Information Technology Integration for Socio-Economic Development

Titus Tossy
Mzumbe University, Tanzania

A volume in the Advances in Human and Social Aspects of Technology (AHSAT) Book Series

IGI GLOBAL
DISSEMINATOR OF KNOWLEDGE
www.igi-global.com

Published in the United States of America by
IGI Global
Information Science Reference (an imprint of IGI Global)
701 E. Chocolate Avenue
Hershey PA, USA 17033
Tel: 717-533-8845
Fax: 717-533-8661
E-mail: cust@igi-global.com
Web site: http://www.igi-global.com

Library of Congress Cataloging-in-Publication Data

Names: Tossy, Titus, 1978-
Title: Information technology integration for socio-economic development / Titus Tossy, editor.
Description: Hershey : Information Science Reference, 2016. | Includes bibliographical references and index.
Identifiers: LCCN 2016017802| ISBN 9781522505396 (hardcover) | ISBN 9781522505402 (ebook)
Subjects: LCSH: Economic development--Technological innovations. | Technological innovations--Social aspects. | Internet in education.
Classification: LCC HD82 .I3413 2016 | DDC 303.48/33--dc23 LC record available at https://lccn.loc.gov/2016017802

This book is published in the IGI Global book series Advances in Human and Social Aspects of Technology (AHSAT) (ISSN: 2328-1316; eISSN: 2328-1324)

British Cataloguing in Publication Data
A Cataloguing in Publication record for this book is available from the British Library.

Advances in Human and Social Aspects of Technology (AHSAT) Book Series

Ashish Dwivedi
The University of Hull, UK

ISSN: 2328-1316
EISSN: 2328-1324

Mission

In recent years, the societal impact of technology has been noted as we become increasingly more connected and are presented with more digital tools and devices. With the popularity of digital devices such as cell phones and tablets, it is crucial to consider the implications of our digital dependence and the presence of technology in our everyday lives.

The **Advances in Human and Social Aspects of Technology (AHSAT) Book Series** seeks to explore the ways in which society and human beings have been affected by technology and how the technological revolution has changed the way we conduct our lives as well as our behavior. The AHSAT book series aims to publish the most cutting-edge research on human behavior and interaction with technology and the ways in which the digital age is changing society.

Coverage

- End-User Computing
- Human Development and Technology
- Cyber Bullying
- Computer-Mediated Communication
- Technoself
- Human-Computer Interaction
- Cyber Behavior
- Technology and Social Change
- Information ethics
- ICTs and social change

IGI Global is currently accepting manuscripts for publication within this series. To submit a proposal for a volume in this series, please contact our Acquisition Editors at Acquisitions@igi-global.com or visit: http://www.igi-global.com/publish/.

The Advances in Human and Social Aspects of Technology (AHSAT) Book Series (ISSN 2328-1316) is published by IGI Global, 701 E. Chocolate Avenue, Hershey, PA 17033-1240, USA, www.igi-global.com. This series is composed of titles available for purchase individually; each title is edited to be contextually exclusive from any other title within the series. For pricing and ordering information please visit http://www.igi-global.com/book-series/advances-human-social-aspects-technology/37145. Postmaster: Send all address changes to above address.

Titles in this Series

For a list of additional titles in this series, please visit: www.igi-global.com

Handbook of Research on Human-Computer Interfaces, Developments, and Applications
João Rodrigues (University of Algarve, Portugal) Pedro Cardoso (University of Algarve, Portugal) Jânio Monteiro (University of Algarve, Portugal) and Mauro Figueiredo (University of Algarve, Portugal)
Information Science Reference • copyright 2016 • 663pp • H/C (ISBN: 9781522504351) • US $330.00 (our price)

Human Development and Interaction in the Age of Ubiquitous Technology
Hakikur Rahman (BRAC University, Bangladesh)
Information Science Reference • copyright 2016 • 384pp • H/C (ISBN: 9781522505563) • US $185.00 (our price)

Examining the Evolution of Gaming and Its Impact on Social, Cultural, and Political Perspectives
Keri Duncan Valentine (West Virginia University, USA) and Lucas John Jensen (Georgia Southern University, USA)
Information Science Reference • copyright 2016 • 456pp • H/C (ISBN: 9781522502616) • US $190.00 (our price)

Handbook of Research on Human Social Interaction in the Age of Mobile Devices
Xiaoge Xu (Botswana International University of Science and Technology, Botswana)
Information Science Reference • copyright 2016 • 548pp • H/C (ISBN: 9781522504696) • US $325.00 (our price)

Defining Identity and the Changing Scope of Culture in the Digital Age
Alison Novak (Rowan University, USA) and Imaani Jamillah El-Burki (Lehigh University, USA)
Information Science Reference • copyright 2016 • 316pp • H/C (ISBN: 9781522502128) • US $185.00 (our price)

Gender Considerations in Online Consumption Behavior and Internet Use
Rebecca English (Queensland University of Technology, Australia) and Raechel Johns (University of Canberra, Australia)
Information Science Reference • copyright 2016 • 297pp • H/C (ISBN: 9781522500100) • US $165.00 (our price)

Analyzing Digital Discourse and Human Behavior in Modern Virtual Environments
Bobbe Gaines Baggio (American University, USA)
Information Science Reference • copyright 2016 • 320pp • H/C (ISBN: 9781466698994) • US $175.00 (our price)

Overcoming Gender Inequalities through Technology Integration
Joseph Wilson (University of Maiduguri, Nigeria) and Nuhu Diraso Gapsiso (University of Maiduguri, Nigeria)
Information Science Reference • copyright 2016 • 324pp • H/C (ISBN: 9781466697737) • US $185.00 (our price)

We dedicate this book to the scholars who contributed to this volume and to all researchers who are engaged in examining the integration of Information Technology for Socio-Economic Development.

Editorial Advisory Board

Table of Contents

Detailed Table of Contents

When Information and Communication Technologies (ICTs) are rapidly revolutionizing, online learning is undergoing dramatic change as powerful new content types, technologies, and services are being adopted for education and training. Traditional learning needs to be supplemented by virtual learning to meet the rapid population growth and counter the lack of traditional higher education institutions. This is particularly true in least developed countries faces critical challenges to meet the new demands in higher education with its ever increasing population, remote and scattered areas, non-availability of resources (including infrastructure, few institutions, few experts or teachers, etc.), and limited funding. While it has been known that developed country are enjoying technology dance for so long time, least developing countries are either not or just joining the dance especially in e-learning. In other hand, there are enough literatures on the e-learning strategic alliances or collaborations/partnerships in developed countries for the purpose of dominating in Education delivery worldwide, and there is no or little literature about existence or non-existence of strategic alliances or collaboration on the e-learning development in the least developed countries. This is why this paper brought forward, enlightening the non-existence or existence of e-learning development strategic alliances, which has been sought to solve dual problem, scarcity of resources and increase the certification and recognition of online degrees in developed countries.

This paper examines e-learning adoption in Higher Education Institutions (HEIs). The paper examines whether there is a difference between the Private and public HEIs e-learning adoption. The rationale for the examination stands from the fact that scholars are of option that there is different between private and public HEIs e-learning adoption. From an empirical survey conducted in Tanzania, this paper evaluates both staffs and students on the current situations, factors affecting, similarities and differences of e-learning adoption between private and public HEIs. The research reveals that, despite of the government efforts to support e-learning initiative by introducing national ICT polices to boost the level of adoption; very few private HEIs have adopted e-learning compared to public HEIs. The limited initial investment, lack of expertize, lack of guaranteed electrical power, poor strategic change management plans, and lack of innovative ideas are the barriers to e-learning adoption.

The subject of Information Technology (IT) adoption and use has been going on for some time. In particular, to electronic commerce (e-Commerce) adoption and use, the invention of Web 2.0 presents new technological features for potential and current e-Commerce adopters as well as new challenges.

With Web 2.0, customers' perceptions, preferences and decisions are not only based on information presented on e-Commerce websites, but are also influenced by content generated by people on social networks and interactive e-Commerce websites. This poses the following question: how can Small, Medium and Micro-Sized Enterprises in the tourism sector (tourism SMMEs) keep up with these technological advancements given their limited resources? Accordingly, this book chapter proposes a framework that shows challenges and incentives (critical success factors) of e-Commerce, identifies e-Commerce platforms tourism SMMEs should adopt in order to maximise benefits and outlines what tourism SMMEs should expect from their e-Commerce platforms.

The E-learning provision in East Africa including Tanzania, Kenya and Uganda is rapid gaining its way. E-learning providers are all groups tasked with e-learning activities including design, delivery, implementation and others. The e-learning providers are surrounded by issues and challenges (main concerns). The purposed of this study was to identify and examine the main concern amongst e-learning providers in East Africa (Tanzania, Kenya and Uganda). In this study the Classic Grounded Theory Research Methodology was used in order to identify the main concern of e-learning providers. In the course of repeated encounters with e-learning providers, this study found the concept of Cultivating Recognition to emerge as the main concern or core variable amongst these e-learning providers. The core variable within the Classic Grounded Theory research methodology is the main entity that accounts for most of the variation in the data. The core variable of the theory, the basic social psychological process of Cultivating Recognition is characterized by Legitimizing and Credentializing. The process of legitimizing involves convincing the e-learning stakeholders that the e-learning programs and projects will be delivered in a timely, valid and sustainable manner. Legitimizing process is achieved through Collaborating (reciprocal and asymmetric), Referral Networking and Strategic alliancing. Credentializing aims to enhance the stakeholders' belief in the e-learning providers' competence prior to the provision of the e-learning programs and projects. While Credentializing is achieved through Endorsementizing, Result Orienting and Prioritizing Duties, Result Orienting itself is achieved through Visualizing, Professionalizing, Focalizing and Role Delineating. This study will help e-learning providers and other groups to cultivate recognition and pave a way for their success in implementation of e-learning.

ICT plays a key role in socio-economic development and ensuring sustainable livelihoods. The purpose of this chapter was to assess the contribution of ICT for sustainable livelihood in Kilosa district. Data were obtained from both primary and secondary sources of data. Structured questionnaires were used in collection of primary data. The collected data was analyzed using SPSS. A sample size of 60 respondents was interviewed. The study determined various contributions resulting from using ICT by rural people

Preface

In an increasingly connected world, ICTs are playing a major role in enabling change. Globalization has altered the landscape of trade, employment, communication and many other areas. Social-development change is no exception to this trend. One of the major challenges facing the world, especially developing countries today is how to take full and smart advantage of quite spectacular and far-reaching advances in science and technology not only in promoting national economic development, but particularly addressing the needs of poor and marginalized sections of society. This effect can be clearly seen in the recent turmoil in the developing countries, in which governments, economies and social mores have changed in ways which have drawn the attention of the world. Because ICT use correlates with many of these developments, and in some cases may cause such changes, understanding the role of ICT in socio-economic development in both developed and developing countries is of great importance.

This Handbook for Research on ICT for Socio-Economic Development discusses and address the role which ICTs are playing in selected, high-impact sectors, such as governance, economic competitiveness, social relationships, and education. It focuses on the required actions needed to enhance the impact of ICT on socio-economic development, in such a way to catalyze positive change in the development challenges. The editors will seek chapters that address different aspects of ICT for Socio-Economic Development (ICT4SED), ranging from the role of ICT in sustainable development, which has been increasingly acknowledged by a variety of stakeholders, including governments, the private sector and civil society, the socio-economic impact of e-services, the role of social media, and the increasing relevance of mobile technologies. Additionally, the handbook will fosters the importance and need of international cooperation for use and promotion of Information and Communication Technologies for socio-economic Development (ICT4SED) trying to bridge the digital divides within countries, regions and the world. Finally, the handbook will provide a platform for sharing experiences and best practices around the world on the ICT for socio-economic development.

This volume brings together a wide range of research on the past, present, and future of the international trend toward the field of Information Technology Integration for Socio-Economic Development (ITi4SED) globally, while providing for further research opportunities in this dynamic field. It is hoped that this text provides the resources necessary for policy makers, technology developers and managers to adopt and implement Information Technology Integration for Socio-Economic Development (ITi4SED). The volume will benefit many groups, not limited to Policy makers, academicians, researchers, advanced-level students, technology developers, and government officials will find this text useful in furthering their research exposure to pertinent topics in Information Technology Integration for Socio-Economic Development (ITi4SED) and assisting in furthering their own research efforts in this field. As a result, this volume contains research in various topics including:

- ICT and Agricultural Development
- ICT for Socio-Economic Development Adoption
- ICT for Socio-Economic Development Platforms
- Social Networking and Politics
- -ICT and Social Networking
- ICT for Sustainable Economic Development
- ICT for Socio-Economic Development in Developing countries
- Barriers to ICT for Socio-Economic Development
- Case Studies on ICT for Socio-Economic Development
- Legal Considerations in ICT for Socio-Economic Development
- Ethical Considerations in ICT for Socio-Economic Development
- ICTs and Industrial Development
- Challenges of ICT for socio-economic development
- Research methods in ICT for socio-economic development
- ICT policy and strategy
- ICT in education
- ICT in healthcare
- ICT in governance
- Mobile technologies for development
- ICT in sustainable development
- Impact of e-Services on Socio-Economic Development
- The role of social media
- ICT in economic competitiveness
- ICT for persons (women, disadvantaged, etc.)
- ICT in social relationships
- ICT for international cooperation

The First chapter written by Prof. Eugenijus Kurilovas, from Vilnius University Institute of Mathematics and Informatics, Vilnius, Lithuania, Vilnius Gediminas Technical University, Lithuania is about Future School: Personalization Plus Intelligence. Prof. Eugenijus Kurilovas is Senior Research Scientist in Vilnius University Institute of Mathematics and Informatics and Associate Professor in Vilnius Gediminas Technical University. He has published about 100 scientific papers, is a member of 35 scientific journals' and conferences Editorial boards and Programme committees. Dr. E. Kurilovas is honoured TOP 100 Scientist 2014 by International Biographical Centre (IBC), Cambridge, and his biographical records are included in: Who is Who in the World – 2014 (31st Edition) and 2015 (32nd Edition); 2000 Outstanding Intellectuals of the XXIst Century – 2014 (8th Edition); Who is Who in Science and Engineering – 2016-2017 (12th Edition); and Dictionary of International Biography – 2014 (37th Edition) and 2015 (38th Edition) by International Biographical Centre (IBC), Cambridge.

The Second Chapter is authored by Dr.Titus Tossy from Mzumbe University, Tanzania, the founder of Tanzania Institute of Technology and Tanzania Business School is titled Collaborating Partnerships: A Project-Based Legitimizing Strategy amongst East African E-Learning Providers. The chapter examines the phenomenon of e-learning in East Africa. The classic grounded theory methodology was used to carry out the study. Through multiple encounters with 24 e-learning providers, the investigation revealed legitimizing of e-learning projects to be a significant issue for these providers. Developing collaborating partnerships was a key strategy used by the providers to legitimize e-learning projects. The study identified two types of collaborating partnerships: Asymmetrical and Reciprocal. These findings are positioned within literature to highlight the contribution that this paper makes to the discourse on e-learning projects in developing countries

The Third Chapter is on Application of Information Communication Technologies for Agricultural Development through Extension Services: A Review. It is written by Mabe L. K. and O. I. Oladele from the Department of Agricultural Economics and Extension, North –West University Mafikeng Campus, South Africa. This chapter outlines the role that Information Communication Technologies (ICT) play in the global context and in Africa, agricultural extension and Agricultural development. The role of and use of ICT by extension officers, the trends of ICT in agricultural information management, how ICT bridge the digital divide as well as the types of ICT tools used by extension officers such as radio, television, computers and internet. It also gives the perspective about the factors that influences use of ICT by extension officers which are seen as playing an important role human development.

The Fourth chapter titled "A Set of Principles for Doing and Evaluating Classic Grounded Theory Research in Information Systems" was written by the joint efforts of Andy Lowe from Grounded Theory Institute, Mill Valley CA USA and Titus Tossy from Mzumbe University, Tanzania. Grounded theory (GT) is a latent pattern recognition research method discovered by Glaser and Strauss (1967). Due to GTs power and transcendence many research papers across several academic disciplines including Information Systems claimed to have used GT when in fact they have used pseudo GT methods. It is argued in this paper that any other research method which adopts the GT label without following orthodoxy of the authentic GT research method should not be called GT. All of the pseudo GT methods make the false assumption that GT is a sub set of Qualitative Data Analysis. This is a false assumption because authentic GT can use either quantitative or qualitative data and it is a general research methodology and produces empirically grounded but modifiable propositions. Within the Information Systems (IS) research community it is therefore not surprising that many, who claim to use GT, have used different types of pseudo GT. They have adopted vocabulary of the GT without following its original tenets. This paper explains how authentic GT can be carried out in an information systems context by trusting in emergence rather than forcing the data.

The fifth chapter is about Cloud Based Geo-Information: Infrastructure with Big Data Analytics to Support Agriculture Activity Monitoring. It is authored by Shamim Akhter, East West University,Bangladesh. Dr. Md. Shamim Akhter received his Ph.D. in Information Processing from Tokyo Institute of Technology (TokyoTech), M.Sc. in Computer Science and Information Management from Asian Institute of Technology (AIT) and B.Sc. in Computer Science from American International University Bangladesh (AIUB) in 2009, 2005 and 2002 respectively. He joined AIUB as a Lecturer in 2002 and since 2005 he has been an Assistant Professor. He was also a JSPS Post-Doctoral Research Fellow in National Institute of Informatics (NII) from FY 2009-2011, Visiting Researcher in Tokyo Institute of Technology, Japan from FY 2009-2011, Research Associate at the RS and GIS FoS, Asian Institute of Technology, Thailand in 2005 and Global COE Research Assistant from Sep 2007~ Aug 2009 in Tokyo Institute of Technology, Japan. He was awarded "The Excellent Student of The Year, FY2008", Global COE Program, Photonics Integration-Core Electronics (PICE), Tokyo Institute of Technology, Japan and Magna-Cum Laude for academic excellence from American International University Bangladesh in 2002. His research interests are Artificial Intelligent, Evolutionary Algorithms and Models for their Parallelization, Remote Sensing (RS) and GIS applications, High Performance Computing (HPC), Algorithm and Complexity Analysis,

Wireless Communications and Mobile Computing. He is a senior member of IEEE and member of JARC-Net.

The sixth chapter is titled "Profiling Collaborating Partners in E-Learning in Developing Countries". It is written by Titus Tossy from Mzumbe University, Tanzania and Prof. Wallace Chigona from University of Cape Town, South Africa. This chapter enlightening the non-existence or existence of e-learning development strategic alliances, which has been sought to solve dual problem, scarcity of resources and increase the certification and recognization of online degrees in developed countries.

The seventh chapter on Before eGovernance and eGovernment, Back to Basics! The Case of the Caribbean was written by Pearson A. Brome, Department of Government, Sociology and Social Work, The University of the West Indies. Dr. Pearson Broome is a lecturer in Political Science and the Programme Coordinator for the MSc. e- Governance for Developing Countries in the Department of Government, Sociology and Social Work at the University of the West Indies, Cave Hill Campus, Bridgetown, Barbados. He has diverse working experience in the Caribbean and served as a Consultant for UNDP, the Commonwealth Secretariat and CSME Unit, CARICOM Secretariat. His research interests include: EGovernance/eGovernment/ eCommerce/ eDemocracy, with particular emphasis on the theoretical approaches and the institutionalization of ICTs; ICTs and the globalisation process focusing on outsourcing, trade, investment and technology transfer flows; the evaluation of national ICT Policies and National Systems of Innovation, the politics of science and technology policy, deregulation and Telecommunications policy, new public management and the restructuring of state bureaucracies, and public policy and its application to public sector management in developing countries.

The eighth chapter is about A Comparative Evaluation of E-Learning Adoption in Private and Public Higher Education Institutions: A Tanzania Survey. Authored by Mpe Paulo Mwamahusi, Tanzania Institute of Technology, Tanzania and Titus Tossy, Mzumbe University, Tanzania. This paper examines e-learning adoption in Higher Education Institutions (HEIs). The paper examines whether there is a difference between the Private and public HEIs e-learning adoption. The rationale for the examination stands from the fact that scholars are of option that there is different between private and public HEIs e-learning adoption. From an empirical survey conducted in Tanzania, this paper evaluates both staffs and students on the current situations, factors affecting, similarities and differences of e-learning adoption between private and public HEIs.

The ninth chapter is titled "The Impact of Web 2.0 on E-Commerce Adoption and Use by Tourism Businesses – Can SMMEs Play the Trick? A Case of the Eastern Cape Province" by Pardon Blessings Maoneke and Naomi Isabirye from University of Fort Hare, South Africa. This chapter proposes a framework that shows challenges and incentives (critical success factors) of e-Commerce, identifies e-Commerce platforms tourism SMMEs should adopt in order to maximise benefits and outlines what tourism SMMEs should expect from their e-Commerce platforms.

The tenth chapter presents new theory in e-learning, the theory of Cultivating Recognition: A Classic Grounded Theory of E-Learning Providers Working in East Africa by Titus Tossy from Mzumbe University and Prof Irwin Brown from University Of Capetown, South Africa. The E-learning provision in East Africa including Tanzania, Kenya and Uganda is rapid gaining its way. E-learning providers are all groups tasked with e-learning activities including design, delivery, implementation and others. The e-learning providers are surrounded by issues and challenges (main concerns). The purposed of this study was to identify and examine the main concern amongst e-learning providers in East Africa (Tanzania, Kenya and Uganda). In this study the Classic Grounded Theory Research Methodology was used in order to identify the main concern of e-learning providers. In the course of repeated encounters with e-learning providers, this study found the concept of Cultivating Recognition to emerge as the main concern or core variable amongst these e-learning providers. The core variable within the Classic Grounded Theory research methodology is the main entity that accounts for most of the variation in the data. The core variable of the theory, the basic social psychological process of Cultivating Recognition is characterized by Legitimizing and Credentializing. The process of legitimizing involves convincing the e-learning stakeholders that the e-learning programs and projects will be delivered in a timely, valid and sustainable manner. Legitimizing process is achieved through Collaborating [reciprocal and asymmetric], Referral Networking and Strategic alliancing. Credentializing aims to enhance the stakeholders' belief in the e-learning providers' competence prior to the provision of the e-learning programs and projects. While Credentializing is achieved through Endorsementizing, Result Orienting and Prioritizing Duties, Result Orienting itself is achieved through Visualizing, Professionalizing, Focalizing and Role Delineating. This study will help e-learning providers and other groups to cultivate recognition and pave a way for their success in implementation of e-learning.

The eleventh chapter on Assessment of Contribution of ICT for Sustainable Livelihood in Kilosa District by Shirima, C. and Prof. Camilius Sanga from Sokoine University of Agriculture, Tanzania. The purpose of this study was to assess the contribution of ICT for sustainable livelihood in Kilosa district.

The twelfth chapter titled "Doing Classic Grounded Theory Research in Information Systems: Trust in Emergence" by Titus Tossy from Tanzania, Prof. Irwin Brown from University of Cape Town, South Africa and Prof. Andy Lowe from Grounded Theory Institute, USA. Following publication of the original grounded theory by Glaser & Strauss (1967), Glaser (1978) went on to explain in detail how to operationalize the GT method. Despite this, some researchers mistakenly continued to classify GT as a subset of the qualitative data analysis research methodology. In doing so, they reveal that they have misunderstood both the purpose of GT and how to use it correctly. In this paper the author will concentrate on the authentic approach to GT, which is termed "Classic Ground Theory" (CGT), so as to differentiate it from any misconceived remodelled version of the theory. This paper explains how CGT can be carried out in an IS context by trusting in the emergence from the data of the required information, rather than forcing it.

The thirteenth chapter is titled "Information and Communication Technologies (ICTs) for Industrial Development: Challenges and Opportunities." Written by Dr. Felichesmi Selestine Lyakurwa and Dr. Joseph Sungau from Mzumbe University, Tanzania. Both authors are lecturers at Mzmbe University.

The fourteenth chapter is about Technology Access and Research Prolificacy: An Econometric Analysis by Pearson A. Brome. Modern information and communication technologies enable research collaborations that were not conceivable a mere decade ago. Moreover, the diffusion of extant technologies (such as broadband, and cellular communication devices) more fully in both developing and developed nations has afforded more people access to key communications technologies, creating ever increasing networks and communities of researchers and collaborators. The current study seeks to more fully explicate the relationship between technology access, diffusion and research output. The study shall consider the impact of telecommunications tele-density, Broadband penetration, Computer access and ICT investment on research output and patent submissions. Both static and dynamic estimations are conducted, employing the Arellano & Bover systems estimator method. The findings suggest that telecommunications tele-density and ICT investment are key determinants of academic research output. They are however not significant determinants of patent submissions. These findings are robust for both static and dynamic estimations.

SUMMARY

This volume will be used as a platform in the establishment of international conference on ICT integration for socio-economic development to be help every year. We hope readers will be part of the coming conference too. There are so many important issues attached to this subject, many explored by scholarly contributions in this volume, dealing both with pragmatic implementation and with conceptual design, that they cannot be enumerated here. What is certain, however, is something close to the heart of and bringing a smile to the lips of every academic: more research is needed! Toward this end, this volume seeks to make a small contribution.

Mpe Mwamahusi
Tanzania Institute of Technology, Tanzania

Acknowledgment

I would like to gratefully thank all the Editorial Board members and reviewers.

I would also like to thank all the authors who have generously and patiently contributed to this book. With their important, innovative, and exciting work, they have made this hand book possible.

I wish to thank all at IGI Global for their kind and supportive help during the time taken to develop this book, with particular thanks to Lindsay Johnston. His professional assistance and guidance have been much appreciated throughout this endeavour.

Thank you.

Titus Tossy
Mzumbe University, Tanzania

Chapter 1
Future School:
Personalization Plus Intelligence

Eugenijus Kurilovas
Vilnius University, Lithuania & Vilnius Gediminas Technical University, Lithuania

ABSTRACT

What learning content, methods and technologies are the most suitable to achieve better learning quality and efficiency? In Lithuania, we believe that there is no correct answer to this question if we don't apply personalised learning approach. We strongly believe that "one size fits all" approach doesn't longer work in education. It means that, first of all, before starting any learning activities, we should identify students' personal needs: their preferred learning styles, knowledge, interests, goals etc. After that, teachers should help students to find their suitable (optimal) learning paths: learning methods, activities, content, tools, mobile applications etc. according to their needs. But, in real schools practice, we can't assign personal teacher for each student. This should be done by intelligent technologies. Therefore, we believe that future school means personalisation plus intelligence. In this chapter, Lithuanian Intelligent Future School project is presented aimed at implementing learning personalisation and educational intelligence.

INTRODUCTION

In Lithuania, we believe that "one size fits all" approach doesn't longer work in education. We strongly believe that future school means personalisation plus intelligence.

Learning personalisation means creating and implementing personalised learning paths based on recommender systems and personal learning environments suitable for particular learners according to their personal needs. Educational intelligence means application of intelligent technologies and methods enabling personalised learning to improve learning quality and efficiency. Lithuanian Intelligent Future School (IFS) project is aimed at implementing both learning personalisation and educational intelligence.

In personalised learning, first of all, integrated learner profiles (models) should be implemented. After that, ontologies-based personalised recommender systems should be created to suggest learning components (learning objects, activities, methods, tools, apps etc.) suitable to particular learners according to their profiles. Thus, personalised learning paths could be created for particular learners for each

DOI: 10.4018/978-1-5225-0539-6.ch001

topic according to curriculum. A number of intelligent technologies should be applied to implement IFS approach, e.g. ontologies, recommender systems, intelligent agents, multiple criteria decision making methods to evaluate quality and suitability of the learning components etc.

Two years' experience of pilot IFS implementation in Lithuania has shown that this personalised intelligent approach is useful and effective. Currently, over 20 primary and secondary schools are involved in this pilot Lithuanian IFS project coordinated by Vilnius University researchers. We create suitable intelligent methods and technologies and validate them at pilot schools in order to create well working intelligent learning system for schools.

IFS approach is based on experience gained in "Future Classroom" projects funded by European Commission and co-ordinated by European Schoolnet (EUN). These flagship European projects in the area of ICT in education were: (1) iTEC (Innovative Technologies for an Engaging Classroom) funded by 7th Framework Programme (7FP) and implemented in 2010–2014; (2) LSL (Living Schools Lab), 7FP, 2012–2014, and (3) CCL (Creative Classrooms Lab), Life Long Programme (LLP), 2013–2015. In Lithuania, these projects were coordinated by the author.

This Chapter is aimed to present several intelligent methods and technologies to improve learning quality and efficiency. They were implemented in IFS pilot schools and validated under the umbrella of EU "Future Classroom" projects.

The Chapter is organised as follows. Section 1 is introductory. Section 2 presents the related research. Section 3 presents integrated IFS learning profile (model) approach based on students' learning styles. In Section 4, we provide intelligent tool based on creating ontologies to interconnect students' learning styles (according to Honey & Mumford (1992) model) with learning activities, methods, learning objects types, and mobile apps. Section 5 presents IFS knowledge-based recommender system to integrate Web 2.0 tools into learning activities according to students' VARK (Visual, Aural, Read/write and Kinaesthetic) learning styles. Discussion and conclusion are provided in Section 6.

RELATED RESEARCH

Personalisation and Learning Styles

Learning personalisation and related issues were very popular in scientific literature in recent years (Bennane, 2013; Yoo et al., 2013; Kim & Lee, 2013; Wallden & Makinen, 2014; Troussas et. al., 2014).

The overview of literature shows that there has not been a concrete definition of personalisation so far. The main idea is to reach an abstract common goal: to provide users with what they want or need without expecting them to ask for it explicitly (Mulvenna et al., 2000). From the educational point of view, personalisation attempts to provide for an individual tailored products, services, information, etc. A more technical standpoint to personalisation is linked with the modelling of Web objects (products and pages) and subjects (users), their categorisation, organising them to achieve the desired personalisation. According to Sampson (2002), personalisation provides training programmes that are customised to individual learners, based on an analysis of the learners' objectives, current status of skills / knowledge, learning style preferences, as well as constant monitoring of progress. Online learning material can be, then, compiled to meet personal needs, capitalising on reusable learning objects.

User-centric environments are learning environments where the learner takes responsibility for his / her own learning, and the instructor acts as the "guide on the side", rather than a "sage on the stage" (Sampson, 2002).

The concept of personalised learning becomes increasingly popular. It advocates that instruction should not be restricted by time, place or any other barriers, and should be tailored to the continuously modified individual learner's requirements, abilities, preferences, background knowledge, interests, skills, etc. The personalised learning concept signifies a radical departure in educational theory and technology, from "traditional" interactive learning environments to personalised learning environments. According to Sampson et al. (2002), some of the most prominent characteristics of this shift can be summarised as follows:

1. While "traditional" learning environments adopt the one-to-many learning mode, personalised learning environments are based on the one-to-one or many-to-one learning concept (i.e. one, or many tutors for one learner);
2. Traditional learning environments usually pose a number of constraints in relation to the learning setting; personalised learning environments, on the other hand, facilitate learning independent of time, location, etc.;
3. Traditional learning environments are usually being designed for the "average learner"; while, in personalised learning environments, the learning material and sequencing, learning style, learning media, etc., depend on the individual learner's characteristics, i.e. background, interests, skills, preferences, etc.;
4. In traditional learning environments, the curriculum, learning units, etc., are determined by the tutor, while in personalised learning settings, they are based on the learner's requirements (self-directed learning).

The Educational Perspective

According to Schunk (1996), the concept of personalised learning builds mainly on the cognitive and constructivist theories of learning. Instructional principles of cognitive theories argue for active involvement by learners, emphasis on the structure and organisation of knowledge, and linking new knowledge to learner's prior cognitive structures. Constructivist instructional theory, on the other hand, implies that instructional designers determine which instructional methods and strategies will help learners to actively explore topics and advance their thinking. Learners are encouraged to develop their own understanding of knowledge.

Several research efforts have been devoted in the identification of the dimensions of individual differences. One of the most prominent research areas in this context concerns the learning styles and learning differences theory, which implies that how much individuals learn has more to do with whether the educational experience is geared towards their particular style of learning. Learning styles are strategies, or regular mental behaviours, habitually applied by an individual to learning, particularly deliberate educational learning, and built on her / his underlying potentials. Learners are different from each other, and teaching should respond by creating different instruction for different kinds of learning. Learners also differ from each other in more subject-specific aptitudes of learning, e.g. some being better at verbal than numerical things, others vice versa. Learning styles have been at the centre of controversy for several decades now, and there is still little agreement about what learning styles really are. There are numerous methodologies and tools that attempt to categorise people according to differences in learning and cognitive styles. The most well-known of these efforts include the Myers-Briggs Type Indicator (Keirsey, 1998); Multiple Intelligences (Gardner, 1999); Auditory, Visual, Tactile / Kinaesthetic Learning Styles (Sarasin, 1998); Grasha-Riechmann student Learning Style Scales – GRLSS (Grasha, 1996); Kolb Learning Styles Theory (Kolb, 1985); Felder and Silverman Index of Learning Styles (Felder, 1996); and Honey and Mumford Learning Styles (Honey & Mumford, 1992).

According to Sampson et al. (2002), in order for these methodologies and tools to be effectively applied, we need to be able to (1) accurately classify each learner according to a selected taxonomy of individual differences, and (2) determine which are the characteristics of the learning environment that are appropriate for this category of learners.

The Technological Perspective

Several notions are used to define personalised virtual learning environments.

They are as follows:

According to Sampson et al. (2002), intelligent learning environments (ILEs) are capable of automatically adapting to the individual learner, and therefore constitute the most promising technological approach towards the realisation of the personalised learning concept. An ILE is capable of automatically, dynamically, and continuously adapting to the learning context, which is defined by the learner characteristics, the type of educational material being exchanged, etc.

According to Brusilovsky et al. (1998), Adaptive Educational Hypermedia (AEH) is a relatively new direction of research within the area of adaptive and user model-based educational applications AEH systems build a model of the individual user / learner, and apply it for adaptation to that user. AEH can be categorised with respect to several dimensions.

The first question to pose about a particular AEH system is: what aspects of the student working with the system can be taken into account when providing adaptation? To which features – that can be different for different students (and may be different for the same student at a different time) – can the system adapt? Generally, there are many features related to the current context of the student work and to the student as an individual which can be taken into account by an AEH system. The features that are used by existing systems are: student's goals, knowledge, background, hyperspace experience, and preferences. Student's knowledge, which is most commonly used in educational systems, is usually represented by an overlay model based on the structural model of the subject domain, which, in turn, is usually represented as a network of domain concepts. Sometimes, a simpler stereotype student model is used, which distinguishes several typical "stereotype students". The student's current goal is usually modelled in a similar manner. That is, the system supports a set of possible student goals, and an overlay student goal model is used to predict the current goal. Student's background and hyperspace experience is also usually modelled through overlay models, while student's preferences are usually either specified by the student, or are deduced by the accumulation of several student models in a group student model.

Another important question concerning AEH systems is: what can be adapted by the system? Which features of the system can differ for different students? What is the space of possible adaptations? The adaptations in AEH systems may include the content of the hypermedia pages (adaptive presentation), as well as the links included in each page (adaptive navigation support). The former case can be further decomposed into adaptive multimedia presentation, and adaptive text presentation, which is most commonly used. The latter case includes direct guidance (providing the "next" node to follow), adaptive sorting of links, adaptive hiding of links, adaptive annotation of links, and / or map adaptation. The last broad categorisation of AEH systems concerns how adaptation can help, i.e. the methods and techniques of adaptation, for content adaptation and adaptive navigation support. Concerning the methods for content adaptation, the most popular one is to hide from the student some parts of the information about a particular subject which are not relevant to the student's level of knowledge about this concept. Following another approach, which has been termed prerequisite explanation, before presenting an explanation of a concept, the system may insert explanations of all its prerequisite concepts which are not sufficiently known to the student. Alternatively, following the explanation variants method, the system may store several variants for some part of the page content, and the student gets the variant that corresponds to his / her student model; or the system may sort the fragments of information about a concept, and present the information that is most relevant to the student's knowledge.

Concerning the techniques for content adaptation, one can distinguish between the conditional text technique, where all possible information about a particular concept is divided into several chunks of text, each one associated with a condition concerning the student's knowledge of the domain, and only the chunks for which the condition is true are presented to the student; or the stretch-text technique, where particular "hot-words" are associated with some text, which is "collapsed", or "un-collapsed" according to the student's knowledge. The most powerful adaptation technique for content adaptation is frame-based adaptation, where all the information about a particular concept is represented in form of a frame, and special presentation rules are used to select which information within a frame will be presented, according to the student's knowledge. Finally, concerning the methods for adaptive navigation support, we can distinguish between global guidance, local guidance, local orientation support, global orientation support and management of personalised views.

Personalised e-learning systems deal with appropriate personalisation techniques in order to maximise the effectiveness of learning (Juskeviciene & Kurilovas, 2014).

Learning Components and Intelligent Technologies

Learning Objects

Learning Object (LO) is referred here as "any digital resource that can be reused to support learning" (Wiley, 2000). The fundamental idea behind LOs is that instructional designers can build small (relative to the size of an entire course) instructional components that can be reused a number of times in different learning contexts. In addition, LOs are generally understood to be digital entities deliverable over the Internet and any number of people can access and use them simultaneously. Moreover, those who incorporate LOs can collaborate on and benefit immediately from the new versions.

According to Kurilovas and Dagiene (2010), the concept of LOs has gained wide spread acceptance in the world of e-learning. The main purpose of LOs is to provide a modularised model, based on the standards that enhance flexibility, platform independence, and reuse of the learning content, as well as providing a higher degree of control for teachers and learners. The definition and meaning of the term "learning object" varies considerably between different actors and communities as well as over time. The idea of LOs is to organise the digital learning content into small, fairly context-independent chunks that can be assembled, disassembled and combined in different ways and in different learning contexts. LOs from different vendors can be combined with one another to form modules that can be used in a specific learning context. In engineering terms, the learning process can be considered as a life cycle of creation and existence of knowledge.

The Ecological Approach

The ecological approach was first proposed by McCalla (2004) in the context of providing a more flexible and powerful way to mark up and use LOs to achieve various pedagogical goals. In the ecological approach the e-learning system keeps a learner model for each learner, tracking characteristics of the learner and information about the learner's interactions with the LOs they encounter. After a learner has finished with a LO, the LO is associated with an instance of the learner model that records the learner's state(s) and activities while interacting with the LO. According to Brooks and McCalla (2006), the information in such a learner model instance can include:

- Information about the learner, including cognitive, affective and social characteristics and the learner's goal(s) in accessing the LO
- Information about the learner's perspectives on the LO itself, including the learner's feedback on the content, or the learner's knowledge of the content (as determined, for example, by a test administered during the learner's interactions with the LO)
- Information about how the learner interacted with the LO, including observed metrics such as dwell time, the number of learner keystrokes, patterns of access, etc.
- Information about the technical context of use, including characteristics of the learner's software and hardware environment
- Information about the social context of use, including links to the learner model instances attached to LOs previously encountered by the learner.

The Standardisation Perspective

During the past few years, a number of international efforts have been initiated for defining specifications and standards which can facilitate reusability in learning technologies. The main initiatives in the area are the IEEE LTSC (Learning Technologies Standards Committee, http://ltsc.ieee.org), the IMS (Instructional Management Systems) Global Learning Consortium Inc (http://www.imsproject.org), the European CEN/ISSS Learning Technologies Workshop (http://www.cenorm.be/isss/Workshop/lt), and the US ADLnet (Advanced Distributed Learning Network, http://www.adlnet.org). These efforts have already resulted in a number of specifications for e-learning applications and services. However, the current versions of these specifications do not support personalised learning.

In particular, today we can describe in a common way LOs, e.g. through the IEEE Learning Objects Metadata (LOM) specification. Further in the Chapter, we'll refer to the popular European application profile of LOM, i.e. LRE APv4.7 (2011). We can also describe learner characteristics in a common format, e.g. through the IMS Learner Information Profile (LIP) specification. Moreover, we can describe learning packages (i.e. collections of LOs) in a common format, though the IMS Content Packaging (CP) specification. However, the current version of this specification facilitates only the definition of simple, table of contents-like structures. As a result, an e-learning system importing a content package can only present the same information to all learners, thus personalised, on-demand learning cannot be supported.

In this context, a number of international efforts have been initiated for the extension of the current versions of these specifications, to allow the definition of rules which determine which (different) parts of learning packages should be selected for different learner categories. One such approach is carried out in the context of the European KOD "Knowledge on Demand" project (http://kod.iti.gr/). The KOD project works on an extension of the CP specification (the knowledge packaging format), so that it can enable the definition of adaptation rules, which specify which parts a learning package should be selected for different learner categories. As a result, the KOD e-learning system (or any e-learning system which is compliant with the KOD knowledge packaging format) can import knowledge packages, disaggregate them, interpret the rules included in them, and present different "knowledge routes" to different learners, according to their individual profiles. Moreover, since the "adaptation logic" (adaptation rules) behind adaptive educational content are represented in a common format, adaptive educational content can be easily interchanged and reused, thus reusability for personalised, on-demand access can be supported.

Another issue in the standardisation of learning material is to provide an efficient way to search and browse various LOs according to individual requirements. A web-course search engine has been developed, which is a user-friendly, efficient and accurate assistant for the learners to get what they want from the vast ocean of LOs being developed all over the world. The system uses Metadata specifications to record and index various LOs, and lets the searchers and the resources "communicate" with each other. Following the Metadata specifications, the system collects exact information about educational resources, provides adequate search parameters for search, and returns evaluative results. With intuitive interfaces, the learners can find the appropriate LOs to suit their needs.

Further in the Chapter, another approach to personalise learning is presented. It is based on semantic search of LOs according to students' learning styles, preferred learning activities, and teaching / learning methods.

Intelligent Pedagogical Agents

Researchers agree that intelligent pedagogical agents could help to personalise learning, but there is no real agreement on what an agent is. Agents' abilities vary significantly, depending on its roles, capabilities, and environments.

According to Aroyo and Kommers (1999), in order to describe these abilities, different notions of agents have been introduced. Intelligent agents are introduced by most of the researchers with four major concepts defining their behaviour:

1. Autonomy,
2. Responsiveness or reactiveness,
3. Pro-activeness and
4. Social ability.

There is also a strong notion on the characteristics of agents, which refer to adaptiveness, pro-activity and intentionality. There are also various taxonomies created for agents, e.g. collaborative, interface, mobile, information, reactive, hybrid, and smart agents. In this context, intelligent agents have been associated with a variety of functions, for example, personal assistants, information managers, information seekers, planning agents, co-ordination agents or collaborative schedules, user representatives, and so forth. Their application in the educational field is mostly as personal assistants, user guides, alternative help systems, dynamic distributed system architectures, human-system mediators, and so forth.

Because pedagogical agents are autonomous agents, they inherit many of the same concerns that autonomous agents in general must address. It has been argued that practical autonomous agents must in general manage complexity. According to Sampson et al. (2002), they must exhibit robust behaviour in rich, unpredictable environments; they must co-ordinate their behaviour with that of other agents, and must manage their own behaviour in a coherent fashion, arbitrating between alternative actions and responding to a multitude of environmental stimuli. In the case of pedagogical agents, their environment includes both the students and the learning context in which the agents are situated. Student behaviour is by nature unpredictable, since students may exhibit a variety of aptitudes, levels of proficiency, and learning styles.

Web 3.0 and Ontologies

There is a lack of the research on using so-called Semantic Web (or Web 3.0) for intelligent semantic search of the LOs suitable for the particular learning styles. If there should exist the qualitative technologies for semantic intelligent search of the relevant learning content on the Web, the learners should get the additional possibility to use this suitable content in their virtual learning environments (VLEs).

According to Berners-Lee et al. (2001), the Semantic Web is not a separate Web but an extension of the current one, in which information is given well-defined meaning, better enabling computers and people to work in cooperation.

Two important technologies for developing the Semantic Web are XML and RDF. XML allows users to add arbitrary structure to documents without saying what these structures mean. RDF allows meaning to be specified between objects on the Web and was intentionally designed as a metadata modelling language.

A third important aspect of the Semantic Web is a set of ontologies. According to Gruber (1993), an ontology is a specification of a conceptualisation. It describes the concepts and relationships of some phenomenon in the world. According to Mohan & Brooks (2003), by using well-defined ontologies on the Web, it is possible for computers to meaningfully process data since there is a common understanding of terms used and the relationships between these terms.

The Semantic Web is concerned about the meaning of all kinds of information (such as LOs) on the Web.

Recommender Systems

E-learning environment is a computer-based online learning system. It supports learner by providing opportunities to learn at the time and location according to the learner's choice and allows his/her interactions with other learners, as well as ensures access to a wide range of learning resources and tools. The e-learning environment, that provides a set of personalisation functionalities such as personalised learning plans and learning materials, and is capable of initialising the interaction with learners by providing suggestions to online learners, is called a personalised e-learning environment. One of the major challenges involved in developing personalised systems is to achieve effective personalisation functionalities such as personalised content management, the learner's model, and adaptive instant interaction. Adaptive systems, autonomous intelligent agents and recommenders provide an important technology for accomplishing personalisation in e-learning systems. Adaptive and intelligent technologies can enhance different sides of learning.

Adaptive hypermedia systems build a model of the goals, preferences and knowledge of each individual user, and use this model throughout the interaction with the user, in order to adapt to the needs of that user (Brusilovsky, 1996). The goal of intelligent agents and recommenders is the use of knowledge about the domain, the learner, and the learning process to support personalised learning, i.e. to improve the learning activities.

Recommender systems (as a kind of services in the e-learning environment) can provide personalised learning recommendations to learners. Recommender systems are information processing systems that gather various kinds of data in order to create their recommendations. The data are primarily about the items (objects that are recommended) to be suggested and the users who will receive these recommendations (Ricci et al., 2011, p.7).

The data can be formalised in a domain ontology, thus the knowledge about a user and items becomes reusable for people and software agents (Li et al., 2007; Youn & Mcleod, 2006; Wang & Huang, 2013; Vesin et al., 2012; Vesin et al., 2013; Chen et al., 2012).

Ontologies can support the definition of such components of recommender system (Buriano et al., 2006):

- The context features and the candidate items
- The output
- The recommendation process, and
- Functional modules

Such information modelling can contribute to tailor the right information to a user and thus facilitate the user-system interaction.

The suggestions relate to various decision-making processes.

There are different types of recommender systems, based on the recommendation approaches. Rule-based reasoning method is the fundamental in knowledge-based recommendation systems (Husain & Dih, 2012; Jadhav & Sonar, 2009). It is deductive and the knowledge base comprises the knowledge that is specific to the domain of the application (Jadhav & Sonar, 2009). The Rule-based reasoning system represents knowledge of the system in terms of a set of rules (facts) (Kapoor & Sharma, 2010). These rules are in the form of IF THEN rules such as: "IF some condition THEN some action". If the 'condition' is satisfied, the rule will take the 'action'.

In order to construct rules of learning tools (i.e. Web 2.0 tools) recommendation, research on the relationship between these tools and the learning process (i.e. existing learning tools selection models presented in literature), was conducted. Results of this analysis are as follows:

According to Conole and Fill (2005), learning activity as an action which occurs within a context with a set of associated attributes and addresses a set of learning outcomes. These outcomes can be achieved through a sequence of tasks and associated roles adopted by the learners which might call upon a set of tools and resources. The toolkit developed by authors offers appropriate media based on the media category (e.g. Adaptive, Productive), the desired learning outcomes (e.g. understand, demonstrate, design, produce, appraise) and strategies (tasks).

Bower et al. (2010) think that for the proper tool selection it is needed to consider the following four element groups of the learning design:

- The overarching learning goals and objectives (outcomes).
- The type of content in terms of knowledge to be represented and cognitive processes in which students are expected to engage.
- The type of pedagogy to be applied (transmissive, dialogic, constructive, co-constructive).
- The preferred modalities of representation (text, image, audio, and/or video), the type of collaboration, and the level of synchronicity required.

Then, based on these elements, matched to the Web 2.0 tools potential to satisfy these requirements suitable Web 2.0 tools can be selected.

According to Starkey (2011), for the proper tool selection it is needed to combine digital tools use (accessing information, presenting, processing information, gaming or interactive programs) with the aspects of learning (doing, thinking about connections, thinking about concepts, criticising and evaluating, creating knowledge, sharing knowledge).

Khalid et al. (2012) consider that in the view of the proper tool selection it is needed to map tools with the problem-and-project-based learning activities that include the tasks (e.g. brainstorming, argumentation, literature storing) performed by learner to achieve learning outcomes in a learning environment while interacting with people or resources.

According to Vega-Gorgojo et al. (2008), selection of suitable tool for learning activity accomplishment needs to map a particular tool with a desirable task to perform. The authors state that the tool supports one or more learning tasks (person-based, group-based or computer-system-based) performed by an actor (person, group or computer system) who can play some role (e.g. editor, communicator, publisher). Tasks are divided into five types: Perception, Construction, Communication, Computation and Information Management. The realisation of these tasks may require an artefact (e.g. document, vote) as input or may produce an artefact (e.g. message, calendar) as output.

INTEGRATED LEARNER MODEL

In Intelligent Future School (IFS) project, first of all, integrated learner model should be created.

User model is representation of information about an individual user. User modelling is the process of creating and maintaining an up-to-date user model, by collecting data from various sources, which may include: (1) implicitly observing user interaction and (2) explicitly requesting direct input from the user (Brusilovsky & Millan, 2007).

Regarding the information contained in the user model, there are identified six features:

1. Knowledge
2. Interests
3. Goals
4. Background
5. Individual traits and
6. Context of work.

The user profiling process generally consists of three main phases (Gauch et al., 2007):

1. Information collection: the process that is used to gather raw information about the user. Depending on the information collection process selected, different types of user data can be extracted;
2. User profile construction (from the user data). There is a variety of ways in which profiles may be represented and constructed;
3. Information in the user profile exploitation. It is done by technology or application in order to provide personalised services.

User profile construction is typically either knowledge-based or behaviour-based (Middleton et al., 2009). Knowledge-based approaches use static models of users and dynamically match users to the closest model and that of behaviour-based apply the user's behaviour as a model, commonly using machine-learning techniques. To obtain behaviour-based knowledge, behavioural logging is used, and knowledge-based approaches often use interviews and questionnaires for gathering knowledge.

In e-learning systems, learner's profiles can be modelled (for personalisation purposes) by several techniques (Dagger et al., 2002) as follows:

- Creating fixed stereotypes: learners are categorised, and the system customises their performance based on the category that has been set for the learner
- Constructing learner's knowledge: a model of the learner's knowledge is constructed on the concept-by-concept basis and updated with the progress of the user through the system
- Combined techniques: the learners are categorised by stereotype initially and then this model is gradually modified, based on learner's interaction with the system.

User profiles can be represented as sets of weighted keywords, semantic networks, weighted concepts, or association rules (Gauch et al., 2007). The keyword-based profiles are often represented as vectors of weighted features.

As mentioned before, one of the features of contemporary e-learning is personalisation because learners should be treated as individuals with differences such us learning styles. The term 'learning style' refers to the concept that individuals differ with regard to which mode of instruction or study is most effective to them (Pashler et al., 2008). Various learning style models are developed. In the work of Coffield et al. (2004), there is a detailed analysis of the best known learning style models. In Popescu (2009), a summary of learning preferences extracted from learning style models (by the main features) is presented. These preferences can be, for example, a specific manner of approaching a learning task, learning strategies activated in order to fulfil the task, preferable ways of gathering, organising, or thinking about information.

The learning style of the student also started to be taken into account, as being one of the individual traits that play an important role in learning. Learning style designates everything that is characteristic to an individual when she/he is learning, i.e. a specific manner of approaching a learning task, the learning strategies activated in order to fulfil the task.

Learning styles represent a combination of characteristic cognitive, affective and psychological factors that serve as relatively stable indicators of how a learner perceives, interacts with, and responds to the learning environment (Keefe, 1979).

Learning styles model systems differ in several aspects: underlying learning style model, diagnosing method (implicit or explicit), modelling techniques (rule-based approach, data mining, machine learning techniques), number of modelled student characteristics besides learning preferences (knowledge level, goals) and the type, size and conclusions of the reported experiments.

In what follows we will focus on the methods used for learner modelling and we classify the systems in two categories:

1. Those that use questionnaires for identifying the learning style and
2. Those that use students' observable behaviour.

The first adaptive educational systems that dealt with learning styles as adaptation criterion relied on the measuring instruments associated to the learning style models for diagnosing purposes. The main advantage of this method is its simplicity: the teacher / researcher only has to apply a dedicated psychological questionnaire, proposed by the learning style model creators. Based on the students' answers to the questions, a preference towards one or more of the learning style dimensions can be inferred.

The main disadvantages of this questionnaire-based approach are:

- Some of the measuring instruments used could not demonstrate internal consistency, test–retest reliability or construct and predictive validity, so they may not be totally reflective of the way a particular student learns (Coffield et al., 2004);
- It implies a supplementary amount of work from the part of the student, who has to fill in questionnaires at the beginning of the course (which sometimes may include over 100 questions, as in case of the Herrmann's Whole Brain Model (Herrmann, 1996));
- It can be easily "cheated" by the students, who may choose to skip questions or give wrong answers on purpose;
- There can be non-intentional influences in the way the questions are formulated, which may lead the students to give answers perceived as "more appropriate";
- It is difficult to motivate the students to fill out the questionnaires; especially if they are too long and the students are not aware of the importance or the future uses of the questionnaires, they may tend to choose answers arbitrarily instead of thinking carefully about them;
- It is static, so the student model is created at the beginning of the course and stored once and for all, without the possibility to be updated.

A method of improving this approach is to give the student the possibility to modify her/his own profile, if she/he considers that the one inferred from the questionnaire results is not appropriate (does not correspond to the reality). This is called an "open model" (scrutable and modifiable) approach and it is used either in conjunction with the questionnaires or instead of them. This direct access of students to their own learner model has several advantages: it provides an increased learner control, it helps the learners develop their metacognitive skills and it also offers an evaluation of the quality of the model created by the system (Kay, 2001).

The main disadvantages of this approach are that it increases the cognitive load of the student and that it must rely on the self-evaluation of a student who might not be aware of her/his learning style.

There is also a second category of systems, which use an implicit and/or dynamic modelling method. Three different approaches have been identified in this respect:

- Analyse the performance of the students at evaluation tests – a good performance is interpreted as an indication of a style that corresponds to the one currently estimated and employed by the system, while a bad performance is interpreted as a mismatched learning style and triggers a change in the current learner model
- Ask the students to provide feedback on the learning process experienced so far and adjust the learner model accordingly
- Analyse the interaction of the students with the system (browsing pattern, time spent on various resources, frequency of accessing a particular type of resource etc.) and consequently infer a corresponding learning style.

Sometimes, these systems use a mixed modelling approach: they first use the explicit modelling method for the initialisation of the learner model and then the implicit modelling method for updating and improving the learner model.

Most of the existing systems treat learning styles in isolation of the rest of the features in the student profile (knowledge, interests, goals). The ideal is to integrate all these features in a more comprehensive and representative learner profile.

The IFS approach is as follows:

1. Selecting good taxonomies (models) of learning styles, e.g., (Felder & Silverman, 1988), (Honey & Mumford, 2000), the VARK style (Fleming, 1995)
2. Creating integrated learning style model (ILSM) which integrates characteristics from several models. Dedicated psychological questionnaire(s)
3. Creating open learning style model
4. Using implicit (dynamic) learning style modelling method
5. Integrating the rest features in the student profile (knowledge, interests, goals)

For example, the system presented in (Sangineto et al., 2008) is based on Felder-Silverman learning style model, and uses fuzzy values to estimate the preference of the student towards one of the four categories (Sensing-Intuitive, Visual-Verbal, Active-Reflective, Sequential-Global). Initially, the system offers to the learner the possibility to use the Soloman and Felder's (1998) psychological test or to directly set the values of the category types, choosing an estimated value for each category (using a slider-based interface).

INTERCONNECTION AND ONTOLOGIES

The second stage of IFS project is establishing interconnections of learning components (e.g. learning methods, activities and LOs) with students' learning styles.

In this Section, we consider interconnections of learning components with Honey & Mumford (1992) learning styles. This model is selected because of its popularity in scientific community. Honey and Mumford (1992) identified four distinct learning styles or preferences: Activist, Theorist; Pragmatist and Reflector.

Literature review has shown that from standardisation perspective the current versions of LOs specifications do not support personalised learning. On the other hand, Web 3.0 approach based on RDF, XML and ontologies is suitable to personalise LOs. Therefore, further on we'll concentrate on creating the ontologies to semantic search of LOs suitable to particular learners.

Interconnections of Learning Components with Learning Styles

Let us analyse personalisation of LOs in terms of providing the learners of particular learning style with LOs suitable for teaching / learning methods that could be part of their preferred learning activities. To achieve this goal, one should interconnect learning styles, learning activities, teaching / learning methods, and LOs types (see Table 1).

Table 1. Interconnection of learning styles, learning activities, teaching methods, and LOs types (according to Kurilovas et al., 2014)

Learning Styles (Honey & Mumford, 1992)	Preferred Learning Activities (Honey & Mumford, 1992)	Suitable Teaching/Learning Methods (iCOPER D3.1, 2009)	Suitable Learning Object Types (LRE APv4.7, 2011)
Activists are those people who learn by doing. Have an open-minded approach to learning, involving themselves fully and without bias in new experiences	Brainstorming, problem solving, group discussion, puzzles, competitions, and role-play.	3 Times 3 Things Learnt, Active Learning, Blogging, Brainstorming and Reflection, Competitive Simulation, Constellation, Creative Workshops, Creation of Personalized Learning Environments, Cultura, Cultural Awareness, E-Portfolio, Exercise Unit, Games Genre, Presenting Homework, Image Sharing, In-class Online Discussion, Mini Conference, Modeling, Online Reaction Sheets, Online Training, Peer Assessment, Process-based Assessment, Process Documentation, Project-based Learning, Resource-based Analysis, Role Play, Student Wiki Collaboration, World Café, WebQuest	Application, Assessment, Broadcast, Case study, Drill and practice, Educational game, Enquiry-oriented activity, Experiment, Exploration, Glossary, Open activity, Presentation, Project, Reference, Role play, Simulation, Tool, Website
Reflectors learn by observing and thinking about what happened. They prefer to stand back and view experiences from a number of different perspectives, collecting data and taking the time to work towards an appropriate conclusion	Paired discussions, self-analysis questionnaires, personality questionnaires, time out, observing activities, feedback from others, coaching, and interviews.	3 Times 3 Things Learnt, Blogging, Brainstorming and Reflection, Constellation, Creative Workshops, Creation of Personalized Learning Environments, Cultura, Cultural Awareness, E-Portfolio, Exercise Unit, Presenting Homework, Image Sharing, In-class Online Discussion, Listen–Do–Reflect, Ten-Plus-Two Variation, Mini Conference, Modeling, Online Reaction Sheets, Online Training, Peer-to-Peer Teaching, Peer Assessment, Process-based Assessment, Process Documentation, Resource-based Analysis, Student Wiki Collaboration, World Café, WebQuest	Application, Assessment, Broadcast, Case study, Demonstration, Glossary, Guide, Presentation, Reference, Textbook, Website
Pragmatists need to be able to see how to put the learning into practice in the real world. Abstract concepts and games are of limited use unless they can see a way to put the ideas into action in their lives. They are experimenters, trying out new ideas, theories and techniques to see if they work	Time to think about how to apply learning in reality, case studies, problem solving, and discussion.	3 Times 3 Things Learnt, Active Learning, Blogging, Brainstorming and Reflection, Competitive Simulation, Constellation, Creative Workshops, Creation of Personalized Learning Environments, Creation of Personalized Learning Environments, Cultura, Cultural Awareness, E-Portfolio, Exercise Unit, Games Genre, Presenting Homework, Image Sharing, In-class Online Discussion, Listen–Do–Reflect, Ten-Plus-Two Variation, Mini Conference, Modeling, Online Reaction Sheets, Online Training, Peer-to-Peer Teaching, Peer Assessment, Process-based Assessment, Process Documentation, Project-based Learning, Resource-based Analysis, Role Play, Student Wiki Collaboration, World Café, WebQuest	Application, Assessment, Broadcast, Case study, Course, Drill and practice, Educational game, Enquiry-oriented activity, Experiment, Exploration, Glossary, Guide, Open activity, Presentation, Project, Reference, Role play, Simulation, Tool, Website
Theorist learners like to understand the theory behind the actions. They need models, concepts and facts in order to engage in the learning process. Prefer to analyse and synthesise, drawing new information into a systematic and logical 'theory'.	Models, statistics, stories, quotes, background information, and applying theories.	3 Times 3 Things Learnt, Blogging, Constellation, Creative Workshops, Creation of Personalized Learning Environments, Cultura, Cultural Awareness, E-Portfolio, Exercise Unit, Presenting Homework, Image Sharing, In-class Online Discussion, Listen–Do–Reflect, Ten-Plus-Two Variation, Mini Conference, Modeling, Online Reaction Sheets, Online Training, Peer-to-Peer Teaching, Peer Assessment, Process-based Assessment, Process Documentation, Resource-based Analysis, Student Wiki Collaboration, World Café, WebQuest	Application, Assessment, Broadcast, Course, Demonstration, Glossary, Guide, Presentation, Reference, Textbook, Website

As it was mentioned before, we'll use Honey and Mumford (1992) research on learning styles and preferred learning activities.

From standardisations perspective, one of the last and very popular initiatives in the area are standardisation of LOs types according to European LRE APv4.7 (2011) (i.e., a popular European LOM standard application profile), and vocabulary of teaching / learning methods suggested by iCOPER D3.1 (2009).

We'll explain some of these interconnections further on using the example of problem solving activity.

Case Study of Problem Solving Learning Activity

Let us analyse the case study of "Problem Solving" activity (A1) on how to create an online course (i.e., sequence of works / methods).

First of all, we should interconnect A1 with suitable iCOPER teaching / learning methods and LRE AP LO types. According to iCOPER D3.1 (2009) teaching / learning methods vocabulary, A1 suitable methods are as follows:

M2: Active Learning: Learners propose, plan, execute, and evaluate a project that requires the application of knowledge from educational science. The setting of the project is the learners' actual work environment. Guidance is provided both at the working place and at the distance university;

M3: Blogging: Students download and set-up a Wordpress blog. They create custom categories, change themes, and install plug-ins. Practical work with the blog is to identify, describe and reflect upon how blogs can be used in educational contexts. Students use the blog to document the personal learning experience while working with the assignment;

M4: Brainstorming and Reflection: Brainstorming and Reflection is used to introduce new concepts and definitions that "seem familiar" to students but are not (yet) exactly and accurately defined, by brainstorming, raising ideas by shouting, and discussing the issues in the whole group;

M14: Presenting Homework: For each lesson a student prepares the homework for presentation in front of the class (e.g. via Power point and beamer, overhead transparency, etc.). The presentation should include both the description of the process and the final product of the homework. During the presentation students and teacher can ask questions and discuss or contribute alternative solutions and strategies;

M17: In-Class Online Discussion: A synchronous online discussion to be held in regular workshops to discuss a pre-determined question reflecting the course readings and lectures;

M21: Modelling: The main goal of the modelling teaching method is to train learners to solve problems autonomously;

M22: Online Reaction Sheets: Reaction sheets are used to collect students' feedback and to share it between all participants. The reaction sheets are likely to have an influence on the follow-up units;

M25: Peer Assessment: Peer assessment is based on the idea that students assess their peers' work in order to enhance students' interpretation and reflection;

M29: Resource-Based Analysis: Divided in groups, students work and compete to analyse a design problem, managing their available resources;

M31: Student Wiki Collaboration: Students work in small groups on a particular problem. All details of the work and the procedure are organized by themselves. Their results shall be published in a Wiki. Finally, short presentations face-to-face are given. Participation is voluntary.

According to LRE APv4.7 (2011) LOs types vocabulary, A1 suitable LOs types are as follows:

T1: Application: Computer software designed to enable end users to perform a specific task or group of tasks;
T12: Glossary: A resource that is a collection of specialized terms and their meanings usually arranged in a specified order;
T16: Presentation: Information organized and delivered (often by using specific presentation software) by an instructor in order to inform a group about a topic;
T18: Reference: A resource such as a database, dictionary, encyclopedia, glossary etc. that is a general source of information or which provides specific information on a topic or activity;
T21: Textbook: Resource providing comprehensive materials for specific topics (use for: chapters or other typical book components);
T22: Tool: Editors and other kinds of programs for producing something. Editors can process e.g. text or pictures and they can be used for creating and editing other LOs. Tools can also perform calculations or conversions;
T23: Website: A collection of web pages (containing a collection of hyperlinked documents and files) typically identified by a common IP address on the World Wide Web on the Internet;
T24: Social Media.

There are several sub-activities used in the problem solving activity A1. They are:

SA1: Discuss the problem-scenario in the group, which promotes communication skills and cooperative learning;
SA2: Brainstorm ideas to cross the learning boundaries promoting creative learning, and knowledge integration;
SA3: Identify the learning issues for research, which promotes active learning and critical thinking;
SA4: Research to construct the action plans, which promote new knowledge development;
SA5: Report the research findings to the group, which promotes peer-to-peer learning complete the final products such as project, research report or presentation, which promotes independent study and self-directed working skills in the teamwork.

Interconnections between these sub-activities (SA) and teaching / learning methods (M) are as follows:

SA1 <=> M17;

SA2 <=> M3 and M4;

SA3 <=> M2 and M22;

SA4 <=> M21, M29 and M31, and

SA5 <=> M14 and M25

Figure 1. Interconnection between teaching/learning methods (M) and LOs types (T) for SA2 sub-activity (according to Kurilovas et al., 2014)

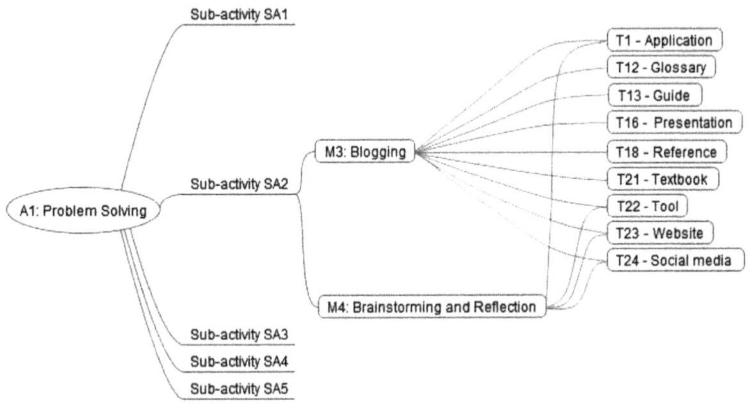

Figure 1 shows interconnection between teaching / learning methods (M) and LOs types (T) for the case of Problem Solving Activity's A1 sub-activity SA2:

Indeed, sub-activity A2 "Brainstorm ideas to cross the learning boundaries, which promotes creative learning, and knowledge integration" can be realised by M3 "Blogging" and M4 "Brainstorming and Reflection" teaching / learning methods, while the other iCOPER methods are not directly applicable to implement A2. In its turn, suitable LOs to implement the M3 "Blogging" method are T1 "Application", T12 "Glossary", T13 "Guide", T16 "Presentation", T18 "Reference", T21 "Textbook", T22 "Tool", T23 "Website", and T24 "Social media". In its turn, M4 "Brainstorming and Reflection" method could be implemented using different means to enhance brainstorming of ideas and its further reflection are applications (T1), tools (T21), websites (T23), and social media (T24).

Figure 2 presents sets portrait between the sets of (1) learning activities, i.e. Problem Solving (A1), (2) sub-activities SA1…SA5, (3) teaching-learning methods M2…M31, and (4) suitable LO types T1…T24:

Figure 2. Sets portrait between the sets of learning methods (M) and LOs types (T) for problem solving activity A1 (according to Kurilovas et al., 2014)

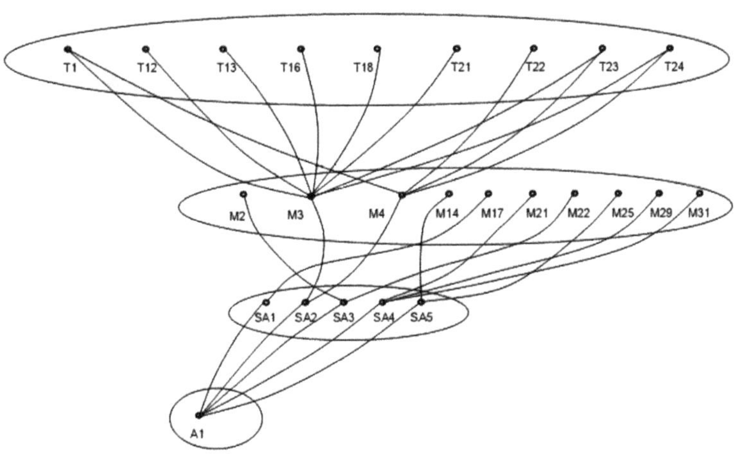

Ontologies

The sets portrait of the presented interconnections should be followed by the appropriate ontology. The ontology is the main tool to create learners' personalised learning environments consisting of LOs, suitable teaching / learning methods, and activities according to their preferred learning styles. The ontology should help the learner to find suitable LOs according to preferred learning methods / activities, and vice versa, and thus to personalise learning.

The sets portraits between the sets of learning activities (A), learning methods (M), and LOs types (T) presented in Table 1 and Figures 1-2 are suitable to apply in the ontologies and, accordingly, in creating search engines to find LOs suitable to particular learners (i.e. learning styles) by implementing semantic search using presented activities / sub-activities and methods, and vice versa. Let us provide some examples to illustrate this.

Ontology Example 1: Query for finding suitable learning activities (i.e. "Problem Solving" activity could be found using "Blogging" teaching / learning method) (see Figure 3).

Figure 3. Ontology example 1: query for finding suitable learning activities by methods (according to Kurilovas et al., 2014)

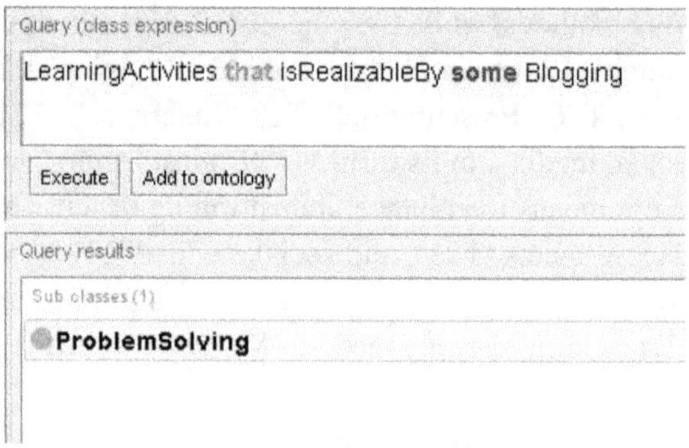

Figure 4. Ontology example 2: query for finding suitable teaching/learning methods by LOs types (according to Kurilovas et al., 2014)

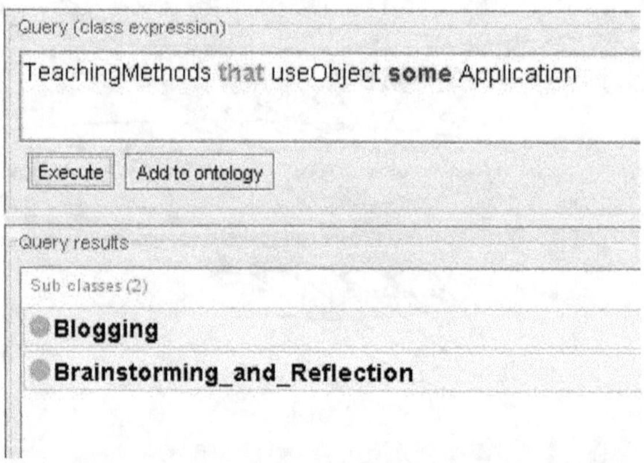

Ontology Example 2: Query for finding suitable teaching / learning methods (i.e. "Blogging" and "Brainstorming and Reflection" teaching / learning methods could be found using suitable Application LO type) (see Figure 4).

RECOMMENDER SYSTEM

The following stage of IFS project is creating recommender system to recommending particular learners suitable learning components.

In this Section, we consider the case of Web 2.0 tools recommender system for learning activities of VARK learning styles' students according to Juskeviciene and Kurilovas (2014).

Web 2.0 tools can be defined as web applications that facilitate participators' information sharing, interoperability, user-centred design, and collaboration on the World Wide Web (Kolesinski et al., 2013, p.16).

There are plenty of Web 2.0 tools and examples of their use in education. Most of the researches focus on e-learning systems personalisation functionalities such as personalised learning plans and learning materials. Although in practice and in many sources of literature the need for the application of Web 2.0 tools in education is highlighted, however there is lack of clear methods how these tools could be applied in learning for a higher learning quality, and there is a lack of e-learning systems that implement these methods. Also, these tools are not always suitable for the task to which they are applied and they are not always properly used for a specific purpose. Furthermore, most of the e-learning systems are focused on the course rather than on the learner, which means that they do not satisfy users. In order to address these problems, learning environments must be more adapted to the learner.

Learning styles model VARK designed by Fleming (2006) is an acronym made from Visual, Aural, Read/write and Kinaesthetic. These modalities are used for preferable ways of learning (taking and giving out) information. Visual learners prefer to receive information from depictions in figures: in charts, graphs, maps, diagrams, flow charts, circles, hierarchies, and others. It does not include pictures, movies and animated websites that belong to Kinaesthetic. The aural perceptual mode describes a preference for spoken or heart information. Aural learners learn best by discussing, oral feedback, email, chat, discussion boards, and oral presentations. Read/write learners prefer information displayed as words: quotes, lists, texts, books, and manuals. The kinaesthetic perceptual mode describes a preference for reality and concrete situations. They prefer videos, teaching others, pictures of real things, examples of principles, practical sessions, and others. Multimodals are those learners who have preferences in more than one mode.

When planning the learning process, first of all we have to set the learning objective (or outcome). One of the basic and essential elements within the education community is a classification of learning objectives (especially, cognitive) in education, presented by Bloom et. al. (1956) is Bloom's taxonomy. The learning objectives can be achieved through a sequence of tasks, performed by the learner. These tasks form a learning method.

The Learning 2.0 allows learners to use Web 2.0 tools, to take an active role in the learning process, and to determine their own learning strategies according to their interests and needs. Active learning methods and Web 2.0 tools can enhance the learning effectiveness. However, it is important to consider each learner's individual preferences. Personalisation deals with this kind of problem. This work takes into account the learner's preferences to the educational material (based on VARK learning styles model) and communication modes.

The proposed method for Web 2.0 tools integration into the learning activities was implemented in prototype of the system. This prototype has been developed following the working principles of the knowledge-based recommender system. The domain knowledge was conceptualised in the ontology. Sub-section 3.1 presents the basic elements of developed ontology. The proposed method is introduced in Sub-section 3.2.

Description of the Basic Elements of the Proposed Method for Integrating Web 2.0 Tools in Learning Activities

With the view to find a particular Web 2.0 tool suitable for learner (for learning activity accomplishment), a link between the tool and the learning activity must be identified. This relationship can be established by interconnections between the defined tool and activity elements.

The learning activity is defined as consisting of the following elements:

1. Learning Activity (what action a learner performs).
2. Content (which object a learner manages).
3. Interaction (with whom a learner interacts).
4. Synchronicity (at what time a learner performs the intended action).

Web 2.0 tool is defined as set of universal functions. This universal function is defined as consisting of the following elements:

1. Function (what action can be performed by using a tool).
2. Artefact (which object (information format based on VARK) can be managed by using a tool).
3. Interaction (what kind of interaction the tool enables).
4. Synchronicity (at what time the intended action is enabled by a tool to take place).

The learning activities and functions of tools are classified mainly based on the Conole & Fill (2005) media taxonomy. These types and particular elements are presented in Table 2.

Table 2. Learning activities and web 2.0 tool function types (according to Juskeviciene & Kurilovas, 2014)

Type	Learning Activities	Subtype (1 -8)	Web 2.0 Tool Function
Narrative	Revise	1 - View	Explore (Read, view, listen)
Information management	Find	2 - Search	Search
	Collect	3 - Host	Host (Store), Syndicate
Productive	Prepare	4 – Create	Create (draw, write, record, edit)
Communicative	Present	5 – Share	Share, publicise
	Dispute	6 – Discuss	Communicate
Imitative	Role play	7 - Imitate	Simulate (Game simulation)
	Observation	8- Model	Model (Phenomenon modelling)

Thus, Web 2.0 tools could be divided based on their usage possibilities, managed objects, communication form, and sort of imitation process into three groups as follows:

1. Artefacts management (including narrative, productive and information management),
2. Communication, and
3. Imitation tools.

For domain knowledge conceptualisation we have used OWL DL (one of the three species of OWL) which can be viewed as expressive Description Logics, with an ontology being equivalent to SHOIN(D) Description Logic knowledge base (Horrocs et al., 2003) and being decidable (Motic et al., 2005).

Also, in the current study, we develop an ontology mainly based on the (Fernández et al., 1999) METHONTOLOGY method which proposes such activities: (1) the specification activity (states why the ontology is being built, what its intended uses are and who the end-users are), (2) the conceptualisation activity (developing an ontology conceptual model), (3) the formalisation activity (transformation of the conceptual model into a formal or semi-computable model), (5) the implementation activity (creating computable models in the ontology language), and (6) the maintenance activity (evaluation, corrections and updates).

During the specification activity, Uschold and King (1995) propose the strategy for identifying the main concepts in the ontology: a top-down approach where the main abstract concepts are identified and then specialised into more specific concepts; and (Gruninger & Fox, 1995) suggest to identify a set of competency questions that must be answered by a model built based on the ontology. The ontology must be able to provide a vocabulary for expressing these questions. Axioms in the ontology should be able to characterise the answers to the competency questions. Ontology evaluation can be seen as evaluation of the properties (e.g., language conformity (Syntax) and consistency (Semantics)) of ontologies generated by development tools (Sure et al., 2003, p.7).

We have defined the following components in the domain ontology visualised with Protégé 4.3 ontology editor: (1) Concepts (Main Classes) (Figure 5), and (2) Relationships between Concepts (Properties) (Figure 6).

Figure 5. The main ontology classes (according to Juskeviciene & Kurilovas, 2014)

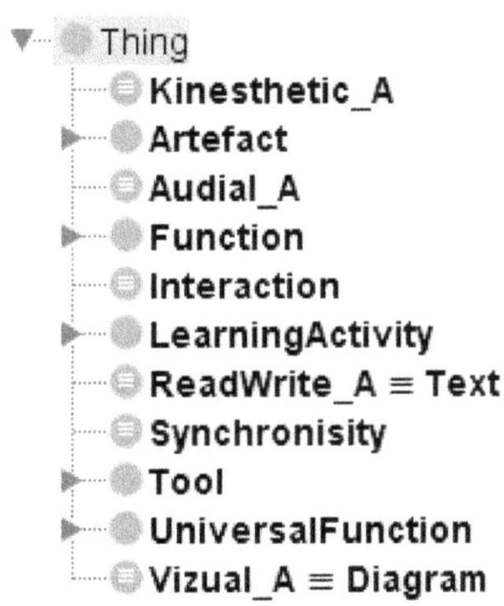

Figure 6. Ontology properties (according to Juskeviciene & Kurilovas, 2014)

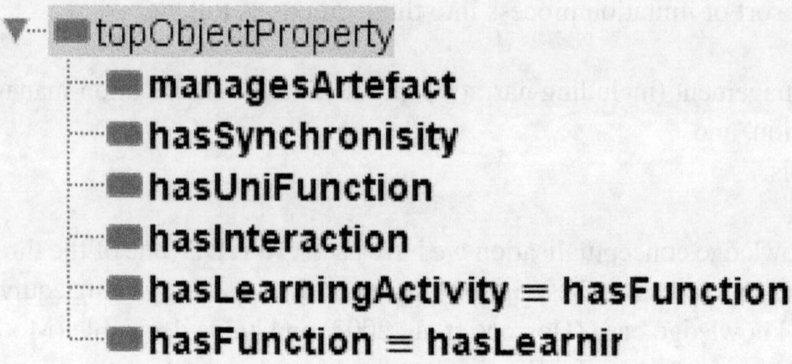

Table 3. Domain and range of properties (according to Juskeviciene & Kurilovas, 2014)

Property	Domain	Range
hasUniFunction	*Tool*	*UniversalFunction*
hasFunction = hasLearningActivity	*UniversalFunction*	*Function*
managesArtefact	*UniversalFunction*	*Artefact*
hasInteraction	*UniversalFunction*	*Interaction*
hasSynchronisity	*UniversalFunction*	*Synchronisity*

Domain and range of properties are presented in Table 3.

The Steps of the Method for Integrating Web 2.0 tools in Learning Activities

These steps are as follows:

- Identification of student's learning style (i.e. preferences of the learning content and communication modes).
- Selection of the learning objective and the learning method.
- Determination of the elements of the learning activities according to chosen learning method.
- Determination of universal function elements of each Web 2.0 tool.
- Finding the link between Web 2.0 tool and learning activity elements.
- Selection of a suitable Web 2.0 tool based on specified elements: Action, Interaction, Synchronicity. Artefact is determined based on individual learning style.

These steps are described below.

An official VARK learning styles model questionnaire was used to diagnose particular learner's preferences. The users can select none or more than one answer to each question, relying on individual perception. Thus, as a result the preferences of learning content format are found.

When a learner is planning the learning process, first (s)he sets the learning objective. The learning method can be selected to achieve this objective. The learning method can be seen as a set of learning activities. Each learning activity can consist of four elements as described in Sub-section 3.1. Each

Web 2.0 tool is analysed as a tool consisting of universal functions. Tool class is described by using closure axiom (i.e., universal restriction that acts along the property to say that it can only be filled by the specified fillers) due to Open world assumption in OWL. These universal functions can consist of four elements as described in Sub-section 3.1.

Interconnections between Web 2.0 tool and learning activity elements are settled as follows:

- Learning Activity and Function
- Content and Artefact
- Interaction and Interaction
- Synchronicity and Synchronicity

With the view to select a suitable Web 2.0 tool based on specified elements the developed ontology was used. Queries could be performed only using a classified ontology. The OWL consistency checker Pellet was used for this purpose as well as for ontology Consistency checking, Concept satisfiability, and Realisation (Sirin et al., 2007).

Also, the DL query tab of the Protégé ontology editor was used to answer the competency questions. The query language (class expression) supported by this plug-in is based on the Manchester OWL syntax, a user-friendly syntax for OWL DL that is fundamentally based on collecting all the information about a particular class, property, or individual into a single construct, called a frame. The example of a formalised competency question: "What tool is suitable for a visual learner individually to view the learning content?" in Protégé 4.3 ontology editor is presented in Figure 7.

This query in the SHOIN(D) descriptive language is described as follows:

$\text{Tool} \sqcap \exists \text{hasUniFunction.(hasFunction.}\{\text{show\_play}\} \sqcap \exists \text{managesArtefact.Vizual\_A} \sqcap \text{hasInteraction.}\{\text{Ind}\})) = \{\text{Lucidchart, Creately, Gliffy}\}$

Furthermore, this query could be written as follows:

$\text{Tool} \sqcap \exists \text{hasUniFunction.(hasLearningActivity.}\{\text{view}\} \sqcap \exists \text{managesArtefact.Vizual\_A} \sqcap \text{hasInteraction.}\{\text{Ind}\})) = \{\text{Lucidchart, Creately, Gliffy}\}$

Such a variety of question expressions is possible due to the statements equivalence made on the properties (e.g. $\text{hasFunction} \equiv \text{hasLearningActivity}$) and equality asserted between individuals (e.g. $\{\text{show\_play}\} \equiv \{\text{view}\}$) in the developed ontology.

Figure 7. The example of query (according to Juskeviciene & Kurilovas, 2014)

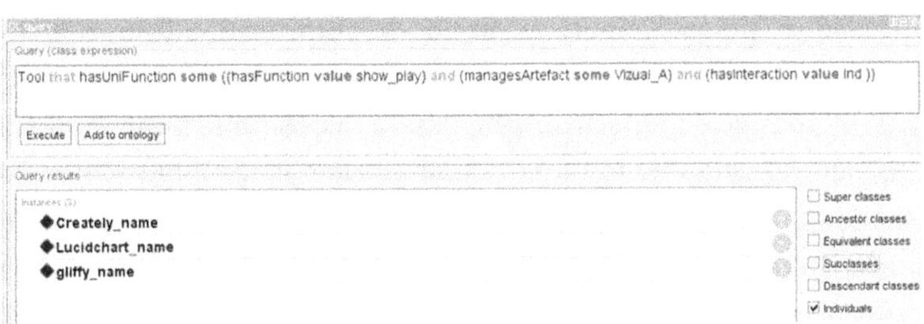

Figure 8. Scheme of the recommender system (according to Juskeviciene & Kurilovas, 2014)

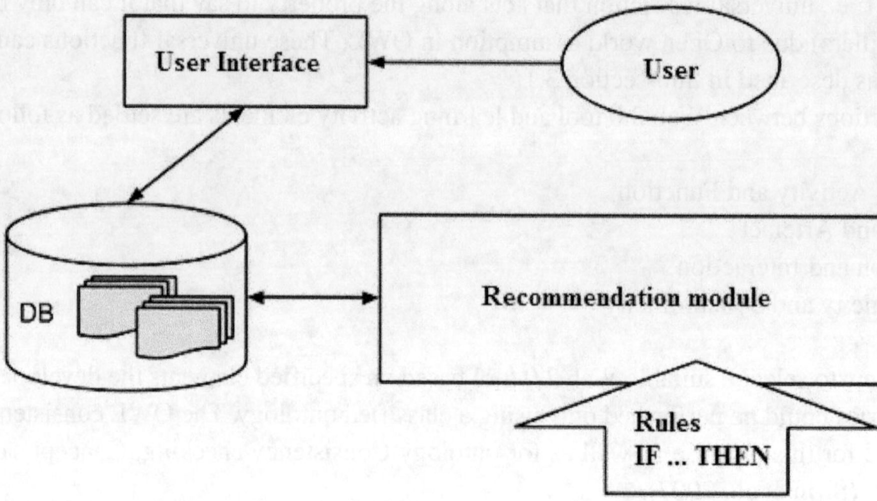

Prototype of the Recommender System

The developed prototype of the knowledge-based recommender system (Figure 8) implements the method proposed completely.

The system uses the background data before the recommending process begins. These data are in the learner's profile (in DB). The profile has been modelled by creating fixed stereotypes. Thus, the learners (USERS) were attached to a particular category based on the VARK learning styles model.

Also, the user's profiles were represented as sets of weighted keywords. Each keyword refers to the learning style (visual, aural, read/write, kinaesthetic), and the weights (numerical representations) indicate its importance in the profile.

The information needed for the system as input data (trough USER INTERFACE) was provided by a user by selecting the learning objective and the learning method as well as by specifying learning activities, i.e. desirable synchronicity and interaction.

The background and input data were combined and processed by programs (in the RECOMMENDATION MODULE) based on the IF…THEN rules. If all the interconnections between the tool and learning activity elements have been found then the appropriate tool (which has the specified properties) was recommended

While performing the expert evaluation of the system prototype based on the method proposed it has been determined that (Juskeviciene & Kurilovas, 2014):

- The results provided by the system are accurate – recommended tools are suitable to implement desirable learning activities, i.e. the tools proposed provide a possibility to perform a desirable pedagogical activity in time and to participate in it for a desirable number of learners, and artefacts managed by the tools are suitable for a particular learner according to his/her learning style.
- The system performance is complete and suitable, i.e. it completely corresponds to the method proposed and it operates as it was planned – the system recommends tools in a personalised way.

- Using the created recommender system, it is possible to find Web 2.0 tools suitable for personalising learning quicker than without the system.
- The method and prototype proposed are qualitative with regard to accurateness, completeness and efficiency criteria – the system is 81.4% of the absolute quality.

CONCLUSION

In this Chapter, the author proposed the Intelligent Future School (IFS) approach to personalise learning using intelligent technologies.

IFS approach consists of the following stages:

1. Creating learners' models (Section 3) by:
    a. Selecting suitable taxonomies (models) of learning styles
    b. Creating integrated learning style model which integrates characteristics from several models. Dedicated psychological questionnaire(s)
    c. Creating open learning style model
    d. Using implicit (dynamic) learning style modelling method
    e. Integrating the rest features in the student profile (knowledge, interests, goals)
2. Creating interconnections and ontologies to establish suitability of LOs, teaching / learning methods, and learning activities to particular learning styles:

The novel sets portrait analysing the interconnections between students' learning styles, their preferred learning activities, relevant teaching / learning methods, and LOs types is presented in the Chapter. The well-known European standardised vocabularies of teaching / learning methods and LOs types were used to establish these interconnections. Problem solving activity example is provided to illustrate the model. The sets portrait of these interconnections is followed by the appropriate ontology to create learners' personalised learning environments consisting of LOs, suitable teaching / learning methods and activities according to their preferred learning styles. The ontology presented should help the learners to find suitable LOs according to their preferred learning methods / activities, and vice versa, and thus to personalise learning.

3. Creating recommender system:

In the research presented, while analysing Web.2.0 tools and recommender systems, the following conclusions and results have been obtained:

- In this work, the Web 2.0 classification method that helps a user to learn effectively in the Web 2.0 tools-rich environment has been proposed. With regard to the usage possibilities and managed objects, Web 2.0 tool could be divided into three groups as follows: (1) artefacts management tools, (2) communication tools, and (3) imitation tools. In this way it is possible to group Web 2.0 tools according to their usage possibilities and managed objects, their communication form and sort of imitation process.

- It is proposed to personalise learning according to learner's preferences in terms of learning content and communication modes according VARK learning styles model theory.
- Recommender systems are suitable to personalise learning by proposing the learning tools according to the learner's preferences. Knowledge-based recommender systems propose the elements associating their features with user's features described in the user profile that could be created in the e-learning system according to the learning styles theory. The domain knowledge and user profile used for recommending could be described by ontology, thus enabling knowledge reusability suitable both for human beings and program modules.

The method for Web 2.0 tools integration into the learning activities was created based on the appropriate research and analysis and on recommender systems application to personalise learning as well as the practical experience gained on using these tools.

The advantages of the method proposed are as follows:

- It helps to select Web 2.0 tool for learner suitable to implement desirable learning activity according to his/her learning style, i.e. learner's preferences in terms of learning content and collaboration form. Thus, learning process is personalised.
- Its elements are described by the ontology in formal way, and this enables to use domain knowledge in recommender system based on this method.

The method and prototype proposed are qualitative with regard to accurateness, completeness and efficiency criteria – the system is 81.4% of the absolute quality.

REFERENCES

Aroyo, L., & Kommers, P. (1999). Intelligent Agents for Educational Computer-Aided Systems. *Journal of Interactive Learning Research*, *10*(3–4).

Bennane, A. (2013). Adaptive Educational Software by Applying Reinforcement Learning. *Informatics in Education*, *12*(1), 13–28.

Berners-Lee, T., Hendler, J., & Lassila, O. (2001). The Semantic Web. *Scientific American*, *284*(5), 34–43. doi:10.1038/scientificamerican0501-34 PMID:11396337

Bloom, B. S., Engelhart, M. D., Furst, E. J., Hill, W. H., & Krathwohl, D. R. (1956). *Taxonomy of educational objectives: Handbook I: Cognitive domain*. New York: David McKay.

Bower, M., Hedberg, J. G., & Kuswara, A. (2010). A framework for Web 2.0 learning design. *Educational Media International*, *47*(3), 177–198. doi:10.1080/09523987.2010.518811

Brooks, C., & McCalla, G. (2006). Towards flexible learning object metadata. *International Journal of Continuing Engineering Education and Lifelong Learning*, *16*(1/2), 50–63. doi:10.1504/IJCEELL.2006.008917

Brusilovsky, P. (1996). Methods and techniques of adaptive hypermedia. *User Modeling and User-Adapted Interaction*, *6*(2–3), 87–129. doi:10.1007/BF00143964

Brusilovsky, P., Eklund J., & Schwarz, E. (1998). Web-based education for all: a tool for development adaptive courseware. *Computer Networks and ISDN Systems, 30*(1-7), 291–300.

Brusilovsky, P., & Millan, E. (2007). User Models for Adaptive Hypermedia and Adaptive Educational Systems. In P. Brusilovsky, A. Kobsa, & W. Neidl (Eds.), Lecture Notes in Computer Science: Vol. 4321. *The Adaptive Web: Methods and Strategies of Web Personalization* (pp. 3–53). Springer. doi:10.1007/978-3-540-72079-9_1

Buriano, L., Marchetti, M., Carmagnola, F., Cena, F., Gena, C., & Torre, I. (2006). The role of ontologies in context-aware recommender systems. In *Proceedings of 7th International Conference on Mobile Data Management (MDM) 2006*. IEEE. Available at: http://creative.eun.org/

Chen, Y., Wu, C., Guo, X., & Wu, J. (2012). Semantic Learning Service Personalized. *International Journal of Computational Intelligence Systems*, *5*(1), 163–172. doi:10.1080/18756891.2012.670528

Coffield, F., Moseley, D., Hall, E., & Ecclestone, K. (2004). *Learning styles and pedagogy in post-16 learning. A systematic and critical review*. Learning and Skills Research Centre.

Conole, G., & Fill, K. (2005). A learning design toolkit to create pedagogically effective learning activities. *Journal of Interactive Media in Education*, *2005*(1), 9. doi:10.5334/2005-8

Dagger, D., Wade, V., & Conlan, O. (2002). Towards a standards-based approach to e-Learning personalization using reusable learning objects. In *World Conference on E-Learning in Corporate, Government, Healthcare, and Higher Education.*

EUN. (n.d.). Available at http://www.eun.org/

Felder, R.M. (1996). Matters of Style. *ASEE Prism, 6*(4).

Felder, R. M., & Silverman, L. K. (1988). Learning and teaching styles in engineering education. *English Education*, *78*(7), 674–681.

Fernández, M., Gómez-Pérez, A., Pazos, A., & Pazos, J. (1999). Building a Chemical Ontology Using METHONTOLOGY and the Ontology Design Environment. *IEEE Intelligent Systems*, *14*(1), 37–46. doi:10.1109/5254.747904

Fleming, N. (2006). Teaching and learning styles: VARK (2nd ed.). Academic Press.

Fleming, N. D. (1995). I am different; not dumb. Modes of presentation (V.A.R.K.) in the tertiary classroom. In A. Zelmer (Ed.), *Research and development in higher education.Proceedings of the1995 annual conference of the higher education and research development society of Australia* (HERDSA).

Gardner, H. (1999). *Intelligence Reframed: Multiple Intelligences for the 21st Century*. Basic Books.

Gauch, S., Speretta, M., Chandramouli, A., & Micarelli, A. (2007). User profiles for personalized information access. In The adaptive Web (pp. 54–89). Springer Berlin Heidelberg. doi:doi:10.1007/978-3-540-72079-9_2 doi:10.1007/978-3-540-72079-9_2

Grasha, A. (1996). *Teaching with Style*. Alliance Publishers.

Gruber, T. R. (1993). A translation approach to portable ontologies. *Knowledge Acquisition*, *5*(2), 199–220. doi:10.1006/knac.1993.1008

Gruninger, M., & Fox, M. (1995). Methodology for the Design and Evaluation of Ontologies. In *Proceedings of the IJCAI'95.Workshop on Basic Ontological Issues in Knowledge Sharing*.

Herrmann, N. (1996). *The Whole Brain Business Book*. McGraw-Hill.

Honey, P., & Mumford, A. (1992). *The Manual of Learning Styles* (3rd ed.). Maidenhead, UK: Peter Honey.

Honey, P., & Mumford, A. (2000). *The learning styles helper's guide*. Maidenhead, UK: Peter Honey Publications Ltd.

Horrocks, I., Patel-Schneider, P. F., & Van Harmelen, F. (2003). From SHIQ and RDF to OWL: The making of a web ontology language. *Web Semantics: Science, Services, and Agents on the World Wide Web*, *1*(1), 7–26. doi:10.1016/j.websem.2003.07.001

Husain, W., & Dih, L. (2012). A Framework of a Personalized Location-based Traveler Recommendation System in Mobile Application. *IJMUE*, *7*(3), 11–18.

iCOPER D3.1. (2009). *ICOPER Report on the Standardized Description of Instructional Models. Deliverable 3.1*. Available at: http://nm.wu.ac.at/nm/icoper

iTEC. (n.d.). Available at: http://itec.eun.org/web/guest

Jadhav, A., & Sonar, R. (2009). An integrated rule-based and case-based reasoning approach for selection of the software packages. In Information Systems, Technology and Management (pp. 280–291). Springer Berlin Heidelberg. doi:doi:10.1007/978-3-642-00405-6_30doi:10.1007/978-3-642-00405-6_30

Juskeviciene, A., & Kurilovas, E. (2014). On Recommending Web 2.0 Tools to Personalise Learning. *Informatics in Education*, *13*(1), 17–30.

Kapoor, B., & Sharma, S. (2010). A Comparative Study Ontology Building Tools for Semantic Web Applications. *International journal of Web & Semantic Technology, 1*(3), 1–13.

Kay, J. (2001). Learner Control. *User Modeling and User-Adapted Interaction*, *11*(1/2), 111–127. doi:10.1023/A:1011194803800

Keefe, J. (1979). Learning style: an overview. In *NASSP's Student Learning Styles: Diagnosing and Prescribing Programs*.

Keirsey, D. (1998). *Please Understand me II: Temperament Character Intelligence*. Prometheus Nemesis Books.

Khalid, M. S., Rongbutsri, N., & Buus, L. (2012). *Facilitating adoption of Web tools for problem and project based learning activities*. Academic Press.

Kim, D., & Lee, J. (2013). Development of Intelligent Instruction System for Mathematical Computation. *Informatics in Education*, *12*(1), 93–106.

Kolb, D. A. (1985). *LSI Learning- Style Inventory*. McBer & Company.

Kolesinski, M. T., Nelson-Weaver, E., & Diamond, D. (2013). *Digital Solidarity in Education: Promoting Equity, Diversity, and Academic Excellence through Innovative Instructional Programs*. Routledge.

Kurilovas, E., & Dagiene, V. (2010). *Evaluation of Quality of the Learning Software. Basics, Concepts, Methods, Monograph*. Saarbrücken, Germany: LAP LAMBERT Academic Publishing.

Kurilovas, E., Kubilinskiene, S., & Dagiene, V. (2014). Web 3.0 – Based Personalisation of Learning Objects in Virtual Learning Environments. *Computers in Human Behavior*, *30*, 654–662. doi:10.1016/j.chb.2013.07.039

Li, L., Tang, S., Fang, L., Xiao, R., Deng, X., Xu, Y., & Xu, Y. (2007). Voeditor: a visual environment for ontology construction and collaborative querying of semantic Web resources. In *31st Annual International Computer Software and Applications Conference* (COMPSAC). IEEE. doi:doi:10.1109/COMPSAC.2007.237 doi:10.1109/COMPSAC.2007.237

LRE APv4.7. (2011). *Learning Resource Exchange Metadata Application Profile v.4.7.* Available at: http://lreforschools.eun.org/c/document_library/get_file?p_l_id=10970&folderId=12073&name=DLFE-1.pdf

LSL. (n.d.). Available at: http://lsl.eun.org/

McCalla, G. (2004). The ecological approach to the design of e-learning environments: purpose-based capture and use of information about learners. Journal of Interactive Media in Education.

Middleton, S. E., Roure, D. D., & Shadbolt, N. R. (2009). Ontology-based recommender systems. In S. Staab & R. Studer (Eds.), Handbook on Ontologies, International Handbooks on Information Systems (pp. 779–796). Springer. doi:doi:10.1007/978-3-540-92673-3_35 doi:10.1007/978-3-540-92673-3_35

Mohan, P., & Brooks, C. (2003). Learning Objects on the Semantic Web.*Proc. Int'l Conf. Advanced Learning Technologies*. doi:10.1109/ICALT.2003.1215055

Motik, B., Sattler, U., & Studer, R. (2005). Query answering for OWL-DL with rules. *Web Semantics: Science, Services, and Agents on the World Wide Web*, *3*(1), 41–60. doi:10.1016/j.websem.2005.05.001

Mulvenna, M. D., Anand, S. S., & Buchner, A. G. (2000). Personalization on the Net using Web mining: Introduction. *Communications of the ACM*, *43*(8), 122–125. doi:10.1145/345124.345165

Pashler, H., McDaniel, M., Rohrer, D., & Bjork, R. (2008). Learning styles concepts and evidence. *Psychological Science in the Public Interest*, *9*(3), 105–119. PMID:26162104

Popescu, E. (2009). Diagnosing students' learning style in an educational hypermedia system. In Cognitive and Emotional Processes in Web-based Education: Integrating Human Factors and Personalization. IGI Global.

Ricci, F., Rokach, L., & Shapira, B. (2011). Introduction to recommender systems handbook. Springer US. doi:doi:10.1007/978-0-387-85820-3 doi:10.1007/978-0-387-85820-3

Sampson, D. (2002). Current and Future Research and Technology Developments in e-Learning. *Industry and Higher Education Journal, 16*(4).

Sampson, D., & Karagiannidis, C., & Kinshuk. (2002). Personalised Learning: Educational, Technological and Standardisation Perspective. *Interactive Educational Multimedia*, *4*, 24–39.

Sangineto, E., Capuano, N., Gaeta, M., & Micarelli, A. (2008). Adaptive course generation through learning styles representation. *Journal of Universal Access in the Information Society*, *7*(1), 1–23. doi:10.1007/s10209-007-0101-0

Sarasin, L. (1998). *Learning Style Perspectives: Impact in the Classroom*. Atwood Publishing.

Schunk, D. (1996). *Learning Theories: An Educational Perspective*. Prentice Hall.

Sirin, E., Parsia, B., Grau, B. C., Kalyanpur, A., & Katz, Y. (2007). Pellet: A practical owl-dl reasoner. *Web Semantics: Science, Services, and Agents on the World Wide Web*, *5*(2), 51–53. doi:10.1016/j.websem.2007.03.004

Soloman, B., & Felder, R. M. (1998). *Index of learning styles questionnaire*. Available at: http://www.engr.ncsu.edu/learningstyles/ilsweb.html

Starkey, L. (2011). Evaluating learning in the 21st century: A digital age learning matrix. *Technology, Pedagogy and Education*, *20*(1), 19–39. doi:10.1080/1475939X.2011.554021

Sure, Y., Staab, S., & Studer, R. (2003). On-to-knowledge methodology (OTKM). In Handbook on Ontologies. Springer.

Troussas, C., Virvou, M., & Alepis, E. (2014). Collaborative Learning: Group Interaction in an Intelligent Mobile-Assessed Multiple Language Learning System. *Informatics in Education*, *13*(2), 279–292. doi:10.15388/infedu.2014.08

Uschold, M., & King, M. (1995). Towards a Methodology for Building Ontologies. *Proceedings of the IJCAI'95.Workshop on Basic Ontological Issues in Knowledge Sharing*.

Vega-Gorgojo, G., Bote-Lorenzo, M. L., Gomez-Sanchez, E., Asensio-Perez, J. I., Dimitriadis, Y. A., & Jorrn-Abellan, I. M. (2008). Ontoolcole: Supporting educators in the semantic search of CSCL tools. *Journal of Universal Computer Science*, *14*(1), 27–58.

Vesin, B., Ivanović, M., Klašnja-Milićević, A., & Budimac, Z. (2012). Protus 2.0: Ontology-based semantic recommendation in programming tutoring system. *Expert Systems with Applications*, *39*(15), 12229–12246. doi:10.1016/j.eswa.2012.04.052

Vesin, B., Ivanović, M., Klašnja-Milićević, A., & Budimac, Z. (2013). Ontology-based architecture with recommendation strategy in java tutoring system. *Computer Science and Information Systems*, *10*(1), 237–261. doi:10.2298/CSIS111231001V

Wallden, S., & Makinen, E. (2014). Educational Data Mining and Problem-Based learning. *Informatics in Education*, *13*(1), 141–156.

Wang, H. C., & Huang, T. H. (2013). Personalized e-learning environment for bioinformatics. *Interactive Learning Environments*, *21*(1), 18–38. doi:10.1080/10494820.2010.542759

Wiley, D. A. (2000). *Connecting Learning Objects to Instructional design Theory: a definition, a Metaphor, and a Taxonomy*. Utah State University. Available at: http://www.reusability.org/read/

Yoo, S., Kim, V., & Park, S. (2013). An Educational tool for Browsing the Semantic Web. *Informatics in Education*, *12*(1), 143–151.

Youn, S., & McLeod, D. (2006). *Ontology Development Tools for Ontology-Based Knowledge Management. Encyclopedia of E-Commerce, EGovernment and Mobile Commerce*. Idea Group Inc.

Chapter 2

Collaborating Partnerships:
A Project-Based Legitimizing Strategy amongst East African E-Learning Providers

Titus Tossy
Mzumbe University, Tanzania

Irwin Brown
University of Cape Town, South Africa

ABSTRACT

This chapter examines the phenomenon of e-learning in East Africa. The classic grounded theory methodology was used to carry out the study. Through multiple encounters with 24 e-learning providers, the investigation revealed legitimizing of e-learning projects to be a significant issue for these providers. Developing collaborating partnerships was a key strategy used by the providers to legitimize e-learning projects. The study identified two types of collaborating partnerships: Asymmetrical and Reciprocal. These findings are positioned within literature to highlight the contribution that this paper makes to the discourse on e-learning projects in developing countries.

INTRODUCTION

E-learning is variously defined as the use of information and communication technologies or ICTs (e.g. computers, the Internet and mobile devices) to facilitate teaching and learning (Cowie & Sakui, 2013; Mayer, 2003). Naidu (2006) draws a distinction between four types of e-learning – individualised learning that is online (i.e. Internet-based); individualised learning that is offline (e.g. materials delivered via CD or DVD); group learning that is asynchronous (e.g. via email lists), and group learning that is synchronous (e.g. video conferencing). E-learning in this paper encompasses all these different types.

Demonstrating the legitimacy of e-learning has been a relevant and persistent concern to researchers and practitioners alike for several years (Kilby, 2001; Lisewski & Joyce, 2003; Yick et al., 2005). This concern has arisen as a result of several factors. For example, predictions were made at the turn of the century, suggesting that e-learning would substantially reduce, if not replace, face-to-face learn-

DOI: 10.4018/978-1-5225-0539-6.ch002

ing environments (Carliner, 2002; Njenga & Fourie, 2010). This prediction has not been fulfilled, with many e-learning projects drastically scaled down or abandoned (Carliner & Shanks, 2007). There has furthermore been an unusually high rate of failure associated with e-learning projects, which has lead to the credibility of these projects being questioned (Keegan et al., 2007). It has also been widely assumed that e-learning results in better learner performance than traditional learning modes (Piccoli et al., 2001). Research evidence has not always supported this assumption (Piccoli et al., 2001; Njenga & Fourie, 2010), highlighting that performance is not necessarily improved simply by changing the media of instruction (Mayer, 2003). Qualifications achieved through purely online learning programmes have faced questions of legitimacy relative to qualifications achieved through traditional face to face learning or blended learning (Adam & DeFleur, 2006; Ndume et al., 2008). Finally, the e-learning profession is at a fairly nascent stage of development, and has therefore faced challenges of "academic legitimacy" within higher educational institutions (Lisewski & Joyce, 2003; Yick et al., 2005).

The context of this study is the developing countries of East Africa. In this context, it has been argued that African universities are not fully prepared to participate in the evolution of ICT for teaching and learning due to a myriad of problems, including lack of legitimacy of some online learning programmes, low investment in ICT infrastructure and the high cost of connectivity and bandwidth (Chacha, 2009; Lwoga, 2012; Sife et al. 2007). Gunga and Ricketts (2007) suggest that collaborative public-private partnerships between stakeholders such as telecommunications providers, e-learning vendors, government and educators may help overcome some of these challenges, as does Lwoga (2012). Collaborative partnerships are touted too by Clarke (2003) as a means of sharing resources and good practice. An example of such a partnership is provided by Sife et al. (2007) who report on an e-learning partnership between universities in Africa and Europe. Not much research in the East African context has focused on the issue of how e-learning projects are legitimized. This paper therefore reports on a legitimizing strategy used by e-learning providers working in East Africa.

The findings reported here arose from a broader grounded theory study whose purpose was to inductively develop a theory to explain the core concerns of e-learning providers working in East Africa, and how these concerns are resolved. The focus on e-learning providers was justified, as much e-learning research has presented either a student or institutional view. An inductive approach was deemed most appropriate, in particular the classic grounded theory methodology (CGTM) (Glaser & Holton, 2004). By following the CGTM principles and procedures, legitimizing of e-learning projects emerged as one of the key concerns for e-learning providers working in East Africa.

This paper is laid out as follows - after this introduction, the CGTM is explained in further detail, since it was foundational to how the research was carried out. Following this, existing literature on the key concepts of the paper are presented. The findings from the empirical data are then presented and discussed in comparison to literature before the paper is concluded.

CLASSIC GROUNDED THEORY METHODOLOGY

In this study we chose to use classic grounded theory methodology (CGTM) by Glaser and Strauss (1967). CGTM has been elaborated and refined by Glaser over the years (Glaser 1978, 1998, 2006). Lowe (1996, p3) describes CGTM as being designed to *"develop and integrate a set of ideas and hypotheses in an integrated theory that accounts for behaviour in any substantive area"*.

Three reasons informed the choice of CGTM. Firstly, CGTM claims to deliver the main concerns of stakeholders involved with some phenomenon. In this study the main concerns of e-learning providers in East Africa were the focus of attention. Secondly, CGTM fits the purpose of the study, which is to provide understanding about a phenomenon which has hitherto received sparse in-depth academic research attention - in this case e-learning in the context of East Africa, from a providers' perspective. Thirdly, CGTM provides a flexible set of inductive strategies for collecting and analysing data. CGTM is rarely used in IS research (Matavire & Brown, 2013), but exemplars do exist of this type of work in IS (Adolph et al., 2012; Volkoff et al., 2005).

Glaser (1992) and Glaser and Strauss (1967) outline the key distinguishing characteristics of the CGTM research methodology as: (a) Simultaneous involvement in the data collection and analysis phases of research; (b) Developing analytic codes and categories from data, not from preconceived hypotheses; (c) Constructing middle-range theories to understand and explain behaviour and processes; (d) Memo-writing, that is, analytic notes to explicate and fill out categories; (e) Making comparisons between sets of data, data and concept, and between different concepts (f) Theoretical sampling, that is sampling for theory construction to check and refine conceptual categories, not for representativeness of a given population; and (g) Delaying the substantive literature review until after the emergence of a core variable.

Stages for Doing CGTM

Throughout the process of CGTM research a researcher needs to maintain transparency, through following all CGTM stages and procedures (see Table 1) outlined by Glaser (1978), including:

Table 1. Stages for doing CGTM

Stages for Doing CGTM	Definition/Main Activities	Reference
Theoretical Sampling	A form of sampling in which it is both inappropriate and impossible, prior to doing the research, to state exactly what data will be needed, it cease once the main concerns of the respondents is discovered (or emerged).	Glaser (1978, 1998, 2006).
Theoretical Memo-writing	the theorising write up of ideas about codes and their relationships as they strike the analyst while coding. Used continually throughout the CGTM process.	Glaser (1978:83)
Substantive Coding	These are the very first attempts at highlighting those data which the analyst believes may have importance for the research beyond the simple description of the context of the data, they are *"gerunded"* and converted to verb ended in "ing" to introduce momentum into the data.	Glaser (1978, 1998, 2006)
Theoretical Coding	A process of generating a conceptual code that arises from the synthesis of the substantive codes. It is a shift from low level context based description to higher level conceptual abstrations	Glaser (1978)
Applying Constant Comparison Method	This process involves comparing data incidents for similarities and differences. As concepts emerge and are named these are compared to other incidents in data, leading to the definition of properties of a category	Glaser (1992), Locke(2001)
Sorting	This is an iterative process of arranging the emergent concepts in the theory around the core variable	Glaser (1978)
Recognising the Emergence of the Core Variable	It is the stage at which you the core variable is recognized with its associate sub-core variables feeding into it.	Glaser (1978, 1998).

Generating Theoretical Sensitivity - the essence of theoretical sensitivity is "the ability to generate concepts from data and to relate them according to normal models of theory in general, and theory development in sociology in particular" (Glaser & Holton, 2004, p23).

During these stages, the order of doing them will change according to what emerges from the data. This kind of flexibility in terms of sequence and process is necessary in order to allow for the emergence of theory from data.

RESEARCH DESIGN AND APPROACH

Research Context

The East African countries targeted for this study were Kenya, Uganda and Tanzania with special focus on the e-learning providers. E-learning providers were operationalised to include higher education institution (HEI) academic staff, educational technologists, programmers, consultants, and management personnel. Sample of E-learning providers, the Higher Education Institutions (HEIs) visited are as listed in Table 2.

Historically the three countries have been economically and diplomatically linked since 1919 and enjoy historical, commercial, industrial, cultural and other ties (EAC, 2001). Included in the process of integrating the East African Community region was the establishment of Inter-university Councils for East Africa (EAC, 2001) for the purpose of broadening and increasing interaction among Higher Education Institutions (HEIs) in the region. During the process of regional integration, HEIs in the region were experiencing significant transformation. For example, due to a series of financial crises in many East African countries during this time, the HEIs in the region could not be adequately financed to cater for the ever-increasing student enrolment (Chacha, 2009). During this period donors began to reduce funding. For example, the World Bank reduced the proportion of funds allocated to HEIs in African countries from 17% between 1985-1989 to a mere 7% between 1995 and 2001 (Chacha, 2009). There continues to be a serious shortage of public funds for expansion, especially for physical infrastructure, teaching materials and e-learning infrastructure (Chacha, 2009; Lwoga, 2012). HEIs in Africa get trapped

Table 2. Institution visited and e-learning status

Name of E-Learning Provider	Country	E-learning Status (Yes/No)
University of Dar Es Salaam	Tanzania	Yes, blended & Pure Online
Open University of Tanzania	Tanzania	Yes, pure online
Dodoma University	Tanzania	Yes, blended
University of Nairobi	Kenya	Yes, blended & Pure Online
DeAfrica	Kenya	Yes, pure online
Makerere University	Uganda	Yes, blended & Pure online
Kampla International University	Uganda	Yes, blended
University of Arusha	Tanzania	Yes, blended
Virtual University of Uganda	Uganda	Yes, Pure Online
Virtual University of Africa	Tanzania, Kenya and Uganda	Yes, Pure Online

in sourcing funding and consultants from developed countries. Because of these problems e-learning in East Africa has manifested either slow delivery or unrealized potential, and challenges to legitimacy.

Data Collection

The data in this study was drawn from e-learning providers working in East Africa. By following CGTM the number, scope and range of those being interviewed were determined in an on-going process based on the emergent theory.

A total of 24 e-learning provider respondents from Tanzania, Kenya and Uganda were interviewed on several occasions (see Appendix 1). Those interviewed included e-learning directors, programme directors, content developers, lecturers and IT managers. All had several years of experience with e-learning. As per Glaser (1978), no notes or tape recordings were used in any of the respondent encounters. There were no detailed semi-structured interview questions, save for the two guiding questions, namely – "*What is your main concern with e-learning?*" and "*How is this main concern resolved?*" The principal author of this paper conducted all the interviews, and concentrated in the initial conversations on establishing rapport and making the respondents feel comfortable and unthreatened. These were deliberate strategies recommended by Glaser (1978). The rationale for this was that the respondent encounters should be more like normal conversations where both parties feel sufficiently relaxed to participate in a candid exchange of ideas. Immediately following each respondent encounter the researcher made very detailed notes that were later converted into theoretical memos.

Apart from interviews, e-learning providers were observed during their daily activities, and thoughts on these observations informed the writing of theoretical memos. A review was also done of written existing documents related to the e-learning providers such as those dealing with e-learning contexts and trends, performance ratings, program logs, tally sheets, and other existing indicators. Much of this data was sourced from the respective country education authorities (such as Ministry of Higher Education, Higher education accreditation authorities, etc.) universities and other governmental and non-governmental sources. All three sources formed the back-drop for writing memos. The memos were hence not tied back to specific "direct quotes" used by specific respondents, as with typical qualitative data analysis techniques.

Data Analysis

The CGTM stages as outlined above were followed in the analysis. Some details concerning the execution of each stage follows. *Theoretical sensitivity* was enhanced through extensive reading on CGTM, and relevant theories in Sociology and Information Systems (IS) specifically. Sampling was driven by the development of the theory (*theoretical sampling*). After fracturing the data collected, detailed *memo-writing* took place (See Appendix 2 for some excerpts). The memo writing enabled the principal researcher to conceptualise the boundaries and properties (characteristics) of each code and illuminated gaps in the emerging theory, thus highlighting where to sample next and for what theoretical purpose (Glaser, 1978). These memos were updated and progressively refined throughout the process of data collection and analysis. Hence they present the final integration of all forms of data collected and analysed from the multiple interviews, observations and documentation. *Substantive coding* resulted in several codes emerging, as labelled in Appendix 2. Illustrative direct quotes from subjects were not the basis for linking codes to evidence, but rather the memos, which were formulated based on the interviews, observation and documentation. *Theoretical coding* was executed by integrating substantive codes into theoretical codes to form the grounded theory. The *Constant comparison method* was employed throughout to arrive at the substantive and theoretical codes. Data incidents were constantly compared with each other, and with other concepts, including those from literature leading eventually to the definition of codes and their properties as in the memos in Appendix 2. Through the process of *sorting* the various substantive codes were established, and re-arranged into theoretical codes in accordance with the data. Initially there were two candidate *core variables* identified. Through subsequent data collection and analysis it was resolved which was the core variable.

LITERATURE REVIEW

Theoretical sensitivity accounted for the ability to generate concepts from the data and to label the concepts appropriately (Glaser, 1978). The literature related to the data-generated concepts of *legitimizing* projects and the key strategy for achieving this – *collaborating partnerships* – is presented in this section. Collaborating partnerships were of two types, *asymmetric* and *reciprocal*, hence literature on these characteristics will also be presented. This literature review was undertaken subsequent to the concepts having emerged from data. It is presented prior to the results to provide the necessary understanding of the emerged concepts, and their relevance to the phenomenon under investigation.

Legitimizing Projects

Legitimizing of projects has been deemed critical in a variety of contexts, wherein there are diverse and multiple stakeholders (Aaltonen, 2013; Kumar et al., 2005). Drawing from organisational literature, Aaltonen (2013) adopts the definition of legitimacy from Suchman (1995), i.e. "Legitimacy is a generalized perception or assumption that the actions of an entity are desirable, proper, or appropriate within some socially constructed system of norms, values, beliefs, and definitions". This definition is applicable to project contexts as well, since projects are in essence temporary forms of organisation (Aaltonen, 2013).

Much of the literature on project legitimacy draws from structuration theory (Giddens, 1984) which highlights three types of structures – structures of signification (reflecting meanings attributed to phenomena); domination (concerning how power is exercised); and legitimation (related to what is taken as legitimate by agents in a particular context) (Giddens, 1984). The concept of legitimation is very much aligned with the definition of legitimacy proffered by Schuman (1995). They both take legitimacy to be concerned with "acceptable social practice" (Hussain et al., 2004:409), rather than codified rules, regulations and policies that are cast-in-stone. Manning (2008), by applying a structurational perspective to project activity concludes that attributes of specific projects need to be legitimized according to the structures of the various organisational and inter-organisational contexts in which the project is embedded. These structural contexts are themselves produced and reproduced as the project is enacted.

Since the implementation of IS leads to organisational, cultural and process changes, legitmizing of IS projects has received significant of research attention so as to understand how to manage the attendant resistance to change, and potential lack of acceptance of an IS (Brown, 1995; Bitektine, 2008). Keable et al. (1998) highlight a legitimacy gap between IS professionals and users, which could lead to project failure. This legitimacy gap arises as a result of differing views between IS professionals and users on the legitimacy of the consequences associated with IS implementation. Klein and Hirschheim

(1989) relate legitimation directly to IS success or failure and in so doing define legitimization as "a social process by which institutions, practices and ideas gain social acceptance" (Klein & Hirschheim 1989: 7). Kohli and Kettinger (2004) identify how legitimization can be achieved in situations where the intended users of an IS are autonomous professionals. Hussain et al. (2004) used structuration theory and activity theory to come up with a Legitimation Activity Model consisting of several stages. These stages include: (1) formulating the desired legitimation structure (e.g. implementation of a new IS); (2) identifying the key stakeholders who may provide legitimacy to the project; (3) learning the norms and structures of these stakeholders; (4) comparing the desired legitimation structures with those of the stakeholders; (5) planning how to close the gap (6) executing the gap-closing strategy; (7) evaluating whether legitimation has been achieved; (8)re-evaluating if legitimation is not achieved. Various other strategies for achieving project-based legitimacy have been proffered. Aaltonen (2013) for example draws from Suchman (1995) who suggest strategies for achieving legitimation through either conforming to, selecting and/or manipulating the environment. Mpazanje et al. (2013) demonstrate the importance of legitimizing a project to ensure IS success in developing countries.

In the e-learning literature several issues have been raised related to legitimacy. For example reports of e-learning project failure have lead to questions of e-learning legitimacy (Keegan et al., 2007). Concerns around the academic legitimacy and standing of e-learning professionals and institutions have been highlighted (Lisewski & Joyce, 2003; Yick et al., 2005), as well as the legitimacy of online degree offerings (Adam & DeFleur, 2006; Ndume et al., 2008). In the developing countries context challenges have been reported around the funding and resourcing of e-learning projects due to questions of e-learning legitimacy (Andersson et al., 2009). While all these issues are pertinent, there appears to be a dearth of studies that have looked specifically at how e-learning projects are legitimised in developing countries, and what this might mean to e-learning providers. This paper seeks to fill this gap.

Collaborating Partnerships

Partnerships are recognised as an important vehicle to achieve mutually beneficial goals in a variety of contexts, including business, politics, government, community, society or any combination thereof. Partnerships may vary from merely informal agreements to full-fledged mergers and acquisitions (Minshall et al., 2010). The degree of interaction between partners may also vary depending on partnership purpose, defined roles, responsibilities, level of integration and so on. Inter-personal behaviour in a partnership can be characterised as having two dimensions – assertiveness and cooperativeness (Cosier & Ruble, 1981). Collaborating behaviour is both assertive and cooperative, which distinguishes it from other behavioural types such as accommodating (unassertive, cooperative) or compromising (intermediate in both assertiveness and cooperativeness) (Cosier & Ruble, 1981). Collaboration hence goes beyond simply cooperating or working together towards some goal (Jamal & Getz, 1995). Thomson and Perry (2006:23) define collaboration as "... a process in which autonomous actors interact through formal and informal negotiation, jointly creating rules and structures governing their relationships and ways to act or decide on the issues that brought them together; it is a process involving shared norms and mutually beneficial interaction". Gray (1985:64) defines a collaborative partnership "as a voluntary pooling of resources (labour, money, information, etc.) by two or more parties to accomplish collaborative goals". When characteristics of collaboration and partnership are synthesised, synergy is created that leads to successful ventures (Saltiel, 1998). There are a myriad of benefits to collaborative partnerships including legitimization of partners, as for example, when a small firm partners with a larger firm: The small firm and its products may gain credibility and legitimacy as a result (Blomqvist, 2002). Selin and Chaves (1995) identifies 3 stages through which collaborative partnerships proceed including (1) problem-setting, (2) direction-setting, (3) structuring.

The importance of collaborative partnerships to e-learning have been highlighted by Clarke (2003) who cites benefits such as resource-sharing, tool-sharing and development of good practice. To address the e-learning challenges in Africa, such as lack of ICT infrastructure and weak ICT policies, collaborative public-private partnerships have also been recommended (Gunga & Ricketts, 2007). Telecommunications service providers, vendors of e-learning products, educators and government leaders when partnering together can make a difference (Gunga & Ricketts, 2007). Sife et al. (2007) describe such a partnership between universities in Tanzania, South Africa and Europe.

In this study two types of collaborating partnerships were identified from discussions with e-learning providers – asymmetric and reciprocal. Baldwin (1978) alludes to the distinction between asymmetric and reciprocal power relationships, whereby asymmetric refers to an imbalance in power. Literature on each concept will be presented next.

Asymmetric Collaborating Partnerships

In the inter-organisational realm, collaboration between business partner firms may be asymmetric due to differences in organisational size, power, management, capabilities and organizational cultures (Blomqvist, 2005; Chen & Chen, 2002). Such asymmetric partnerships, it is noted, require both trust and contractual obligations to be successful (Blomqvist, 2005). These conditions may equally apply to intra-organisational partnerships, or partners working on a project, such as an e-learning project.

Reciprocal Collaborating Partnerships

Reciprocal collaborating partnerships refer to situations where partners have equal power (Baldwin, 1978). Harper and Selina (2000) highlight that establishing a reciprocal relationship is an important step in setting up collaborative relationships. Given this literature understanding of these key concepts, the next section now turns to describing the theoretical outcomes of the analysis.

THEORETICAL OUTCOMES

Legitimizing E-Learning Projects

Legitimizing was identified as a main sub-core variable amongst e-learning providers based in East Africa and who participated in this study. The e-learning providers continually used *legitimizing* in order that e-learning projects could be successfully implemented. Based on the data *legitimizing* is defined as *the process of ensuring the e-learning activities, including planning, design, delivery, and evaluation take place and that each e-learning product or deliverable is valid and sustainable.* Thus the validity, in terms of quality and appropriateness for purpose, and the sustainability of the project is ensured and maintained by means of the *legitimizing* process. For e-learning projects to be successfully executed, e-learning providers need to ensure a project is legitimized by all stakeholders (i.e. clients, funders, learning institutions, etc.) and is accorded respect throughout the process. The main symptom and reason for a lack of legitimization revealed in the data is the lack of respect and trust shown to e-learning providers by some of the stakeholders. A legitimizing process is undertaken by e-learning providers in order for them to gain the necessary respect from the stakeholders.

The process of legitimizing is a fairly tricky one for e-learning providers to achieve. Legitimizing can only happen with the approval of respected third parties. These third parties sometimes have difficulty prior to the project's commencement understanding the full implications of e-learning in terms of increasing educational effectiveness and quality. One of the main barriers experienced by these third parties is the inability to understand clearly the technical jargon employed by e-learning providers working in this field. E-learning providers also often find it problematic to communicate clearly and precisely to their partners what it is they are doing. In order to reduce the potential for misunderstanding they have to legitimize their actions. As soon as legitimization has been achieved, the project gains greater acceptance, which in turn leverages future funding.

Collaborating Partnerships

Based on the data collected and analysed legitimizing involves at its core the process of a collaborating partnership. This is a process in which e-learning providers collaborate with other experienced e-learning partners prior to a formal funding submission to donors for their e-learning projects. Such collaboration builds trust, which is a necessary condition for achieving project legitimization. *Collaborating Partnerships* involve creating a working agreement in such a way as to ensure each partner gains from the agreement. The process of collaborating with partners can be either asymmetrical or reciprocal. These two labels were derived by researcher as a result of developing theoretical sensitivity (Glaser, 1978).

Asymmetric Collaborating

Asymmetric Collaborating is a form of collaboration in which the weaker party in the collaboration is able to obtain the benefits that the collaboration brings, whilst skilfully avoiding being dominated by the more powerful partner. However, the weaker partner is potentially always more vulnerable. The more leverage the powerful partner has, the more this partner will strive to ensure that the project is implemented. In *Asymmetric Collaborations* the aim of the weaker partner is to manage to skilfully negotiate for resources and finances from the dominant party in a fair exchange.

During the e-learning implementation process the need for more partners will often arise. This in turn will trigger more collaboration within the e-learning project. Other factors which facilitate the refining of such collaboration can be the rapid changes in technology and the availability of funding and of e-learning experts. A change involving one or more of these factors will lead to the start of a new e-learning collaboration or agreement, or to a process of refining the current e-learning collaborations. This in turn rejuvenates the e-learning project. As soon as the junior e-learning partner is able to stand, the e-learning provider tends to start e-learning projects running in tandem with partnering projects and with new e-learning partners. *Asymmetric Collaborating* tends to be spasmodic when certain relationships create unanticipated problems and necessitate the putting of strategies in place to deal with the unforeseen problems. Asymmetric relationships can also be perpetual. Even in harmonious relationships conflict of interest can exist, or agenda priorities differ. To avoid and deal with these issues it is also quite feasible that *Asymmetric Collaborating* will also have to be perpetual.

Figure 1. Collaborating partnerships: a project-based legitimizing strategy

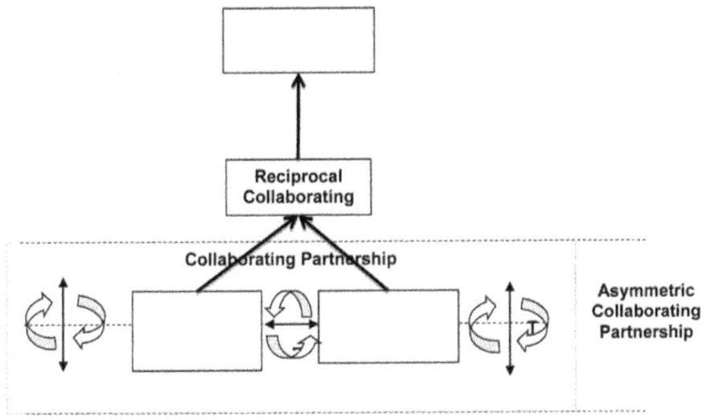

Reciprocal Collaborating

The e-learning providers ideally should attempt to transform the *Asymmetric Collaboration* to one characterised by a reciprocal relationship wherever possible. The reason for this is that asymmetric relationships tend to be limited or are reduced to being contractual relationships whereas reciprocal relationships always have a greater potential for developing beyond a single project.

Reciprocal Collaborating is defined as a process of providing mutual benefit for the parties in terms of sharing power and status. The process involves two parties coming together to achieve agreed objectives by contributing different and unique skills and investment to the partnership. For e-learning providers to share the right of ownership of e-learning projects equally, they have to manoeuvre themselves into a position of parity. To ensure this, e-learning providers design and sign various agreements and memoranda of understanding. All parties are obliged to sign such agreements to enter into the collaboration. The parties coming into a *Reciprocal Collaboration* expect to have equal power and status. If one partner gains a higher status and greater power than the other in the collaboration, the collaborating process will change to an asymmetric collaboration. The collaboration tends to be entered into by the partners in order to secure funds, strengthen the e-learning programs and as a means to achieve legitimacy. As soon as such goals are achieved, the process ends. At this point the relationship tends to evolve into new e-learning collaborations as the situation changes. Figure 1 illustrates the process.

DISCUSSION OF FINDINGS

In this discussion section, the theoretical outcomes arising from the data are compared, contrasted and integrated with literature to highlight the paper's contribution. Concerns around legitimacy of e-learning have abounded in literature. These have included legitimacy concerns with respect to the e-learning profession, virtual institutions, programmes, qualifications, and projects (Adam & DeFleur, 2006; Keegan et al., 2007; Lisewski & Joyce, 2003; Yick et al., 2005). This study has revealed achieving project-based legitimacy as a key concern to e-learning providers working in East Africa. The credibility and respect afforded e-learning providers is a characteristic property of project-based legitimacy in this study. This finding resonates with the notion of "academic legitimacy" proffered by Adam and DeFleur (2006), but overall the perspective of legitimacy discovered here is much wider and is focused more on the e-learning project per se.

Not much research in e-learning has addressed the project-based dimension of legitimacy, a dimension that has received some attention in the project management literature (Aaltonen, 2013), and more specifically IS project management literature (Hussain et al., 2004). In the developing countries context Mpazanje et al. (2013), by employing the lens of Actor Network Theory, describe a process that was carried out to ensure a project was legitimized. They fail however to explicitly refer to this process as the concept of project-based legitimization. By providing an explicit label for this phenomenon, this paper contributes to theoretically elaborating on legitimization of e-learning projects in developing countries.

The literature on project–based legitimizing draws quite liberally from structuration theory, making reference to the concept of legitimation as a means of theorising the process (Hussain et al., 2004; Manning, 2008). In this study, collaborating partnerships, inclusive of asymmetric collaboration was found to be a key strategy to achieving legitimization. The asymmetric aspect of the collaborating partnerships reflects asymmetry in power between partners, and is similar to the structure of domination in structuration theory. According to Giddens (1984) there is interaction between structures of domination and legitimation. This observation from structuration theory provides backing for the link between asymmetric collaboration and project–based legitimization as revealed in this study.

Prior studies have alluded to achieving legitimacy as one of the benefits of collaborative partnerships (Blomqvist, 2002). In this study it was demonstrated that collaborating partnerships can in fact be a key strategy for achieving project-based legitimacy, and is not merely just one of many benefits.

Blomqvist (2005) identified the need for both trust and signed contracts in order for collaborative partnerships to be successful. In this study, it was found too that signed contracts and memoranda of understanding were a feature of the collaborating partnerships, and that trust was a characteristic built through reciprocal relationships.

Table 3. E-learning collaborating partnership stages, sequence, main activities, and output of each stage

Sequence	E-Learning Collaborating Partnership Stage	Main Activities	Output
1	Determination of Key e-learning Drivers	• Determining the key drivers to partner	• Key drivers listed including lack of resources, etc.
2	Problem Setting	• Consensus is reached on partnership operations. • agreement signing	• Problem identified • Agreed document on how to solve the problem
3	Direction-Setting	• Establish goals to achieve the common vision recognized Stage 2. • Set ground rules for the operation of the partnership • Identify and appreciate a sense of common purpose.	• Goals, vision and rules for partnership • Type of Partnership
4	Structuring	• Formalising and Institutionalizing • Devising a regulatory framework to guide future collective actions	• Formal Relationships, roles, tasks, monitoring and control systems • Signed Agreement
5	Achievement of Outcome	• outcomes of the process are realized	• Program Modules • No. Of students • Offices • Income

Prior research has mentioned the presence of both asymmetric and reciprocal collaborative partnerships, in particular in inter-organisational contexts (Blomqvist, 2002). This paper goes further in showing how the type of collaborating partnership in an e-learning project is dynamic, and may change between asymmetric and reciprocal over time, and in response to events that occur.

The importance of collaborative partnerships to e-learning have been highlighted by Clarke (2003) and Gunga and Rickett (2007). This study goes further by showing how collaborating partnerships can become a key strategy for legitimizing e-learning projects.

Drawing from existing theory on collaborative partnerships, such as Selin and Chaves (1995), and relating it with the findings of this study, the evolution of e-learning collaborating partnerships can be thought of as proceeding through 5 stages – (1) determination of key e-learning drivers; (2) problem-setting, (3) direction-setting, (4) structuring and finally (5) achievement of outcomes. Before the detailed discussion on the five stages see the summary of main activities in each stage in Table 3.

At the *determination of key e-learning drivers*, there exists a wide range of these key drivers, or pressures to form partnerships in East Africa, including scarce resources, common vision, incentives, existing networks, high population growth, lack of qualified human resources and lack of physical infrastructure (Gunga & Ricketts 2007; Ndume et al. 2008; Sife et al. 2007; Sifuna 2007).The collaborating institutions operate within societal environments. These societal environments exert competitive, technological, political, social and economic forces and pressures to increase and strengthen collaborating partnerships (Selin and Chaves 1995). Gunga and Ricketts (2007) argue that partnerships bring together innovative minds to help overcome the various e-learning challenges. They also avoid dual, or separate, problem solving within the same country and in the same context. From these key drivers, partnerships evolve sequentially through a problem setting, direction-setting and structuring phase (Selin & Chaves, 1995).

At the *Problem-setting* stage of the collaborating partnerships, the institutions in such partnerships appreciate the interdependencies existing among them. This is because they come to realize that the action to take in solving a particular problem, or problems, needs collective efforts among them. It is during this stage that consensus is reached on partnership operations. This will include such actions as agreement signing. At this stage the partners and stakeholders begin to mutually acknowledge the issue, or issues, which bring them together. Positive perception in terms of the unforeseen benefits will increase the strength of the partnership.

At the *Direction Setting* stage the partners tend to establish goals to achieve the common vision recognized in the problem setting stage. The established goals tend at this stage to be measureable and responsive to the perceived benefits stipulated at the problem setting stage. The partners set ground rules for the operation of the partnership including joint information searches, exploring options and organising sub groups of operations. The direction setting stage is also the stage where partners begin to identify and appreciate a sense of common purpose. The joint information search helps strengthen the future of the partnership.

The *Structuring* stage involves formalising and institutionalising the collaborating partnership and devising a regulatory framework to guide future collective action (Gray, 1989). For the partnership to survive stakeholders' interactions need to be managed in a systematic manner. At this stage, the legal formalisation of the collaborating partnership is developed and put in place. This involves signing various agreements, assigning roles, and finally reaching formal agreements to monitor and assure collective compliance to the goals of the group (Selin & Chaves, 1995). At the end of this stage all formal relationships, roles, tasks and monitoring and control systems are in place. Some of the roles are assigned to special committees, such as a procurement committee, online content committee, etc.

It is at the last stage in the process of the collaborating partnership, the *Outcome* stage, that the outcomes of the process are realized (Selin & Chaves, 1995). These outcomes may be visible and tangible products of the partnership and may include program modules, offices, and numbers of students registered for the program modules. For the collaborating partnership to flourish there needs to be communication throughout the collaborating process. As the various forms of feedback continue, the collaborating partnership process becomes cyclical to the benefit of the partners and of the partnership. This means that a continuous process of evaluation and re-evaluation becomes one of the main purposes and characteristics of the partnership, thus constantly improving the quality of deliverables, broadening the scope of the partnership, and admitting more partners into the collaborating partnership. These outcomes reflect that the e-learning project has been legitimized.

CONCLUSION

This paper was part of a wider study that sought to develop a theory explaining the core concerns of e-learning providers working in East Africa. Legitimizing e-learning projects emerged as one of the key sub-variables associated with the wider theory, and so was the focus of attention in this paper. Establishing collaborating partnerships was identified as a major strategy used for achieving project-based legitimization for e-learning providers. Collaborating partnerships were found to be of two types – asymmetric and reciprocal. Asymmetric collaborating partnership explains how the weaker party in the collaboration is able to obtain the benefits that collaboration brings whilst skilfully avoiding being dominated by the more powerful partner. A reciprocal collaborating partnership is the process of generating mutual benefit for the parties in the partnership through the sharing of power and status. Partnerships at various times evolve between the two types, and are not static. These partnerships may exist between stakeholders internal to an organisation, as well between internal and external entities.

Integrating these findings with other theory on partnerships such as that of Selin and Chaves (1995) leads to a model of evolutionary collaborative partnership that explains how partnerships begin with the identification of key e-learning drivers. These partnerships then evolve sequentially through problem setting, direction-setting and structuring phases, before final outcomes are achieved. In the case of e-learning providers in East Arica, the desired outcome is legitimization of e-learning projects.

Future research can build on this study by gathering empirical evidence to validate that the phases of e-learning drivers, problem-setting, direction-setting, and structuring are followed in achieving the outcome of legitimization. How these phases are executed in the light of asymmetric and reciprocal collaborating partnerships can be elaborated on.

Overall this study has made contributions to knowledge by providing insight in to a key concern of e-learning providers in East Africa - that of legitimising their e-learning projects. How this concern is resolved through collaborating partnerships is also explained. The study makes a methodological contribution by employing classical grounded theory methodology. Most theory-building research in information systems employs various evolved forms of grounded theory methodology. Few have attempted to make use of the original formulation by Glaser and Strauss (1967), which offers opportunity for further insights in to IS phenomena, based on the experiences of IS professionals and other relevant actors.

ACKNOWLEDGMENT

Dr. Andy Lowe is acknowledged for his guidance regarding Classic Grounded Theory research methodology and for being the methodological coach for the broader study.

REFERENCES

Aaltonen, K. (2013). The establishment of legitimacy: The case of international projects. *International Journal of Managing Projects in Business*, *6*(1), 13–35. doi:10.1108/17538371311291008

Adams, J., & DeFleur, M. H. (2006). The acceptability of online degrees earned as a credential for obtaining employment. *Communication Education*, *55*(1), 32–45. doi:10.1080/03634520500343376

Adolph, S., Kruchten, P., & Hall, W. (2012). Reconciling perspectives: A grounded theory of how people manage the process of software development. *Journal of Systems and Software*, *85*(6), 1269–1286. doi:10.1016/j.jss.2012.01.059

Andersson, A. S., & Grönlund, Å. (2009). A conceptual framework for e-learning in developing countries: A critical review of research challenges. *The Electronic Journal of Information Systems in Developing Countries*, *38*(8), 1–16.

Baldwin, D. A. (1978). Power and social exchange. *The American Political Science Review*, *72*(4), 1229–1242. doi:10.2307/1954536

Bitektine, A. (2008). Exploring slow institutional variables in healthcare context: A study of electronic health record in Canada. *ASAC*, *29*(28), 15–39.

Blomqvist, K. (2002). *Partnering in the dynamic environment: The role of trust in asymmetric technology partnership formation*. (Doctoral Thesis). Lappeenranta University of Technology.

Blomqvist, K., Hurmelinna, P., & Seppänen, R. (2005). Playing the collaboration game right—balancing trust and contracting. *Technovation*, *25*(5), 497–504. doi:10.1016/j.technovation.2004.09.001

Brown, A. D. (1995). Managing understandings: Politics, symbolism, niche marketing and the quest for legitimacy in IT implementation. *Organization Studies*, *16*(6), 951–969. doi:10.1177/017084069501600602

Carliner, S. (2002). *Designing e-Learning*. Alexandria, VA: American Society for Training and Development.

Carliner, S., & Shank, P. (2007). *The e-Learning Handbook: Past Promises, Present Challenges*. San Francisco: Wiley & Sons.

Chacha, F. (2009). Annual E-learning Report on East Africa.*Annual East African University Council Meeting*, Arusha, East African University Council.

Chen, H., & Chen, T. J. (2002). Asymmetric strategic alliances: A network view. *Journal of Business Research*, *55*(12), 1007–1013. doi:10.1016/S0148-2963(02)00284-9

Clarke, C. (2003), *Towards a Unified E-learning Strategy*. A consultant document submitted to the Department for Education and Skills, United Kingdom. Available online at http://www.education.gov.uk/consultations/downloadableDocs/towards%20a%20unified%20e-learning%20strategy.pdf

Cosier, R. A., & Ruble, T. L. (1981). Research on conflict-handling behavior: An experimental approach. *Academy of Management Journal*, *24*(4), 816–831. doi:10.2307/256179

Cowie, N., & Sakui, K. (2013). It's never too late: An overview of e-learning. *ELT Journal*, *67*(4), 459–467. doi:10.1093/elt/cct037

EAC. (2001). *Treaty Establishing the East African Community*. Available at http://www.eac.int/treaty/

Giddens, A. (1984). *The Constitution of Society: Outline of the Theory of Structuration*. Cambridge: Polity Press.

Glaser, B. (1978). *Theoretical Sensitivity: Advances in the Methodology of Grounded Theory*. Mill Valley, CA: Sociology Press.

Glaser, B. (1992). *Emergence Vs Forcing: Basic of Grounded Theory Analysis*. Mill Valley, CA: Sociology Press.

Glaser, B. (1998). *Doing Grounded Theory: Issues and Discussions*. Mill Valley, CA: Sociology Press.

Glaser, B. (2006). *Doing Formal Grounded Theory: A proposal*. Mill Valley, CA: Sociology Press.

Glaser, B., & Holton, J. (2004). Remodelling Grounded Theory. *Forum Qualitative Sozial Forschung*, *5*(2), 1–17.

Glaser, B., & Strauss, A. (1967). *The Discovery of Grounded Theory: Strategies for Qualitative Research*. Hawthorne, NY: Aldine De Gruyter.

Gray, B. (1985). Conditions facilitating interorganizational collaboration. *Human Relations*, *38*(10), 911–936. doi:10.1177/001872678503801001

Gray, B. (1989). *Collaborating: Finding Common Ground for Multiparty Problems*. San Francisco: Jossey-Bass.

Gunga, S., & Ricketts, I. (2007). Facing the Challenges of E-learning Initiatives in African Universities. *British Journal of Educational Technology*, *38*(5), 896–906. doi:10.1111/j.1467-8535.2006.00677.x

Harper, G. W., & Salina, D. D. (2000). Building collaborative partnerships to improve community-based HIV prevention research: The university-CBO collaborative partnership (UCCP) model. *Journal of Prevention & Intervention in the Community*, *19*(1), 1–20. doi:10.1300/J005v19n01_01

Hussain, Z., Taylor, A., & Flynn, D. (2004). A case study of the process of achieving legitimation in information systems development. *Journal of Information Science*, *30*(5), 408–417. doi:10.1177/0165551504046725

Jamal, T. B., & Getz, D. (1995). Collaboration theory and community tourism planning. *Annals of Tourism Research*, *22*(1), 186–204. doi:10.1016/0160-7383(94)00067-3

Keable, M., Landry, M., & Banville, C. (1998). The legitimacy gap between IS professionals and users. *Communications of the ACM*, *41*(5), 146–156. doi:10.1145/276404.276406

Keegan, D., Lõssenko, J., Mázár, I., Michels, P. F., Paulsen, M. F., Rekkedal, T., ... Zarka, D. (2007). *E-learning initiatives that did not reach targeted goals*. Bekkestua: NKI Forlaget.

Kilby, T. (2001). The direction of web-based training: A practitioner's view. *The Learning Organization*, *8*(5), 194–199. doi:10.1108/EUM0000000005912

Klein, H. K., & Hirschheim, R. A. (1989). Legitimation in information systems development: A social change perspective. *Information Technology & People*, *5*(1), 29–46.

Kohli, R., & Kettinger, W. J. (2004). Informating the clan: Controlling physicians' costs and outcomes. *Management Information Systems Quarterly*, *28*(3), 363–394.

Kumar, R., Rangan, U. S., & Rufín, C. (2005). Negotiating complexity and legitimacy in independent power project development. *Journal of World Business*, *40*(3), 302–320. doi:10.1016/j.jwb.2005.05.006

Lisewski, B., & Joyce, P. (2003). Examining the five-stage e-moderating model: Designed and emergent practice in the learning technology profession. *Alt-J*, *11*(1), 55–66. doi:10.1080/0968776030110106

Locke, K. (2001). *Grounded Theory in Management Research*. London: Sage.

Lowe, A. (1996). *An Explanation of Grounded Theory*. Glasgow, UK: University of Stratchlyde. Unpublished manuscript.

Manning, S. (2008). Embedding projects in multiple contexts–a structuration perspective. *International Journal of Project Management*, *26*(1), 30–37. doi:10.1016/j.ijproman.2007.08.012

Matavire, R., & Brown, I. (2013). Profiling grounded theory approaches in information systems research. *European Journal of Information Systems*, *22*(1), 119–129. doi:10.1057/ejis.2011.35

Mayer, R. (2003). Elements of a science of learning. *Journal of Educational Computing Research*, *29*(3), 297–313. doi:10.2190/YJLG-09F9-XKAX-753D

Minshall, T., Mortara, L., Valli, R., & Probert, D. (2010). Making "asymmetric" partnerships work. *Research-Technology Management*, *53*(3), 53–63.

Mpazanje, F., Sewchurran, K., & Brown, I. (2013). Rethinking information systems projects using Actor-Network Theory–Perspectives from a developing country. *The Electronic Journal of Information Systems in Developing Countries*, *58*(7), 1–37.

Naidu, S. (2006). *E-learning: A Guidebook of Principles, Procedures and Practices*. New Delhi: CEMCA.

Ndume, V., Tilya, F., & Twaakyondo, H. (2008). Challenges of adaptive e-learning at higher learning institutions: A case study in Tanzania. *International Journal of Computing and ICT Research*, *2*(1), 47–59.

Njenga, J. K., & Fourie, L. C. H. (2010). The myths about e-learning in higher education. *British Journal of Educational Technology*, *41*(2), 199–212. doi:10.1111/j.1467-8535.2008.00910.x

Piccoli, G., Ahmad, R., & Ives, B. (2001). Web-based virtual learning environments: A research framework and a preliminary assessment of effectiveness in basic IT skills training. *Management Information Systems Quarterly*, *25*(4), 401–426. doi:10.2307/3250989

Saltiel, I. M. (1998). Defining collaborative partnerships. *New Directions for Adult and Continuing Education*, *1998*(79), 5–11. doi:10.1002/ace.7901

Selin, S., & Chaves, D. (1995). Developing a Collaborative Model for Environmental Planning and Management. *Environmental Management*, *19*(2), 189–195. doi:10.1007/BF02471990

Sife, A., Lwoga, E. & Sanga, C. (2007). New technologies for teaching and learning: Challenges for higher learning institutions in developing countries. *International Journal of Education and Development Using ICT*, *53*(2), 57-67.

Sifuna, D. (2007). The challenge of increasing access and improving quality: An analysis of universal primary education interventions in Kenya and Tanzania since the 1970s. *International Review of Education*, *53*(5-6), 687–699. doi:10.1007/s11159-007-9062-z

Suchman, M. (1995). Managing legitimacy: Strategic and institutional approaches. *Academy of Management Review*, *20*(3), 571–610.

Thomson, A. M., & Perry, J. L. (2006). Collaboration processes: Inside the black box. *Public Administration Review*, *66*(1), 20–32. doi:10.1111/j.1540-6210.2006.00663.x

Volkoff, O., Strong, D. M., & Elmes, M. B. (2005). Understanding enterprise systems-enabled integration. *European Journal of Information Systems*, *14*(2), 110–120. doi:10.1057/palgrave.ejis.3000528

Yick, A., Patrick, P., & Costin, A. (2005). Navigating distance and traditional higher education: Online faculty experiences. *International Review of Research in Open and Distance Learning*, *6*(2).

APPENDIX 1: RESPONDENT DETAILS

Table 4.

Respondent	Country	Position	Experience in E-Learning	Specialty
X_1	Tanzania	E-learning Director & Content Developer	17 years	IT & Management
X_2	Tanzania	E-learning Program Director	12 years	ICT
X_3	Kenya	Executive Director & Chairman	23 years	ICT and Management
X_4	Kenya	E-learning Lecturer	14 years	ICT & E-learning Developer
X_5	Kenya	Distance Learning Director	17 years	ICT & Management
X_6	Kenya	E-learning Content Developer	11 years	IT & Programming
X_7	Uganda	E-learning Centre Manager	20 years	ICT & Programming
X_8	Uganda	E-learning Director	27 years	IT Management
X_9	Uganda	E-learning Coordinator	6 years	ICT
X_{10}	Tanzania	Associate Director (E-learning)	9 years	IT & Management
X_{11}	Kenya	IT manager	13 years	ICT
X_{12}	Kenya	E-learning Lecturer	5 years	IT
X_{13}	Uganda	E-learning Coordinator	11 years	IT
X_{14}	Kenya	Lecturer	7 years	IT
X_{15}	Tanzania	Lecturer	6 years	ICT
X_{16}	Tanzania	Educational Technology Director	21 years	Educational Technology
X_{17}	Tanzania	Online Content Developer	16 years	Programming
X_{18}	Tanzania	Online Program Coordinator	10 years	IT
X_{19}	Tanzania	ICT Director	18 years	ICT & Management
X_{20}	Tanzania	ICT Online Program Administrator	15 years	Management
X_{21}	Tanzania	Online Content Developer	7 years	IT
X_{22}	Kenya	Program Manager	16 years	IT & Management
X_{23}	Uganda	Program Manager	12 years	ICT
X_{24}	Uganda	Center Manager	19 years	ICT & Management

APPENDIX 2: EXCERPTS OF MEMOS

Memo: Legitimizing

Definition (from the Data)

E-learning professionals find it problematic to communicate exactly what they are doing to their collaborative partners. These problems are caused by the barrier of IT technical jargon which the client partner rarely understands. To reduce the potential for misunderstanding they have to legitimize their actions. The process of legitimization is achieved by a combination of the e-learning professional's qualifications, past experiences and endorsed referral from other clients.

As soon as the legitimizing has been achieved, the project gains credibility and more people tend to help it and get funding. For example, an E-learning Director felt disrespected during the launch of the e-learning institution. Most people viewed him as wasting his time. Just after securing endorsement from the Ministry of Education [*Collaborating*], he was able to legitimize the program, and more people were now coming to join the program. He worked to legitimize his profession to ensure respect and recognition.

Properties of Legitimizing

- Increased respect from top managers.
- Synchronizing of the project with other projects.
- Acceptability and recognition of the project.

Memo: Asymmetric Collaborating

Definition (from the Data)

This concept explains how the weaker party in a collaborative partnership is able to obtain the collaborative benefits whilst skilfully avoiding being dominated by the more powerful partner. In all collaborations each partner will have slightly differing agendas. The weaker partner is potentially always more vulnerable.

The more powerful partner has an edge in ensuring the project is implemented. An example of an asymmetric collaboration is when a university is collaborating with the government. The government is always more powerful than a university. In asymmetric collaborations the weak partners skilfully manages to extract resources and finances from the dominant party without being overwhelmed. The e-learning professionals attempt to transform the asymmetric collaboration to a reciprocal relationship wherever possible.

Properties of Asymmetrical Collaborating

- Being recognized and increased credibility.
- Increased and known intellectual property over ownership.
- New Operation expansion.
- Quality control and assurance.

- Project sustainability and growth.
- Attracting more donors and funding.

Memo: Reciprocal Collaborating

Definition (from the Data)

Two parties come together to obtain agreed objectives by contributing different and unique skills and investments. The parties tend to share equally the right of ownership of e-learning projects. It is mutually beneficial between the parties or done in return. No party is under the other party and they have equal status and power to manage the project. To ensure there is equal status and power, e-learning providers have been designing and signing various agreement and memorandum of understanding. All parties do sign such agreements to enter into the collaboration. For example, during establishment of an e-learning institution, after the project had been endorsed by the Minister of Education, every partner was supposed to sign an agreement to ensure mutual benefits for all.

Properties of Reciprocal Collaborating

- Referral networks creation.
- Degree of mutual benefits.

Chapter 3

Application of Information Communication Technologies for Agricultural Development through Extension Services: A Review

L. K. Mabe
North-West University – Mafikeng, South Africa

O. I. Oladele
North-West University – Mafikeng, South Africa

ABSTRACT

This chapter outlines the role that Information Communication Technologies (ICT) play in the global context and in Africa, agricultural extension and Agricultural development. The role of and use of ICT by extension officers, the trends of ICT in agricultural information management, how ICT bridge the digital divide as well as the types of ICT tools used by extension officers such as radio, television, computers and internet. It also gives the perspective about the factors that influences use of ICT by extension officers which are seen as playing an important role human development.

INTRODUCTION

ICT and the Global Context

According to Roy (2005), the use of Information Communication Technologies is expanding rapidly. ICTs comprise of a diverse set of technological tools and resources to create, disseminate, sore and manage data and information. Traditional ICT tools such as television, radio and telephone have proven their effectiveness in promoting development in marginalized areas. The emergence of computers, the internet and wireless communication technology, along with powerful software for processing and integrating

DOI: 10.4018/978-1-5225-0539-6.ch003

text, sound and video into electronic media comprise modern ICTs. The spread of the global electronic network of computers popularity referred to as the internet and wireless telephony has generated an unprecedented global flow of information, product, capital and ideas. Furthermore Adesope, Asiabaka and Agumagu (2007), states that the advent of the computer gave a new direction to the information and communication technology industry. This is evident in the use of technologies such as electronic mail (e-mail), electronic commerce (e-commerce) and more recently mobile phone, which Salihu (2000) referred to as Palm assistant, because of its palm-size nature. Information communication technologies such as electronic mail and electronic conferencing provide scientists, administrators and information staff with rapid and reliable communication, while increasing productivity and decreasing communication costs by reducing the physical means of communication channels (Kerrigan, Lindsey & Novak, 1994).

ICT and Africa

Agriculture has and will be the backbone of the African economy contributing to the livelihoods of rural communities. The prospects of African globalization cannot be separated from the nature of its historical incorporation in the world economy, the ensuing commodity based export structure, unequal terms of trade and declining position in the world economy. Globalization demands driven re-casting critical questions on the role of state and NGO's to steer economic development through the use of ICTs. In Africa, most farming communities rely on the public Agricultural Extension services for technical farming advice and information. Information communication Technology can be the instrument for stimulating the growth and by ushering in unique forms of national and transnational exchange relationships between producers and consumers, for instance, the scope of widening markets, profitability and investments and the choice of diverse range products, services and facilities (Roy, 2005).

Africa, like the rest of the world, needs to attain the United Nation's Millennium Development Goals (MDGs) and the World Food Summit (WFS) goals to reduce the number of hungry people from 790 million to 400 million by 2015. To reduce poverty by 50% by 2015, the African economies need to grow at 7% per annum. Another target requires that a national strategy for sustainable development be in place in every country to ensure the reversal of current trends in the loss of environmental resources at both national and global levels by 2015. Several factors are responsible for the numerous challenges facing the continent. First, there has been underinvestment in the rural areas. Inadequate access to markets and unfair market conditions also continue to be major stumbling blocks. These challenges, coupled with the crippling impact of limited access to advanced technologies, weak infrastructure, high production and transport costs, present the image of a continent that will be poverty stricken for many years to come (ECA, 2005).

Richardson (1997) points out that ICTs are considered to be drivers of change for rural and agricultural development, as they are efficient tools for reaching rural and remote communities and for improving agricultural productivity. Due to this factor, ICTs can speed up the extension of development services in areas that include healthcare, education and agriculture. Furthermore, ICTs can be instrumental in strengthening partnerships and in providing a framework for shared learning. It is not surprising, therefore, that ICTs have led to increased use of a networked information environment and the development of platforms for better sharing and exchange of information and knowledge (Van Audenhove, 2003).

ICTs, Agricultural Extension, and Agricultural Development

Agricultural Extension and advisory services in the field of agricultural development has seen many changes in the past decades. The first trend is the change in philosophy and rationale behind extension services towards more participatory approaches and participatory technology development (PTD), a shift has been made in thinking about the role of extension services in broader information and knowledge sharing processes (Leeuwis, 2004). A second trend in extension services, situated at a less conceptual level and a more professional level, seems to be the development towards 'extension-plus' which means that extension and advisory services plays an expanded role because extension has long been primarily associated with 'transferring technologies to farmers' (Sulaiman, 2003).

According to Munyua (2008), Information and knowledge play a central role in rural agricultural development. In describing the role of women and men in agricultural and rural development, information and knowledge are essential if women and men are to respond to opportunities and major challenges of the coming century. In emphasizing the role of information and knowledge, information and knowledge is described as prime productive resources. The Food and Agriculture Organization of the United Nations (2000) advances a corresponding view to this perspective in its assertion that "information and knowledge play a key role in ensuring food security and sustainable development". World Bank (2002, cited in IICD 2006), also takes the position that "information and communication technologies are a key input for economic development and growth".

Munyua (2008) states that Africa has a cumulative body of agricultural information and knowledge, know-how and practices that need to be shared continent-wide and applied for improved livelihoods and sustainable development. Although there is a lot of information relevant to small-scale farmers, this is scattered and hard to find. Most of the institutions implementing initiatives on ICTs and small-scale agriculture in Africa have developed local agricultural information and knowledge resources in English, and in some cases, in local languages. Bertolini (2004) notes that most farmers in Africa rely on intermediaries such as nongovernmental organizations (NGOs), extensionists and producer organizations to access and exchange information on innovations in crop varieties, pest management, soil fertility, weather forecasting and irrigation among others. Wider access to such information and in a more cost effective manner ought to be given priority.

There seems to be recognition that extension should merge with, or at least should incorporate, other fields such as improved access to markets, research, advice, credit, infrastructure, farmer organization development and business development services. "Research and extension strategies should emerge out of a broader livelihood analysis organized through a wider consultative exercise." Next to these developments in fundamental thinking in the academic and professional worlds, a third global economic trend from public services to private services can be identified. Although, for instance in India, the state Department of Agriculture (DoA) continues to dominate extension provisions, initiatives towards privatization and public-private partnerships do exist (Sulaiman, 2003).

Effective agricultural development requires access to information on all aspects of agricultural production, processing and marketing and it seems likely that if anything this need is increasing (Jones, 1997). ICT is already showing the potential to play an important role in the delivery of this information to this sector in both developed and developing countries. In most cases the base technology is universal, rather than being specific to agriculture, and hence usage evolves from existing designs and practices (Zijp, 1994).

The agricultural communications industry will change in the future because the agricultural industry itself is changing dramatically. There is more information but perhaps less knowledge than ever before in agriculture and that is a huge opportunity for extension officers. At the same time, the emergence of the internet as a communications medium, as well as the convergence of many other traditional communications vehicles like print, radio, television and database marketing, has a direct impact on extension officers approach their jobs (Boone, Meisenbach & Tucker, 2000).

According to Boone et al. (2000) today agriculture holds a unique position in society. As more and more of the world's population moves from rural to urban areas, agriculture as a way of life is changing. Although its role in food production has not changed, however, and more and more of the world's food production is centered on agriculture with fewer and fewer people engaged in production farming. The gap between rural and urban society has narrowed. Media and popular culture have changed the way that farmers work, live and interact with other parts of society. Depending on the commodity in which a farmer specializes, information and technology is critical for efficiency and fiscal success. Technology is not limited entirely to the production aspects of agriculture. Farmers, as has much of society have adopted personal computers technologies into their farm operations; therefore computer technology is built into most equipment in use on farms today. Specialty crops and specialty farming operations are also tied into advanced technology.

Role of ICTs in the Extension Service

Information and communicative technologies are key enablers of globalization. They allow for the efficient and cost-effective flow of information, products, people and capital across national and regional boundaries. ICT is not a panacea for rural development problems, but it has the potential to help the rural poor to leaping some of the traditional barriers to development, by improving access to information, expanding the market base, enhancing employment opportunities and making government services work better (Swanson & Rajalathi, 2010).

According to Leeuwis (2004),the conventional method of training and information dissemination through the extension channel has been based on cascade model: master trainers are trained, they in turn train the trainers who go to the field to conducting training of extension officers / workers; extension agents carry information to the final recipients. This model has obvious disadvantages such as loss of information, degradation of quality of training, long lead times and high costs. In fact, the lead time of such a training cycle is so long that by the time one training cycle is completed, the second is overdue. There is therefore a great need to speed up the extension process and enhance its quality. It is in this area where the use of ICTs amongst extension officers plays an important role and with the Participatory Programme Extension Approach, the gap that exists in the information flow between the end users will be addressed.

Effective ICTs strategies must provide the basic infrastructure for connectivity and access, develop human capacity, offer affordable demand-driven ICT services and involve local stakeholders and beneficiaries in project design and implementation. Liberalization of the telecommunications sector can dramatically expand connectivity and reduce costs, but the strategic government investment or regulatory measures are needed in disadvantaged areas where the private sector does not provide services. Community- based telecenters offer a way of providing affordable access to ICT services in rural areas and this will assist agricultural extension officers to disseminate required farming information that will increase the agricultural productivity (van Veldhuiven, et al, 1997).

The diffusion of ICTs is fundamental to enable developed and developing countries to stimulate socio-economic development and usher in structural change. The term digital divide has been coined to capture the bias in the patterns of access to ICT in terms of the levels of use on the internet (region, sector, socio-economic group), the ownership of computers or public facilities, and the availability of telecommunications (telephone lines), the nature of use (e-mail, using the World Wide Web), and the institutions (public and private) offering ICT facilities (Roy, 2005)

ICTs are in addition are considered to be drivers of change for rural and agricultural development as they are efficient tools for reaching rural and remote communities and for improving agricultural productivity. Due to this factor, ICTs can speed up the extension of development services in areas that include healthcare, education and agriculture. Furthermore, they can be instrumental in strengthening partnerships and in providing a framework for shared learning (Richardson, 1997).

Uses of ICT's by the Extension Service

The ICT sector as a whole (telecommunications, broadcasting, computer hardware and software, and related technologies) has emerged as a strategically important sector driving social and economic change. The paradigms shift in development concept towards participation and sustainability coupled with revolutions in the information and communication technology has provided opportunities for extension and rural communities to move into the information age. Extensive review of literatures in agricultural development showed that there is a direct relationship between research and the utilization of research findings. This is a strong linkage complimented by flawless information flow enhanced by the effective use of ICT by the extension services significantly boost agricultural production and improve rural livelihoods in developing countries. Front line extension workers, who are the direct link between farmers and other actors in the agricultural knowledge and information system, are well positioned to make use of ICT to access expert knowledge or other types of information that could facilitate the accomplishment of their day-to-day activities (Aboh, 2008).

The important role played by agricultural extension services in providing linkages and support to agricultural research information and technology transfer for farmers and farming communities has been crucial to agricultural success. Strong criticism of public agricultural extension services has circulated in recent years. This criticism is due to agricultural extension top-down approach, which has been supply- driven, technically weak, patronizing, and catering only to large farmers and providing insufficient coverage of and contacts with farmers (FAO,2004).

IFAD (2002) reached the conclusion that extension services in Africa have failed to address the needs of small-scale farmers. In another study, Richardson (2006) argues that agricultural extension services that provide agricultural information do not work effectively in Africa. These shortfalls may be due to changes in the extension process that have resulted in the shift to the facilitation and brokerage of information, communication and advocacy services. This range of services, meant to improve rural livelihoods, can benefit from the applications of ICTs. It has been found for instance that the technologies could be used to restructure extension services. For example, they can be used to disseminate information and advisory services to farmers to improve their efficiency. More specifically, agricultural information on input supply, use, product marketing, and value addition could be transmitted through ICTs, thus cutting down the number of visitations by extension workers. ICTs could also free time for extension workers if greater use is made of telecentres and other public access points.

It is true that public extension services unfortunately have been ineffective in reaching farmers and farming communities with information and technologies needed to ensure food security and sustainable development. Farmers have managed to obtain information from other sources where agricultural extension services have been insufficient. These preferred sources include other farmers, private sector agricultural marketing and sales providers (of seeds, fertilizers and agrochemicals) or from NGOs and civil society organizations. There is substantial literature providing evidence to this fact. What is clear from many studies, however, is the continued need for information by farmers and farm communities- thus necessitating continued improvement in ways and means to provide the extension support that farmers and farm communities require and demand.

According to FAO, (2004) given the reality of an interconnected world, the recent trend currently confronting agricultural extension globally are also manifesting regionally and nationally. These include the responses to the effects of globalization, privatization, increasing commercialization of agriculture, increased democratization and participation, environmental degradation, increasing incidences of natural disasters, improved communications including the availability of information technology, the need to address the HIV/AIDS epidemic, preferences for multidisciplinary approaches and holistic development and criticism of extension itself. The faces of globalization and privatization on extension, for instance, are probably most pronounced in the highly competitive, market – driven orientation of agribusiness. However, most extension services are ill prepared to deal with the changes due to extension agents' lack of knowledge and skills in those areas.

The twin challenge for developing country agriculture, then, is to harness the power of ICTs to compete in complex and rapidly- changing regional and global markets while empowering poor smallholders with information and communication assets and services that can increase their productivity and income and protect their food security and livelihood.

New technologies, new applications and new business modes and their wider availability and affordability are enabling widespread innovation in addressing the needs of the poor farmer and farming communities. The uses of ICTs are driving changes in agriculture that make it even more urgent to understand their potential and pitfalls. The dramatic and rapid expansion in global information and communications networks in the past decade, and the concomitant proliferation of new information and communication devices, applications, services and business models pose significant new opportunities and new challenges for developing subsistence agriculture and for rural poor populations. Globalised food markets have both opened new market opportunities for developing country agriculture and created new risks and volatility, including new challenges to food security, and risks of marginalization from agricultural value chains that increasingly depend on technical sophistication for speed, customization and food safety.

Agriculture Technology Dissemination

According to (Nagasri, 2000; Aizaki, Nakashima, Ujie, Takeshita & Tahara, 2010), all agricultural extension and farmer programs face major challenges which are ensuring cost effective outreach, designing solutions tailored to needs of individual farmers and cultivating an image that is farmer friendly. Any change in technology, the economic impact of ICT occurs through improvements in efficiency and increasing productivity. This can take place in different ways including improving efficiency in resource allocation, reducing transaction costs, and technical improvements that result in an outward shifting of the production function. In particular, through the provision of information from a source that is relative affordable, accessible and broadly available, ICT can contribute to the reduction of uncertainty in activities and transactions, reduce the extent to which markets are thin, missing or incomplete, and reduce the extent to which information asymmetries can be exploited by the relatively informed to extract rent when transacting with the relatively uninformed.

Most of the farming communities, particularly in the rural areas, do not have access to the huge knowledge base acquired by agricultural extension centers and businesses. ICT can provide vital access to information, markets by connecting the rural poor and marginalized to the worlds information resources and opportunities. The inequality in opportunities presented by ICTs is widest between commercial and subsistence farmers, urban and rural groups, rich and poor, men and women, and the educated and uneducated. Despite this, ICTs use in rural areas is increasing, such as the internet and cell phones and the individual, community ad national benefits they bring by making information available at the fingertips are forever emerging. A productive sector depends on a fruitful and fair interaction between the diverse actors-communication and information flows are critical to this process. The technology and socioeconomic context of the farmers determines the selection of ICT in terms of content, media and form of communication.

According to FAO (2006), there are numerous instances where improved production and market information is important to farmers who are often a particularly vulnerable group. These might include extension and research on adoption of new crop varieties, mechanization, pests and weed control, processing and the care of livestock. A further observation was made by FAO (2006) that, technology and what can be accomplished with it has implications for rural communities and producers of all sizes, whether these are larger commercial producers who need to understand global market situations that affect them or subsistence producers concerned with local input markets. This is particularly relevant in the agriculture sector which is an activity that is often highly dependent on externally determined requirements. Government, parastatals and private sector agri-business frequently regulate commercial agricultural production by placing requirements on quality, safety, logistical arrangements and even quotas. Inputs to the agricultural sector may be similarly affected, including seed, fertilizers, pesticides and herbicides as well as livestock feed and veterinary services (FAO, 2006).

The development of agricultural producers can be enhanced by using ICT is though what is known as site-specific management, also called precision agriculture. This refers to a knowledge intensive management strategy that involves the application of information technology to crop production. The literature provides many examples of ICT applications in this domain, including uses in the application of chemical samples, application of fertilizers, application of herbicides, application of liquid fertilizers, application of pesticides, and so forth, as well as activities for efficient resource management such as livestock movement regulations (FAO, 2006).

Farmer Education

According to FAO (2004) strengthening and upgrading rural education and training is a sure means for addressing the knowledge, skills and information gaps that currently exist for farmers as well as extension agents. Today's farmers seem to be more innovative and extension services have become laggards. The important role that rural education institutions can play in this is vital and will require policy changes and commitment and support by governments, donors and communities, as rural schools and institutions are often given lower priority compared to urban institutions and are thus under - resourced and under-utilized. Strengthening of education's linkages with extension is a needed component.

Extension personnel are unable to creatively respond to the change taking place in the environment and remain duplication and tradition bound. Farmers need dynamic information relating to agricultural rural development. There is presently a gap between farm families need by way of generic and dynamic information in what the conventional extension agencies are able to provide. Therefore, to satisfy the need s of farmers and farming communities, information and communicative technologies would be very effective. ICTs can play a major role in updating the farming community by ensuring information flow through agri-websites, enabling the industry to take an updated, uniform database of farming practices to the individual villages, with the data customized to individual agro-climatic zones (Bhatnagar & Schware, 2000; Acker & Gasperini, 2008).

There is potential to bring about transformation in agriculture, through the enhancement of education and research. Perhaps the most straight-forward way in which this can occur is through the conventional agricultural extension system. Traditional Training and Visit extension is a comparatively costly approach requiring the preparation, printing and dissemination of training material, large numbers of trained extension officers who carry the messages to be conveyed, and the risk that messages may become distorted when they are eventually conveyed. Extension officers who are connected through ICT will be better able to update their knowledge on a continuous basis than in the past, avoiding the criticism that the information provided by these services is often irrelevant or out-of-date. This approach does not require any ICT capacity on the part of the farmer, and as a result, may be relatively simple to implement in many countries. However when farmers are digitally literate a range of new opportunities become available.

According to FAO (2006), ICT can be used for agricultural research surveys and censuses completing a 'virtuous circle' of information exchange and can also be used to build what have been termed Knowledge Digital Libraries which can include the collection of indigenous knowledge about crops and cultivation practices specific to local contexts. Through participatory research strategies that place farmer's and communities at the centre of the research process, a network for small scale development projects can foster the exchange of information, experiences, expertise and solutions to technical problems and adapt the often generic nature of information accessed by ICT to local conditions. Such appropriately supported ICT strategies have the potential to enhance access to information and information technologies as well as access to channels and other modes of communication.

Prize Realization

The benefits of using ICTs can be broadly categorized as personal and commercial. On the personal front ICTs can be used to enable farmers to keep in touch with their relatives and friends whereas commercially, farmers can use them to obtain information on the prices of agricultural inputs and marketing their products. Most farmers sell their produce to middlemen or in the nearest markets where the middlemen decide on price. The farmers have virtually no interaction with the buyer nor do they know the prices ruling at nearby markets. By making commodity prices and market information on a real time basis available on the internet, the farming community can be provided with choices that they lack today. This will ensure better price realization and stimulate a drive towards better productivity. The role of extension services becomes increasingly important in ensuring that the use of ICTs is more relevant and effective.

Furthermore the use of ICTs amongst extension officers can also enormously benefit farmers who grow cash crops by providing forecasted information on future process of commodities. This will prevent the tendency of farmers to jump a decision on the basis of ruling price levels and later discover that the process have crashed when they are ready to sell their produce at the end of the season. Information on likely future process of commodities can avert this disaster to a large extent (Nagasri, 2000)

According to Rivera and Sulaiman (2009) in India, Reuters Market Light provides information on market prices, weather conditions, agricultural policy news, and tips on farming cycles via SMS to fee-paying subscribers in India. Information can be personalized based on the type of crop, region of the country, and local language. The service employs a staff of 300 full-time content professionals to provide news and data on more than 250 crop types, 1,000 markets, and weather forecasts for 2,500 locations.

Back-Up Services

An important task in the process of development is that of carrying information on the latest agricultural concern to rural areas. It is not sufficient for extension officers to provide farmers exhaustive information through web portals, their specific queries have to be replied though on-line chats, they have to be provided with early warning systems regarding pest onset and weather forecasts to their local areas. Digital data storage appeared on the scene much later than long distance communication. While telephones and telegraphs revolutionized the transfer of information, traditional analog methods like pen and paper were still considered sufficient enough to keep track of production. Bookkeeping had become an art by then, and farmers simply did not have enough time to learn the ins and out of computerization.

However, as the volume of information increased dramatically over the years, the limitations of old methods became increasingly clear. They first became apparent in central organizations like Ministries of Agriculture which had to keep track of production on an exceedingly vast scale and maintain the information indefinitely. The vulnerability of paper and the massive amounts of record to track meant that digital storage quickly went from being a novelty to a necessity. The use of ICTs amongst extension officers will attract large number of farmers; lot of new industries interested in providing services like transportation ad mechanized farm operations like pest control will get created (Gurstein, 2000).

Enhancing Agricultural Production

Swanson and Rajalahati (2010) reported that most countries have somewhat similar agricultural extension systems, but most systems differ in terms of their respective management structure, including how extension priorities are set, how extension programs are planned and by whom, and how extension programs are actually organized and implemented. Farmers especially emerging ones often face threats from soils, drought, erosion and pests. Key areas where ICTs can help improve this is by providing up to date information about pest and disease control, early warning systems, new varieties, new ways to optimize production and regulations for quality control.

Capacity Building and Empowerment

The meaning of the term capacity building seems to vary according to the user, but there seems to be no doubt that ICTs are capable of helping to achieve it. There are many definitions, but they all carry the same theme concerning developing an organization's or individual's core skills and capabilities to help them achieve their development goals. This definition suits the context of ICTs as it assumes knowledge of the existence of development goals, without which ICTs are unlikely to be of much value (Marker, McNamara & Wallace, 2002).

It is widely accepted today that the information society is going to lead the knowledge society where individuals as well as institutions are valued and judged according to what they know and how much they know. The farming population needs new knowledge and new skills to understand, to feel at ease with, to take advantage of, to benefit from, and to operate ICTs efficiently. The sped of change of ICTs means that acquisition of this new knowledge and skills needed to operate ICTs is becoming a never ending process.

Marker et al. (2002), further states that capacity building also relates to the accumulation of social capital, which refers to those features of social organization such as networks, norms, and social trust that facilitate coordination and cooperation for mutual benefit and establishing networks for mutual benefit that can be nurtured and extended through the use of ICTs. ICTs can help create and sustain online and offline networks that introduce and interconnect people who are working toward similar goals. Many NGOs can use ICTs as an advocacy tool to champion their courses.

ICTs can be used strengthen communities and farmer organization to strengthen their own capacities and better represent their constituencies when negotiating input and output prices, land claims, resource rights and infrastructure projects. Rural communities are able to interact with others via the use of ICTs which reduces social isolation that they would otherwise be facing. Besides, ICTs are able to make processes like law-making and land title approvals more transparent (FAO, 2004).

In the past senior subject-matter specialists were needed to provide technical support to field extension staff about key crops, livestock, fisheries, and other commodity areas. In the future, it may be possible to make more of this technical and marketing information available electronically to the field extension staff, given the complexity of problems being addressed. The goal should be to enhance the capacity of the field extension staff to make available a wide range of technical and market information to the different group of farmers being served, especially small-scale men and women farmers through the use of ICTs (Swanson et. al, 2010).

According to Flor and Hazelman (2004) in Thailand, it is mandatory for the field extension staff of the DOAE to undergo training on Basic Computer Applications and Web Tools. The DOAE ICT Center conducts both training courses on a regular basis. However, secondary data from the 2001 UNDP Human Development Report reveal that computer literacy in the rural areas is still quite low. Generally, Thai extension workers are computer literate and proficient in using the World Wide Web. Their role may go beyond plain users but as contributors to and developers of online knowledge products. Thus, training programs on Web technologies (i.e., Web writing, html, and interface design) might be in order.

Knowledge Base

A knowledge base is a portal –type application that can centralize many major sources of information into one searchable reference library. The sources of information could include government websites, universities, document archives, the Internet, economic reports, historical data and others. The information may have different security levels depending on who is looking at the information, but in general would be open to browsing or searching by topics and keywords. Internet information would be filtered and categorized for more accurate and valuable results. Information in the knowledge base include: technology, seeds, livestock, veterinarian resources, best farming practices and natural language sources (Richardson, 2006).

AGRICULTURE KNOWLEDGE MANAGEMENT

Knowledge sharing, exchanging and dissemination are elements in a broader theme which is knowledge management. The central purpose of knowledge management is to transform information and intellectual assets into enduring value. The basic idea is to strengthen, improve and propel the organizations by using the wealth of information and knowledge that organizations and its members collectively possess. It has been pointed out that a large part of knowledge is not explicit but tacit. This is true for knowledge in agriculture where a lot of good practices are transferred without being well documented in books, papers or extension documents. To manage the knowledge properly, ICTs are needed. In effect, there are many information technologies that can be used for knowledge management (Said, Rafea, El-Beltagy & Hassan, 2009).

Content management system in its wider sense including data bases and multimedia, *is* the core technology of information and knowledge management. This technology can be used in different applications: Building a national agriculture research information system (NARIS) which needs to include research outcomes, projects, institutions and researchers in every country, and a regional research information system that works as a portal for all the NARIS. An example NARIS has been developed at the agriculture research center in Egypt, developing an information system of indigenous agricultural practices can enable researchers to examine this knowledge and decide on its usefulness for sustainable development. Such a system will also keep this knowledge for future generations before it disappears as a result of advanced technologies, developing an information system recording matured technologies that on a trial basis have proven successful and success stories that have achieved economic growth will strengthen the interaction between inventors and innovators. This will lead to an innovation-driven economic growth paradigm, storing and retrieving images, videotapes and audiotapes related to different agricultural activities (Said et al. 2009).

Said, et al. (2009) further states that geographic information systems (GIS) are needed to store databases about natural resources with a graphical user interface that enables users to access these data easily using geographical maps. Furthermore decision support system techniques are needed in many applications such as simulating and modeling methods can be used to build computer systems that can model and simulate the effect of different agricultural production policies on the economy and the environment to help top management make decisions, using expert systems technology to improve crop management and track its effect on conserving natural resources is elaborated in This technology may also be appropriate for keeping indigenous knowledge. Expediting the expert systems development by generating agriculture specific tools to overcome the well-known problem of knowledge acquisition and enhancing the explanation capabilities of expert systems developed for agriculture.

Agricultural Information Needs

The effectiveness of information transfer process depends on the effectiveness of the communication system through which the transfer is to take place, and that, in turn, depends on whether the target audience has access to the medium and can understand the message which it conveys (Feather,1994). In light of the constraints of human and financial resources and geographical distances, the media and ICT are being promoted as valuable tools in the delivery of extension information. Increasingly, ICT is being regarded as a tool for sustainable development and poverty reduction. The extent and rate of change now occurring in the development of ICTs have opened the way for significant change in crop production management, agricultural decision-making and information dissemination. The farmers may depend on extension personnel to get the proper advice to cultivate the crop. The information needed relate to different schemes, crops, technologies, seeds, fertilizers, pesticides, availability of fertilizers, seedlings, biopesticides, soil fertility, pest and disease diagnosis and many more.

The agricultural marketing information is essential for farmers to increase their profits. Information such as price details of seeds, fertilizers, pesticides and availability of these products in the market enables the farmer to take decision in choosing right seed, fertilizer and pesticide required for the better farming. The vital information that flows from the agricultural policy maker's desk, such as fixation of procurement price, procurement targets and policy relating to exports helps to farmers to get maximum profits.

In India, Reuters Market Light provides information on market prices, weather conditions, agricultural policy news, and tips on farming cycles via SMS to fee-paying subscribers. Information can be personalized based on the type of crop, region of the country, and local language. The service employs a staff of 300 full-time content professionals to provide news and data on more than 250 crop types, 1,000 markets, and weather forecasts for 2,500 locations (Rivera & Sulaiman, 2009).

The delivery of information, technical advice and agricultural skills training for farmers rests with national extension services. However, in the Pacific extension activities are typically given a low priority. Extension work continues to face challenges because of the scarcity in human, financial and physical resources. The need 'to do more for less' is a realty, and partnering with the media can help enhance extension services (SPC, 2011).

Agriculture Support

Research suggests that increasing agricultural productivity benefits the poor and landless through increased employment opportunities. As the vast majorities of poor people lives in rural areas and derive their livelihoods directly or indirectly from agriculture, support for farming is a high priority for rural development. ICTs can deliver useful information to farmers in the form of crop care and animal husbandry, fertilizer and feedstock inputs, drought mitigation, pest control, irrigation, weather forecasting, seed sourcing and market prices. Other uses of ICTs can benefit farmers, enabling them to participate in advocacy and co-operative activities.

Vulnerable and Marginalized Groups

According to Marker et al. (2002), women in developing countries particularly, face difficulties in using ICTs, as they tend to be poorer, face greater social constraints and are less likely to be educated or literate than men. They are likely to use ICTs in different ways, and have different information requirements, to men. Women are less likely to be able to pay for access to ICTs, either because of an absolute lack of funds or because they lack control of household expenditure. Constraints on women's time or their movement outside of the home can also reduce their ability to access technologies. Such groups usually require special assistance and attention in order to benefit from programmes that are targeted at poor people.

Munyua (2000) further notes that, in any farming system it is important to recognize the various roles of men, women, youth, and children. Women in rural areas have very little access to information. They are mostly poor, illiterate, and unable to afford even the very most basic forms of ICTs, such as radios and telephones. Nevertheless, rural women actively seek and disseminate information. So ICTs must be appropriate to enable women to gain access to information efficiently and cost-effectively.

Gender considerations have, however, not received the attention they deserve in the design and use of information services or in the application of ICTs. Therefore the types of information required to meet the needs women express, to determine the information gaps, and to select the ICTs and services to best close these gaps must be identified.

The lack of reliable and comprehensive information for rural female farmers is a major hindrance to agricultural development. They require information on agricultural inputs; market prices; transportation systems; product potential; new environmentally sound production techniques and practices; new agricultural technologies; new markets; food processing and preservation; decision-making processes; the resource base; trade laws; and trends in food production, demand, and processing. Women also need to exchange indigenous knowledge. However, most available local information is packaged in a raw form and therefore difficult to access or use .The situation is compounded because women do not know where to find this information. Moreover, current mass media and communication systems have not been used to maximum effect in development. Information should be accessible to female farmers at selected sites, with various ICTs to facilitate easy access to relevant information and information exchange to meet the information needs of rural female farmers (Munyua, 2000).

Trends in Agricultural Information Management

According to Ballantyne (2009), the world of agricultural information and communication is changing, as agricultural innovators become active creators and managers of information and knowledge and information managers become innovators. These effort calls for greater investment in knowledge creation, information access, and the wider use of information technologies. There is an exploration on what the renewed interest might mean for information and communication specialists working in agricultural environment. Starting from an "innovation systems" perspective which looks across the agricultural information for development landscape and highlighting some items on the agenda of information and communication specialists working area. The widening recognition of the value of farmers knowledge, growing used of information and communication information and communication technologies to enable different agricultural development activities, concerns to ensure that public investment result in public goods whose benefits can travel, related efforts to ensure that agricultural content is open and accessible, the discovery and increasing use of a 'social' web, and some emerging new roles for agricultural library information and information centers to meet changing demands.

Virtual Community

Satyanarayana, Reddy, Balasubramani, Sharma, and Bhaskar (2008) defines a virtual community, e-community or online community as a group of people that primarily interact via communication media such as letters, telephone, email or usenet rather than face to face. If the mechanism is a computer network, it is called an online community. Virtual and online communities have also become a supplemental form of communication between people who know each other primarily in real life. Many means are used in social software separately or in combination, including chartrooms and forums that use voice, video and/ or text.

Virtual Information Networks

A number of virtual information networks with the objective of linking agricultural institutions for facilitating better access to information resources have been initiated successfully. While some of these network agricultural institutions within a country, link to agricultural institutions around the globe.

Agriculture Network Information Center

Agriculture Network Information Center (AgNIC) is an Internet based network of public and private agricultural libraries and information centers, coordinated by the National Agriculture Library (NAL), USA. The network aims to provide global access to agricultural information. AgNIC Members represent 40 land-grant universities and other national and international partners including universities and research institutions, government agencies, and a non-profit organization. Through its website the network provides access to a network of electronic sources on research and teaching in agriculture, food, renewable natural resources, forestry, and physical and social sciences. AgNIC is a distributed discipline-oriented source of agricultural information in electronic form on the Internet. The goals are to: identify major collections of agriculture-related information; provide mechanisms to facilitate access/ retrieval from these information resources; create mechanisms to encourage organizations to collaborate in creating/ using AgNIC. AgNIC partners select important information sources for inclusion in the system (Satyanarayana et al. 2008).

Services include: Resource database with web sites, image collections, lists of publications, documents, databases, and other resources; Calendar of events which includes meetings, seminars, national and international symposia and conferences, conventions, and workshops in agricultural and related sciences; News items; Specialized Services – such as Plant Disease Announcements; discussions on emerging plant diseases around the world; Expertise where specialists respond to individual questions. There are partnerships between libraries and United States Department of Agriculture (USDA) Cooperative Extension programs; between libraries and academic departments within colleges; between states and between technologists and librarians. Member participants take responsibility for small segments of agricultural information and develop Web sites and reference services in their specific subject areas. Nearly all participating AgNIC institutions have developed partnering relationships with a variety of internal and external institutions, groups, and agencies to develop content and tools for their respective Web sites.

Agricultural Libraries Network

The Agricultural Libraries Network (AGLINET) coordinated by the Food and Agriculture Organization of the United Nation (FAO), is a world-wide, voluntary network of international agricultural libraries with regional/country coverage and other comprehensive or specialized subject resource collections. All member libraries provide, upon request, access to the literature originating in the country or region or for a given specialization. AGLINET aims at combining resources for mutual and rational use through delivery of primary documents, by means of inter library loan provision of reproductions, bibliographic information; with appropriate regional and subject specialization. Member libraries include Agricultural libraries of national or regional importance with comprehensive collections and strong regional coverage; Libraries in special subject fields within the broad domain of agriculture with worldwide coverage.

AGLINET Network members include Libraries from Argentina, Australia, Belarus, Belgium, Benin, Botswana, Brazil, Bulgaria, Brazil, Canada, China Costa Rica, Czech republic, Denmark, Egypt, Estonia, Ethiopia, Finland, France, Germany, Hungary, India, Indonesia, Italy, Japan, Jordan, Kenya, Korea, Latvia, Lithuania, Malawi, Malaysia, Mexico, Netherlands, Nigeria, Norway Philippines, Poland Portugal, Russia, Slovenia, South Korea, Sprain, Sri Lanka, Sweden Syria, Thailand, UK, Uruguay, USA, West Indies and Zambia (Satyanarayana et al. 2008).

International Information System for Agricultural Sciences and Technology

The Food and Agriculture Organization (FAO) of the United Nations initiated agricultural information networking through AGRIS. Established in 1974, AGRIS facilitates information exchange and brings together world literature dealing with all aspects of agriculture and related subjects. AGRIS is a cooperative system in which 240 national, international and intergovernmental centers in participating countries input references to the literature produced within their countries and draw on the information provided by the other participants. The new vision of AGRIS, in response to the needs includes: a decentralized approach with greater emphasis on national partnerships, improved linkages between the AGRIS network and other FAO initiatives; focus on management of documents in agricultural science and technology in full text; information about activities, organizations, and people in agricultural science and technology (FAO, 2006).

Agrigate

Coordinated by the National Library of Australia is a project of the libraries of the Universities of Melbourne, Adelaide and Queensland, and the Commonwealth Scientific and Industrial Research Organization (CSIRO) in Australia. This is a subject information gateway for resources, both online and offline, in agricultural research. The purpose is to support the identification and dissemination of quality research materials selected by an editorial review process consisting of members of the agricultural research community. The content covers agriculture and related areas of value to the Australian agricultural research community. The resources selected have been reviewed by subject specialist librarians. The majority of resources identified in the database are available online.

AGROWEB CEE Network

Satyanarayana et al. (2008) states that AGROWEB CEE network has been established by organizations and individuals involved in agricultural and rural development in Central and Eastern Europe. This collaborative network is maintained by representatives of participating countries, and facilitated by International Association of Agricultural Information Specialists (IAALD) and FAO. Twenty-five countries from the Balkan region, New Independent States, Baltic States, Caucasus region, Central Asia, and European Union are participating in the network. National web pages have been established as portals for the countries in the region, which provide access to information about agriculture-related institutions - ministries, libraries, information centre's, universities, research centre's, NGOs, agro-marketing organizations and other organizations. The Agro Web national portals provide Internet links, to these national institutions and to other relevant national web sites. In addition to the national portal pages, cross-cutting subject areas have been identified and a number of thematic sectors of the network are being coordinated by regional focal points.

Asia-Pacific Association of Agricultural Research Institutions (APAARI)

Information exchange among the Asia-Pacific national agricultural research systems is one of the primary objectives of the Asia-Pacific Agricultural Research Information System (APARIS). Components of APARIS include Management Information System (MIS) Tools viz.: Regional Research Networks (RRN) Database, NARS Database, etc.; information on regional events; access to scientific publications generated by agricultural research in the region; electronic Forums to facilitate dialogue among stakeholders of Agricultural Research for Development (ARD) in the region on issues of strategic importance; gateway/portal service - portal to RRNs, websites of NARS Institutions in the Asia/Pacific region, web-enabled information on key topics/themes of ARD; Knowledge Networks on ARD in the APAARI region; information dissemination through APAARI web site and publications. The web site has linkages to various national, regional and international institutions, networks, and other organizations (Satyanarayana et al. 2008).

The Consultative Group on International Agricultural Research

According to FAO (2006), The Consultative Group on International Agricultural Research (CGIAR) is a research network of fifteen international agricultural research centers. The 15 centers supported by the CGIAR are independent institutions, each with its own charter, international board of trustees, director general, and staff. The institutions are Africa Rice Center (WARDA), Biodiversity International, Centro Internacional de Agricultura Tropical (CIAT), Center for International Forestry Research (CIFOR), International Maize and Wheat Improvement centre (CIMMYT), International Potato center (CIP), International Center for Agricultural Research in the Dry Areas (ICARDA), International Crops Research Institute for the Semi-Arid Tropics (ICRISAT), International Food Policy Research Institute (IFPRI), International Institute of Tropical Agriculture (IITA), International Livestock Research Institute (ILRI), International Rice Research Institute (IRRI), International Water Management Institute (IWMI), World Agro-forestry Center (ICRAF) and World Fish Center (WFC).

Thirteen of the 15 CGIAR Centers are located in developing countries and the knowledge gathered is shared and disseminated across the world. The CGIAR website gives access to a Virtual Information Center and Library. One can tap into agricultural information databases, including online libraries of the CGIAR Centers and the Core Collection Database. The CG Library gives access to databases and e-journals and facility to go directly to the full text of publications. One can search CGIAR libraries/ other agricultural libraries/by specific topics. The Virtual Information Center provides information on various topics mentioning the source against each topic. Each institutional library is the source for/ and responsible for providing information on specific subject areas viz.,: Agriculture in the Dry Areas-ICARDA library; Agriculture in semi arid tropics- ICRISAT library; Agro-forestry- ICRAF library; Aquaculture and fisheries -WFC library; Food policy- IFPRI library, Agricultural biodiversity – IPGRI library; Water Management-IWMI library; Wheat - CIMMYT library(FAO,2006).

DAINet: German Agricultural Information Network

The German Agricultural Information Network (DAINet) is a catalogue of Internet sources in the field of nutrition, agriculture and forestry. DAINet was established in 1995 to guide people interested in subject specific information to sources on the Internet. DAINet has about 7,500 references to information sources and about 800 web links to agriculture-related organizations. Three sections cover, 'information', 'service' and 'dialogue'. The 'information section' organized by subjects, structures worldwide existing agricultural information under 'subject fields', 'user groups' and 'topics'. The 19 'subject fields' cover various topics of agriculture. The division into five user groups caters to the specific needs of farmers, economists, scientists, educators and journalists. The 'service section' is structured according to types of data. The 'dialog section' is used for professional communication through Internet. 'Information' and 'service' sections offer links to databases and web pages and also give access to newsgroups and mailing lists (Satyanarayana et al. 2008).

ELIN: Entomology Library and Information Network

According to Satyanarayana et al. (2008), the ELIN project hosted by the Natural History Museum, London, aims to establish a global Entomology Library and Information Network to provide entomologists with a co-ordinated and structured electronic information resource including a gateway to a wide range of entomological information. A mailing list was launched in 1998 and has over 100 members from libraries and information centre's from around the world, which has encouraged international communication between specialists working with entomological information. The ELIN project provides up-to-date, detailed information about entomological libraries around the world. ICTs.

Information Network on Post-Harvest Operations

Information Network on Post-Harvest Operations (INPhO) is an FAO databank project of the Post-Harvest Management Group. It is an international collaborative effort by the FAO, GTZ and CIRAD which aims to support the collection and dissemination of information on proven technologies and products in post-harvest systems. Components of the network include a comprehensive collection of information on post-harvest issues, communication/interactive services; links with other databases. The website includes full text documents of training and technical publications, country profiles, crop profiles (main focus is on cereals and grains, fruits and vegetables, oil seeds, roots and tubers). Recipients of this information include people working in the agricultural production and marketing sector viz., producers, researchers, policy makers, private investors and donors (FAO, 2006).

Philippine Agricultural Libraries and Information Services Network

Philippine Agricultural Libraries and Information Services Network (PhilAgriNet) aims to create and maintain a central electronic database of Philippine technical agricultural literature and make this accessible to agricultural scientists worldwide. Membership is open to agricultural institutions. The database intends to cover all technical publications on agriculture, generated by member institutions, both published and unpublished, and written in any language or dialect, regardless of format. Affiliation is open to libraries of Government agencies and corporations engaged in agricultural research; state colleges and universities, whose curriculum includes agriculture and independent or private agricultural agencies.

Virtual Extension Research Communication Network (VERCON)

Virtual Extension and Research Communication Network (VERCON) established by FAO aims to use the potential of the Internet and apply it to strengthening and enabling linkages among the research and extension components of the national agricultural knowledge and information system. VERCON aims at improving linkages between and within agricultural research and extension institutions through human and technological components. The human component is a network of staff of research and extension institutions, faculty of agricultural education, NGO workers, in some cases agro producers committed to strengthening collaboration, communicating, sharing information and supporting improved agricultural production. In order to link the human component there is the technological component, which allows members of the network to communicate and develop, share, store and retrieve information (FAO, 2006).

Network members can engage in two-way horizontal communication to address problems, discuss solutions and coordinate local, regional and national agricultural activities. An illustration of how VERCON functions: A farmer visits an extension office with an insect that is causing heavy damage to his cotton crop. The extension worker identifies the insect by consulting a database of digital photographs of various cotton pests that have been photographed in the area and scanned. Using a chat room, a number of integrated pest management specialists are consulted simultaneously about how to accurately identify and control the pest. Relevant information is quickly compiled from research results, pest distribution maps, publications and the photo file into a Fact sheet using a template. A message with the fact sheet attached is sent to all extension officers in the region, warning about the presence of the pest and providing recommended integrated pest management measures. This is communicated to farmers through IPM farmer field schools, facilitated by extension.

In 2000, the Virtual Extension and Research Communication Network (VERCON) project was funded by the FAO Technical Cooperation Program (TCP) to develop a Web-based information system to strengthen the link between research and extension. The Egyptian Government established a pilot VERCON to support research and extension interaction and the flow of information between research, extension and farmers in Egypt. This network has been extended to include other stakeholders, and other services through a project funded by Italian Debt Swap Program and executed by FAO in collaboration with Central Laboratory for Agricultural Expert Systems (CLAES). Several expert systems have been made available on this network in addition. In collaboration with other stakeholders, CLAES has developed three regional expert systems for wheat, faba and barley. Furthermore CLAES also developed the National Agricultural Research Management Information System (NARIMS) through a project funded by FAO/TCP (FAO, 2003).

Web-Based Information Service for Agricultural Research for Development

Web-Based Information Service for Agricultural Research for Development (WISARD) is a web-based information platform that provides searchable information on experts, organizations, outputs and projects in the fields of ARD, Natural Resource Management (NRM) and Sustainable Development (SD) from the mid-nineties till date. The system can be used at organization, network, national and international levels. WISARD allows decentralized data input and management through focal points at organization, network or national level. A module has been added for depositing outputs of research and another is under development for 'Technologies and Best Practices'. Present partners include among others, CGIAR, IPM-Europe, and Rice-Wheat Consortium for the Indo-Gangetic Plains, FAO, donor agencies, international and national NGOs and European agencies and national agricultural research systems of India, Pakistan, Nepal and Bangladesh.

Bridging the Digital Divide with Technology

The emergence of digital divide lies in the acknowledgement of the existence of an environment that sustains, unequal and unaffordable access to digital and network resources; enormous disparities in the ICT endowment/infrastructure and the lack of suitable local digital content. Digital divide cuts across international borders and goes beyond the gap between developed and developing nations to include the unequal participation of women, youth and marginalized communities, as well as the lack of local culture and content. The concept applied to reflect the disparity in access to information that is growing within nations, namely, between the rich and poor and between urban and rural communities. In recent years, as ICTs have become a more important part of the global information economy, more attention has been given to the gap in access to ICTs between developed and developing nations, and also within countries (Alao, 2010).

E-Agriculture

The FAO defines e-Agriculture as an emerging field in the intersection of agriculture informatics, agricultural development and entrepreneurship, referring to agricultural services, technology dissemination and information delivered or enhanced through the Internet and related technologies. More specifically, it involves the conceptualization, design, development, evaluation and application of new (innovative) ways to use existing or emerging information and communication technologies (ICTs)" (FAO 2006). E-Agriculture is intended to promote the integration of agricultural stakeholders and technology with multimedia, knowledge and culture, and aims to improve communication and learning processes.

According to FAO (2006) e-Agriculture is an emerging field focusing on the enhancement of agricultural and rural development through improved information and communication processes. E-Agriculture also addresses the 8th Millennium Development Goal, and the target to make available the benefits of new technologies, especially information and communication technologies to the fight against poverty. More specifically, e-Agriculture involves the conceptualization of design, development, evaluation and application of innovative ways to use information and communicative technologies in the rural domain, with a primary focus on agriculture. All stakeholders of agriculture industry need information and knowledge about the phases of agriculture, for example: water management, fertilizer application, food safety and food marketing and how to manage them efficiently. Any system applied for getting information and knowledge for making decisions in any industry should deliver accurate, complete, concise information in time or on time. The information provided by the systems must be in user friendly form, easy to access, cost effective and well protected from unauthorized accesses. ICTs can play a significant role in maintaining the above mentioned properties of information.

The World Summit on the Information Society (WSIS) Plan of Action emphasizes two key areas on e-Agriculture, namely:

- Ensuring the systematic dissemination of information using ICTs on agriculture, animal husbandry, fisheries, forestry and food in order to produce ready access to comprehensive, up-to-date and detailed knowledge and information, particularly in rural areas.
- Establishing public-private partnerships that seek to maximise the use of ICTs as tools to improve production, marketing and food safety standards (FAO 2006).

Electronic Publishing in Agriculture

Satyanarayana et al. (2008) states that Electronic Publishing (e-Publishing) is quickly becoming an important part of publishing mainstream. Recent innovations in this area have made it possible to publish on the information super highway. E-document can be accessed at the computer. It makes much easier for readers to search the information. It is very quick and easy for a reader to browse the table of the contents of previous issues, to jump directly to a particular section of a document or even particular section of an article. E-Publishing (EP) difference lies in the new levels of value it provides through features not possible in traditional media. EP products may differ to an even greater degree than print products. Nevertheless, there are some common features to distinguish EP from print Publishing in terms of value to end-users.

Satyanarayana et al. (2008) further states that EP products create additional value for the user with regard to following three dimensions: content availability, content transparency and interactivity and content format. Content availability means that EP products can be delivered and accessed with more independence of time and place than can be traditional print products and that their delivery is less limited with regard to quantity. Content availability includes: time of delivery - available any time; location of delivery - consumption anywhere; amount of information - end of traditional boundaries imposed by paper volume and price. Content transparency and interactivity refers to new tools and opportunities concerning information navigation The main features are: Interactivity- contextual hyperlinks open new dimensions of information retrieval and lead to new types of information behavior: browsing, etc. the possibility to integrate content and services; and search tools across one or thousands of documents - interactive information processing.

For example, Electronic Publishing in Agriculture in India, The Directorate of Information and Publications of Agriculture, (DIPA), New Delhi, is the official publication wing of the ICAR through which the research and other activities are revealed to the world. DIPA brings out a variety of publications in English and Hindi languages for the use of scientists, researchers, students, policy planners, extension personnel, farmers and the general public. The e-publications of DIPA include - "Handbook of Horticulture", ICAR Research Projects Information – Research Project Files (RPF) Database, ICAR Vision 2020 Document etc. Some other important institutions bringing out e-publications in Agriculture in India include, National Institute of Agricultural Extension Management (NIAEM), Indian Agricultural Research Institute (IARI), New Delhi, Indian Farmers Fertilizer Cooperative Limited (IFFCO), Krishak Bharati Co-operative Limited (KRIBHCO), National Bank for Agricultural and Rural Development (NABARD) National Agricultural Co-operative Marketing Federation of India Ltd. (NAFED) Indian Agricultural Statistics Research Institute (IASRI), and State Agriculture Universities of Tamilnadu, Uttarakhand, Punjab, Haryana and Andhra Pradesh. Two major open universities namely Indira Gandhi National Open University (IGNOU), Delhi and Yeshwantrao Chavan Maharashtra Open University (YCMOU), Nashik and National Institute of Agricultural Extension Management, MANAGE, have taken open and distance learning in Agriculture on a national scale, and all these institutions are in the process of bringing out a number of e-publications in Agriculture (Satyanarayana et al. 2008).

E- Commerce

Harris (2004) indicates that m-commerce using wireless mobile phones and PDAs to sell goods and services has been growing rapidly. It is the case however, that most countries cannot make payments online due to national ICT policies, security risks and inadequate infrastructure such as low bandwidth to support e-commerce. Other bottlenecks have been created by inadequate skills to use e-commerce, the legal status of e-mail communication and the high telecommunication costs. Not surprisingly therefore, in Africa, e-commerce has been insignificant in the agricultural sector. Thus, a number of transactions have to be conducted off-line (Tregurtha & Vink 2002). Some of the reasons given for not adopting e-commerce include: lack of applicability and little incentive to change business models when returns are not clear; poor trust in e-transactions; privacy considerations; shortage of ICT-skilled labour (Qiang, Clarke & Halewood, 2006).

The findings of a survey on e-commerce that was conducted in South Africa there are a number of challenges to using e-commerce such as: the need for face-to-face interaction; data privacy and security issues; hesitance of customers to use the technology; inadequate staff with e-commerce expertise; costs of implementing e-commerce; unwillingness to make organisational changes; low levels of Internet usage; non-supportive business laws; inadequate legal protection (Qiang et al, 2006).

Types of ICT Tools Used by Extension Service

Increasing diversity and complexity of global agriculture, and the seed of global innovation in ITC tools, services and business models, requires in turn a flexible approach to policies and investment choices that focuses not on 'choosing winners' among ICT options but on maximizing conditions for local innovation and creative public –private partnerships to expand ICT access and services. According to Obayelu and Ogunlade (2006), ICTs consist of various collections of resources and technical tools that are used for connecting, spreading, storing and managing information. ICT represents the collection of hardware and software that is used for producing, preparing, transferring and storing data via devices such as computers, radios, televisions, etc., and it includes an extensive scope of traditional and modern media.

In general, ICTs can be classified into three groups: New ICTs: This group consists of computers, satellites, one-on-one connections, wireless phones (mobile), the internet, e-mail, the web, internet services, video conferences, CD-ROMs, personal computers (PC), distance control systems, informational-geographical systems, global positioning systems (GPS), electronic cameras, databases, etc. The hidden concept behind these technologies is that they are not automatically considered to be new, but their common and inexpensive availability has resulted in them being regarded as new; Old ICTs: This group consists of radios, televisions, telephones, telegraphs, audio and video cassettes, films and slides. This group of technologies have been used for several decades; Very Old ICTs: These groups of technologies have been used for several centuries and includes newspapers, books, photo albums, posters, theater, human interactions, markets and plays (Obayelu & Ogunlade,2006).

According to FAO (2006), combining traditional information and dissemination methods with new forms of extension of extension delivery will promote information access by farmers and farming communities. The ICT tools that have great potential for use in agricultural extension include radio, television, telephone (GSM), the web, search engines, packet digital assistants, cameras, video, email, computer, contact data bases and system, CD-ROM, DVD, current awareness, group ware, rural radio, etc. regardless of the tool, the focus must be on the people who will use the technologies and the content rather than the technologies themselves. Farmers depend on multiple sources of information and technologies. In most societies, face- to- face contact is still the preferred mode of communication. Given the large populations to be served by extension agents and the dispersed and distant locations, face-to-face communication is not always possible and thus to combine strategies and communication channels to include traditional information sources, such as radio, print, video and television, together with new forms of delivery such as through ICT.

Rivera and Sulaiman (2009) states that different approaches to delivering farm extension services via ICT channels in a sustainable (i.e. without on-going donor support, although often with government support) and scalable ways (i.e., to reach tens of thousands or millions of farmers) have been identified. Often, more "high tech" ICT channels such as mobile phone services are combined with an older ICT channel, like radio, that has been used well for many years in sub-Saharan Africa and for decades in the US and elsewhere. Strengthening the use of ICT to close the gap between those who have access to extension services and those who do not, along with the strategies use of traditional media, is an opportunity that needs to be harnessed. Use of ICTs is now increasingly available and becoming more affordable. Linking ICT tools and other traditional mediums is also advocated, with choices dependent upon the local situation and realities. ICT provide information access and networking, which are two areas where the current extension systems are weak- between research and extension and between extension and farmers.

According to Lie (2006) the renewed interest in ICTs for agricultural development is remarkable and triggers the question: "What is the difference between the classic mass media attempts and the 'new' attempts to use ICTs as instruments for bringing about change?" The first thing that seems to be different relates to the shift in paradigmatic thinking about change and development in general and the role of ICTs in development in particular. This change is often framed under the heading 'from modernization to multiplicity'. The new paradigm emphasizes non-linear bottom-up perspectives and local cultural relevance. The second thing that seems to be different and is more specific is that the new ICTs, such as internet and mobile phones are not only and primarily used to reach the masses as the mass media were supposed to do. The new ICTs can also cater for small and specific audiences, which opens up possibilities for sharing more relevant and even tailor made information.

As the mass media mainly aimed at national development, the new ICTs can also be appropriate instruments for aiming at local, community levels. The implementation of any technology needs to be participatory and sustainable in its very nature. Although difficult to implement, the new ICTs have the potential of becoming a more democratic medium than traditional media such as national television that is often operating in a national power driven environment. So, an important difference between then and now seems to lie in the possibilities of the technology itself. The possibilities of the ICTs have changed towards more flexibility as it is related to speed, convergence of media forms; interactivity and specificity (Lie, 2006).

Furthermore, Lie (2006) states that in the area of 'agricultural extension systems' or 'agricultural knowledge and information systems' (AKIS) one of the changes taking place is the integration of new information and communication technologies. The mainstream focus of bringing ICTs to rural areas seems to be lying on (multi-purpose) telecentres or information kiosks as they are called in India. Two mainstream ICTs in the debates seem to be the use of the internet (including email) and the use of mobile phones, although mobile phones seem to get far less attention. Besides these two 'new' ICTs, radio remains one of the most important communication and information technologies for rural areas, but also the overall paradigmatic thinking about agricultural extension has changed.

The present systems no longer think in terms of 'adoption and diffusion of innovations', but talk about 'agricultural knowledge and information systems', emphasizing social learning and negotiation, participation and interactivity, the sharing—instead of transmissions—of all kinds of knowledges, knowledge markets and networks and continuous change. Within this paradigm shift from 'extension' to 'communication for change' in the agricultural sector, there is also a growing recognition of the importance of cooperation and integration of the natural sciences and the social sciences, in case of the communication sciences. Without intrinsic trans disciplinary cooperation, ICTs would only have a limited chance of being successful in an appropriate and sustainable way (Lie, 2006).

Radio

Radio has achieved impressive results in the delivery of useful information to farmer and farming communities. One of its strengths is its umbiquity. For example, a recent survey of 15 villages in Nepal found radios in every village, with farmers listening to them while working in their fields. Another survey of 21,000 farmers enrolled in radio-backed farm forums in Zambia found that 90 percent found radio programmes relevant and more than 50 percent credited the programmes and forums with increasing their crop yields (Dodds, 1999).

Community radio projects have indicated how extension officers and farming communities can use ICTs tools appropriate for their own purposes. For example, in Nepal, two-community radio stations were established, radio Lumbini in Manigram in western Nepal and Radio Madan Pokhara in Palpa District. Village development committee of the communities held a license and a community group holds the other. Both services have proven to be very popular. Ownership of radio receivers in the coverage areas has shown an increase of 68 percent in the use of community radio station for agricultural information. Programmes include valuable development messages; the Kothmale community radio station in Sri Lanka *p*rovides an interface between smallholder farmers in remote regions. In daily agricultural programmes presenters accepts requests for information from community members and browse the Internet live in response to listeners' requests, the information accessed is explained, and communities are enabled to develop their own websites which are then hosted on the station's server and searches the internet for answers, which it then broadcasts on air (UNDP-APDIP, 2004).

For example, in Ghana the community radio station provides agricultural programmes intended to develop and assist farmers in accessing information about the agricultural practices. Agricultural programmes are prepared by community radio extension officers who visit communities to discuss problems and priorities and record discussions with local experts. They enable smallholder farmers to hear people within their own communities discuss the issues in their local languages. *Arid Lands Information Network-East Africa (ALIN-EA), partnered* with World Space Foundation, ALIN-EA uses digital satellite radio broadcasting to provide Web-based information for smallholder farmers in Ethiopia, Kenya, Tanzania and Uganda. In Kenya, Mali Shambani is a weekly hour-long radio program featuring agricultural news and responding to the business and market access needs of small, rural farmers. The program covers a wide range of topics, including market prices and trends, farming techniques, weather and seasonal issues, financing opportunities, inputs, land use, and quality standards. Each program also offers an interactive call-in component where farmers are given the opportunity to pose agricultural questions to a panel of experts either via phone or SMS (FACET, 2010). Furthermore, Rivera and Sulaiman

(2009) states that, Farmer Voice Radio (FVR) is a radio extension service currently operating in Kenya, Malawi, Tanzania, Mali, Ghana and Zambia that targets smallholder farmers. FVR's radio extension agents pro-vide regular, on-site extension support to a small group of pre-selected farmers, which are then documented and broadcast via radio.

Television, Videocassette, VCD, and DVD

The UNDP-APDIP (2004) reports that television (TV) is commonly cited as having considerable development potential. It provides programmes with moving pictures and sounds otherwise unavailable to smallholders, using standardized and readily available technologies. Television programmes can be pre-recorded or live and interactive (with smallholders calling in), although the latter may have to be asynchronous over greater distances and different time zones. Television programmes can now be Internet streamed. Videocassettes, VCDs and DVDs can be shared between individuals, families and community groups for self-paced learning. Production and material rights acquisition costs can be high and time consuming to maintain.

There are examples that exist of television usage for agricultural educational purposes. The most notable example of TV used for agricultural development and growth comes from China with its TV University and agricultural TV station. In Viet Nam, two universities in the Mekong Delta Region work with the local TV station to broadcast weekly farmers' workshops that are watched by millions of farmers and rural communities (UNDP-APDIP, 2004). Similarly, the South African Broadcasting Cooperation (SABC) broadcast a multidisciplinary Agricultural TV programme in partnership with the Department of Rural Development and Land Reform. The Agri TV program broadcast information related to agriculture, rural development and land reform. It also aims to inform and keep up to date with the latest developments and trends, while educating by means of practical information on topics ranging form vegetable production, animal health to irrigation technology.

Video and Digital Video Cameras

Video and digital video cameras are emerging as important ICT tools for extension services. Rural smallholders can visually record and communicate problems to subject specialists. Digital photographs and video clips can be distributed and exchanged online through email or instant messaging, or by regular mail, enabling interactive and collaborative problem solving. Use of these applications should grow as costs reduce, and as digital video and operating systems become more standardized and available. Commonwealth of Learning Media Empowerment (COLME), highlighted that in Ghana and the Caribbean, working with in-country agencies, COLME identifies rural community needs and trains extension workers in shooting and editing videos using local content. The tapes and/or broadcasts serve large numbers of farmers, thus offsetting reducing extension services. The programme provides training in literacy for smallholders and women in Ghana, and in agribusiness and environmental sustainability in the Caribbean. For example in India, Digital Green disseminates targeted agricultural information via digital media to small-scale and marginal farmers. The system includes a digital video database that is produced for farmers by farmers. Participating villages are provided with a TV, DVD player and camcorder operated by local NGO staff and managed by farmers, along with DVDs that are shipped to the village. Nightly viewings are set up on a rotating basis around different areas of the village for small groups of 10 to 20 farmers (Rivera & Sulaiman, 2009; FACET, 2010).

Telephone

In many countries, access to the telephone system is a vital element that government's attempts to implement various policies to offer affordable telephone service but unfortunately some countries lack sufficient telephone lines. In Bangladesh, the Grameen Bank, a village-based micro-finance organization leases Grameen hand phones or cellular mobile phones to successful farmers and community members, this practice has delivered significant benefits to the farmers and rural community dwellers. The phones are mostly used for exchanging price, business and health related information. They have generated information flows that have resulted in better prices for outputs and inputs, easier job searches, reduced mortality rates for livestock and poultry, and better returns on foreign-exchange transactions. Phone owners also earn additional income from providing phone services to others in the community and rural people account for one-fourth of all the phone calls made (Bayes, Von Braun & Akhter, 1999). Bayes et. al (1999) further indicates that for rural communities in general, phones offer additional non-economic benefits such as improved law enforcement, reduced inequality, more rapid and effective communication during disasters and stronger kinship bonding. Furthermore these telephones have perceptible and positive effects on the empowerment and social status of phone-leasing women and their households.

The United Nations Development Programme (Asia-Pacific Development Information Programme, 2004) in collaboration with the Government of India conducted a study on China and found that villages that had telephones as the most basic communications technology, experienced declines in the purchase price of various agricultural commodities and lower future price variability. It also noted that the average prices of agricultural commodities were higher in villages with telephones than in villages without telephones. Vegetable growers said that access to telephones helped them to make more appropriate production decisions, and users of agricultural inputs benefited from a smoother and more reliable supply. Better information also improved some sellers' perception of their bargaining position vis-à-vis traders or intermediaries (UNDP, 2004). Village telephones facilitated job searches, access to emergency medical care and the ability to deal with natural disasters; lowered mortality rates for livestock thanks to more timely advice from extension workers; and improved rates in foreign-exchange transactions (Eggleston, Jensen, & Zeckhauser, 2002). In Nigeria the telephone use in extension delivery even with the recent launch and explosion of the global system of mobile (GSM) as a dividend of democracy is non-existent to very insignificant. The serious limitation of access to reliable telephone lines makes even ordinary intra-and inter-organizational networking for agricultural information exchange a harrowing and frustrating experience (Arokoyo, 2010)

Cell Phones

According to Rivera and Sulaiman (2009) there are currently a wide variety of approaches being tried to use cell phone networks to enhance extension services. Few have managed to scale to millions of farmers and it is not clear yet due to the paucity of research and the relatively short time periods of the implementations which approaches have the most impact. Clearly, a variety of approaches (or combinations of approaches) may prove successful. The approaches are varying in several ways: whether text or voice is used and, for text, if searches or queries are made via SMS or more advanced data queries which require a more capable and expensive phone hand-set or whether farmers use the mobile phone services directly or via a mediator such as a farm extension worker.

Whether cell phones enable one or two-way communication, whether farmers or intermediaries are "pull" information where farmers initiate the query, or information is "pushed" to the farmer or intermediary via a text message, a voice phone call or voicemail. This capability to enable two-way exchanges (synchronous or a synchronous) is an important enhancement to farm extension service delivery because it allows the service deliverers to receive frequent feedback regarding the most important information for farmers and how well they understand the information being delivered to them (Rivera & Sulaiman, 2009). In the Pacific, mobile phones offer another potential for extension workers, with estimates of up to 50% of Pacific Islanders having access to mobile phones. Mobile phones could help extension workers disseminate information on crop pricing or pests and diseases to farmers. In Tonga, a pilot exercise is underway using mobile phones to link farmers and extension officers. The United Nations Conference on Trade and Development (UNCTAD) is helping develop an efficient market information system and a user-friendly platform to disseminate information via mobile phones. UNCTAD supports the establishment of a regional information hub that would be based at SPC (http://www.spc.int/lrd/index.php).

Public Address Systems

Public address systems (PAS) commonly found in China and Vietnam are used to deliver public information, announcements and daily news. A community in VietNam is planning to augment its PAS, by connecting it to the internet so as to obtain useful information for broadcasting (UNDP, 2004). Public address systems which are localized are more popular than the radio which is technically simpler and less expensive. However, research on less developed communities suggests that the telephone and radio remain the most important to provide direct access to ICT tools for changing the lives of the farmers and rural people (Heeks, 1999).

Computers and the Internet

According to Aboh (2008), the use of computers and Internet as ICT tool will help extension officers to analyze data, write scientific papers, prepare plan of work, draw budgets, produce curricula and handouts and reach out to farmers easily. The Internet has grown to be the world's most important communication medium. It is a reasonably inexpensive, fast, two-way medium and a powerful tool for storage, retrieval, and dissemination of information. It is also good for publishing. The Internet can introduce new information resources and open new communication channels for rural farmers. Extension officers can send the document through which could serve as a subject of demote in an internet discussion group, and they could be placed on a website for viewing .Computer is a group of integrated parts that have the common purpose of performing the given operations and can be defined broadly as any of a class of man-made devices or systems that can modify data in some meaningful way whereas the internet is a loose amalgam of computer networks connecting thousands of sites and millions of users all over the world (FACET, 2010).

Asia-Pacific Development Information Programme (APDIP) conducted by UNDP (2004) in collaboration with the Government of India, APDIP on computers and the internet are commonly made available to rural communities in the form of community-based telecentres. Telecentres provide shared access to computers and the internet and is the only realistic means of doing things for rural communities. Although telecentres come in many guises, with the two key elements being public access and a development orientation. This characteristic distinguishes telecentres from internet or cyber cafes.

Internet cafe or cyber cafe is a place where people can get connected to the Internet while eating or have a cup of coffee using publicly accessible computer. While the main activity in the cyber cafe is the Internet, such as email, newsgroup and web site surfing, other applications usually available are office suite to type document, spreadsheet, games, such as online games and other utilities such as printing service, scanning and digital photo service. Alao (2010) indicates that cyber cafes can be a useful device in fostering development through ICTs, but the difference is crucial, because development-oriented telecentres embody the principle of providing access for a purpose and that of implementing a development agenda. To achieve this development objective, telecentres perform community outreach services in order to determine the types of information used to foster development activities. Computer literate telecentres staff acts as intermediaries between farmers and rural community members who may not be familiar with ICTs and the information services that they require.

According to UNDP-APDIP (2004) telecentres can provide a range of ICT-based services from which they can earn an income, such as telephone use and faxing, photocopying and printing, email and word processing. This helps with financial self-sustainability, which telecentres are often required to attain, although some argue that rural people should not pay ICT based development services provided as a public service, rather like libraries. The results of experiments with telecentres mixed functions have demonstrated considerable benefits for their target audiences; others are struggling with fragile connectivity and uncertain communities and very few have achieved self-financing sustainability. With Kiosks, telecentres and knowledge centers', rural farmers acquire skills to improve their products better by making them appeal to local and global markets. They have the access to promote their products, handle simple transactions such as orders, and seek opinions to make decisions for their own development.

Telecentres and knowledge centers" have a relative positive impact on rural communities in terms of promoting development, political awareness and social change (UNDP, 2004). For example, the International Rice Research Institute (IRRI) provides a comprehensive resource links with the IRRI Rice Web through the *Rice Knowledge Bank*, which provides information on practices in the field; research findings; technology transfer methods and support skills; training materials; statistics and other data; a discussion area; and access to other agricultural sites. In Kenya, the farmers helpline operated by KenCall, a for-profit call center, is a real-time call center service staffed by agricultural experts that provides agricultural information, advice and support to small holder farmers over the phone, using voice and voice call-back to farmers, not SMS. The service has received funding from GSMA and the Rockefeller Foundation (FACET, 2010).

SIMputers

According to Lie (2006) the Simputer is a low cost portable alternative to PCs, by which the benefits of IT can reach the common man. Simputers are handheld computers similar to personal digital assistants (PDA). However, they are fitted with SIM cards for data transmission and reception. Simputers may be used by an extension worker to provide pricing and market information to farmers received from a central server located in the dzongkhag. The Simputer itself is invented, developed and produced in India. It has a special role in the third world because it ensures that illiteracy is no longer a barrier to handling a computer. The key to bridging the digital divide is to have shared devices that permit truly simple and natural user interfaces based on sight, touch and audio. The Simputer meets these demands through a browser for the Information Markup Language (IML). IML has been created to provide a uniform experience to users and to allow rapid development of solutions on any platform. This ICT has been tested successfully in Thailand by the Bank for Agriculture and Agricultural Cooperatives.

By being inexpensive, portable, running on the open source operating system Linux, having network (including internet) facilities, having text-to-speech facilities, being able to run on three AAA batteries, and being robust, the original idea was that it could be capable of playing a role in closing the digital divide. The Simputer was intended to be 'an access device for the masses' and aimed at farmer communities in rural areas. Critiques have been skeptical about these intended target groups. Technical issues as well as economic and socio-cultural issues remain unclear. There were power problems, sales and production costs problems, but also, only a few projects have been recorded that actually use the Simputer in rural change projects.

Bridging the digital divide by using the Simputer that runs on Linux, means introducing ICTs and providing infrastructure (wireless or connected to desktops). From an end-user perspective, open source software is not always an interesting option. Although it is possible to do basic text editing and to work with a spreadsheet on the Simputer, it is not comparable with the major Desktop/notebook applications (Lie, 2006). In July 2003, it was reported that 600 Simputers had been sold and 1,500 to 2,000 Encore Simputer machines were out in the market in September 2003 (Ganapati, 2003). According to Fonseca & Pal "there is limited evidence of end-user consumer purchases in rural India." The only information that seems to be available on existing projects that use the Simputer is the information that is provided by PicoPeta, one of the organizations involved in the manufacture of the Simputer (I4D, 2004). These projects involve fields such as spot billing in electricity metering and micro financing in Karnataka, India. With the coming of the new Amida model of the Simputer in April 2004, the market also seems to have shifted a little bit to the urban rich and urban young who cannot afford to buy an expensive PDA, but can afford to buy a Simputer. For them the Simputer could be an interesting alternative as the price is more interesting. This shift in markets is emphasized by the way the Simputer is advertised (Lie, 2006).

Intranet

These processes link rural radio stations with each other and with the internet. A team of communication experts specifically trained for information gathering and dissemination on the internet, by adapting information to the local context, radio scripting and interviewing techniques, manages the network of stations thus facilitating external as well as location specific information.

Web-Based Information and Learning Resources

Web-based "learning objects"-individual, self-contained and often customizable information or instructional packages with text, images, animations or movie clips-are rapidly growing in number and availability. They allow farmers and rural communities to pursue their own learning routes and interests, but there are issues of cost, computer literacy and Internet access. Smallholders also have to be able to plan and manage their own learning, discriminate between reliable and dubious source material, and draw their own conclusions from the available resources. Program developers must be capable of using advanced authoring tools to customize courses to individual needs, while achieving economies of scale by designing for large learning groups.

CD-ROM and DVD-ROM

Compact Disc-Read Only Memory (CD-ROM) and Digital Versatile Disc or Digital Video Disc- Read Only Memory (DVD-ROM) are ICT tools with storage medium that, together with a microcomputer, offers rapid access to a very large volume of data, including text, audio, graphics, animation, slides and video in computer-accessible learning and demonstration documents. They are considerably cheaper to mail than print material, but farmers and smallholders must have access to computers and printers and be able to afford the connection, printing and other costs. They are also interactive and provide farmer and extension feedback. Programs can be multimedia, combining text, sound, graphics and video, and can provide simulations, tests and other active learning tasks (Zijp, 1994).

Agricultural development depends on rural communities and farmers having access to information they need and can use effectively. Yet poor communications systems make transmitting timely, accurate information to agricultural extension agents, health workers, and others, a difficult task. Smallholder farmers must be computer literate, or helped in using the technology and computer access is an issue, but mobile computing with laptop and notebook computers is a promising solution for low-income countries (Zijp, 1994). According to Zijp, (1994) CD-ROM is used across all sectors, including agriculture, natural resource management, medicine and social sciences. Most applications involve storage of large amounts of material which can be easily retrieved. CD-ROMs can provide ready access to a vast amount of information on local pests and plant diseases, which can be used to support decision making in cop management and integrated pest management, thereby increasing productivity and reducing the environmental impact of agriculture. They can also delver crop protection information efficiently and effectively to extension agents and other intermediaries, as well as to farmers themselves. For example, CTA- EU selected agricultural information services in Cameroon, Kenya, Mali, Papua New Guinea, Trinidad and Tobago, Western Samoa, Zaire, Zambia and Zimbabwe to participate in a pilot project aimed at improving national self- sufficiency in the dissemination of scientific agricultural information and to introduce computer technology in developing country information services in order to encourage the production of local databases.

Organizations which disseminate scientific agricultural information, such as ministerial libraries, universities and regional development centers, were provided with the necessary CD- ROM equipment and commercially available databases. Up to two weeks of onsite training were given as well. Six to twelve months after installation, evaluation of the project found that the equipment had been used 50 percent of the time for CD-ROM application and the remaining time was put toward word processing, local databases, inputting into international databases and online access to remote databases (Zijp, 1994).

GIS/Decision Support Systems

According to Munyua (2008) Geographical Information System (GIS) is an information technology that links activities in the field and the office, and allows for comparisons between different types of agricultural data. It helps to establish relationships between data sets and is important in the production of maps as well as charts that present results in visual form. In addition, the spatial variability that it captures allows for unique handling of certain locations. It also helps in the analysis of post harvest variation in crop yield measures. This is done by analysing yields of previous years to guide future management decisions. The GIS details include maps and additional information associated with coordinates and time, and these are all linked in a computer database that could be analysed further. In general, a GIS technology provides a holistic view of the production system and helps make precision agriculture practical and useful even for small-scale farmers. GIS accounts for its significance in the following initiatives:

In Kenya, the International Livestock Research Institute (ILRI) has used the GIS technology and remote sensing in its Small Dairy Project. It has also employed the technology to map fences in its livestock-wildlife ecosystem initiative; furthermore, the Kenya Agricultural Commodity Exchange (KACE) plans to use a GIS in the near future to develop commodity patterns and seasonal patterns. The information obtained can then be used to produce a map of the country, depicting the seasonality of different crops, the nature of what is being produced and their geographical coverage, among others. In some cases, the data will help in the estimation of supplies and expectations, apart from assisting in the selection of markets before harvesting and in the provision of dispersion distribution and productivity patterns. Finally, the forecasting of supplies and market potential that the technology makes possible is viewed as critical by KACE in its preparation of export plans and local market sales. DrumNet in Kenya plans to apply GIS and has acquired GPS technology to help map out where farmers are located, the collection points, and easy ways of accessing transportation, cell phone accessibility, roads and cell phone connectivity, among others. The International Small Group and Tree Planting Alliance (TIST) is training farmers in Kenya and Uganda in community-based reforestation projects to use PDAs and Global Positioning System (GPS) technology to gather reforestation data, which is then uploaded to an online database (Munyua, 2008).

Munyua (2008) further indicates that in South Africa, the GenARDIS project has used GIS applications in natural resources management in the Roiboos tea lands and in exploring the spatial dynamics of gender in rural areas.

Mobile Mapping

Mobile mapping is a concept and system with the ability to collect field data, including unique geospatial time tags and attributes, for integrating into/updating a GIS. Mobile mapping essentially remembers what has been recorded, when it was recorded and where it was recorded, and allows for addition or expansion of the stored data. Thus, mobile mapping provides the freedom to collect data anytime, anywhere and in any manner (GIS Development 2006).

Factors Influencing Use of ICT by Extension Officers

Agricultural Extension systems in most developing countries are under – funded and have had mixed effects. Much of the extension information has been found to be out of date, irrelevant and not applicable to small farmers needs, leaving such farmers with the very little information or resources to improve their productivity. ICT helps the extension system in re- orientating itself towards the overall agricultural development of small production systems. With the appropriate knowledge small- scale producers can even have a competitive edge over larger operations. When knowledge is harnessed by strong organizations of small producers, strategic planning can be used to provide members with at least- cost inputs, better storage facilities, improved transportation links and collective negotiations with buyers (Amutha, 2010).

Extension services combine educational methodologies such as farmer field schools and demonstration plots as well as advice from extension agents. They include the dual challenges of informing farmers as well as keeping extension agents themselves well trained with access to back up experts for tough problems. They are delivered by public and private entities via extension agents. National ministries of agriculture have hundreds and sometimes thousands of extension agents, and large buyers often have their own private agent networks. Still farmers often go years without having direct contact with an extension agent. To increase access to these critical services, both public and private provider shave been using a variety of delivery "channels," such as radio, print media and, increasingly, ICT-enabled channels, some of which are often combined. Mo-bile phone networks are the ICT channel most often used, given the recent dramatic increase in access to such networks, even in rural areas, and affordability even for the poor (Rivera & Sulaiman, 2009).

The information and communication intensity of global agriculture has increased dramatically, as globally – traded commodities and the demand of a growing global middle class for higher value-added, more diversified food products necessitate more time –urgent and contextualized information about crop varieties, supporting technologies, market trends, trading partners and the competition. Poor smallholder farmers face an increased risk of being left et the "bottom of the value chain", seeking to earn income from commodity crops that are increasingly volatile in price while being faced with substantial information disadvantages (impediments and costs imposed in part by physical isolation, weak infrastructure and asymmetric access) that continue to limit their leverage at market and their ability to innovate in response to changing demand.

Some prior studies have demonstrated that demographic and characteristic background such as age, gender, ethnicity, marital status, level of education, prior experiences with computers and the Internet influence the ICT and or e-Learning adoption. Agricultural extension depends to a large extent on information exchange between and among farmers on the one hand, and a broad range of other actors on the other, has been identified as one area in which ICTs can have a particularly significant impact. There is growing recognition that farmers and members of rural communities have needs for information and appropriate learning methods that are not being met. In the midst of this change, extensionists are grappling with the question of how best to harness ICTs to improve rural livelihoods. Meera et al (2004) had noted that as a result of the emerging new paradigm of agricultural development, old ways of delivering important services to citizens are being challenged; traditional societies are also being transformed into knowledge societies all over the world.

Age of Extension Officers

Flor and Hazelman (2004) states that in Indonesia most extension workers based in the local governments are beyond 40 years old and awaiting retirement. They belong to the old school of information and communication. A training course that covers design, development and utilization of information and communication media by agricultural extension workers is required. Although this course would include sessions on digital multimedia production, it should likewise incorporate interfaces with conventional and traditional media as well as alternative last mile linkage strategies. In Indonesia, there are eight thousand researchers compared to 35 thousand extension workers, most of who are now under the payroll of local governments. It is mentioned that innovative agricultural technologies are housed in the AIATs which now carry part of the agricultural extension burden. However, the AIATs are staffed by researchers and not by extension workers.

Gender Composition of Extension Officers

About 94 percent of agricultural extension agents worldwide are male, multiple studies have reported problems with male extensionists delivering services only to men, or focusing narrowly on traditional roles when delivering programmes to women (FAO, 2006). A holistic approach to agricultural extension today goes beyond technology transfer for major crop and livestock production systems. It also includes goals for human capital development, in terms of enhancing the management and technical skills of farm households relating to production and postharvest handling of high-value crops, livestock and fisheries, sustainable natural resource management, family health and nutrition, and leadership and organizational skills, in addition to social capital development, that is, organizing producer groups (Swanson 2008). As tools embedded within existing social and gender relations, information and communication technologies (ICTs) are perceived, used, and appropriated in ways that are subject to asymmetries in power, privilege, opportunity, and choice.

Several studies including the CTA-sponsored one on rural women's access to public and private extension services have revealed that the village level extension agent is the most effective source of information for farmers but certainly not the most efficient in terms of cost and coverage (Arokoyo, 2010). According to FAO, (2006) the number of female extension officers has increased, although their representation continues to remain extremely low, ranging from 3.1% of the extension staff to 22%. While increased number of women agents does not necessarily guarantee that more women farmers will be reached. Several reports indicate that female extension agents are more inclined to contact female farmers and are more likely to address issues of concern to them, such as subsistence production and household food security. At the same time, women farmers may be more inclined to attend a demonstration when a woman gives it. The situation is still similar in South Africa. The agricultural sector in the Northwest Province is clearly a male dominated sector with more than 80% of the labour force being males. According to the Gender Focal Points, 2003 the total number of extension officers in the department was 939, with the total number of males being 602 (64%) as compared for women 337 (36%)(Mabe, 2005). Modern ICT including internet and web technology is needed to make these systems available regionally and globally. Accessing information communication technologies will bring a wealth of information to all agriculture stakeholders including women farmers in rural and urban areas and will help in overcoming the existing digital divide.

Education Level and Years of Experience of Extension Officers

Adesope et al. (2007) revealed that there was a significant but negative relationship with information technology needed. This implies that the higher the educational qualification the less the information technologies needed and the lower the educational qualification the more the information technologies needed. It is possible that respondents with higher qualifications may have too many responsibilities to grapple with, hence may not have the time to appreciate more need for Information technologies. Agricultural extension has the ability to solve many of the challenges that agriculture is facing, but it needs to be made attractive. Indeed, the nature of agricultural extension itself is challenging since it involves working with the farming community (which is not easy as we need to understand rural life and human behavior), have a good knowledge on agriculture (pest and disease management, market prices and opportunities, other stakeholders in agriculture and their role etc.) and have specific skills (good communication skills, listening skills, be a trainer and learner at the same time etc.).

Access to Allocated Vehicle for Extension Work

Given the problems that extension agents face in facilitating direct contact with farmer clients and with researchers due to the physical distances involved and the lack of transportation needed for their mobility, the application of ICT offers excellent possibilities. Utilizing ICT for strengthening research-extension-farmer linkages must, however, move from pilot testing phases to integration into research-extension systems and beyond the urban focus. Major factors to also address include the creation of an enabling environment, development of infrastructure, development of software and information content, making ICT sustainable, building capacities, combining its use with traditional media and learning from success stories (APO, 2002).

Average Distance between Extension Officer and Clients

Given the problems that extension agent's face in facilitating direct contact with farmer clients and with researchers due to the physical distances involved and the lack of transportation needed for their mobility, the application of ICT offers excellent possibilities. Utilizing ICT for strengthening research –extension-farmer linkages must, however, move from pilot testing phases to integration into research- extension systems and beyond the urban focus. Major factors to also address include the creation of an enabling environment, development of infrastructure, development of software and information content, making ICT sustainable, building capacities, combining its use with traditional media and learning form success stories (FAO, 2006). Mmbengeni and Mokoka (2002) indicated that extensionists in Limpopo Province, South Africa experience a serious lack of transport during the execution of their day-to-day duties. This constraint has a direct influence on the ability of extensionists to reach farmers on a regular basis, on the ability of extensionists to facilitate communities and the provision of transport for farmers during farmers-day and training events.

Number of Farmers Covered by Extension Officers

Several studies including the CTA-sponsored one on rural women's access to public and private extension services (Arokoyo, Chikwendu & Ogunbameru, 2002) have revealed that the village level extension agent is the most effective source of information for farmers but certainly not the most efficient in terms of cost and coverage. At the inception of the state-wide ADPs in 1980 the extension agent: Farmer (EA: Farmer), ratio ranged "between" 1:2000 to 1:3000. This was expected to come down to "between" 1:800 to 1:1000 by the project completion date and the withdrawal of World Bank support. This target was never achieved. A recent field survey by NAERLS and PCU (2002) showed that the EA: Farmer ratio was "between" 1:848, in Ogun State in the South- West Ecological Zone to 1:1650 in Katsina State in the North-West Ecological Zone. This is similar to the finding by Arokoyo et al (2002), of ratio of "between" 1:1000 and 1:2000 in the public extension service, and worse for the women-in-agriculture program, which was found to have several extension blocks expected to be filled by female extension agents, vacant. It is clear therefore, that no matter however effective, extension delivery through the village extension agent can neither be efficient nor cost-effective for a developing country like Nigeria, with a population of about 150.00 million, majority of who are involved in agriculture (70 – 80 percent) and illiterate. The Extension Service therefore must be appropriately supported with the use of ICTs (Arokoyo, 2010).

According to Glendenning et al, (2010) in India, the extension staff of the DoA operates at the district and block levels, which are administrative subdivisions. A block is a subdivision of a district. The numbers and capacity of the staff vary greatly throughout the country, but it is well acknowledged that staff numbers are low. The ratio of staff to farmers varies widely across the country with 1:300 in Kerala and 1:2,000 in Rajasthan. In Ghana the ratio of extension workers to farmers is estimated to be 1 to 2000, but in recent years the number of agricultural extension workers has been going down, while the numbers of farmers is growing. This means the role of the extension workers in the dissemination of knowledge within rural communities is diminishing. The use of ICTs to support the work of the Extension Officers might be beneficial for the development of the agricultural sector (Akkermans, 2010).

In agriculture extension service and the Veterinary services Units play a major role in improving agricultural production. To ensure their effective operation, the district has been delineated into 22 operational areas. Currently there are 13 extension officers in the district giving an extension officer farmers ration of 1: 5000, which is high compared with the acceptable ratio of 1: 400. This indicates that the district has a serious problem as far as the number of extension officers in concerned. A survey of the extension service unit reveals that, the unit faces a number of problems, which have inhibited its ability to reach farmers effectively which include: poor condition of roads leading to some farming communities, inadequate means of transport, delays in the payment of staff allowance, extensive coverage areas for field officers, high extension officer farmer ratio, lack of motivation for field staff.

In Papua New Guinea, the extension officer to farmer disparity points to challenges in delivering information. The extension farmer ratio is one extension officer per 3,600 people, and in some parts of the northern Solomon Islands this ratio is 1 per 14,000. For medium size islands such as Samoa and Tonga, the ratio is closer to 1 per 800 people. Extension work is further hampered by a lack of transport, or lack of fuel for transport.

Contact with Farmers

According to Asian Productivity Organization (2002) given the complex nature of agriculture and the challenges being faced, the use of multidisciplinary approaches involving a mix of specializations is becoming more and more the norm in addressing information delivery today. Technical specialists along with social scientists and communications and information specialists are combining expertise to address farmers' and farming communities' problems and needs. Studies have shown that farmers depend on multiple sources for information and technologies. In most societies, face-to-face contact is still the preferred mode of communication. Given the large populations to be served by extension and the dispersed and distant locations, face-to-face communication is not always possible and thus needs to combine strategies and communication channels to include traditional information sources, such as radio, print, video and television, together with new forms of delivery such as through ICT.

Due to the limited number of extension workers, farmers are using the phone to seek advisory services. Extension information can also be shared through a website. In addition, PDAs could be used to collect and disseminate critical agricultural information that farmers need. The shift to these technologies would cut down face-to-face contacts drastically and reduce telecommunication and transport costs. Farmers would, however, require training and a change of attitude to be able to access the ICT driven information and advisory services. Of importance is the fact that the applications of ICTs in extension services would provide for more content diversification and ensure a wider reach. In India for instance, the e-choupal model has made it possible to reach farmers through public access centres that include telecentres, knowledge centres and access points (Munyua, 2008).

Area Covered by Extension Officers

Several research studies conducted on extension organizations have revealed that the delivery of goods is effective when the grass roots extension worker covers a small area of jurisdiction with multiple purposes or broad basing. The existing system of large jurisdiction, each with a narrow range of activities, is less effective. Broad basing agricultural extension activities, developing farming system research and extension, having location specific modules of research and extension, and promoting market extension, sustainable agricultural development, and participatory research are some of the numerous areas where ICT can play an important role. However, broad basing requires grass roots workers to be at the cutting edge of extension and master of many trades, which is not really possible. The emergence of such knowledge workers will result in the realization of the much talked about bottom-up, demand driven technology generation, assessment, refinement and transfer. ICTs can help by enabling extension worker gather, store, retrieve and disseminate a broad range of information needed by farmers, thus transforming them from extension workers into knowledge workers. ICT has many potential applications in agricultural extension. It can bring new information services to rural areas where farmers, as users, will have much greater control than before over current information channel. Access to such new information sources is a crucial requirement for the sustainable developing of farming systems. If extension agents will effectively and efficiently use ICT, it will make their work easier and moreover, facilitates their information delivery and after extension services (Amutha, 2010).

Effect of E-Readiness of Extension Officers

Swatman (2006) clarify that e readiness is defined as being prepared mentally or physically for some experience or action. In terms of e-Learning, Borotis and Poulymenakou (2004) defined e-Learning readiness as the mental or physical readiness of an organization for some e-Learning experience or action. According to Arokoyo (2010) most of the nation's agricultural research and extension system (NARES) members now have computers for information and data management, most of the computers have neither telephone nor internet access. Even ordinary email access becomes a big deal for a whole national Research Institute. The low computer literacy level of the managers and staff makes the ICT-readiness of the system very low. Kauffman and Kumar (2005) introduce three stages of diffusion of ICT at the country level of analysis are the ICT readiness stage, the ICT intensity stage and the ICT impact stage. In the first stage of ICT readiness, they argue that when the technology is new to a country or a region, the readiness of its people to adopt it is a crucial issue. On the other hand, Kaur and Abas (2004) noted that ICT readiness assessment allows one to design comprehensive e-Learning strategies and effectively implement ICT goals. Hence, generally ICT readiness assessments help a country's leaders to measure and plan for ICT integration, focus their efforts and identify areas where further attention is required (Krull, 2003).

FAO (2006) stated that several studies have revealed that lack of access to information hampers the performance of extension activities. In most cases, it was found that extensionists have little or no access to sources of agricultural information. Although most extensionists are aware of the large amount of information available on the internet, they do not have access to internet facilities causing constraints not only in terms of access to information, but also in terms of communication (i.e. e-mail).

Constraints Faced by Extension Officers

According to Raabe (2008), information flow within the public sector moves linearly, with content focusing on the transfer of technology for increasing crop production. A wider definition of agricultural extension, beyond improving crop productivity, has not been embraced. Information flow is supply-driven and not needs-based or area-specific so farmers see the quality of the information provided by the public extension staff as a major shortcoming. This is due to the static and inflexible nature of the organization, where a top-down hierarchical approach continues. Sulaiman, (2003) further states that access to extension is also an issue, because of the low level of outreach by public extension services. This is partly due to the public staff being overburdened with implementing state and centralized schemes, which are also not easily modified to suit local needs and conditions.

There are also insufficient funds for operational costs, training, and capacity development, which limits the activities and continual development of the extension staff. Sulaiman, Hall, and Suresh (2005) indicate that of the required 1.3 million to 1.5 million extension personnel, there are only about 100,000 on the job. At the state level, the various line departments have been criticized for working in isolation, with weak linkages and rare partnerships, which limits information flow.

Furthermore, the research–extension link has been criticized for not absorbing or using feedback from farmers and extension staff. Extension personnel and farmers are passive actors, and scientists have limited exposure to field realities. The various components of the public-sector extension system suffer from duplication of programs, without convergence. While Agricultural Technology Management Agency (ATMA) is pushed as the platform through which the multiple agencies can converge, the implementation difficulties are proving too great for effective integration, with shortages of both personnel and funds (Glendenning, Babu & Asenso-Okyere, 2010).

Effects of ICT Use on Information Access by Extension Officers

Information has been identified as one of the resources required for the improvement of agricultural production (Aina, 1995). It is said to be a resource that must be acquired and used in order to make an informed decision. Those who possess appropriate and timely information will make a more rational decision than those without. The various agricultural information users populations can be categorized as policy makers and planners; researchers; extension staff; educators and students; agro-based industries and services staff; and farmers. It was also observed that each of the sectors mentioned above contribute directly to the improvement of agriculture hence relevant information provided to each category of these user populations will contribute positively to the development of agriculture in Africa. Information is the potential ingredient for the advancement of agriculture. Without information the agricultural sector may degenerate into extinction. Information provides the basis for scientific innovations in agriculture. Agricultural information has been defined as all published or unpublished knowledge on all aspects of agriculture.

Application of ICT in Other Countries

The applications of ICT for agriculture were reviewed in this section with specific examples from countries in Africa, Asia and Europe.

India

The e- Choupal and Satcom were reviewed form India. Swanson and Rajalahati (2010) reported that e-Choupal (which means gathering place in Hindi) and a profitable rural transformation initiative in India, is a web based information and procurement tools for Indian farmers that places computers with internet access in rural farming communities to have access to information and an e-commerce hub. e-Choupal was developed by the private sector in India, and has enabled farmers from selected parts of the country to gain access to up- to- date price and marketing information, as well as to engage in online sales of agricultural products and online purchases of inputs. This privately financed ICT system is owned and managed by ITC Limited and it currently has over 10,000 eChoupal sites where farmers regularly engage in online sales and purchases on a fee – for- service basis. Given the success of this system, the Government of India is now establishing similar common service centers (CSC) whereby rural people and other parts of the country can gain access to similar technical and marketing information, including online transactions. Currently, there are 6,500 functioning CSC's and the plan is to scale this number up to 100,000 CSC's serving the 638,000 rural villages across India. In addition, some block-level extension offices that received World Bank support also have PC's with Internet access: therefore, farmers in these areas cam also gain access eChoupal and other technical and market information.

The e-choupals serves as both a social gathering place for exchange of information which delivers real-time information and customized knowledge to improve farmer's decision-making ability, thereby better aligning farm output to market demands; securing better quality, productivity and improved price discovery. This model helps to aggregate demand in the nature of a virtual producers' co-operative, in this process facilitating access to higher quality farm inputs at lower costs for farmers. The e-choupal initiative also creates a direct marketing channel, eliminating wasteful intermediation and multiple handling, thus reducing transaction costs and making logistics efficient.

This system has also catalyzed the rural transformation that has helped to alleviate rural isolation, create more transparency for farmers, and improve their productivity and income. e-Choupal has helped to empower farmers in India with real time information on weather and prices so that they can insulate themselves from unpredictable climatic conditions and price fluctuations at markets.

Bhatnagar and Schware (2000) also reported that India was amongst the first countries to explore the use of Satellite Communication technology (Satcom) to support development and today its space efforts are recognized the world over for their wide spectrum of application relevant to societal benefit. Satcom offers the unique capability of being able to simultaneously reach out to large numbers spread over large distances even in the remotest corners of the country. It is a very strong tool to support development education. Satcom has been used both in broadcast mode as well as an interactive mode to reach out to rural audiences at large, and to conduct training programmes for extension staff as well as rural population who are participating in the implementation of development actives.

Sri Lanka

The Department of Agriculture is presently moving towards introducing electronic media for dissemination of information to extension agents as well as farmers. Cyber Extension is one such initiative where farmers could make use of e-mail and the internet facilities to access information and communicate, particularly with the Department of Agriculture (DOA) to get advice and assistance to solve their farming problems. Furthermore the Audio –visual Center of the DOA has initiated a toll-free telephone service for the farmers to directly contact agricultural researchers for obtaining advice to their problems. This service is becoming popular among farmers but publicity has to be given to reap the full of benefit of it (Mudannayake, 2006).

China

According to Ke and Zhang (1999), subsistence farmers do not need to interact significantly with people beyond their own families. Yet in raising output above the subsistence level, communication is important. Despite low levels of ICT in rural China, studies have shown that ICT, including the telephone and internet have positive impacts on farm households in terms of industrial patterns, market expansion, sale and supply channels, education and healthcare. For example, duck farmers in the Hubei province now sell their ducks directly to customers through a telephone order system. With access to telephones, the potential exists to eliminate intermediaries who do not add value to products. Furthermore in Haikou village in the suburbs of the city of Wuhan is well known for producing balsam pears. Although farmers had bumper harvests for years prior to installing telephones, they did not profit from them because local intermediaries had forced prices down. After the installation of telephones in 1998, farmers were able to contact vegetable dealers in their neighboring province and sell their products at much higher prices. This experience led to the installation of 41 telephones in this village within two months. Agricultural universities and research institutes have become pioneers in providing internet- based-education for farmers. The Capital Farmer Network (www.capitalfarmer,net) is a popular web site that specializes in agricultural technology and provides free online seminars to farmers on new agricultural techniques. Lack of internet infrastructure, however, means that few farmers benefit from these services (Ke & Zhang, 1999).

Ireland

According to Acker and Gasperini (2008) the use of ICT has improved safety of animals in Ireland, in 2001, the foot and mouth crisis in Surrey brought back memories of the fears to many Irish farmers, but on this occasion ICT has been able to provide them with some reassurances. The very fact that Northern Ireland can prove that no animals from Surrey have entered it in recent times shows that the benefits of ICT in action. All cattle entering the island of Ireland, on both sides of the border, are tracked and monitored with software used to trace their movement.

A major cattle movement is operated through a monitoring system which is a great help in ensuring safety, as it ensures traceability of cattle from birth to death. ICT is enabling both government and industry to reassure all stakeholders be they consumer of farmer that agricultural products in Ireland are safe to use. The department of Agriculture plans to roll out the monitoring system to cover the movement of ll animals by the end of 2008.

Israel

According to Gelb et al (2008), the collaborative contribution of extension and research to Israel's agricultural productivity and profitability dates back to the late 19th century. It facilitated technological innovation in agriculture, transfer of information, training of farmers and sector connectivity. These were in effect a dictate of continuous deteriorating agricultural terms of trade, crippling water shortages and severe climate and soil constraints. The Israeli Extension Services (IES) strategic decision to adopt Information and Communication Technologies significantly enhanced Extension's capabilities and scope of activities. ICT empowered extension's ability to develop, introduce and adopt innovative agricultural technologies and collaborate with research and international partners. Adoption of ICT as one instance of technological innovation dramatically improved the transfer and management of information, production chain efficiencies and integration within and with the agricultural sector. These were and are critical Israeli Extension success factors enabling research, services and farmers' ability to sustain a profitable, thriving agriculture sector. The use of ICTs contributed to rural viability and a model for adoption of technological innovation.

Burkina Faso

According to Munyua (2008) the Business Intelligence Trade Points model in Burkina Faso is used in the agro-business sector to address the socio-economic exclusion of small-scale farmers. The project collects and disseminates market information for local producers and facilitates trading through e-commerce which is growing gradually.

Egypt

In 1987, officials at the Egyptian Ministry of Agriculture and land reclamation recognized expert systems as an appropriate technology for speeding development in the agricultural sector. To realize this technology, in 1989, the ministry initiated the Expert Systems for Improved Crop Management Project (ESICM) in conjunction with the Food and Agriculture Organization of the United Nations (FAO) and the UNDP. The project began in mid-1989 and the CLAES joined the Agricultural Research Center (ARC) in 1991. Through the development, implementation and evaluation of knowledge-based decision support systems, CLAES is helping farmers throughout Egypt optimize the use of resources and maximize food production. A dozen expert systems have been developed for horticulture and field crop management (Rafea, 1999).

Jamaica

RU Network in Jamaica is a pilot project of the Caribbean Agricultural Advisory Service (CAIS), which uses "Information Cafés" with internet connection to exchange knowledge and experience between rural communities, researches and extension services. It provides CD-ROM based information and printers, scanners, digital cameras and audio recorders for capturing, storing and distributing local information.

Mauritius

In Mauritius, there is a mobile service used in Agricultural Extension is the "SMS Disease Alert" by the Agricultural Research and Extension Unit (AREU). This service is about sending SMS to registered planters when there is a disease outbreak on a particular crop. The service is beneficial to the planters as they can take precautions to prevent the disease to affect their crops.

Rwanda

In Rwanda the Sustainable Partnership for Rural Enterprise Agricultural Development (SPREAD) programme is using ICTs to sell coffee internationally and to monitor the international coffee auction through an innovative partnership that links SPREAD, Texas &M University and Michigan State University. This trading model facilitates direct access to new markets and was instrumental in introducing international quality standards on coffee auctions via the Internet (Munyua, 2008).

Senegal

Manobi project uses cell phones with wireless access protocol (WAP) and short messaging service (SMS) technology to provide members of Senegalese fishing communities, and fruit and vegetable grower farming communities, with up to date weather and market price information. The goal of users is to secure higher prices from middlemen and improve the timing of entry to markets when demand is high and supply is low. Users are also able to reduce spoilage by locating buyers while goods are still fresh. The project has also contributed to the expansion of mobile telephone infrastructure in targeted towns. The service integrates a variety of existing technologies, including voice telephone, web and personal data assistants (PDAs) to create a real time data and information processing network (Batchelor, Evangelista, Hearn, Sugden & Webb, 2003).

South Africa

According to Bua news (2010) Limpopo Department of Agriculture, budgeted R6 million in 2010 financial to install technology –based information systems that will enable extension officers to provide relevant and accurate information to farmers' instantly. The Department of Agriculture introduced the e- Agriculture project, which aimed at improving emerging farmers' access to information and markets through the use of information communication technologies such as internet and SMS messages. Since the e-Agriculture project was launched in October 2010, 215 farmers and 16 extension officers have been trained in basic ICT skills using Digital Doorways, which are robust computer stations with multiple screens that provide internet access to rural communities.

Uganda

Information and Communication Technology has a role to play in any country's development. Like other countries' Uganda has recognized the potential and enabling elements of information and communication technologies as a tool for social and economic development. Although there has not been a comprehensive survey of human resources in ICT in Uganda, it is fairly obvious that various levels of skills are required and the existing training institutions are meeting some of the training needs. There is a need to assess the national requirement for ICT skills, establish how much is available and then determine the best strategy of meeting the appropriate ICT skills requirement. Communication is one of the basic human rights and access to communication channels should not be left entirely to market forces. This creates a situation of information dualism, with a minority urban information rich and the rural majority being information poor. Conducive environment for investments in communication infrastructure should be prompted to narrow the information gap between the urban and rural areas (Ministry of Works, Housing and Communications, 2003).

Zambia

Agriculture is the economic backbone for many Zambians especially in the rural areas: as such it plays an important role in the social and economic development of the country Kasonde (2009). According to the National Information and Communication (ICT) Policy, the Zambian government has committed itself in making sure that ICTs are integrated in the agricultural sector reform process in order to contribute to the social and economic revival of the country. This will be achieved by working closely with the local community so that they develop human capacity to manage and maintain ICT equipment, create integrated agricultural information systems on agro technologies and techniques, pricing and market information, for all agro products in order to provide strategic information for farmers, governments and other stakeholders at national, provincial and districts level. ICTs should be undertaken to create awareness campaigns for all types of farmers in the use of traditional and new ICT tools at all levels, develop and promote ICT skills development among agricultural extension workers and farmers, develop a monitoring and evaluation system for the conversation and sustainable utilization of natural resources in agricultural production systems.

Zimbabwe

Gudza (2010) notes that podcasting has been adopted as a media to disseminate agricultural information amongst rural farmers. This is an innovative way of using ICTs to disseminate information to those areas that have limited road networks and poor communication infrastructure. Podcasts are being used as an alternative to distribute audio and video content particularly on the internet. Podcasting project has already taken off in Mbire district and it can help to transform the lives of small scale farmers despite of lack of electricity and poor communications. Traditionally the agricultural extension officers who work under Agricultural Research Extension Service (AREX) would provide agricultural information to farmers orally and through brochures, posters and pamphlets even though the local people did not have an input in the development of the content. However with podcasting the local people are able to develop their own content because the concept is dependent upon people, technology and content.

According to Gudza (2010) the podcasts' are in local or indigenous languages and all stakeholders collaborate to develop the podcasts. Theses podcasts will have information covering a variety of themes on crop and livestock production and management. The podcasts are played on MP3 players which are powered by rechargeable batteries. These podcasts are indispensible tools in disseminating knowledge on agriculture to disadvantaged farmers. This project is collaboration between government, non-governmental organizations like practical action and local community leaders. The project has made the farmers more knowledgeable and it has helped them to generate and share knowledge that improves productivity through using ICTs.

Sudan

Golder, an international company, has been working with ASBNACO - a Sudan-based company that manages the Agadi Farm and the Arab Authority for Agricultural Investment and Development (AAAID), to introduce commercial mechanised farming. As part of this project, the GPS and the GIS technology have been introduced to help improve farm productivity while conserving the environment. This has been achieved through an acquired auto-steer tractor, fitted with a GPS satellite guidance system that controls tractor steering. PA has helped to cut down planting time on the Agadi farm by 60%, compared with the time taken for a similar task in two previous seasons.

Spatial yield monitoring, using GPS field monitors have also been installed in combine harvesters to monitor yield variations within each field and produce yield maps so that farmers can identify high and low productivity areas on their farms. The maps also guide fertilizer, seed and agro-chemical applications. Controlled traffic farming (CTF) has also been introduced in the project, using GPS-based self-steering tractors that limit movements of tractors to selected rows within a field to reduce soil compaction and erosion. In addition to these changes, a GIS-based farming information and management system is in the process of being developed, while plans are underway to introduce infra-red photography and Variable Rate Application (VRA) technology. These systems and technology would help to identify weed infestations, water stress, and crop pest outbreak areas and make it possible to select appropriate chemical applications, thus reducing the wastage normally associated with conventional blanket spraying. It has been observed that VRA enables the efficient use of fertilizers, chemicals and seed based on the analyses of previous seasons' production and the chemical content of soils. Furthermore, it also prevents unnecessary groundwater contamination (AAAID, 2006).

Key challenges faced by the Agadi Farm include the need to ensure reliable operation and maintenance of highly specialized equipment under harsh and remote conditions; demands associated with providing training in the use of GPS equipment and GIS software for local technicians; intricacies associated with assessing the adoption of precision farming systems or their integration into current practices. It is anticipated that the wide acceptance of PA in Africa will enable the potential of PA to be fully exploited. However, the technology requires adequate funding for high-tech farming applications, which is beyond the financial reach of small scale farmers as individuals (AAAID, 2006).

Factors Influencing Application of Information Communication Technology in Agricultural Extension Services

The theoretical framework explores research on the use of ICT by extension officers in different parts of the world. This is the view of establishing a conjectural link between several variables have been used to determine ICT use, constraints to and effect of e- readiness among extension officers. Ahmadpour, Mirdamadi, Hosseini and Chizari (2010), noted that variables such as age, gender, education level, the number of farmers covered by extension officers, number of years of experience as an extension officer, e readiness, access to allocated vehicle for extension work, and studying for a higher degree have impact on the use of ICT's by extension officers.Using descriptive statistics and (Binary) Logistic model, Ndag, Sanusi and Aigbekan (2008) found that the proportion respondents who did not have the requisite knowledge on computer use was slightly higher the North-central (57.14%) than in the South-west (55.71%). Age and ownership of a personal computer (PC) were significant factors ($p< 0.1$) determining the probability of ICT use to search for information by respondents in North-central Nigeria.

Hosseini, Niknam and Chizari (2009) categorized factors influencing ICT use by extension officers into six namely: organizational, technical, financial, social, regulatory and human challenges, ordered by their impact. Based on the findings, organisational challenges are the most important issue. Technical challenges are always potentially troublesome in the application of ICT's. Bheenick et al.., (2003) and Schimtz (2003) stated, for extension organization, in order to fulfill its function, has to overcome technical challenges along with human, regulatory, financial and social challenges as main constraints to apply ICTs. The findings about financial challenges is in accordance with those of Khan (2001); Kushner et. al., (2004) and Stribhadung (2006) which shows that cost of buying and maintaining the system would have effect on the application of ICT. The importance's of technical factors have been pointed out by several authors as Barajars et al. (2000), Surry (2002), Bheenick et al. (2003), Ebadi (2005), and Castels (1996). The findings also reflect an important fact that positive attitude and knowledge and skills of experts directly impacts the application of ICT. This is in corroborating with the findings of Lynch (2001). The results of the study pointed out to this fact that, there was relationship between social factors and application of ICT by the extension service in Iran. The findings are in accordance with studies by Sullivan (2000), Samak (2006), and Tyan (2003). ICTs have tremendous potential in Iran and it is evident now that ICT's will change the way people live, work and learn. In this regard, agricultural extension has the opportunity to harness the effective use of ICTs.

Adesope, Asiabaka and Agumagu (2007), examined the effect of personal characteristics of Extension managers and supervisors on their information technology needs in the Niger delta area of Nigeria. The result showed that there is no significant difference in the Information Technology needs of both categories of respondents. Findings also revealed that number of associations belonged to, educational qualification, training, category of organization were the personal characteristics that significantly influenced information technologies needed by extension managers and supervisors. Lashgarara, Mirdamadi and Hosseini (2010), noted that Information and communications technologies (ICTs) represent an important strategy that can be used in attaining food security in Iran. The results showed that, according to the experts' point of view, the situation of food security in rural Iranian households was medium, but that ICTs could play an important role in improving the situation. The results further showed that increasing food production, transferring of new methods and technologies, improving interactions and communications, providing information about cultivation and harvest, facilitators and content of old technologies account for 71% of the variance of the food availability of rural Iranian households. Moreover, the path analysis technique demonstrated that the improvement of interactions and communications had the greatest influence on determining the casual model of improving the food availability of rural Iranian households.

Aboh (2008) noted and examined the frequency of ICT tools usage by agricultural extension agents in Imo State, Nigeria. Findings of the study revealed that only mobile phone and computer were frequently used by respondents. Though ICT tools were not frequently used for extension services and this has grave implications for agricultural development. Based on the findings of the study, the following recommendations were made: there should be adequate provision of computers for all the extension agents. Adequate provision should be made for the maintenance of the ICT facilities and systems, provision should be made in providing training programme for the extension agents in order to update their knowledge on the latest and new information technology, government to establish policy programme as means of bridging the digital divide, creating digital opportunity for various users of extension services in the country. ICT enabled extension service must be able to define its key stakeholders and audience and jointly determine their needs.

Purnomo and Lee (2010) noted that in Indonesia, the perception of agricultural extension officers', readiness and barriers towards implementation of ICT programme, the assessment model of e-LRS to measure the readiness of ICT programme implementation. The first finding reveals that they perceived that three out of the four factors of readiness as positive. The e-LRS assessment revealed that they perceived farmer readiness as lowest and thus considered it as a barrier. Also, the technological and organizational cultures were also seen as the main barriers of ICT programme implementation, and demographic variables such as regency and age, must also be considered when ICT programmes are implemented.

Oladele, (2011) compared the effect of Information Communication Technology on Agricultural Information access among researchers, extension agents, and farmers in South Western Nigeria, and noted that a significant difference in the effect of ICT on agricultural information access among researchers, extension agents, and farmers . The study concluded that access to agricultural information through ICT will continue to improve, since perceptions are overtly positive among researchers, extension agents, and farmers. Agwu, Uche-Mba and Akinnagbe (2008) ascertained that the level of usage as well as constraints to the use of information communication technologies (ICTs) among major stakeholders in the agricultural development process in Abia and Enugu States of Nigeria. The findings indicate that majority of researchers; extension workers and farmers had high, moderate and low knowledge levels of available ICTs, respectively. Out of 24 information communication technologies researchers utilized more ICT than extension officers and farmers. The study revealed major constraints to the use of ICTs to include lack of sufficiently trained computer personnel, lack of confidence in operating modern ICTs, erratic and fluctuating power supply, poor finance, lack of internet access in the rural areas and high cost of ICTs hard and software's, among others. The study concludes with the need to have a National Agricultural Extension Policy on the use of ICTs with major emphasis on access, availability and use.

Akparo, Okon and Inyang, (2007) analysed constraints affecting the utilization of Information and Communication Technologies (ICT) for agricultural extension activities by agricultural extension officers in Nigeria's Niger Delta Region. The findings revealed important specific constraints which include: poor ICT infrastructure development, high cost of broadcast equipment, high charges for radio/ television presentations, high cost of access/interconnectivity and electricity power problems. The use of Factor Analysis aided to crystallize identified constraints into three factors of 'poor enabling environment', 'lack of accesses and 'dissemination of unrelated information'. Adebayo and Adesope (2007) examined awareness, access and utilization of ICT among female researchers and female extensionists in South Eastern Nigeria. Information collected showed that female researchers and female extensionists are aware of ICT; both categories of respondents know how to access Internet on their own, but do not have adequate access to IT. The types of ICT needed by female researchers and female extensionists include; World Wide Web, Electronic Mail, Electronic Spreadsheet, Word Processing, CD-ROM, Use of Projector, Use of computer, Web Design, Chartroom. Salau and Saingbe (2008) indicated that in Nigeria researchers had greater access and utilization to ICT facilities than extension workers. The level of education positively influenced the level of utilization of ICTs while years of working experience had negative influence. The key problems militating against the use of ICTs in the area were poor access to ICT facilities, lack of computer knowledge, low income and poor power supply. It was recommended that agricultural organizations should install all necessary ICT facilities in their establishments and provide training opportunities for their staff. Constant power supply to both urban and rural communities should be considered a fundamental human right and treated as such.

REFERENCES

Aboh, C, L. (2008). Assessment of the frequency of ICT tools usage by Agricultural Extension agents in Imo State, Nigeria. *Journal of Agriculture and Social Research*, *8*(2).

Acker, D., & Gasperini, L. (2008). Education for rural people: What have we learned. *Journal of International Agricultural and Extension Education*, *15*(1), 28–30. doi:10.5191/jiaee.2008.15103

Adebayo, E. L., & Adesope, O. M. (2007). Awareness, access and usage of information and communication technologies between female researchers and extensionists. *International Journal of Education and Development using Information and Communication Technology, 3*(1), 85-93.

Adesope, O. M., Asiabaka, C. C., & Agumagu, A. C. (2007). Effect of personal characteristics of extension managers and supervisors on information technology needs in the Niger Delta area of Nigeria. *International Journal of Education and Development Using ICT*, *3*(2).

Agwu, A. E., Uche-Mba, U. C., & Akinnagbe, O. M. (2008). Use of Information Communication Technologies (ICTs) among Researchers, Extension Workers and Farmers in Abia and Enugu States: Implications for a National Agricultural Extension Policy on ICTs. *Journal of Agricultural Extension*, *12*(1), 37–49.

Agwu, E. A., Uche-Mba, U. (2010). *Congruency, agreement and accuracy among researchers, extension workers and farmers on the role of ICTs in Nigeria's agricultural development*. Available at iaald2010.agropolis.fr/proceedings/national-experiences.html

Ahmadpour, A., Mirdamadi, M., & Hossein, J. (2010). Factors influencing the design of electronic system in agricultural extension. *American Journal of Agricultural and Biological Sciences*, *5*(2), 122–127. doi:10.3844/ajabssp.2010.122.127

Aizaki, H., Nakashima, Y., Ujie, K., Takeshita, H., & Tahara, K. (2010). Influence of Information about Agrochemicals provided via Internet –enabled mobile phones on Consumer Valuation of food. *Journal of Agricultural & Food Information*, *11*(3), 209–211. doi:10.1080/10496505.2010.490509

Akkermans, N. (2010). *The Role of ICTs in Knowledge Sharing within Rural Communities in Ghana, Internship MA International Relations*. University of Groningen.

Arokoyo, T. (2010). *ICTs in the transformation of Agricultural Extension: The case of Nigeria.* Available at www.fao.org

Arokoyo, T., Chikwendu, D., & Ogunbameru, K. (2002). *A study of the access of Rural Women to Public and Private Extension Service in Nigeria*. Report of a study commissioned and funded by CTA.

Asian Development Bank. (2004). Building e-community centers for rural development. Report of the Regional Workshop, Bali, Indonesia.

Asian Productivity Organization (APO). (2002). *Report of a study meeting on Integration of Agricultural Research and Extension.* APO.

Ballantyne, P. (2009). Accessing, Sharing and Communicating Agricultural Information for Development: emerging trends and issues. *Sage Journal*, *25*(4), 260-271. Available at http://www.sagepublications.com

Batchelor, S., Evangelista, S., Hearn, S., Sugden, S., & Webb, M. (2003). *ICT for development-Contributing to the Millennium Development Goals: Lessons learned from Seventeen infoDev Projects*. Washington, DC: World Bank. Available on line: www.infodev.org/symp2003/publications/CaseStudies.pdf

Bayes, A., Von Braun, J., & Akhter, R. (1999). *Village Pay Phones and Poverty Reduction: Insights from a Grameen Bank, Initiative in Bangladesh*. ZEF Discussion Papers on Development Policy No. 8. Centre for Development Research. Available on line: http://www.apdip.net/publications/iespprimers/eprimer-pov.pdf

Bertolini, R. (2004). *Strategic thinking: making information and communication technologies work for food security in Africa*. Available online: http://www.ifpri.org/pubs/ib/ib27.pdf

Bhatnagar, S., & Schware, R. (2000). *Information and communication technology in development: Cases from India*. Sage Publications India Pty, Ltd.

Bheenick, K., & Brizmohunr, G. (2003). *The scope of information and communication technology application in agriculture extension in Mauritius*. Available on line: www.Uom.ac.mu/faculty/foa/ais/amas/htm

Boone, K., Meisenbach, T., & Tucker, M. (2000). *Agricultural Communications: Changes and Challenges*. Iowa State University Press.

Boone, K.M., & Paulson, C.E., & Barrick, R.K. (1993). Graduate Education in Agricultural Communication: The need and Role. *Journal of Applied Communications*, *77*(1), 16–26.

BUA News. (2010). *South Africa: Limpopo farmers get ICT boost*. Author.

Chizari, M., Lashgarara, F., Mirdamadi, S. M., & Hosseini, S. J. F. (2009). Role of information and communication technologies in improving food availability of Iranian rural households. *International Journal of Education and Development using Information and Communication Technology*, *5*(3), 193-205.

Dodds, T. (1999). *Non-Formal and Adult Basic Education Through Open and Distance Learning in Africa*. Academic Press.

Economic Commission for Africa (ECA). (2005). *Emerging issues in science and technology for Africa's development: science, technology and innovation for meeting key MDGs*. Addis Ababa: ECA Sustainable Development Division.

Eggleston, K., Jensen, R., & Zeckhauser, R. (2002). Information and Communication Technologies, Markets and Economic Development. In *The Global Information Technology Report: Readiness for the Networked World* (pp. 62–63). New York: University Press.

Engelhard, R. (2001). *Wireless: a help line for agricultural development*? Report of the 4th Consultative Expert meeting of CTA"s Observatory on ICTs. CTA Working Document-8024. Available: http://www.agricta.org/observatory

FACET. (2010). *ICT to enhance farm extension services in Africa*. Available on line: www.facet.org

FAO. (2000). *The role of information and communication technologies in rural development and food security*. Rome: FAO. Available on line: http://www.fao.org/sd/CDdirect/CDre0055.htm

FAO. (2003). *Virtual extension and research communication network.* Available on line: http://www.fao.org/sd/2003/KN10053_en.html

FAO. (2004). *Institution Building to strengthen Agriculture Extension.* 27th FAO Regional Conference for Asia and the Pacific, Beijing, China. Available on line: www.fao.org

FAO. (2006). *The Internet and Rural and Agricultural Development – An Integrated Approach.* Food and Agriculture Organisation. Available on line: http://www.fao.org/docrep/w6840e/w6840e05.htm

Flor, A. G., & Hazelman, M. (2004). *Regional Prospects and Initiatives for Bridging the Rural Digital Divide.* AFITA/WCCA Joint Congress on IT in Agriculture. Available online www.upou.edu.ph/downloads/2004AFITA.pdf

Gelb, E., Gal, B., & Wolfson, D. (2008). *Information and Communication Technologies(ICT) for Agricultural Extension –An Overtime Israeli Perspective.* Israel: Ministry of Agriculture and Rural Development.

GIS Development. (2006). *GIS application.* Available online: http://www.gisdevelopment.net/application/agriculture/overview/agrio0011dhtm

Glendenning, C. J., Babu, S., & Asenso-Okyere, K. (2010). *Review of Agricultural Extension in India, Are farmers' information needs being met?* IFPRI Discussion Paper 01048. Eastern and Southern Africa Regional Office.

Gudza, L. D. (2010). *Podcasts can inform poor farmers.* Available on line http//www.scidev.net/en/agriculture-and-environment/farming-practices/opinions/podcasts-can-inform poor farmers.html

Gurstein, M. (2000). Community Informatics: Enabling Communities with information and communication technologies. Idea Group Publishing. doi:doi:10.4018/978-1-878289-69-8 doi:10.4018/978-1-878289-69-8

Harris, R. (2004). *Information and communication technologies for poverty alleviation.* Asia-Pacific Development Information Programme.

Heeks, R. (1999). *The Tyranny of Participation in Information Systems: Learning from Development Projects.* Institute for Development Policy and Management, University of Manchester. Available on line: http://knownetgrin.honeybee.org/honeybee.html

Hosseini, S. J. F., Niknami, M., & Chizari, M. (2009). To determine the challenges in the application of ICT's by the agricultural extension service in Iran. *Journal of Agricultural Extension and Rural Development, 1*(1), 27-30.

International Fund for Agricultural Development (IFAD). (2002). *Challenges in rural development and poverty alleviation in Eastern and Southern Africa: the role of civil society and development institutions.* Available on line: http://www.ifad.org/ngo/events/pf/reports/session3.pdf

International Institute for Communication and Development (IICD). (2006). *ICTs for agricultural livelihoods: impact and lessons learned from IICD supported activities.* The Hague: IICD.

Jones, G. E. (1986). Investing in Rural Extension: Strategies and Goals. Elsevier Applied Science Publishers.

Kasonde, T. (2009). *Zambia: Promoting Agriculture through the use of ICT's.* Available on line: www.lusaka times.com

Kauffman, R. J., & Kumar, A. (2005). *A Critical Assessment of the Capabilities of Five Measure for ICT Development.* University of Minnesota.

Ke, B., & Zhang, X. (1999). *The development of ICT and its impacts on rural economy in China.* Beijing: RCRE.

Kerrigan, K., Lindsey, G., & Novak, K. (1994). Computer networking in International Agricultural Research Experience of the CGNET. New Information technologies in Agriculture. *Quarterly Bulletin of the International Association of Agricultural Information Specialists, 39*(1-2), 182-193.

Khalil, M., Dongier, P., & Qiang, C. Z. W. (2009). Overview in Information and communications for development (IC4D) Information and communications for development (IC4D). Washington, DC: The World Bank.

Khan, B. H. (2001). *A framework for e-learning.* LTI Magazine.

Krejcie, R. V., & Morgan, D. W. (1970). Determining sample size for research activities. *Educational and Psychological Measurement, 30*, 607–610.

Kushner, J., & Chong, P. (2004). *Conducive environments for promoting community e-centers.Asian Development Bank Regional Workshop*, Bali, Indonesia.

Lashgarara, F., Mirdamadi, S. M., & Hosseini, S. J. F. (2010). Determining the causal model role of ICT in improving food availability of Iranian rural households. *International Journal of Advanced Science and Technology, 24*, 27–40.

Leeuwis, C. (2004). *Communication for Innovation: Rethinking Agricultural Extension* (3rd ed.). Blackwell Science Ltd.

Lie, R. (2006). *Background paper: ICTs for Agricultural Development. An exercise in Interdisciplinarity, A final report.* Hampton Press.

Lynch, M. M. (2001). *The online educator: a guide to creating the virtual classroom.* London: Routledge.

Mabe, L. K. (2005). *The Role of Women in Agriculture and other non-farm activities in three regions of the North-West Province.* (MSc. Thesis). North-West University, Mafikeng Campus.

Marker, P., McNamara, K., & Wallace, L. (2002). *The Significance of Information and Communication Technologies for Reducing Poverty, The Think! Programme.* Development Policy Department, Department of International Development.

Meera, S. N., Jhamtani, A., & Rao, D. U. M. (2004). Information and communication technology in agricultural development: A comparative analysis of three projects from India. *Agricultural Research and Extension Network, 135.*

Ministry of Works. (2003). *Housing and communications.* Uganda: National ICT Policy.

Mmbengeni, M. E., & Mokoka, M. S. (2002). *Provincial Report on Education and Training for Agriculture and Rural Development in Limpopo Province.* Limpopo Department of Agriculture.

Mudannayake, I. (2006). *Forging relationships between Libraries and Extension Services for Improved Access to Agriculture Information: A case study in Sri Lanka.* World Library and Information Congress: 72nd IFLA General Conference and Council, Seoul, Korea.

Munyua, H. (2000). Application of ICTs in Africa's Agrcultural Sector: A gender perspective. International Development Research Center (IDRC).

Munyua, H. (2008). *Report on ICTs and small - scale Agriculture in Africa: A scoping study*. IDRC.

Nagarsi, K. (2000). *Assess of creative potential and practice of extension personnel*. New Delhi: IARI Pusa.

Ndag, I., Sanusi, R. A., & Aigbekaen, E. O. (2008). Comparative analysis of Information and Communication Technology (ICT) use by agricultural extension workers in South-west and North-central Nigeria. *Proceedings of the 19th Annual International Information Management Association.*

Obayelu, A., & Ogunlade, I. (2006). Analysis of the uses of Information & Communications Technology for Gender Empowerment & Sustainable Poverty Alleviation in Nigeria. *International Journal of Educational Development.*

Oladele, O. I. (2011). Effect of Information Communication Technology on Agricultural Information Access Among Researchers, Extension Agents, and Farmers in South Western Nigeria. *Journal of Agricultural and Food Information, 12*(2), 167- 176.

Purnomo, S. H., & Lee, Y. H. (2010). An Assessment of Readiness and Barriers towards ICT Programme Implementation: Perceptions of Agricultural Extension Officers in Indonesia. *International Journal of Education and Development using Information and Communication Technology, 6*(3), 19-36.

Qiang, C. W., Clarke, G. R., & Halewood, N. (2006). The role of ICT in doing business. 2006 information and communications for development: global trends and policies. Washington, DC: World Bank.

Raabe, K. (2008). *Reforming the agricultural extension system in India: What do we know about what works where and why?* IFPRI Discussion Paper 00775. Washington, DC: International Food Policy Research Institute.

Rafea, A. (1999). *Managing agricultural knowledge: role of information and communication technology*. American University in Cairo.

Richardson, D. (1997). *The internet and rural and agricultural development – an integrated approach*. Rome: FAO. Available online: http://www.fao.org/docrep/w6840e/w6840e00.htm

Richardson, D. (2003). *Agricultural extension transforming ICTs? Championing universal access* (Background Paper for CTA's ICT Observatory). Available on line: http://www.cta.int/fr/content/download/2792/20425/version/1/file/backgroundpaper2003.pdf

Richardson, D. (2005). How can agricultural extension best harness ICTs to improve rural livelihoods in developing countries. In E. Gelb & A. Offer (Eds.), *ICT in Agriculture: Perspectives of Technological Innovation*. Jerusalem: Hebrew University of Jerusalem, Centre for Agricultural Economics Research.

Richardson, D. (2006). *ICTs – Transforming Agricultural Extension.* Report of the 6th Consultative Experts Meeting of CTA's Observatory on ICT's, The ACP-EU Centre for Agricultural and Rural Cooperation (CTA).

Rivera, W. M., & Sulaiman, R. V. (2009). Extension: Object of reform, engine for innovation. *Outlook on Agriculture, 38*(3), 267–273. doi:10.5367/000000009789396810

Roy, S. (2005). *Globalisation, ICT and Developing Nations: Challenges in the information age*. New Delhi, India: Sage Publications Ltd.

Said, A., Rafea, A., El-Beltagy, S., & Hassan, H. (2009). Automatic Generation of Explanation for Expert Systems implemented with Different Knowledge Representations. *WSEAS Transactions on Systems, 8.*

Salau, E. S., & Saingbe, N.D. (2008). Access and Utilization of Information and Communication Technologies (ICTs) Among Agricultural Researchers and Extension Workers in Selected Institutions in Nasarawa State of Nigeria. *PAT, 4*(2), 1-11. Available online: www.patnsukjournal.com/currentissue

Satyanarayana, K. V., Reddy, M. N., Balasubramani, N., Sharma, V. P., & Bhaskar, G. (2008). *National Institute of Agricultural Extension Management*. Available at www.manage.com

Schmitz, J. G. (2003). Agricultural Extension on the Web. University of Illinois.

Secretariat of the Pacific Community (SPC). (2011). *Pacific Community*. Retrieved from: http//www.spc.int

Sulaiman, R. (2003). *Agricultural Extension: Involvement of Private Sector*. Mumbai, India: National Bankfor Agricultural and Rural Development.

Swanson, B. E. (2008). Global review of good agricultural extension and advisory service practices. Rome: Food and Agriculture Organization of the United Nations.

Swanson, B. E., & Rajalahti, R. (2010). Strengthening Agricultural Extension and Advisory Systems: Procedures for Assessing, Transforming and Evaluating Extension Systems. The International Bank for Reconstruction and Development/The World Bank.

Tregurtha, N., & Vink, N. (2002). *E-commerce and the South African horticultural export industry: current status and future directions*. Available on line: http://www.gapresearch.org/production/EcommercefinalreportAug2002.pdf

UNDP. (2001). *Information Communications Technology for Development. Essentials ICTD 31. Synthesis of Lessons Learned*. Evaluation Office N0. 5. Retrieved from http://www.undp.org/eo

United Nations Development Programme. (2004). *Asia-Pacific Development Information Programme (UNDP-APDIP)*. Available on line: www.apdip.net

Van Audenhove, L. (2003). *Towards an integrated information society policy in South Africa: an overview of political rhetoric and policy initiatives 1994 – 2000*. Cape Town: HSRC Publishers. Available on line: www.hsrcpress.ac.za

Van Veldhuizen, L., Waters-Bayer, A., & De Zeeuw, H. (1997). *Developing Technology with Farmers: A Trainers Guide for Participatory Learning*. Zed Books Ltd.

Zijp, W. (1994). Improving the transfer and use of agricultural information: A guide to information technology. The International Bank for Reconstruction and Development.

Chapter 4
A Set of Principles for Doing and Evaluating Classic Grounded Theory Research in Information Systems

Andy Lowe
Grounded Theory Institute, USA

Titus Tossy
Mzumbe University, Tanzania

ABSTRACT

Grounded Theory (GT) is a latent pattern recognition research method discovered by Glaser and Strauss (1967). Due to GTs power and transcendence many research papers across several academic disciplines including Information Systems claimed to have used GT when in fact they have used pseudo GT methods. It is argued in this paper that any other research method which adopts the GT label without following orthodoxy of the authentic GT research method should not be called GT. All of the pseudo GT methods make the false assumption that GT is a sub set of Qualitative Data Analysis. This is a false assumption because authentic GT can use either quantitative or qualitative data and it is a general research methodology and produces empirically grounded but modifiable propositions. Within the Information Systems (IS) research community it is therefore not surprising that many, who claim to use GT, have used different types of pseudo GT. They have adopted vocabulary of the GT without following its original tenets. This paper explains how authentic GT can be carried out in an information systems context by trusting in emergence rather than forcing the data.

INTRODUCTION

There are number of theories from other disciplines such as social sciences and others being adopted for use in Information Systems (IS) (Gregor, 2006; Truex et al, 2006). While some are adopted as they are, some are modified and some are mixed to fit within the IS discipline. Such, combinations erode

DOI: 10.4018/978-1-5225-0539-6.ch004

the known procedures and methods. In other words, there is likely repeating of the same mistakes made from the original disciplines (Truex et, 2006). One of these theories is the Classic Grounded theory, which was developed from the background of Sociology discipline (Glaser & Strauss, 1967). In the recent years, classic grounded theory (CGT) has been used in information systems researches, and various Information Systems researchers claim to use CGT methodology in them researches. Unfortunately, many of the claims of the use of GT research by IS researchers should be more accurately described as pseudo GT methods. There are many researchers do mix CGT with other theories such as Quantitative Data Analysis (QDA). This is noted as wrong, as Glaser (2003:1-2) argues that "Mixing *QDA and GT downgrades or erodes the GT goal of conceptual theory through the process of default remodelling*". CGT becomes remodelled as another QDA method with all its descriptive baggage. This is because there are confusions on doing and evaluating classic grounded theory. This paper fills this gap, by outlining necessary way for doing and evaluating classic grounded theory in Information Systems. It also processes the principles for Classic Grounded theory research.

The paper has three main sections; a clarification as what grounded theory is, an overview of the prior use of the grounded theory method amongst the IS research community and how to do authentic grounded theory research in an IS context.

CLARIFICATION OF THE MEANING OF THE GROUNDED THEORY RESEARCH METHODOLOGY

Classic Grounded theory methodology is faithful to the original formulation and follows all the original tenets of Grounded theory by Glaser and Strauss (1967). It has been further elaborated by Glaser (1978, 1992, 1998, 1999a, 1999b, 2001, 2003, 2005, 2006, 2008, 2009, 2011). Classic Grounded theory methodology is a general inductive research method designed to revel deep seated latent patterns of human behaviour and how the main concerns are being continually being resolved [Glaser, (1998)]. Glaser (1978:93) argues "the goal of GT is to generate a theory that accounts for a pattern of behaviour which is relevant and problematic for those involved... the goal is not voluminous description, not clever verification"

CGTM provides IS researchers with a means to build theory relevant to the discipline as in Matavire and Brown (2011). Lyytinen and King (2004) argues that building theory for the field would help strengthen it. In the IS discipline opportunities exist to build theory in both emerging fields and well-researched ones as outlined by Matavire and Brown (2011). As Glaser (1992) states, in domains where a large body of knowledge already exists, new concepts and categories may not emerge, but CGT will still contribute to a better understanding of the basic social processes at play. Many authors refer to having used classic grounded theory in their studies, when in fact what they mean is that they have been influenced by the idea of a grounded theory without following all the procedures (Bryant, 2002). In this paper we caution against such incorrectly labelled GT protagonists as pseudo GT They simply use Classic Grounded theory vocabulary and avoided following all the tenets of the method. As Glaser (2009) explains in his rebuttal of Bryant and Charmaz (2007) that these researchers only use the words of GT without the essence of GT as a result force the data rather than let emerge. The seeds of the major source of confusion concerning the use of the GT method can be directly traced to the publication by Corbin and Strauss (1990). When this book first appeared Glaser wrote to his old collaborator Anselm Strauss and asked that he either withdraw the book or publish it under another title. Strauss declined to do this so Glaser wrote a book Glaser (1992) with the same title as the Corbin & Strauss (1990) giving line by explanations as where their book was both incorrect and misleading. The major source of disagreement between Glaser (1992) and Corbin and Strauss (1990) is concerned with how the GT researcher should make the transition from low level substantive coding to higher level theoretical coding. The authentic way to do this is by making use of a combination of constant comparison and sorting so that the way in which the lower level substantive codes relate to each other can emerge as a theoretical coding outcome. Whereas the Corbin and Strauss (1990) chose to introduce an intermediate stage between substantive and theoretical coding called axial coding. The axial coding approach encourages the researcher to force the data into deductively derived patterns [selected by the researcher] rather than allow the deep seated patterns to emerge. The Corbin and Strauss (1990) gained in popularity because it gave a short cut to GT researchers but in so doing it also killed any prospect of revealing the core variable. It is not surprising that those researchers who were unaware of this unauthorised corruption of the GT by Corbin and Strauss (1990) believed they were doing GT when in fact they were doing a form of Qualitative Data Analysis. GT is not a subset of Qualitative Data Analysis because it is designed to use any kind of data including quantitative data. QDA is a legitimate research method but it should never be confused with GT because GT is a general research method.

An Overview of the Prior Use of the Grounded Theory Research Methodology in Information Systems

It is not surprising that many in the IS research community have largely misunderstood the different between authentic GT and pseudo GT research methods. This is because the confusion caused and spread by Corbin and Strauss (1990) still persists. There are two main reasons for its continued persistence. Firstly, pseudo GT methods present the attractive illusion that they can also deliver all that the authentic GT method can do without the challenges that are necessarily present in authentic GT. Secondly, authentic GT gives the appearance of being quite easy to do and has great appeal because it can reveal previously unknown deep seated latent patterns of human behaviour.

The authentic GT method places several difficult demands that those doing GT which the pseudo GT research methods do not. GT researchers have to have the twin characteristics of a high tolerance for uncertainty and developed skills in conceptualisation. Pseudo GT research methods let the researcher off the hook by forcing the data instead of allowing it to emerge.

There is range of papers claiming to use classic GT in information systems. As Matavire and Brown (2011) argued that some papers used in information systems used classic grounded theory. However, most of such papers does not follow the classic GT tenets.

How to Do Authentic Grounded Theory Research

Avoidance of Researcher Contamination

Authentic GT research focuses on the issues of concern of those being researched rather than issues of interest in the professional research community. GT should not concern itself with issues of professional interest for any constituency. To do so would preconceive the outcome of the research. The purpose of the GT method is to discover robust but modifiable hypothesis which are grounded in data not in the professional interest concerns. GT's other objectives are twofold. Firstly, is to discover the main concern of those being researched. Secondly, to reveal how these main concerns are being continually resolved. The researcher is advised not research contexts with which they have great familiarity. This is because it is extremely difficult to do GT research when the researcher perspective is already contaminated by his prior knowledge. Where it is not possible for the GT researcher to not have any prior knowledge of the research context a technique called "preunderstanding" needs to be employed. This means that the researcher, before doing any research, writes down all the issues that he believes are likely to be of concern to the respondent constituency involved in the GT research project. The effect of this is to sensitise the GT researcher both to his own "preunderstanding" and also to new issues which might arise.

Awareness of the Dangers of Using Pseudo GT Research Methodologies

The GT research method has been compromised by many researchers due their misunderstanding of what the method entails. Even in the International Journal of Information Systems there is evidence of confusion as what the GT research method actually is. Walshaw (2006:324) states that:

"Orlikowski (1993) describes the use of GT including open and axial coding to generate a set of concepts from data".

Walshaw (2006) is quoting from a researcher [Orlikowski] who does not understand that axial coding plays no part at all in the authentic GT method. The axial coding concept was introduced by Corbin& Strauss (1990). Glaser (1992) exposed this approach as a means of prematurely forcing the data into patterns dictated by the researcher rather than the data itself. At the heart of the authentic GT method the researcher has to trust in emergence. This means that if the latent pattern is genuine it will be there whether the researcher reveals it or not. By following the tenets of the authentic GT research method the deep seated latent will be revealed without any need for premature forcing.

Walshaw (2006:325) incorrectly suggests that the concepts which arise in the GT method are chosen by the GT researcher himself:

"In terms of learning from the data itself, grounded theory offers one way of doing this, although the coding is a subjective process, because the researcher chooses the concepts to focus on"

The GT researcher does not choose the concepts to focus on. Correct use of the authentic GT method means that the researcher first develops substantive codes from the low level description of the respondent narrative. Then the GT method creates theoretical codes which are a conceptualisation as to how the substantive codes relate to each other as a modelled, interrelated, multivariate hypothesis which accounts for the main concern of those being researched. Forcing the data by the researcher is not grounded theory. Instead the grounded theory researcher has to make use of a combination of constant comparison analysis and sorting from which the core variable emerges. This is a mainly inductive rather than a deductive process.

Epistemological Stance of Classic Grounded Theory

Authentic GT or Classic grounded theory methodology (CGTM) is known to be ontological and epistemological free research methodology, as Christiansen (2005:81-84) argues that classic GT is an "ontology free and epistemology free methodology". Glaser (1978) reiterates that ontological assumptions have no place in the use of Classic GT since GT is grounded in data. Such argument would consider epistemology (the theory of knowledge pertaining to knowing or 'recognizing'), as a mental activity and therefore having little place in the application and use of CGTM. To elaborate more on " CGT being free from ontological and epistemological assumptions" Glaser (2004:76) argues that "It is only a question of applying a rigorous and systematic method for discovering and explaining these patterns. Thus, *just do it*." Furthermore, Glaser (1978:45) explains that in the CGTM methodology all ontological and epistemological suppositions are irrelevant and are not useful. As Glaser (1978:45) states "... logical conjectures or preconceptions that pre-frame the research, and pre-framing has to be avoided in any CGTM work, in order to keep close to what the data has to say, and to discover and name emergent latent patterns in the data."

The CGTM approach minimises the influence and potential distorting effect of preconceptions, logical deductions and elaborations and ungrounded assumptions (Glaser, 1978). As a result, a researcher using CGTM is free from both ontological & Epistemological assumptions. However, if researchers choose to use other methodologies ontological and epistemological diversity would be acceptable.

The misunderstanding identified is that those papers that were categorized as "GT" were then subdivided into either "interpretative" or "positivistic". Instead of categorizing, it is more important to say that several authors who had claimed to be using the CGT method in fact were either using either the QDA approach [interpretative] or hypothesis testing [positivistic].

Glaser (2004 :2) reminds us what the GT research method is all about "GT is a latent pattern indicator based on index formation. The index formation model in turn is based on accumulation and summing of indicators from [originally] from survey data to generate indexes or concepts is fundamental to GT." Whatever research methodology the researcher decides to employ it is always preferable for the researcher to stay within the tenets of that methodology instead of trying to introduce elements of other methods in order to validate their research.

Basic Philosophy Underpinning Classic GT Methodology

CGTM was founded by Glaser and Strauss in 1967. At the time they considered this new research methodology to be a necessary challenge and alternative to the ungrounded and unquestioned "great men" theories being presented to students at academic institutions: "Many of our teachers converted departments ... into mere repositories of "great man" theories and taught these theories with a charismatic finality that students could seldom resist. Currently, students are trained to master "great-man" theories and to test them in small ways, but hardly to question the theory as a whole in terms of its position or manner of generation…some theories of our predecessors, because of their lack of grounding in data, do not fit, or do not work, or are not sufficiently understandable to be used and are therefore useless in research, theoretical advance and practical application" (Glaser & Strauss, 1967:10-11). Glaser developed and expanded his theoretical position of the sixties in 1998 putting the case for a grounded theory which was more reliable than previous theories, was designed to minimise preconceptions and assumptions and to address "the main concern", "A methodology was needed that could get through and beyond conjecture and preconception to exactly the underlying processes of what is going on so that professionals and laymen alike could intervene with confidence to help resolve the participants' main concern surrounding learning, pain and profit. In this sense, grounded theory has put vested social structures in some jeopardy, because the dependent variables are vital to work with; hence, it is hard to ignore grounded theory results." (Glaser, 1998:5)

Glaser (1998:115) argued that a researcher using Grounded Theory "... tries to understand the action in a substantive area from the point of view of the actors involved. This understanding revolves around the main concern of the participants whose behaviour continually resolves their main concern. Their continual resolving is the core variable". The main objective of CGTM is to discover a robust empirically derived hypothesis or core variable through a specific process. The core variable articulates the main concern of the respondents. It is so called because the core variable is able to explain most of the variations in the data. The particular feature of the CGTM is that it is transcendent of both time and place (Glaser, 1978, 1992, 1998, 2000, 2003, 2008, 2009). According to Glaser this is made possible due to the fact that CGTM delivers a conceptual explanation rather than a low level descriptive context-based narrative. Thus the emerged grounded theory must pay its way by being able to demonstrate that it represents the main concern of those participants who are the subjects of the research. Glaser (1992:15) outlines and explains the four central characteristics of CGT and emphasising the value to the research process of it "modifiability": "There are only four central criteria with which to evaluate CGT: fit, work, relevance and modifiability. If a grounded theory is carefully induced from the substantive area its categories and their properties will fit the realities under study in the eyes of the subjects, practitioners and researchers in the area. If a grounded theory works it will explain the major variations in behaviour in the area with respect to the processing of the main concerns of the subjects. If it fits and works the grounded theory has achieved relevance. The theory itself should not be written in stone or as a 'pet'; it should be readily modifiable when new data present variation in emergent properties and categories. The theory is neither verified nor thrown out; it is modified to accommodate by integration the new concepts. When these four criteria are met, then of course, the theory provides a conceptual approach to action and changes and accesses into the substantive area researched." (Glaser, 1992:15)

Stages and Procedures of the Grounded Theory Research Method

Although there are very specific processes through which the CGTM researcher must go through they do not always occur in a linear format. This kind of flexibility in terms of sequence and process is necessary in order to reveal certain deep seated patterns of human behaviour. The structure of CGTM will be outlined in the following section.

1. Generating Theoretical Sensitivity

It is essential that the researcher, prior to embarking on any CGTM research, develop theoretical sensitivity. None of the advocates of pseudo GT make this a perquisite. What this entails are explained by Glaser (1978). Since there are already in existence literally hundreds of patterns of human behaviour, having a repertoire of these before embarking the research is absolutely essential. Some of the examples of theoretical codes: Causal consequence models (cause, consequences, contexts, contingencies, co variances and conditions); Process models (stages, phases, polarity, progressions, temporal pacing and cycling); Degree models (cutting points, probability, polarity, continuum and intensity); Asymptote models (mathematical models for getting as close as possible); Isomorphism (mathematical models how one established theory can trigger a new theory); Moment capture (business models which explain how many financial services operate)

If armed with a repertoire of theoretical codes, it is easier for the CGTM researcher to start making sense of data.

2. Theoretical Sampling

This is a form a sampling in which it is both inappropriate and impossible, prior to doing the research, to state exactly what data will be needed (Glaser, 1978, 1998, 2006). This is because the sampling process can only cease once the core variable has emerged. The researcher samples data until s/he discovers the main concern of the respondents. The researcher then samples within a chosen group and demographic area for integrating into the research, the newly emerged concept. Once the researcher reaches the point where no further patterns emerge, the data is said to be saturated and the core variable to have emerged (Glaser, 1978, 1998).

According to the CGTM theoretical sampling is a process of data collection for generating theory whereby the researcher simultaneously generates, codes, analyses and synthesizes data and uses this as a basis to decide what data to collect next and where to find them (Glaser, 1978). The researcher begins by selecting a context for the research. A series of short conversations with a small number of respondents is then set in motion. There are two reasons for this. Firstly, a high level of rapport must be established between the researcher and respondent(s) so that trust and respect can develop. This cannot happen easily with single encounters. Secondly, the researcher should not make use of any form of recording or note taking throughout the whole process of conversation/research with the respondent(s) encounter (Glaser, 1978). This both deepens the researcher/respondent trust and also increases the level of the researcher's concentration. Immediately following all encounters with respondents, the researcher should document the interaction using theoretical memos.

3. Theoretical Memos

Theoretical memos as described by Glaser (1978:83) are "the theorising write up of ideas about codes and their relationships as they strike the analyst while coding". They are the means to abstraction and ideation and can be used continually throughout the CGTM process. Initially a theoretical memo may consist of a short sentence but as the analysis proceeds, they are updated and developed and can be several pages long (Glaser, 1998). Theoretical memos are a crucial part of the GT method for six main reasons. Firstly, it is a device which allows the researcher to use both deductive and inductive logic in order to reveal the deep seated latent patterns of human behaviour which the GT method is designed to reveal. Secondly, theoretical memos encourage the researcher to have a dialogue with his data. This means that GT researcher can interrogate the data to assess the voracity of emerging theoretical concepts. Thirdly, the theoretical memo enables the GT researcher to move from low level descriptive coding to higher level theoretical coding. The transition from description to conceptual coding cannot happen without theoretical memoing. Fourthly, theoretical memos are very important because it is the main way in which the GT researcher can understand how to carry out the most appropriate theoretical sampling strategy. Theoretical sampling is necessary in the GT research process because it ensures that the researchers to look beyond the obvious descriptive narrative and understand what are the main concerns of those being researched. Fifthly, the GT method, for the researcher, is a delayed action phenomenon because it is impossible for any researcher to immediately understand all aspects of his data on a single examination of the data. In other words, when the GT researcher is engaged in the simultaneous writing of several theoretical memos it allows the researcher's own though processes to discover insights in partially completed theoretical memos on which he is not currently involved in writing. This is because the mind is highly discursive and is always attempting to make sense of phenomena; but it takes time. Sixthly, when the researcher is engaged in the simultaneous writing of theoretical memos it stimulates the possibility of the cross fertilization of new ideas and concepts. This serendipity stimulates the discovery of new ideas and concepts.

In short any research which claims to have used the GT research method and does not make full use of the theoretical memo will have failed to reveal the main concerns of those being researched.

4. Substantive Coding

These are the very first attempts at highlighting those data which the analyst believes may have importance for the research beyond the simple description of the context of the data (Glaser, 1978:1998). These codes are labelled and often 'gerunded' (Glaser, 1978). In other words, the researcher converts the substantive code label into a verb ending in "ing". The purpose of this is to sensitize the researcher to the processes and patterns which may be revealed at a later stage (Glaser, 1978; 1998; 2006). Gerunding is effective because it introduces momentum into the data (e.g. the emergence of legitimizing and credentializing was due to gerunding)

5. Theoretical Coding

This is a conceptual code. It arises from the synthesis of the substantive codes. To move from substantive codes to theoretical codes is a tricky and often elusive process (Glaser, 1978). This synthesis can be effected by examining the interplay between the theoretical memos. This process is known as the constant comparison method (Glaser, 1978). At this point the researcher has to engage both intellect and intuition to achieve a shift in perspective from low level context based description, to higher level conceptual abstraction.

6. The Constant Comparison Method

When using the constant comparison method, the researcher has to painstakingly look across all data in all theoretical memos to look for various types of indicators which might reveal previously hidden connections and patterns (Glaser, 1978; 1998). There is a hierarchy of processes involved in this research method. It starts with the recording of raw data in the theoretical memos and proceeds to substantive coding which is then fractured into categories and sub categories. These in turn have different properties (Glaser, 1998). The term "properties" is not strictly relevant in the context of explaining the constant comparison method. The expression "properties" is more relevant when writing theoretical memos. If the researcher believes that some elements of an emergent pattern might be on the brink of discovery, he could return to the data [or perhaps the respondent for another encounter] and try and establish what are the properties of a particular concept concerned. For instance, the properties of legitimizing in this research revealed that they were reciprocity, asymmetrical and networking. So properties are another word for characteristics and as such are not exclusively linked with the constant comparative method. Then, both the categories, as well as the properties, are constantly compared across all the theoretical memos. As this process begins to gather momentum theoretical codes begin to emerge. What begins as a property of a category of a substantive code sometimes emerges as a theoretical code in its own right. Only through a process of constantly comparing the data for differences and similarities can theoretical codes emerge (Glaser, 1978). The exceptions to this are those rare occasions when the respondent is an articulate conceptualize and is able to step back from personal context and give an authentic explanation of the main concern (Glaser, 1978, 1998).

7. Sorting

This is an iterative process which occurs at the later stages of the CGTM process. Sorting will ensure that the emerging theoretical codes have earned their place in the emergent process of theory creation (Glaser, 1978). If sorting does not take place several times it is unlikely that the research will move much beyond low level narrative description (Glaser, 1978, 1998). The principle aim of sorting is to ensure that the emerging core variable has been fully saturated and has earned its place in the research process (Glaser, 1978).

8. The Emergence of the Core Variable

The core variable is so called because it is able to account for most of the variation in the data (Glaser, 1978). It is the main concern of the respondents translated and carried on to the conceptual level. There are frequently also sub-core variables which feed into the core variable (Glaser, 1978, 1998).

Respondent Encounters

Classic GT allows collecting data through oral conversations, face to face, online and many others. The most important aspect of the interviews should be short duration, conducted without tape recordings, and followed immediately by detailed memo writing (Glaser & Strauss, 1967; Glaser, 1978). Several interviews rounds are expected. For example, the initial round is recommended to start from the respondent's perspective. Glaser (1978) deliberate recommends the following interviewing strategies: (1) no notes or tape recordings is recommending to be used in any of the respondent encounters (2) no semi-structured interview questions. Instead the researcher concentrated in the initial conversations on establishing rapport and making the respondents feel comfortable and unthreatened. The rationale for such strategies is that the respondent encounters should be more like normal conversations where both parties felt sufficiently relaxed to participate in a candid exchange of ideas. Immediately following each respondent encounter the researcher should create a very detailed notes, on which later is converted into theoretical memos.

The Classic Grounded Theory research method is known as a "delayed action phenomenon" (Glaser, 1978). This is because it takes some time for both the researcher and the respondent to consider what is and has been said. For this reason, it is essential not to attempt to collect all the data first and then proceed to attempt to analyse it. It is essential in CGTM that the researcher follow wherever the data leads. This is because the GT researcher has to be able to move from the low level realm of contextual description to the higher level of theoretical explanation.

Once the researcher had completed at least six initial interviews with different respondents, appointments for subsequent interviews were made with respondents. Prior to the next round of respondent encounters more theoretical memos were written. The focus of the respondent encounters shifted and followed the issues emerging from the most recent iteration of theoretical memos. Although there is no standard template which all theoretical memos must adhere to, there are some basic issues which all theoretical memos must address. If these are not present, there will be no transparency in terms of how the researcher progressed from raw data to an emerged core variable. In order to meet transparency requirements all memos should include: (1)Memo title; (2) Summary of any substantive codes, categories and properties; (3) Conceptual indicators [These are important links that form the bridge from substantive description to conceptual explanation]; (4) Emergent theoretical codes, categories and properties (5) A statement as to the type of data the researcher believes has been given to him or her by the respondent [Baseline, proper lined vague, Zero, or interpreted data]. It actually does not matter when doing CGT what type of data the researcher has because everything is considered to be data. If, for example people, are untruthful, in encounters, this would probably be more interesting and significant for the researcher than if they were being honest. What is of the utmost importance is that the researcher fully understands what type of data he has, rather than worry about accuracy. A research should have: (1) A clear separation of empirical data from conjecture data. Both are important in CGT but they are very different types of data and need to treated and used accordingly; (2) How has the data been fractured? Which conceptual theoretical perspectives has the researcher used to better understand the latent patterns embedded in the data? (3) What links have been highlighted in other memos from the researcher's bank of memos? (4) What unanswered questions arising from the data will the researcher use in his/her next meeting with the respondent? and (5) Have the theoretical concepts in each memo been constantly compared and then sorted and by what process?

Once a critical mass of data had been generated this researcher had to use a combination of constant comparison and analytic induction to tease out the main issues of concern as expressed by the respondents. Two competing main issues or core variables emerged. The researcher then followed the specific guidance given by Glaser (1978:93-100): to re-enter the field and investigate which of the two variables proved to be more 'robust'. This situation occurred when Glaser and Strauss were writing the Discovery of Grounded Theory (1967), where two variable emerged. As Glaser and Strauss (1967) did, the researcher had to re-enter the field to investigate which of the two core variable candidates was more robust. First the researcher returned to the data and laboriously and constantly compared these across all the theoretical memos to see which of the two candidates for the core variable was more prominent. "Cultivating Recognition" began to emerge across all the data whilst the other core variable candidate only featured occasionally. The researcher re-entered the field and met with four respondents from the original cohort of respondents. A total of 101 respondent encounters were theoretically sampled for "Cultivating Recognition". Cultivating Recognition was then instantly acknowledged by the respondents as their key issue or concern. Prior to this final wave of interviews, the respondents had repeatedly been talking about their concern with regard to the lack of respect on the part of some of their clients. Although it worried them considerably it turned out to be the symptom of something much more important and pervasive. It was a symptom of a general lack of recognition. Thus Cultivating Recognition emerged as the core variable. Figure explains how "cultivating recognition" emerged as the core variable.

As has been outlined above, each of the interviewees was re-interviewed several times until latent patterns of behaviour began to emerge. Once these patterns emerged the second stage of interviewing took place with more university professionals as well as other constituencies identified by the interviewees as being important people to speak with concerning this issue. As key concept indicators emerged theoretical sampling was the feature of the second stage of interviewing. The interviewing and memo writing and sorting continued until no newer patterns emerged. When this happens the data is said to be 'saturated': no more data is required because predictable patterns have begun to emerge.

Nature of the Respondent Encounters

The meetings with respondents should be scheduled earlier before meeting them to ensure a respondent is free. Each meeting is introduced with an explanation. Should be made clear to the respondent that the research focus is to reveal, understand and explain what the main issues were which confronted a respondent in the execution of their role. Encourage the respondents to let the researcher know if they felt uncomfortable about any line of enquiry. The absence of any tape recording or note taking during these encounters helps to make the respondents feel more relaxed. Increase the possibility of transforming the nature of the encounters from an interview to a something akin to a normal conversation during which both interviewer and respondent are able to pose questions to each other. All encounters recommended to be kept deliberately brief so that the researcher is able to make detailed notes immediately after the conclusion of each meeting. The purpose of the series of meetings is to discover the respondents' main concerns regarding their professional workings including the challenges, how they solve problems and what they consider to be achievements and successes in the course of their daily working lives. It is considered vital to schedule several meetings with each respondent as this would give each respondent the time and space to reflect subsequent encounters and enabled the researcher to encourage a respondent to think more deeply about what their real concerns were. Following this incremental process of having several interviews with each respondent ensure that the respondent's agenda emerged rather than that of the researcher.

Choice of Interview's Location

The choice of the locations is an important consideration. The idea is to try and ensure that the respondents felt relaxed. Thus a variety of locations, other than their own offices, is highly recommended when doing CGT. These included restaurants, university canteens, meeting rooms, public bars, walking in the street and driving in the car.

Types of Data

As part of the process of the grounded theory [GT] method the researcher has immediately to be aware of the types of data which emerge from conversations with respondents. From a GT perspective there are five types of possible data to be derived in any encounter with respondents:

1. **Base Line Data:** The respondent is being as truthful as possible, as far as the interviewer can gauge.
2. **Interpreted Data:** The respondent is filtering his/her message either to deliberately complicate or simplify her/his responses.
3. **Proper Lined Data:** The respondent deliberately edits the data so that it is not in conflict with the "official" line of argument from his employing authority.
4. **Vague Data:** The respondent deliberately gives an indistinct version of reality to try and confuse the researcher, obfuscate the issues and throw the researcher 'off the scent'.
5. **Zero Data:** The respondent refuses to say anything at all in order to avoid saying anything which he might later regret.

From a GT perspective it does not matter which type or types of data the respondent delivers because each type of data has the potential to communicate something of interest and value to the research.

Awareness of Ethical Issues

The GT research process will not be effective unless there is a very high level of rapprochement between the two parties. The researcher has to establish credibility by offering the respondent anonymity. This is made easier when using the GT research method because none of the respondent encounters are ever tape recorded. Note taking should also be avoided during the research process. This ensures that the respondent encounters take the form of conversations rather than typical interviews that a researcher might when adopting other types of research methods. When the respondents are relaxed and comfortable they will be more likely to reveal issues which are of genuine concern.

Data Analysis

The Grounded Theory research method is a general research method which treats everything as data, whether quantitative or qualitative Glaser (1992). Grounded Theory primarily uses an inductive research approach in which the researcher has to be subordinate to the data. Grounded Theory is not a sub-set of Qualitative Data Analysis [QDA]. This is because QDA values accuracy above all within a narrow research context whereas Grounded Theory is mainly concerned with the discovery of latent patterns of human behaviour which are transcendent of the original context of the data. The following Grounded Theory Research process was followed as outlined by Glaser (1978, 1992): (1) Theoretical sampling and substantive coding; [2] Theoretical coding; [3] Theoretical memo writing; [4] Constant comparison between and within memos; [5] Sorting codes; [6] Data saturation; [7] Emergence of the core variable. Although the above is represented in a linear format, in practice the exact sequence varied according to what patterns seemed to be emerging from conversations with respondents. The reason for this is that GT is a process not a unit based approach. The data was simultaneously, subsequently, and sequentially collected, analysed and synthesised. Once most of the variations in the data were explained in terms of a key concept, the data was said to be saturated and therefore no further need to collect more data since the core variable could be said to have emerged. The sequence of the research process, although as already mentioned, not a linear one, was as follows: (1) Gaining access and carrying out the first few interviews with e-learning providers; (2) GT skill development on the part of the researcher in memo writing, substantive coding and theoretical coding using the data from the series of interviews with the initial group of interviewees; (3) Managing the transition from substantive to theoretical coding, achieved by using a combination of the constant comparison method, cutting point analysis and grounding; (4) The emergence of latent patterns following another series of interviews with e-learning providers The data yielded was sorted so that conceptual indicators could help clarify these emergent latent patterns; (5) Once the data could be said to be saturated the core variable emerged; (6) The core variable was then used as a basis for a comparative literature review; (7) The findings are published in the form of this PhD thesis.

Coding Data for Emergent Fit with Previously Grounded Theory

Authentic Classic GT [as explained by the co-originator of GT Dr Barney Glaser] is a latent pattern identifier. It is not evidence based but instead these patterns are revealed by the discovery of the main concerns of those being researched. These main concerns are continually being resolved by the respondents whether they have been revealed by research or not. The main concerns are not the respondent's narrative but rather their motivations behind their narratives. Because of this the use of the authentic GT method means that emergent theory is transcendent of both time and place. The theoretical explanations of human behaviour which GT reveals will also be applicable across a wide range of different contexts. The product of the any GT research project is a robust hypothesis which can be modified as new data comes to light. As a result, CGTM has two basic models for studies: (1) The first model is by pure discovery of all the concepts by going through the prescribed procedures; and (2) The second model is "starting from scratch" (Glaser, 1978). The first model is the emergent fit model and applies the same procedures as the second model, but does not "start from scratch". It proceeds from a previously generated CGT and its substantive concepts. The second model is especially suited to an advanced user of a generated CGT who would want to scrutinise the extent to which a theory, which has been generated from a given set of data from a given substantive area, fits his/her data from the same or another substantive area. The use of this second model requires a minimum level of familiarity with the CGTM. For advanced users of Classic GTM in business, (e.g. business consultants, who have been trained in the use of the methodology) the modification and refinement of any generated CGT may be applied within any area of study including the area of business. The second model may become an option for a researcher in the business field, as well as a funnelling down on some company-specific issues. Use of this second model needs to be embarked upon with care since potential pitfalls exist. In the case of an analyst who has some basic knowledge on data gathered from research and analysed that s/he already generated and wishes to extend it, or use some grounded formal theory to analyse it, Glaser and Holton (2005:18) sound a warning to those following rigidly a research model through with "pre-emergent analytic thinking" without reassessing and making constant comparisons throughout the process: "The second model looks a bit like deductive, logical elaboration, but it is not, providing the analyst follows the grounded approach. S/he does not start "empty" or "non-preconceived" as in the first model. S/he engages in pre-emergent analytic thinking, and sampling before approaching the field. But once in the field, s/he starts correcting early thoughts and follows the grounding in subsequent theoretical sampling. And s/he ends up as s/he would in the first model, searching for comparison groups, as it becomes clearer and clearer where to go for fit as the theory develops."

The second model equates to a GTM which begins with some pre-framed view on the part of the researcher who starts by trying out a particular set of interrelated concepts, while data is collected and constantly compared. The second model saves time because large amounts of time will have been spent by the researcher using the first model

Criteria for Evaluating Classic GT

The grounded theory research method is a general inductive methodology which can be applied to all types of data, both quantitative and qualitative, or the combination thereof as has been outlined by Glaser (1978, 1998). GT is not a sub set of Qualitative Data Analysis (QDA). Therefore, QDA criteria are inappropriate for the evaluation of research done using the grounded theory research method (Glaser, 1978).

Glaser (1978, 1998, 2000, 2001) gives explicit examples of the four main criteria with which to evaluate grounded theory study: fit, workability, relevance and modifiability. These criteria will be explored in the following section.

1. Fit

If a concept can be said to be a 'fit' it means that the concept adequately reflects the data that it purports to express. The categories of the emerged core variable must fit the data used to create the theory. Ungrounded assumptions and data which have been forced to fit into preconceived concepts derail fit in CGTM, as well as invalidating the theory (Glaser, 1978). Although fit would appear to correspond to positivistic validity, it is fit in terms of CGTM action and usage, not as a result of testing (Glaser, 1998). Unlike research based on preconceptions, such as research conducted in order to test a hypothesis, in CGTM data which cannot be forced are discarded rather than used to correct the emerging categories. With CGTM only data 'sorted' and used for developing theory can be said to constitute a fit (Glaser, 1978, 1998). The fit could be expressed as a 'refit or an 'emergent fit'. During research categories emerge fast, so the need arises to refit them to the data as the research proceeds and to be sure that they fit all of the data they purport to indicate, thus increasing and ensuring transparency. Categories can emerge between data and pre-existent categories (Glaser, 1978).

2. Workability

How the core variable accounts for the respondents' continual resolution of their main concern is a key issue. The emergent GT must clearly explain what is happening, and the process of its happening, and by so doing should be able to predict future behaviour (Glaser, 1978, 1998, 2001).

3. Relevance

How the emerged core variable has been received by the members of the constituency from which the data was drawn is also pertinent. Prior to the research it is very likely that respondents would be unable to articulate their main concern. Once they have read the research they should instantly be able to recognize the emerged core variable as being authentic (Glaser, 1998).

4. Modifiability

The CGT is considered to be modifiable if it is easy for subsequent CGTM researchers be aware of what research has been done so far in the area, and to proceed to modify or refine the theory as they collect and code new data (e.g., code new data for emergent fit), without invalidating the theory.

Certain questions on the issue of transparency when using CGTM have been posed: "How can research using CGT have a chain of 'legitimate', 'accurate' evidence? How can one reconcile this with the approach of not recording or taking notes in interviews? How does one demonstrate, or substantiate the chain of evidence in the process of moving from data to concept/theory?" Such questions assume accuracy in the collection of evidence based data as a requirement for 'accurate', 'reliable', 'legitimised' research (Ndume et al. 2008).

Since CGTM is an inductive research methodology it is appropriate to use a legitimate set of evaluation criteria as outlined by the CGT's co-originator Glaser (1978). However, there must be total transparency when using CGTM so that it can easily and clearly be seen how the emergent core variable arose.

Glaser (2007:41-42) explains the reasons for the persistence of a level of misunderstanding of the CGTM amongst many academics and researchers: they continue to fail to distinguish clearly between "conceptual generalisations" and "descriptive generalisations": "The constant comparative method [at the heart of CGT] was originated in 1967 to generate concepts by constantly comparing indicators of a latent pattern. After several comparisons the result is conceptually naming the pattern and its properties. The category is abstract of time, place and people. The conceptual categories apply with emergent fit. The Formal CGT research generates categories that relate to each other as generalizations. In contrast descriptive comparisons are as old as research itself. They just compare differences and similarities and a sum of description of people's social actions in a unit of time. The concern is always accuracy, that is, the researcher has worrisome accuracy to constantly contend with. Generalisations are difficult to make from one unit to another and even within the unit studied. These are the problems of accuracy of external and internal validity respectively of descriptive generalizations. Conceptual generalizations do not have this validity problem. They just apply within a context applied to with modification. For example, controlling clients by pseudo-friending simply varies for client types and client conditions."

Glaser (2009:15) gives a robust response to the continuing misconceptions regarding CGTM, as articulated by Bryant and Charmaz (2007:120), in particular the "worrisome accuracy" associated with data analysis and the failure of researchers to appreciate the value of the conceptual level of CGTM: "Data worries dominate the GT jargonizing of QDA issues in the Handbook. And why shouldn't it? Worrisome accuracy is the central issue of QDA. QDA research has to yield accurate description, which puts an emphasis on analysing the data used in any research every which way to see what, indeed, is being described and is verifiable. And furthermore how fast will the description get stale-dated and if it can be momentarily discursively generalized. This is a perennial, non-solvable problem. It totally neglects the conceptual level of GT, which is free of and abstract of place, time and people and hence free of worrisome accuracy. Comparisons generate a GT's constant concept modifiability." Transparency of a fully grounded theory will be achieved only when the reader is able to clearly see how the researcher progressed from raw data to the final integrated core variable.

A Set of Principles for Classic Grounded Theory Research

However, Glaser (1978:1-2) specified the criteria for evaluating classic GT, there are persistent confusion among many researchers. This raises the need for setting principles for classic GT research. Principles for evaluating all researchers claim to use CGT in IS. This paper we propose a set of principles for the evaluation of classic grounded theory research in information systems and beyond. The source of these principles includes the Glaser (1978) and the past practice of classic grounded theory research in information systems, sociology and other management disciplines as well as our understanding of the underlying philosophical foundations of classic grounded theory.

As Myers and Klein (2011) citing Klein and Myers (1999) we use the word principles to emphasize that the general principles we propose are offered in the spirit of being fundamental ideas that may be helpful to classic Grounded theory researchers in IS. The principles are delivered from the philosophical writings of Glaser and Strauss (1967) and Glaser (1978, 1992, 2001, 2004, 2005, 2008, 2011). As noted by Myers and Klein (2011:24): "the principles should be helpful because they summarize impor-

Table 1. Summary of principles for authentic GT research

The Fundamental principle of the emergence of core variable This principle suggests that the Authentic GT outcome is a Core Variable which account of most of Data. This forms the basis of the theory to come. This principle is the fundamental to other principles.
The principle of core variable accountability for most of the variation in the data This principle signifies that all is grounded into data. All is about Data
The principle of stages and processes in the discovery of core variable The principles insist in following all stages and processes in discovery of core variable including developing theoretical sensitivity, Identify your substantive area, Collect data, Open code your data as you collect it, Write memos throughout the entire process, Conduct selective coding and theoretical sampling, and Sort your memos and find the Theoretical Code(s) which best organises your substantive codes-- the core variable emerge.
The principle of core variable explains how the respondent's main concerns are continually resolved This principle means that categories must fit the data relevant to respondent. Data should not be forced or selected to fit preconceived or pre-existent categories or discarded in favour of keeping an exact theory intact (Glaser, 1978:4)
The principle of research design involves the discovery of a hypothesis and does not test one • GT exist for discovery of hypothesis not to test. hypothesis should not be forced or selected, rather should be emerged from data- data and only data
The principle of core variable is transcendent of the context of the data The theory should be able to explain what happened, predict what will happen and interpreted what is happening in an area of substantive or formal inquiry (Glaser, 1978:4)
The principle of emerged hypothesis is modifiability as new data emerges in future research This principle allows for ready, quick modification to help explain surprising or new variation- such as developing new condition, almost as new data appears. It is through this quick approach that the tractability of GT over social life is maintained, hence its relevance (Glaser, 1978:5)
The principle of researchers demonstrates full transparent transition from substantive to theoretical codes The principle allows the researcher to build necessary theoretical sensitivity by which they can render theoretically their discovered substantive, grounded categories.
The principle of processes of both sorting and constant comparison. The principle enables a research to discover a rich core variable. Firs analyst starts by comparing incident to incident. Secondly research compare concept to more incidents. Lastly, while the first and second type of comparisons continues throughout the study, the analyst also compares thirdly concept to concept (Glaser 1978:49-54)

tant insights which as a whole are (as yet) not embedded in the current practices of {Classic Grounded theory}, but could improve them if observed more systematically" (see Table 1).

The Principle of the Emergence of Core Variable

This is the goal for doing classic grounded theory. As Glaser (1978:94) argues that the analyst should consciously look for a core variable when coding his data. In other words, while doing all you can do [comparing data, etc.], be alert to the one or two that are core. This means that a research constantly looking for the main theme, for what – in their view-is the main concern or problem for the people in the setting, for what sums up in a pattern of behaviour the substance of the what is going on in the data, for what is the essence of relevance reflected in the data, for gerunds which brings out process and change (Glaser, 1978:94). When researcher starts coding, categories tend to emerge, giving way to finding core categories, the researcher should be alert of these core variable.

Glaser (1978: 95) sum up the criteria by which a researcher can make his judgement as to the core category:

- It is saturated: The core variable takes time to saturated than other variables because it occurs frequently in the data
- It relates meaningfully and easily with other categories-such connections need not be forced
- A core category in a substantive study, has clear and grabbing implication for formal theory
- It carry-through
- It is a dimension of the problem
- It is completely variable
- It is grounded into data

The principle of emergency of core variable suggests that the Authentic GT outcome is a Core Variable which account of most of Data. This forms the basis of the theory to come. This principle is the fundamental to other remaining principles.

The Principle of Core Variable Accountability for Most of the Variation in the Data

Glaser (1978) argues that CGT is grounded into data. Using the principle of emergence of core variable, then should reoccur frequently in the data. As it occurs frequently becomes accountable for most of the variation in the data. Its frequent reoccurrence it comes to be seen as a stable pattern and becomes more and more related to other variables. Glaser (1978:95) argues that if it does not reoccur a lot, it does not mean the category is uninteresting, it may be quite interesting in its own right, but it just means it is not core.

The Principle of Stages and Processes in the Discovery of Core Variable

The principles insist in following all stages and processes in discovery of core variable including developing theoretical sensitivity, identify your substantive area, Collect data, Open code your data as you collect it, Write memos throughout the entire process, conduct selective coding and theoretical sampling, and Sort your memos and find the Theoretical Code(s) which best organises your substantive codes-- the core variable emerge.

The Principle of Core Variable Explains How the Respondent's Main Concerns Are Continually Resolved

This principle means that categories must fit the data relevant to respondent. Data should not be forced or selected to fit preconceived or pre-existent categories or discarded in favour of keeping an exact theory intact (Glaser, 1978:4). The core variable it must be central, that is related to the many other categories and their properties as possible and more than other candidates. This criteria of centrality are a necessary condition to make it core.

The Principle of Research Design Involves the Discovery of a Hypothesis and Does Not Test One

This principle explains how you discovered the core variable. GT exist for discovery of hypothesis not to test. Hypothesis should not be forced or selected, rather should be emerged from data- data and only data.

The Principle of Core Variable Is Transcendent of the Context of the Data

The theory should be able to explain what happened, predict what will happen and interpreted what is happening in an area of substantive or formal inquiry (Glaser, 1978:4).

The Principle of Emerged Hypothesis Is Modifiability as New Data Emerges in Future Research

This principle allows for ready, quick modification to help explain surprising or new variation- such as developing new condition, almost as new data appears. It is through this quick approach that the tractability of GT over social life is maintained, hence its relevance (Glaser, 1978:5).

The Principle of Researchers Demonstrates Full Transparent Transition from Substantive to Theoretical Codes

The principle allows the researcher to build necessary theoretical sensitivity by which they can render theoretically their discovered substantive, grounded categories.

The Principle of Process of Both Sorting and Constant Comparison

The principle enables a research to discover a rich core variable. Firs analyst starts by comparing incident to incident. Secondly research compare concept to more incidents. Lastly, while the first and second type of comparisons continues throughout the study, the analyst also compares thirdly concept to concept (Glaser, 1978: 49-54; see Table 2).

Writing a Paper Using the GT Method Which Focused on Previously Published Research in Information Systems Research

There are a number of ways a GT research project could deal with such a question. Before explaining how to do, it is important to understand that one of the main themes which was present in all of the papers read is the pre-occupation of the researchers into verifying whether the existing "professional concerns" those working in the IT environment established in the literature is also present in their data. All the papers have in greater or lesser ways taken these professional concerns as their hypothesis and tried to understand them by testing the various hypotheses. An approach to doing a GT research paper on this issue could be either make use of 100% secondary data [in other words using the papers as the data] or commissioning a study [using primary data] to discover whatever the actual main concerns of the IT community rather than just testing established hypotheses (Glaser, 2009). The CGT research design would focus on establishing what were the main concerns of members of the IT community rather

Table 2. Overall assessment of the five articles claim to use classic grounded theory

	Scott (2007)	Orlikowski (1993)	Huff and Munro (1985)	Munro, Huff, and Moore (1998)	Lundell and Lings (2003)
1. The principle of the emergence of core Variable (CV)	Temporal Integration	None Emerged	None Emerged	None Emerged	None Emerged
2. The principle of CV accountability for most of data	✓	x	x	x	x
3. The principle of stages and processes in the discovery of CV	✓	No explanation to how they were revealed	No explanation to how they were revealed	x	x
4. The principle of CV explains how the respondent's main concerns are continually resolved	✓	x	x	x	x
5. The principle of research design involves the discovery of CV	✓	Hypothesis testing/ QDA	Hypothesis testing/ QDA	Hypothesis testing/ QDA	Hypothesis testing/QDA
6. The principle of core variable is transcendent of the context of data	✓	None, because no core variable	None, because no core variable	None, because no core variable	None, because no core variable
7. The principle of emerged hypothesis is modifiability as new data emerges in future research	✓	No new concise thypothesis, therefore it is not modifiable	No new concise thypothesis, therefore it is not modifiable	No new concise thypothesis, therefore it is not modifiable	No new concise thypothesis, therefore it is not modifiable
8. The principle of researcher's demonstrate full transparent transition from substantive to theoretical codes	✓	Opaque	✓	✓	✓
9. The principle of process of both sorting and constant comparison	✓	No constant comparision stated	x	x	X
OBSERVATIONS	**USED CGT**	**DID NOT USE CGT**	**DID NOT USE CGT**	**DID NOT USE CGT**	**DID NOT USE CGT**

than testing established concepts which are cited in the literature. The GT research project would first seek to focus on a very specific category of professional IT research community. There several reasons for this. Firstly, for the project to be manageable, a specific community should be selected. Secondly, this avoids the requirement to be representative of the IT community as a whole. Thirdly, since the GT researcher does not need to be representative of a context this issue is not of importance. Finally, what is of the greatest importance is for the GT researcher to understand that the latent patterns which exist within any constituency of IT professional are not only transcendent of the IT professional communities but also of many other similar communities.

Most important to note is that whichever type of data used the GT process would be the same because in GT everything is data. Apply the GT process to the data and a core variable will emerge.

Special Characteristics of Grounded Theory

There are two main "hallmarks" of the GT methodology: (1) "finding the core variable first" and then delimiting the rest of the work to what is closely related to the core variable, and (2) "the special Classic GT form of induction", which is used constantly and in all tasks with the exception of theoretical sampling and recognition of theoretical codes (Glaser, 1978; 1998). Then there is a third element: the pivotal procedure (3) involving "constantly comparing while coding (conceptualizing)", which, in a sense also applies in memo-writing and sorting, and which is always used in conjunction with (2). Then there is a fourth (4): the procedure of "memo-writing" and (5) the procedure of "sorting" (Glaser, 1978). Many other features are involved in a CGTM process, such as "interchangeability of indices" and "explication de text". Glaser (1998:24-25) describes the circumstances of his exposure to 'explication of text' at the University of Paris, for him "a major root in the discovery of grounded theory" and explains the process and purpose of explication of text' in grounded theory. It involves: ""... reading closely line by line to ascertain what exactly the author is saying without imputing what was said, interpreting it or reifying its meaning……Explication de text helps the researcher who is doing constant comparisons to generate concepts that closely fit without imputation as to what is going on in the substantive area, while at the same being able to claim the authorship of the concept he generated" …… So the message I received was to read and compare line by line and then to take credit for one's generated, emerged concepts – not disavow it and give it away...

CONCLUSION

We have explained how a classic grounded theory research study can be carried out in an information systems context in order to elucidate good practice and help prevent the misuse of authentic grounded theory. The paper has responded to call made by Matavire and Brown (2011) to further work can be done particularly to build better understanding of the process and outcomes of following a CGT approach". We have detailed on what an IS researcher should be aware of when using classic grounded theory. The researcher should be aware of pseudo CGT methodologies, epistemological stance of CGT, basic philosophy underpinning CGT, stages and procedures of CGTM, how to conduct interviews with respondent encounters including nature of meetings, choice of location; types of data collected, ethical issues, and special characteristics of CGT. Furthermore, an information systems researcher should maintain a high level of transparency when using CGT. Finally, the researcher should be aware of Data analysis, criteria for evaluating CGT and how to write papers using CGT method which focused on previously published research in IS researches.

ACKNOWLEDGMENT

We thank Dr. Barney Glaser, the co-originator of classic grounded theory methodology, from whom we have learned a lot about doing classic grounded theory research. We also thank various classic GT fellows in various countries around the globe for the promotion of use of classic GT.

REFERENCES

Banville, C., & Landry, M. (1989). Can the field of MIS be disciplined? *Communications of the ACM*, *32*(1), 48–60. doi:10.1145/63238.63241

Baskerville, R., & Pries-Heje, J. (1999). Grounded action research: A method for understanding IT in practice, *Accounting. Management and Information Technology*, *9*(1), 1–23. doi:10.1016/S0959-8022(98)00017-4

Bryant, A. (2002). Re-grounding grounded theory. *Journal of Information Technology Theory and Application*, *4*(1), 25–42.

Chen, W., & Hirschheim, R. (2004). A paradigmatic and methodological examination of information systems research from 1991 to 2001. *Information Systems Journal*, *14*(3), 197–235. doi:10.1111/j.1365-2575.2004.00173.x

Glaser, B. (1964). *Organizational Scientists: Their Professional Careers*. Chicago: Aldine.

Glaser, B. (1965). The Constant Comparative Method of Qualitative Analysis. *Social Problems*, *12*(4), 436–445. doi:10.2307/798843

Glaser, B. (1978). *Theoretical Sensitivity: Advances in the Methodology of Grounded Theory*. Mill Valley, USA: Sociology Press.

Glaser, B. (1992). *Emergence vs Forcing: Basics of Grounded Theory Analysis*. Mill Valley, CA: Sociology Press.

Glaser, B. (Ed.). (1993). *Examples of Grounded Theory: A Reader*. Mill Valley, CA: Sociology Press.

Glaser, B. (Ed.). (1995a). *Grounded Theory 1984-1994, Volume One*. Mill Valley, CA: Sociology Press.

Glaser, B. (Ed.). (1995b). *Grounded Theory 1984-1994, Volume Two*. Mill Valley, CA: Sociology Press.

Glaser, B. (1996). *Gerund Grounded Theory: The Basic Social Process Dissertation*. Sociology Press.

Glaser, B. (1998). *Doing Grounded Theory: Issues and Discussions*. Mill Valley, CA: Sociology Press.

Glaser, B. (1999). *The Discovery Of Grounded Theory: Strategies for Qualitative Research*. Mill Valley, CA: Sociology Press.

Glaser, B. (1999b). *The Future of Grounded Theory, Qualitative Health Research. Sage Publications*, *9*(6), 836–845.

Glaser, B. (1999c). The Future of Grounded Theory, Keynote address at the 4th annual Qualitative Health Research Conference. *The Grounded Theory Review*, *1*(1), 1–9.

Glaser, B. (2001). *The Grounded Theory Perspective: Conceptualization Contrasted with Description*. Mill Valley, CA: Sociology Press.

Glaser, B. (2003). *The Grounded Theory Perspective II: Description's Remodeling of Grounded Theory*. Mill Valley, CA: Sociology Press.

Glaser, B. (2004). *Glaser's explanations on ontology and epistemology*. Academic Press.

Glaser, B. (2004). *Staying open*. Unpublished paper.

Glaser, B. (2005). *The Grounded Theory Perspective III: Theoretical Coding*. Mill Valley, CA: Sociology Press.

Glaser, B. (2005b). *The Roots of Grounded Theory*. Keynote presentation delivered to the 3rd International Qualitative Research Convention, Johor Bahru, Malaysia.

Glaser, B. (2006). *Doing Formal Grounded Theory: A proposal*. Mill Valley, CA: Sociology Press.

Glaser, B. (2008). *Doing Quantitative Grounded Theory*. Mill Valley, CA: Sociology Press.

Glaser, B. (2009). *Jargonizing: Using the grounded theory vocabulary*. Sociology Press.

Glaser, B. (2011). *Getting Out of the Data: Grounded Theory Conceptualization*. Mill Valley, CA: Sociology Press.

Glaser, B., & Holton, J. (2004). Remodeling grounded theory. *Forum: Qualitative Social Research, 5*(2), Article 4. Retrieved from http:// www.qualitative-research.net/fqstexte/2-04/2-04glaser-e.htm

Glaser, B., & Holton, J. (2005). Basic Social Processes. *The Grounded Theory Review*, *4*(3), 1–21.

Glaser, B., & Kaplan, D. W. (Eds.). (1996). *Gerund Grounded Theory: The Basic Social Process Dissertation*. Mill Valley, CA: Sociology Press.

Glaser, B., & Strauss, A. (1965). *Awareness of Dying*. Chicago: Aldine Publishing Company.

Glaser, B., & Strauss, A. (1967). *The Discovery of Grounded Theory: Strategies for Qualitative Research*. Hawthorne, NY: Aldine de Gruyter.

Gopal, A., & Prassad, P. (2000). Understanding GDSS in Symbolic Context: Shifting the Focus from Technology to Interaction. *Management Information Systems Quarterly*, *24*(3), 509–546. doi:10.2307/3250972

Gregor, S. (2006). The nature of theory in information systems. *Management Information Systems Quarterly*, *30*(3), 611–642.

Levina, N., & Vaast, E. (2006). Turning a Community into a Market: A Practice Perspective on Information Technology Use in Boundary Spanning. *Journal of Management Information Systems*, *22*(4), 13–37. doi:10.2753/MIS0742-1222220402

Lyytinen, K., & King, J. (2004). Nothing at the center? Information systems as a reference discipline? *Journal of the Association for Information Systems*, *5*(6), 220–246.

Matavire, R., & Brown, I. (2008). Investigating the use of 'Grounded Theory' in Information Systems research. *Proceedings of the SAICSIT 2008 Conference*.

Matavire, R., & Brown, I. (2011, August). Profiling grounded theory approaches in information systems research. *European Journal of Information Systems*, *30*. doi:10.1057/ejis.2011.35

Munro, M. C., Huff, S. L., & Moore, G. (1985). Information Technology Assessment and Adoption: A field study. *Management Information Systems Quarterly*, *9*(4), 327–340. doi:10.2307/249233

Munro, M. C., Huff, S. L., & Moore, G. (1988). Expansion and Control of End-User Computing. *Journal of Management Information Systems*, *4*(3), 5–27. doi:10.1080/07421222.1987.11517798

Ren, Y., Kiesler, S., & Fussell, S. R. (2008). Multiple Group Coordination in Complex and Dynamic Task Environments: Interruptions, Coping Mechanisms, and Technology Recommendations. *Journal of Management Information Systems*, *25*(1), 105–130. doi:10.2753/MIS0742-1222250105

Truex, D., Holmstrom, J., & Keil, M. (2006). Theorising information systems research: A reflective analysis of the adaptation of theory in information systems research. *Journal of the Association for Information Systems*, *7*(12), 797–821.

Watts, S., & Henderson, J. C. (2006). Innovative IT climates: CIO perspectives. *The Journal of Strategic Information Systems*, *15*(2), 125–151. doi:10.1016/j.jsis.2005.08.001

Zahedi, F. M., Van Pelt, W. V., & Srite, M. (2006). Web Documents' Cultural Masculinity and Femininity. *Journal of Management Information Systems*, *23*(1), 87–128. doi:10.2753/MIS0742-1222230103

Chapter 5
Cloud-Based Geo-Information Infrastructure to Support Agriculture Activity Monitoring

Shamim Akhter
East West University, Bangladesh

Kento Aida
NII, Japan

ABSTRACT

Agriculture activity monitoring needs to deal with large amount of data originated from various organizations (weather station, agriculture repositories, field management, farm management, universities, etc.) and mass people. Therefore, a scalable environment with flexible information access, easy communication and real time collaboration from all types of computing devices, including mobile handheld devices as smart phones, PDAs and iPads, Geo-sensor devices, and etc. are essential. It is mandatory that the system must be accessible, scalable, and transparent from location, migration and resources. In addition, the framework should support modern information retrieval and management systems, unstructured information to structured information processing (IBM Info Stream, text analytic, pig & hive, etc.), task prioritization, task distribution (Hadoop), workflow and task scheduling system, processing power and data storage (Amazon S3 and Google BigTable). Thus, High Scalability Computing (HSC) or Cloud based system can be a prominent and convincing solution for this circumstance.

INTRODUCTION

Agricultural activity monitoring, enclosed quantifying the irrigation scheduling, tracing the soil hydraulic properties, generating the crop calendar, prediction on crop growth in terms of planting date, acreage, planting intensity, water stress, biomass, yield etc., is very important. It can also contribute to better policymaking, timely countermeasures, optimization of water resources distributions, damage assessment and finally to food supply security and stable market. Farmers want to know the above information in a regular basis. Researchers of agriculture also try to analyze various information about crops in order to

DOI: 10.4018/978-1-5225-0539-6.ch005

take measures if they had some problems. Particularly, when an on-going experiment covers large area such as a country, Remote Sensing (RS) plays a vital role by providing useful information over large areas. However, some information, or crop parameters, e.g. ground water level, cropping season time extent, date of emergence of crop, irrigation scheduling cannot be visible directly through RS images, which reflects a practical problem that we cannot generate or observe those parameters from remote places. To collect those data, time by time basis field experiments are required. This is a time consuming, complex and expensive procedure. To overcome such problems, indirect methods such as inverse modeling with crop model can be used to obtain those basic input parameters. One such method is the manual calibration by "trial and error" procedure, which is very subjective and time consuming and uncertainty associated with them cannot be quantified. A more robust way of inverse modeling is to combine the model with optimization algorithm. However, processing the inverse modeling with crop model has a problem in practicality, that is, they require a huge amount of processing times. It becomes necessary to introduce methods for using higher processing power such as High Performance Computing (HPC) technologies. Some protocols or tools have been developed concerning the inverse modeling techniques and their HPC implementation models. However, the interoperability protocol between those agriculture applications and existing remote sensing (RS) image processing software is also necessary to improve practicality.

Inverse Modeling Techniques

Crop models, Soil-Water-Air-Plant (SWAP) (Van Dam et al., 1997) or Decision Support System for Agro technology Transfer (DSSAT) (Tsuji et al., 1994), have capacity to simulate soil, water and crop processes and serve as crop productivity monitoring tool. Crop Assimilation Model (CAM) predicts parameters of crop models with satellite images. A new methodology was developed in (Ines, 2004), CAM-GA, to analyze the crop model (SWAP) parameters assimilation with Remote Sensing (RS) data and that parameters assimilation procedure was optimized by an evolutionary searching technique called Genetic Algorithm (GA). CAM with double layers GA, CAM-DLGA (Akhter et al., 2010b), uses directly visible multi-resolution RS images (ASTER Image Webpage, 2009) (MODIS Image Webpage, 2009) and inversely assimilates to SWAP model data for estimating the non-visible model parameters. Other similar functionality models, e.g., CAM-PSO (Kamble & Chemin, 2006) and CAM-PEST (Dorji, 2003), use different evolutionary searching techniques. However, processing the agricultural information with CAM has a problem in practicality, that is, they require a huge amount of processing times. It becomes necessary to introduce methods for using higher processing power such as High Performance Computing (HPC) technologies.

High Performance Computing Issues

Multi computer based distributed systems (Clusters and Grids) have a large processing capacity for a lower cost; naturally, choice turns towards developing HPC applications. However, it is not an easy job to port CAM in HPC environment. The application performance is significantly affected by the data and task distribution methods on the HPC and developers of agriculture or satellite image processing applications need to solve the problem of both data and task distribution, or how to distribute data and tasks among single or multiple clusters environment. The workload in HPC, the bandwidth, the processors speed, parameters of evaluation methods and data size are additional concerning factors. CAM-GA Model in successfully implemented on cluster computers and the implementation strategies are described in (Akhter et al., 2007a). Experiments and results illustrated that the distributed pixel model is performing well in cluster computing testbed (Akhter et al., 2005) (Akhter et al., 2007b). Interoperability between the agriculture application and existing satellite image processing software is necessary to improve practicality. Although users need to manually extract satellite data from some databases in the existing crop model assimilation works. GRASS GIS is an open source software/tool, which has been used to process satellite images. Inside GRASS, different modules have been developed for processing satellite images. A framework has been developed to enable the GRASS GIS environment for satellite image processing on distributed computing systems, as an example basis (Akhter et al., 2007c). Afterward, the CAM-GA application is imported as a GRASS module. High demand of distributing behavior is appealed inside the GRASS CAM-GA modules. The performance improvement of CAM is discussed in (Akhter et al., 2006) and the work includes parallel methods of CAM-GA. However, the discussion in (Akhter et al., 2006) was limited to parallel approaches only, and their performance analysis was missing. Thus, the performance impact of parallel methods on the Grid was unknown. In particular, the performance was significantly affected by task distribution methods on the Grid, and developers of RS applications need to solve the problem of task distribution, or how to distribute tasks among multiple clusters on the Grid. Three different CAM-GA parallel methods, the pixel distribution, the population distribution, and the hierarchical distribution have been successfully implemented and their performances are compared through the experiments on the real Grid testbed. These methods use GridRPC as the programming framework but ways of task distribution are different (Akhter et al., 2007b). The results show in (Akhter et al., 2007b) that the pixel distribution model exhibits the lowest communication overhead and present the best performance on some setting of the Grid. However, the hierarchical distribution model is the most preferable model to employ more parallelism in the CAM-GA application (Akhter et al., 2010a). Afterwards, the same research has been extended with real experimental data and implemented on real Grid testbed (Akhter et al., 2008).

Web Portal Implementation Issues

It is hard for agricultural researchers to understand the HPC system and to port their agriculture models. Web based technology can contribute an understandable, efficient and effective monitoring system. (Figure 1)

Some researches separately focused on the performance improvement of agricultural model activities or satellite image processing with HPC and very few of them are in the merging domains on satellite image processing, agriculture and HPC. The realistic structure of the distributed agriculture monitoring system with a web portal on HPC and its implementation for larger domain like country or province has been proposed and discussed (Figure 2) with the following schemes:

Processing satellite images automatically through HPC,
CAM HPC implementation with appropriate data and task distribution schemes.

Individually, each step (Figure 1) has been discussed and implemented with different data and application domains in (Akhter et al., 2007c) and (Akhter et al., 2008). GRASS GIS (GRASS GIS, 2009) tool has been used to process the satellite images automatically. CAM has been implemented as a GRASS module. The interconnectivity between the GRASS on HPC platform has been successfully established and then the GRASS CAM module has been implemented in HPC platform with different data and task

Figure 1. Essential Components to Build the Framework

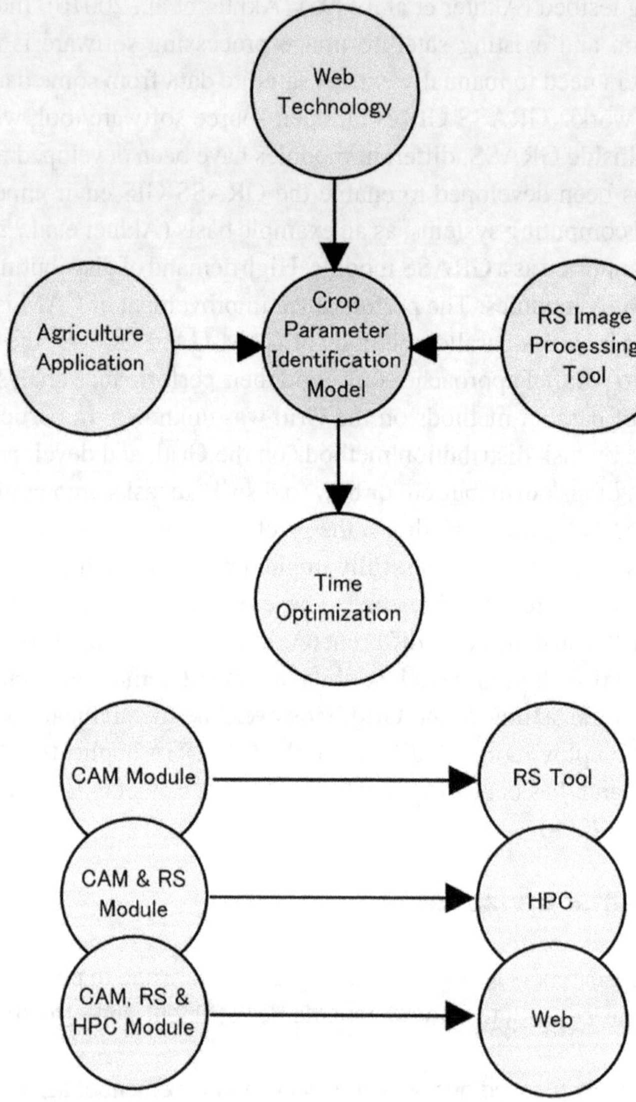

distribution methods. However, their combined framework for modeling the overall distributed agriculture monitoring scheme has not yet been established. Web based portal system on HPC for processing RS images protocol need to be established and that will merge those issues together into a unique platform.

Data Collection and Integration Issues

Data collection, integration and processing appropriate information from data repositories are additional concerns to this research implementation. This research will co-ordinate with large amounts of data. The following figures provide an example of soil map (Figure 3), and the Japan Metrological Agencies web site (Figure 4) to collect metrological data for crop model.

Figure 2. The Realistic Agriculture Framework with Web Portal and HPC

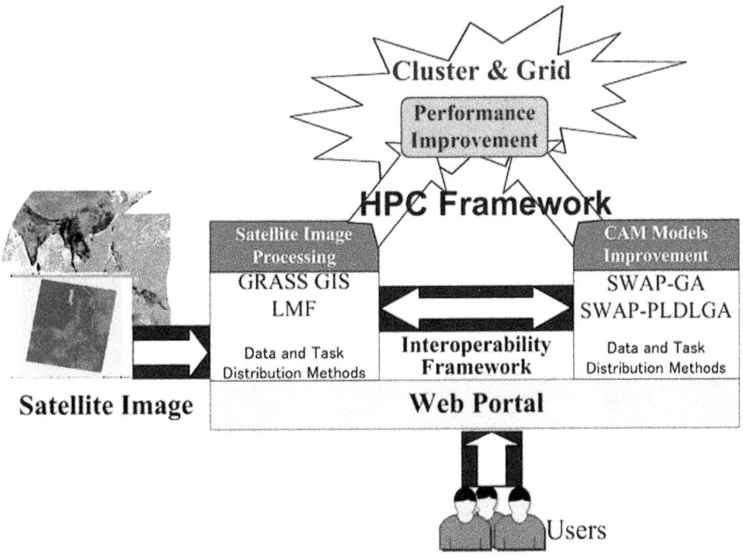

Figure 3. The Soil Map File

However, those data repositories frequently update their resources. Thus, stream computing techniques are required to capture real time unstructured data streams, convert them to the required structure, store in databases and analyze them. Google Big Query (Web5, 2014), IBM Info Sphere Streams (Web1, 2014) with support of Hardoop (Web2, 2014), pig (Web3, 2014) &/ hive (Web4, 2014) tools can be used to solve processing, analyzing and storage problems.

RESEARCH OBJECTIVES

Based on the above observation, large scale agricultural activity monitoring requires to congregate information from Remote Sensing (RS) images and that type of processing takes a huge amount of computational time and optimization on the computational time is a vital requirement. High Perfor-

Figure 4. The Japan Metrological Agencies Web Site

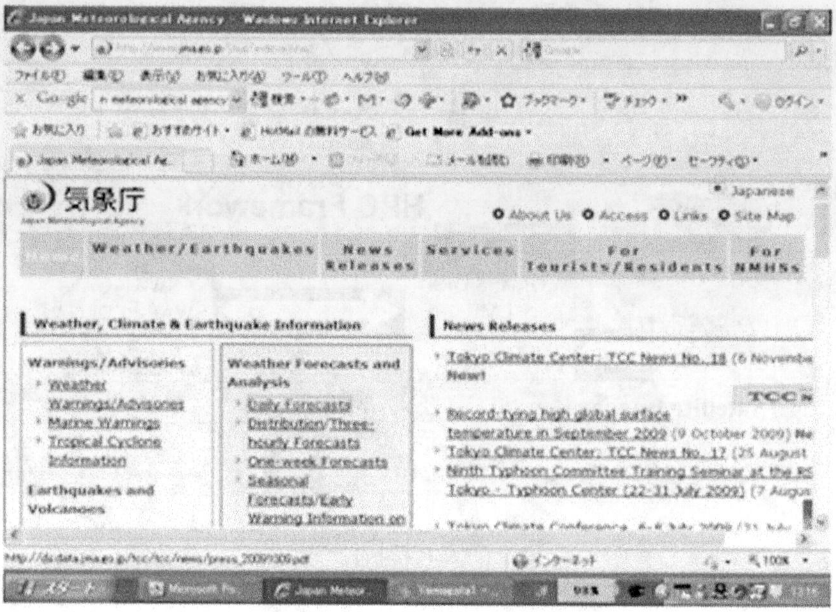

mance Computing (HPC and parallel computing) can be a solution but not enough. Agriculture activity monitoring also needs to deal with large amount of data originated from various organizations (weather station, agriculture repositories, field management, farm management, universities, agencies etc.) and mass people. Therefore, a scalable environment provided with flexible information access, easy communication and real time collaboration from all types of computing devices, including mobile handheld devices, such as smart phones, PDAs and iPads, Geo-sensor devices, such as spectrometer and turbidity sensors, and etc. are essential. It is mandatory that the system must be accessible, scalable, and transparent from location, migration and resources. In addition, the framework should support modern information retrieval and management systems, unstructured information to structured information processing (IBM Info Stream, text analytics, pig&/hive etc.), task prioritization, task distribution (Hadoop), workflow and task scheduling system, processing power and data storage (Amazon S3 and Google BigTable (Pandey et al., 2011)). Thus, High Scalability Computing (HSC) or Cloud based system can be a prominent and convincing solution for this circumstance. In addition, the following researches will include new dimensions to the above research arena:

Real applications like to monitor the agricultural activities over a city or province requires tracing the crop type in the field and for this particular crop what will be the available crop model data, weather and soil data. In the part of satellite image processing requires to solve the crop classification problem, mixed crops in an image pixel problem and calibrations of weather station data with satellite image pixels are vital requirement.

Fully automated data collection, analyze, and storage modules are required. Data security and privacy modules are additional requirements.

Moreover, in a disaster event scenario, the policy makers need to prioritize any new agriculture activity monitoring task over existing priority tasks by preemption a resource from a task and assign the resource to a new prioritized task. Our proposed "MAS Workflow Model and Scheduling Algorithm"

Figure 5. CAM Model Accuracy Framework

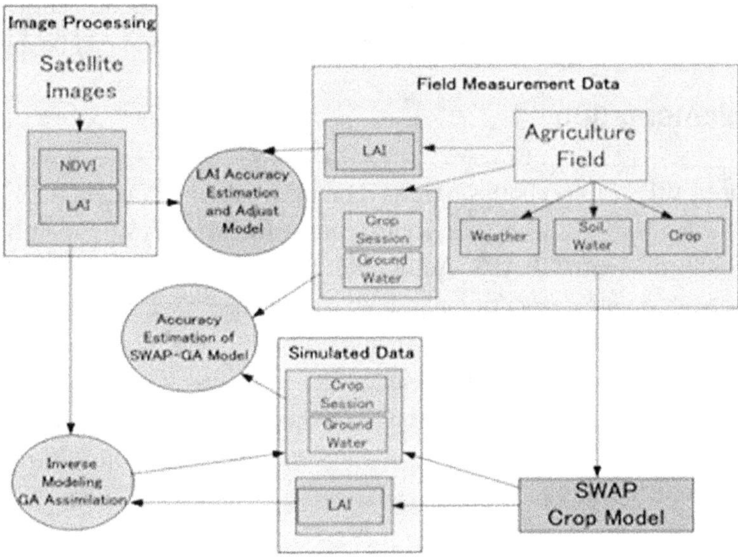

(Habiba and Akhter, 2012) can use as a supporting module on the Cloud based rich portal system to help the policy makers prioritize their tasks in a natural disaster scenario.

RESEARCH METHODOLOGY

Three different methodological issues are traced to implement the whole CAM model. First, the CAM model implementation with real field data, second web portal implementation, third the cloud based framework. Following are the description of the three implementations.

CAM Model Implementation with Real Field Data

To monitor the agricultural activities require tracing the assimilation model accuracy in a small regional crop field with known crop, weather, soil water information, and the remote sensing images.

First, the model is required to proof the workability in a small region, thereafter; the same procedure can be extended to the provincial/state level agricultural activity monitoring. Figure 5 explains the scenario to implement the crop models in small regions.

The Leaf Area Index (LAI) and other necessary crop model information need to be collected from the field experiments and then the traced LAI from satellite images will be calibrated with field LAI.

The crop assimilation model will run with inverse modeling and generate the crop unknown parameters, and then the unknown parameters values will be examined with known field experimental data.

Thus, the model accuracy will be examined and adjust with the experimental results and then extend the model for larger areas directly without being examined with filed experiments anymore. However, the following issues strictly need to be considered to get the model best accuracy: i) appropriately collect the climate, soil and water information, ii) the cropping fields dimension and iii) the availability of RS images. Hyper spectral images will be better option as it covers small dimension as well as high spatial resolution. However, it is relatively costly and thus, provides low temporal resolution. ASTER or EROS satellite images are capable to provide 15m spatial resolution images and can be an alternative option.

Web Portal Implementation

Openlayer (2009) or PyWPS (2009) both supports dynamic mapping concept with interconnection between the image repositories with necessary security infrastructure. PyWPS is relatively new concept and easy to create connection with GRASS. Additionally, it supports dynamic GRASS location creation for a given input image during the execution phase, makes the task easier for sharing input images from another repository. Two separate modules for client and server will be provided. User can select the specific region from a given image through the interface. XML based parser system will trace the user given queries and submit those queries to the server modules. Server module will execute the GRASS based CAM module with user given queries and generate unknown crop information; those are not directly extractable from RS images.

High Scalability Computing (HSC) or Cloud-Based System

The agriculture framework is going to implement with three layers of Cloud environment in a similar fashion of our recent work on Cloud system (Habiba & Akhter, 2013):

Layer 1: PaaS is the preferred model over fully outsourced data processing and handling, presumably gaining support for having clear visibility, ownership and control over all the data. At the same time, system can quickly obtain the benefits of a fully-maintained software solution on a subscription basis. With PaaS system can get full control over data encryption and security.

Layer 2: IaaS is the most important layer. Amazon EC2 can be a suitable candidate as IaaS. The main components of this proposed DMS such as Workflow Engine (WE), Workflow Scheduler (WS), Monitor, Cloud web services as well as temporary data storages are put in this layer.

Layer 3: The proposed web portal is established in SaaS layer. This proposed system will be implemented on a Cloud simulation platform, icanCloud (Nazrov, 2011). After running the model with icanCloud then the full system will be port to the real cloud testbed. Remote Sensing image processing tool ArcGIS (ArcGIS, 2012) is already implemented in Cloud. However, the challenging issue is to port the GRASS open source tool run able on Cloud environment and that solving that interoperability will provide a new research dimension for GRASS researchers.

CONCLUSION

Cloud based system will provide a scalable environment with flexible information access, easy communication and real time collaboration between RS image and agriculture data repositories. The system will be accessible, scalable, and transparent from location, migration and resources. Real time data collection and integration with intelligent system (data frame and algorithm) supports can provide successful automation and data storage. In future those data can be used for any further prediction purpose. The successful implementation of this research can be extended to city or provincial level, helps the policy makers to monitor the on field agriculture behavior and take prompt decision and action regarding any unusual condition. Through the Cloud based web portal (on hadoop infrastructure) the user can select the specific area for running the agriculture model and later the outputs will be generated regarding the specific crop parameters which are not directly visible from RS image. Users don't need to bother about the backend processes like the crop models or HSC or the distribution mechanisms. The expected time for processing the user request will be within few hours range. Multi-users can interact in the system at the same time. So, the portal must be capable to serve multi users processing requests.

REFERENCES

Akhter, S., Aida, K., & Chemin, Y. (2010). GRASS GIS on High Performance Computing with MPI, OpenMP and Ninf-G Programming Framework. *Proceeding of ISPRS 2010*.

Akhter, S., Chemin, Y., & Aida, K. (2007). Porting a GRASS raster module to distributed computing. Examples for MPI and Ninf-G. *OSGeo Journal*, *2*, 36–44.

Akhter, S., Honda, K., Chemin, Y., & Uthayopas, P. (2005). Distributed Pixel Method to speed-up the RS data assimilation of SWAP model. In *Proceedings of the MapAsia conference*.

Akhter, S., Honda, K., Chemin, Y., & Uthayopas, P. (2007). 2007a, Exploring Strategies for Parallel Computing of RS Data Assimilation with SWAP-GA. *Journal of Computer Science*, *3*(1), 47–50. doi:10.3844/jcssp.2007.47.50

Akhter, S., Jangjaimon, I., Chemin, Y., Uthayopas, P., & Honda, K. (2006). Development of a GRIDRPC tool for Satellite Images Parallel Data Assimilation in Agricultural Monitoring. *International Journal of Geoinformatics*, *2*(3).

Akhter, S., Osawa, K., & Aida, K. (2007). Performance Evaluation of Distributed SWAP-GA Models with GridRPC. IPSJ SIG Technical Reports.

Akhter, S., Osawa, K., Nishimura, M., & Aida, K. (2008). Experimental Study of Distributed SWAP-GA Models on the Grid. *IPSJ Transactions on Advanced Computing Systems*, *1*(2), 193–206.

Akhter, S., Sakamoto, K., Chemin, Y., & Aida, K. (2010). Self-organizing GA for Crop Model Parameter Estimation using Multi-resolution Satellite Images. *International Journal of GeoInformatics*, *6*(4).

Apache™ Hadoop. (2014) Retrieved from: http://hadoop.apache.org/

ArcGIS. (2012). *ArcGIS as a System for Emergency/Disaster Management, ArcGIS as a System for Emergency/Disaster Management*. Author.

ASTER Image Webpage. (2009). Retrieved from http://asterweb.jpl.nasa.gov/

Dorji, M. (2003). *Integration of SWAP Model and SEBAL for Evaluation of on Farm, Irrigation Scheduling with Minimum Field Data*. Enschede: ITC.

Google BigQuery. (2014). Retrieved from: https://developers.google.com/bigquery/

Grass GIS. (2009). *Geographic Resources Analysis Support System*. Retrieved from http://grass.itc.it/

Habiba, M., & Akhter, S. (2012). MAS Workflow Model and Scheduling Algorithm for Disaster Management System. *International Conference On Cloud Computing Technologies, Applications And Management* (ICCCTAM-12).

Habiba, M., & Akhter, S. (2013). A Cloud Based Natural Disaster Management System. *The 8th International Conference on Grid and Pervasive Computing* (GPC 2013). doi:10.1007/978-3-642-38027-3_16

Hive. (2014). Retrieved from: http://hive.apache.org/

Ines, A. V. M. (2004). *Improved Crop Production Integrating GIS and Genetic Algorithms.* (PhD Thesis). Asian Institute of Technology (AIT), KhlongLuang, Bangkok, Thailand.

InfoSphere Streams. (2014) Retrieved from: http://www.ibm.com/developerworks/bigdata/streams/

Kamble, B., & Chemin, Y. H. (2006). *GIPE in GRASS Raster Add-ons.* Retrieved from http://grass.gdf-hannover.de/wiki/

MODIS Image Webpage. (2009). Retrieved from http://modis.gsfc.nasa.gov/

Nazrov, E. (2011). *Emergency Response management in Japan*. Final Research report. ASIAN Disaster Reduction Center: FY2011A Program.

OpenLayer. (2009). *OpenLayers: Free Maps for the Web*. Retrieved from http://openlayers.org/

Pandey, S., Karunamoorthy, D., & Buyya, R. (2011). *Workflow Engine for Clouds,CloudComputing: Principles and Paradigms*. New York: Wiley Press.

Pig. (2014). Retrieved from: https://pig.apache.org/

PyWPS. (2009). *Python Web Processing Service*. Retrieved from http://pywps.wald.intevation.org/

Tsuji, G. Y., Uehara, G., & Salas, S. (1994). *DSSAT v3.0*. Honolulu, HI: University of Hawaii.

Van Dam, J. C., Kroes, J. G., Groenendijk, P., Hendriks, R. F. A., & Jacobs, C. M. J. (1997). *SWAP Model*. Retrieved from: http://www.swap.alterra.nl/

Chapter 6
Profiling Collaborating Partners in E-Learning in Developing Countries

Titus Tossy
Mzumbe University, Tanzania

Wallace Chigona
University of Cape Town, South Africa

ABSTRACT

When Information and Communication Technologies (ICTs) are rapidly revolutionizing, online learning is undergoing dramatic change as powerful new content types, technologies, and services are being adopted for education and training. Traditional learning needs to be supplemented by virtual learning to meet the rapid population growth and counter the lack of traditional higher education institutions. This is particularly true in least developed countries faces critical challenges to meet the new demands in higher education with its ever increasing population, remote and scattered areas, non-availability of resources (including infrastructure, few institutions, few experts or teachers, etc.), and limited funding. While it has been known that developed country are enjoying technology dance for so long time, least developing countries are either not or just joining the dance especially in e-learning. In other hand, there are enough literatures on the e-learning strategic alliances or collaborations/partnerships in developed countries for the purpose of dominating in Education delivery worldwide, and there is no or little literature about existence or non-existence of strategic alliances or collaboration on the e-learning development in the least developed countries. This is why this paper brought forward, enlightening the non-existence or existence of e-learning development strategic alliances, which has been sought to solve dual problem, scarcity of resources and increase the certification and recognition of online degrees in developed countries.

DOI: 10.4018/978-1-5225-0539-6.ch006

INTRODUCTION

Least Developed Countries (LDCs or 4th World) represent the poorest and weakest segment of the international community characterized by extreme poverty or low income, the structural weaknesses of their economies and the lack of capacities or human resources related to growth (UN-OHRLLS, 2009; Wikipedia, 2009;). A country may 'graduate' if it passes these criteria. It has been known that LDCs are acute susceptibility to external economic shocks, natural and man-made disasters and communicable diseases. According to UN-OHRLLS (2009), Thirty three countries in Africa are the least developing countries including Angola, Benin. Burkina Faso, Burundi, Central African Republic, Chad, Comoros, Democratic Republic of the Congo, Djibouti, Equatorial Guinea, Eritrea, Ethiopia, Gambia, Guinea, Guinea-Bissau, Lesotho, Liberia, Madagascar, Malawi, Mali, Mauritania, Mozambique, Niger, Rwanda, Sao Tome and Principe, Senegal, Sierra Leone, Somalia, Sudan, Togo, Tanzania, Uganda and Zambia.

The least developed countries are concentrated with high population, lacking physical infrastructure such as schools, hospitals roads, etc which point out that there is necessary for something to be done or try technology as a tool for development. The introduction of e-learning may be of the utmost importance (Juhary, 2005) since least developed countries' population is too high and pressured by education for all by 2015, especially for ages 17-24 attending the higher education. Equipping these youths with the right skills will help to make their countries 'graduate' and not become 'least developed countries'. E-learning is flexible, making our young people study as well as be working (Juhary 2005). Generally E-learning is noted as a tool to cater for the higher population with few physical building for education. While e-learning is so important(Juhary, 2005), yet there are so many challenges(Ricketts & Gunga, 2007; Ndume et all, 2008; Juhary, 2005; Puteh, 2008;) in its development in least developed countries including people's perception and acceptance, Quality and certification of e-learning, intellectual investment, management support, Technology infrastructure, lack of experts and much more are the three pillars of the ICT revolution that is connectivity, capacity and content, are yet to be realized in least developed countries (Ricketts & Gunga, 2007). Partnerships bring together innovative minds and solve such kind of problems (Ricketts & Gunga, 2007) and avoid dual problem solving within the same country. While by partnership other refers to Public-private partnership other refers to Strategic alliance and many more. This paper looks to find out the exactly term been used to refer collaboration within e-learning development, as well as to see the existence and non-existence of these terms in least developed countries.

The next section will reveal the literature review, followed by methodology, discussion and finally conclusion which will suggest further research area to be considered for researchers.

REVIEW OF LITERATURE

E-Learning, Africa, and LDCs

In contrast to traditional modes of education delivery, e-learning carried three important principles: Scalability, access, and timeliness which makes e-learning to be very important for the 21st century. E-learning can be scaled almost infinitely at little additional cost, making it available for access anywhere there is internet connection and continually updated with new information and knowledge relatively cost effectively (Clarke & Herbmens, 2001)

Figure 1. Open, flexible, and distributed e-learning (Khan, 2005)

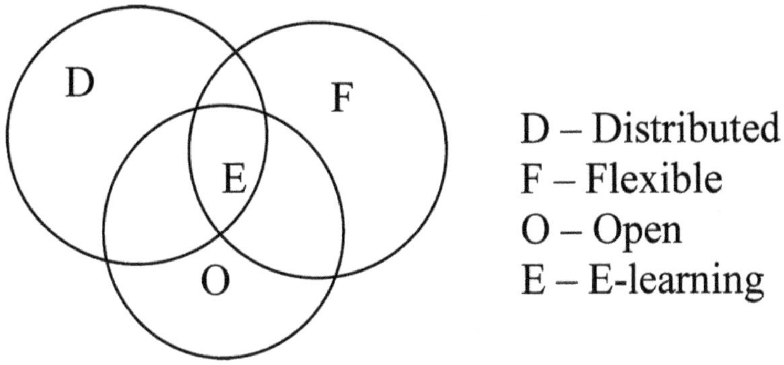

However there are many definitions of e-learning (Henze & Nejdl, 2004), Khan (2005) define e-learning as "an innovative approach for delivering well-designed, learner-centered, interactive, and facilitated learning environment to anyone, anyplace, anytime by utilizing the attributes and resources of various digital technologies along with other forms of learning materials suited for open, flexible, and distributed learning environment". Open learning allows student to learn in their own time, pace and place (Calder & McCollum, 1998) that means if is open and flexible then the student is able to say where, how and when the learning place will take place (Ellington, 1997). In other hand, Saltzberg and Polyson (1995) defined distributed learning as:

An instructional model that allows instructor, students, and content to be located in different, non-centralized locations so that instruction and learning occur independent of time and place...the distributed learning model can be used in combination with traditional classroom-based courses, with traditional distance learning courses, or it can be used to create wholly virtual classroom.

The e-learning has been delivered through internet technologies due to its ability to be a device, platform, time and place independent. To more clarify the terms open, distributed and flexible learning see the Figure 1.

E-learning offers important contributions by accommodating diversity and improving the quality of teaching and learning and to be more open, less constraining and narrowly prescriptive.

Africa is looking at the new century as an era that present challenges of rapid advancement in ICTs, which forces Africa to continually look more closely at the education demands (Juhary 2005). LDCs are experiencing an explosion of technologies including mobile phones sales. The technology explosion, witnessed LDCs benefiting from ICTs and more people joining the dance for example 28% of Tanzanians are connected. In other hand, due to lack of building and with the high populations to attend education, e-learning is sought to be the solution to the problem. According to Unwin (2008) indicates that there is a wide variety of different e-learning practices in Africa; e-learning still very much in its infancy across the most of the content and there is much enthusiasm amongst African people for developing the potential of e-learning in their countries. Regardless of the efforts within countries there are various constraints or challenges in seeking to develop and implement e-learning practices including lack of infrastructure- connectivity, electricity, computers, etc. (Unwin, 2008; Ricketts & Gunga, 2007), appropriate training and capacity development (Ricketts & Gunga, 2007; Unwin, 2008), lack of relevant digital content (Ricketts & Gunga, 2007; Unwin, 2008), cost of implementation (Unwin,2008). The problem in Africa is generally not just the near absence of e-learning programmes but also the inability of students to gain access even to the few that do exist (Ricketts & Gunga, 2007). However due to the emerging technologies of blackberries phones and other technologies access would not such big problem. The e-learning development efforts are so fragmented within the same country, same culture, same economic and same program of study, making the institutions dual solving the problem and scramble against little resources they have, because most of the e-learning development efforts are being done individually (by the institutions) rather than collaborative efforts. For together we can make e-learning happen without scrambling against little resources we have as Ricketts and Gunga (2007) noted that all the e-learning implementation problems can be solved through collaborative efforts, or partnerships such as Public-private partnerships. However it is not clearly known does only one type of partnership exist? Is it possible for other types of partnerships to exist? This is what the paper is trying to discover, read the next section to know about partnerships.

Challenges Facing E-Learning Development

The benefits of integrating e-learning in developing countries are not dispute (Ahmed, 2004; Stepanyan, 2008), it is seen as vital in narrowing the digital divide. While developed countries are enjoying the e-learning dance, due to difference in economic issues and cultural differences, Least Developed Countries are faced by various challenges in joining the dance including: lack/Imperfect ICT infrastructure (Clarke & Gunga, 2007; Stepanyan, 2008; Ahmed, 2004), Lack of human resources-experts/Educators (Clarke & Gunga, 2007; Stepanyan, 2008; Ahmed, 2004), Lack, lack of funding, Lack of recognition of e-learning certificates (Clarke & Gunga, 2007;Ndume et al., 2008; Ahmed, 2004), Intellectual investment (Ndume et al., 2008), management support (Clarke & Gunga, 2007; Ndume et al., 2008), resources availability and accessibility (Ndume et al., 2008), New Culture influence/for learning (Ndume et al., 2008; Stepanyan, 2008; Ahmed, 2004), Language differences and lack of local content (Stepanyan, 2008; Ahmed, 2004), security and quality assurance of online programs (Ndume et al., 2008; Ahmed, 2004), Technology support for e-learning (Ndume et al., 2008; Clarke & Gunga, 2007; Ahmed, 2004), Corruptions (Ahmed, 2004).

Corporate Developments and Strategic Alliances in E-Learning

The 21st century is witnessing extensive corporate developments in e-learning industry, which is hopefully stimulated by the combination of the demand for access to continuous, flexible education and training, and the arrival of the increased bandwidth of more powerful information and communication technologies. Strategic alliances of leading universities, e-learning and technology companies are forming to provide online delivery of sophisticated, media rich, interactive education and training on a global basis (Clarke & Hermens,2001). While other side of the world, Governments are struggling to respond to education for all through e-learning, other part of world(especially developed countries) are using e-learning as a tool to get more profit through offering education globally because they have recognized that education is becoming the World's economy, as Clarke and Hermens (2001) noted that hundreds of e-learning companies are competing in this new marketspace with content, technology and services, for example major companies are establishing their own corporate universities while other higher learning institutions are forming strategic alliances.

As the education and training sector continues to grow it is increasingly subjected to pressures to change (Clarke & Hermens, 2001), especially in least developing countries. Among the factors fueling the drive towards alternative forms of delivery of education, Clarke and Hermens (2001) noted the following:

- *Advance of globalisation with a demand for world class products, services and technical infrastructures;*
- *Emergence of post-industrial information age, with sophisticated communications systems, and the explosive growth and distributed nature of new knowledge;*
- *Demand for greater access to tertiary education and the need for lifelong learning created by rapid changes in the economy;*
- *Growing reluctance on the part of governments to fund the increasing demand for higher education;*
- *Resultant increasing costs of higher education and the growth of the earner-learner market (as distinct from the learner-earner market);*
- *Dissatisfaction of industry with the responsiveness of traditional providers (Cunningham et al., 2000 , p. xv)*

Both in developed and developing nations, universities, companies are wrestling to take advantage of e-learning. Regardless of little resources available, universities and companies in developing countries are scrambling to get a portion of it to build individual e-learning. For example they wrestle against skilled experts, funding, and even medial presence. Most of the fragmented e-learning development efforts finally fail due to lack of continuous funding, certification or public recognization and lack of positive perception from the entire public.

In other hand, knowledge workers require flexible and continual learning rather than occasional, highly structured courses, that is learning anywhere, anytime for anyone (Clarke & Hermens, 2001). In a volatile market environment of often speculative investment (Clarke & Hermens, 2001), with dramatic advances in the technology of delivery, lack of funding, few experts and poor infrastructure there is a need for universities to engage into collaborative kind of e-learning development. Clarke and Hermens (2001) classified this type of collaboration as:

- **Strategic Alliances:** *The sudden emergence of ambitious global networks of leading universities, e-learning companies, and technology companies engaged in strategic alliances to extend online infrastructures nationally or globally.*
- **E-Learning Companies:** *A rapid proliferation of e-learning companies, innovatively disaggregating the education and training value chain by inserting new modes of electronic delivery at critical stages.*
- **Corporate Universities:** *The establishment of a growing number of "corporate universities" by large companies investing in their own delivery systems for education and training;*

According to Rickett and Gunga (2007), the collaboration and partnerships of any kind will enhance the e-learning development. The problem of the digital divide can only be solved by harnessing the power of partnerships (Utsum, 2005): strategic alliances, e-learning companies or corporate universities. Among of the advantages of such partnership include collaboratively addressing educational and developmental issues, strengthening technical and human capacity building for teaching, learning and research, combating lack of funding, experts and the rest (Rickett & Gunga, 2007). Because all higher learning institutions existing in the least developing countries are there for responding to 'education for all' and all are classified as non-profit making, then there is need for them to create strategic alliance to overcome the constraints in e-learning development. Strategic alliance brings together innovative minds including experts from governments, business, civil society, academia and the international organization, and hence successful e-learning in the country. The partnership within same country, culture, environment, economy is very important.

While there are different models of public-private partnerships in education in Africa, there is little institution-institution partnership (Institution strategic alliance). There is need for universities in least developing countries to form strategic alliance to bring together their efforts in tackling the barriers to e-learning development.

Strategic alliance between education institutions will enhance the product offering, expand distribution channels, and explore new market segments. Regardless of these importance, there is little been seen in least developed countries, where universities instead of joining together, they scramble and continue fighting for resources. The opposite is true for developed world, where there are viable strategic alliances efforts to bring e-learning to the world, and have strong immense influence and penetration in e-learning. It is believed that is least developed countries form their own strategic alliance for e-learning development then the accreditation of the new online universities or programs and exclusivity of partnerships will be essential for differentiation among e-learning competitors (Hambrecht & Co, 2000, p. 24).

Examples of Strategic Alliances

Clarke and Hermens (2001) noted that Cataloguing the increasing array of strategic alliances in e-learning is not an easy business, due to the speed of creation of these alliances, the complexity of the relationships between multiple partners, and their tendency to rapidly transform into something different from what was originally conceived (together with a high failure rate and dismantling of alliances, often as part of the restructuring of new strategic alliance formations). However, it is possible to highlight the characteristics of some of the major strategic alliances in e-learning presently. A good example noted by Clarke and Hermens (2001) includes:

- **Unext:** *(*http://www.unext.com*) with headquarters in the USA is a strategic alliance of universities including Columbia, Stanford, Chicago, Carnegie Mellon and the London School of Economics which intends to provide MBA business courses to multinational companies through the newly formed online Cardean University*
- **Fathom:** *(*http://www.fathom.com*) is an e-learning portal established by a group of leading universities and cultural institutions comprising Columbia University, London School of Economics, Cambridge University Press, the British Library, Smithsonian Institution's National Museum of Natural History, and the New York Public Library. These eminent sources will use the portal for aggregating, hosting and distributing content to provide e-learning opportunities on a wide range of subjects to a diverse life-long learning population. Fathom's tools and resources are overseen by an advisory board drawn from its partners, and also offers an access channel for the purchase of books, periodicals, and articles.*
- **Universitas 21:** *(*http://universitas.edu.au *) is company incorporated in the UK with a network of 18 universities in ten countries which collectively enroll approximately 500,000 students each year, employ 44,000 academics and have a combined operating budget of $US9 billion. It is hoped*

this network will enable the pursuit of agendas beyond the individual capabilities of members. Universitas 21 has three different e-learning companies for content, intellectual property rights, and delivery system.

- **Scottish Knowledge:** *(*http://www.scotish-knowledge.com *) is an alliance of Scottish universities and businesses with Scottish Knowledge acting as a broker and global commercial marketing partner. Launched in 1999 it has developed materials as needed, and focused on key markets. It offers modular programmes built from multi-source suppliers to provide tailor-made solutions for clients.*
- **Western Governors University (WGU):** *(*http://www.wgu.edu*) in the USA is a partnership between 40 universities who are sharing resources. The emphasis is upon "competency-based assessment, testing and credentialling performative knowledge rather than discursive or expository knowledge" (Cunningham et al., 2000 , p. 46).*
- **Open University of Catalonia (UOC):** *The UOC is a public-private partnership operating with a network of spin-off and joint venture businesses funded by state and private capital. It employs virtual teachers from universities in Spain and globally, and has grown to 20,000 students and 16 on-line virtual courses at bachelor, masters and doctorate levels. Based in Barcelona it covers Spain, the Spanish speaking Americas, and is moving to provide English language offerings globally.*
- **Pensare:** *Is in a collaborative partnership with over 45 "major brandholders" among academic institutions including Harvard Business School Publishing, Duke University, Wharton School of Business, University of Southern California, corporate universities, and leading business authors. Pensare offers an interactive e-learning solution integrating content, technology and services aiming to create knowledge communities tailored to specific organisations.*

With all of the examples above, very clear that it is possible to develop strategic alliance for e-learning development and avoid dual problem solving and reaching the very large number of students. However it is little or none realized in least developing countries in Africa in such kind of partnership, rather than in the same country more than 3 or more universities scrambling to get their e-learning practices going.

Methodology

Research Instruments

The research methodology employed in this study is literature review from both library and online materials using the keywords: partnership, e-learning, collaboration, least developed countries, strategic alliances. The idea was to find out what are the whole idea about partnership/collaboration/alliance in least developed countries.

Data Analysis

In order to identify the non-existence and existence of e-learning development partnerships/collaborations/alliances in least developed countries, the literatures where thoroughly studied and analysed.

Results and Discussion

E-learning is a key-stone in the creation of the nationalizing environment as a key to progress in the villages and districts. Literatures noted that e-learning has the potential to enable Africa, especially least developed countries to achieve educational for all, because are faced with shortage of trained teachers, lack of physical buildings. In recent years, e-learning has gained strong acceptance in least developed countries. Furthermore, it has been recognized that universities are able to collaborate and form partnerships to combat against challenges facing these countries. E-learning has made a shift from centralised classroom-based education towards distributed e-learning courses (Alonso et al., 2005). E-learning promote team-learning pedagogy (Clarke & Gunga, 2007). E-learning has change teachers role, as they become expert questioner rather than providers of answers and designers of student learning experiences rather than providers of contents (Clarke & Gunga, 2007). Many African student start seeing the important of online education, and expect technology to have a significant role in their learning (Juma, 2003).

As Clarke and Berhemens (2001), there are various types of partnerships can be deployed in least developed countries, including *Strategic alliances, E-learning companies, and Corporate universities.* Strategic alliances enables the leading universities, e-learning companies, and technology companies engaged in collaboration to extend online infrastructures nationally. E-learning companies, profit making companies comes together and innovatively disaggregating the education and training value chain by inserting new modes of electronic delivery at critical stages. Corporate universities, large companies join their hands and form this kind of universities.

Stakeholders' Perceptive

It has been noted that the number of groups (teachers, students, workers, etc.) increasing to realize the potential of e-learning. The only obstacle is the recognization of the online programs by respectively education bodies in the respective countries. This could be facilitated by the fragmented e-learning efforts going on, lowering quality of e-learning programs and lacks trust. Teachers and learners are shifts their preferences from blended learning to purely online learning, especially in the turbulent of few physical buildings and lack of teachers in the sea of advanced technologies. Research by Clarke and Gunga(2007) shows that collaborative efforts(strategic alliance) will increase the awareness of online programs, and combat the connectivity, and other challenges.

Collaboration/Partnership/Alliances in Africa

E-learning networks and collaboration could provide the alternative way for cost sharing, and for leveraging access to education (Ahmed, 2004). Ahmed (2004) revealed some of the types of networks including Twinning, Remote Satellite campus strategy, consortia, spin-offs and outsourcing, and many more. The literature shows the Public-Private Partnership (PPPs) as one of well-known type of partnerships going on in Africa, with the example of African Virtual University, lacking evidences of Institution-Institution Partnership (IIPs) within the same culture, country, political influence and courses. Hence it shows that for the past years, much has been promoted for regional partnerships rather than country partnerships or to say village or district partnerships with the same culture and environment. Utsumi (2005) noted that the only way to face e-learning challenges in Africa is through collaborations. However it did not note which partnership style, but it is clear that any type of the partnership will be valuable tool to both bring e-learning to the hands of our people, and nationalize our programs for better gaining certification and recognition. As Clarke and Gunga (2007) noted that partnerships caters for more learners with a lower outlay on resources, and further the support to teaching, learning, research and administration (Johnson, 2005).

The literature has revealed that Ethiopia under the UniversityNet programme has managed to bring all 12 Ethiopian universities together and networked-however no research done on its success. Under Schoolnet more than 500 secondary schools including the Technical and vocational education and training schools are networked (Reif, 2005) with no evidence if specifically for e-learning. Furthermore, Somali tertiary institutions has joined together and form online distance learning initiatives as partnership to enable students to attain accredited university qualifications through partnerships with other institutions-to safe guard their identity. With this literature it shows there are one type or another type of partnerships for the e-learning development in Africa. However other countries like Tanzania are not joined their efforts to form partnership rather than every institution works in its own. Issues that can be resolved through partnerships are many and include policy formulation for online education, training online educators, content development, network support systems, content delivery and liaison with vendors of VLEs (Clarke & Gunga, 2007).

CONCLUSION

In this study, it has been shown that the only way to face the challenges in e-learning development is through partnerships, at the level of village, district and country region and national. This can be easy because are within same culture, political, economy and the rest, rather than crossing national borders.

In additional it has been revealed that developed countries are much far ahead in collaboration and partnerships in all three types of partnerships than developing countries. However, little has been shown that even least developed nations are forming partnerships- with little literature about their success. The current challenge, therefore is to develop and nurture value-added partnerships in which members have complementary skills and are ready to establish a local presence in Africa (Clarke & Gunga, 2007).

This study has revealed areas of further researches to help least developed countries. First, the research should be done to find out why the universities within same culture, politics, and economic situation are not working together (forming partnerships- strategic alliance) for the success of e-learning Development? Because it is clearly known that the opportunities for using ICT in education in one country differs from those in other due to education system, diverse culture, social norms, change the perception of education and politics (Stepanyan, 2008). Hence if such environments remain constant, we hope there will be partnerships existing. Secondly, it is needed to know for those managed to form partnership in national level, what are the challenges faced? This paper hopes that researchers will come up further this research.

REFERENCES

Ahmed, A. (2004). Making Technology Work for the Poor: Strategies and policies for African sustainable development. *International Journal of Technology Policy and Management*, *4*(1), 1–17.

Beebe, M. (2003). Partnerships, Alliances and Networks for E-learning. In AfricaDotEdu. IT Opportunities and Higher Education in Africa. New Delhi: Tata McGraw Hill.

Calder, J., & McCollum, A. (1998). *Open and flexible learning in vocational education and training*. London: Kogan Page.

Clarke, T., & Hermens, A. (2001). Corporate developments and strategic alliances in e-learning. *Journal of Education and Training*, *43*(4/5), 256–267. doi:10.1108/00400910110399328

Ellington, H. (1995). Flexible learning, your flexible friend. In C. Bell, M. Bowden, & A. Trotts (Eds.), Implementing flexible learning (pp. 3-13). London: Kogan Page.

Henze, N., & Nejdl, W. (2004). A logical characterizing of adaptive education hypermedia. *Journal of New Review in Hypermedia and Multimedia, 10*(1).

Juhary, J.B. (2005). A step towards e-learning: some pedagogical issues. *International Journal of Pedagogies and Learning*, *1*(1), 48-58.

Juma, M. (2003). An evolution of an existing institution. In *The university – Current challenges and opportunities*. Nairobi: African Virtual University. Retrieved February 29, 2009, from http://www.unesco.org/iiep/virtualuniversity/home.php

Ndume, V., Tilya, F., & Twaakyondo, H. (2008). Challenges of Adaptive elearning at higher learning institutions: A case study in Tanzania. *International Journal of Computing and ICT Research*, *2*(1), 47–59.

Puteh, M. (2008). E-learning implementation in Malaysian Universities: The Universiti Teknologi Malaysia Experience. *Proceedings of the 3rd international conferences on e-learning*. University of Cape Town.

Ricketts, I. W., & Gunga, S. O. (2007). Facing the challenges of e-learning initiatives in African universities. *British Journal of Educational Technology*, *38*(5), 896–906. doi:10.1111/j.1467-8535.2006.00677.x

Saltzbert, S., & Polyson, S. (1995). Distributed learning on the world wide web. *Syllabus*, *9*(1), 10–12.

Steiner. (2005). *African tertiary institution connectivity survey*. Nairobi: African Virtual University. Available online at http://www.avu.org/documents/partnership%20connectivity%20report-%20revised%2017-10-04.pdf

Stepanyan, K. (2008). *Challenges and opportunities for practicing e-learning globally*. Buckinghamshire Cilterns University College.

UN-OHRLLS. (2009). *Least Developed Countries: about LCDS*. The United Nations Office of the High Representative for Least Developed Countries. Available online at http://www.unohrlls.org/en/ldc/25/

Unwin, T. (2008). Survey of e-learning in Africa- based on a questionnaire survery of people on the e-learning Africa. *Database*, 2007.

Utsumi, Y. (2005). *Connect the World official launch*. Available online at http://www.itu.int/partners/pdf/SG_preso.pdf

Wikipedia. (2009). *Least developed countries*. Wikimedia Foundation, Inc. Available online at http://en.wikipedia.org/wiki/Least_Developed_Countries

Chapter 7
A Comparative Evaluation of E-Learning Adoption in Private and Public Higher Education Institutions:
A Tanzania Survey

Mpe Paulo Mwamahusi
Tanzania Institute of Technology, Tanzania

Titus Tossy
Mzumbe University, Tanzania

ABSTRACT

This paper examines e-learning adoption in Higher Education Institutions (HEIs). The paper examines whether there is a difference between the Private and public HEIs e-learning adoption. The rationale for the examination stands from the fact that scholars are of option that there is different between private and public HEIs e-learning adoption. From an empirical survey conducted in Tanzania, this paper evaluates both staffs and students on the current situations, factors affecting, similarities and differences of e-learning adoption between private and public HEIs. The research reveals that, despite of the government efforts to support e-learning initiative by introducing national ICT polices to boost the level of adoption; very few private HEIs have adopted e-learning compared to public HEIs. The limited initial investment, lack of expertize, lack of guaranteed electrical power, poor strategic change management plans, and lack of innovative ideas are the barriers to e-learning adoption.

DOI: 10.4018/978-1-5225-0539-6.ch007

INTRODUCTION

The use of ICT in teaching and Learning is popular known as e-learning (Khan, 2005; Twaakyondo, 2008; Wentling, et al. 2000; Galagan, 2000). The adoption of e-learning in Higher Education Institutions (HEIs) has enacted opportunities for education sector (Papalewis, 2003; Woodward & Pilling, 1993). While Mafu (2004) argues that the use of e-learning offers students and teachers learning and teaching opportunities in HEIs, Cruthers (2008) and Watson et el, (2004) claims that e-learning allows remote training participation and support collaborative learning process, it reduce geographical barriers and minimize the tuition cost to affordable level.

The difficulties to share the training materials, higher operation cost, delivery time limitations, impossible to launch new training sessions without physical construction of new buildings, and limitations on the number of students enrolled are the factors promote e-learning adoption in HEIs (Noe, 2005; CIRTL Network, 2012; OPPAPERS, 2012; BLR-Employee Training Centre 2012; Chiemelie, 2012; Scribd, 2005). While Twaakyondo (2008) argues that the colleges and universities in developing countries are scarcely available to customers due to the geographical locations and expensive training cost, Mgendi (2010) and Twaakyondo (2008) argues that there is difference in adoption between public HEIs and private HEIs. In order to examine the e-learning adoption difference, the paper first examines the status of Tanzanian education. Second, the paper examines through an empirical survey, the current situation and factors affecting e-learning adoption. Third, the similarities and differences of e-learning adoption between private and public HEIs are discussed. The paper finally attempts to propose possible solutions that will allow fair play ground for the adoption of e-learning as well as fasten the e-learning adoption.

E-LEARNING

Wentling et al. (2000:5) define e-learning as:

The acquisition and use of knowledge distributed and facilitated primarily by electronic means. This form of learning currently depends on networks and computers but will likely evolve into systems consisting of a variety of channels (e.g. Wireless, satellite), and technologies (e.g. Cellular phones, etc.) as they are developed and adopted. E-learning can take the form of courses as well as modules and smaller learning objects. E-learning may incorporate synchronous or asynchronous access and may be distributed geographically with varied limits of time. (Wentling et al., 2000:5)

E-learning captures a wide range of terms (Albert & Mori, 2001) referred to as *'labels'* which have been used to describe the concept of e-learning. These labels include, but are not limited to *Web Based Learning (WBL), Web Based Instruction (WBI), Web Based Training (WBT), Internet Based Training (IBT), Online Resource Based Learning (ORBL), Advanced Distributed Learning (ADL), Tele-Learning (T-L), Computer-Supported Collaborative Learning (CSCL), Mobile Learning (M-learning or ML), Nomadic Learning, Off-Site Learning* (Collis, 1996; Khana, 2005; Yieke, 2005; Bates, 2001; Dam, 2004; Goodear et al., 2001; Pegler & Littlejohn, 2007; Dabbagh et al., 2000; Barbara, 2002, 2004; Cramer et al., 2000; Salzbert & Polyson, 1995; Schreiber, et al., 1998; Schank, 2001; Howard, 2003; and Singh, 2003). The e-learning term is used interchangeably with other related terms such as online learning, virtual learning, and web-based learning (Twaakyondo, 2004).

While The use of e-learning has the added value of flexibility *(" anywhere, anytime, anyplace"),* E-learning facilitates both learner engagement and the engaging of experiences (Uys, 2004; Meyen, 2000; 2002). Meyen (2002) demonstrate how e-learning helps to overcome the traditional barriers to education delivery. These barriers include lack of physical infrastructure, lack of qualified teaching staff, absence of adequate education budgets, and the failure of traditional pedagogy and curricula. East African countries are characterized by these barriers (Ndume et al, 2008). The failure of the government's efforts in building physical classrooms has created an opportunity for innovative education delivery via e-learning (Yieke, 2005). As Alavi and Leidner (2001) argues that e-learning's importance will grow right across the educational spectrum from primary to HEIs, the e-learning implementation in Tanzania HEIs is taking place despite the various outlined barriers. The e-learning implementation differs from one HEI to another.

TANZANIA HIGHER EDUCATION STATUS

According to TCU (2010), the education sector in Tanzania has grown drastically for the past fifty (50) years; this has been due to an increase in the number of Higher Education Institutions (HEIs). The students' enrolment has increased tremendously since independency. As MoEVT (2011) states that the number of students enrolled in HEIs increased drastically. In 1961, Tanzania had 1,737 students enrolled in 4 HEIs, while in 2011 a total of 244,045 students in 358 HEIs (MoEVT, 2011). This emanated from free markets which encourages establishment of both private and public HEIs, backed by various government policies on education sector such as Vision 2025, ICT Policy and Higher Education Master Plan (HEMP), which enhance the establishment of both private and public HEIs (Maliyamkono, 2006:396-445). Despite the fact that the number of HEIs has increased since 1961, the pace of increase of students compared to overall national population growth doesn't match the enrolment offered by these institutions (Maliyamkono, 2006). This is due to limitation on enrolment capacity, geographical constraints, cost of education, lack of enough infrastructures, lack of qualified personnel and lack of innovative ideas (Chiemelie, 2012). In the light of those challenges, e-learning is sought to be the ultimate solution in which the enrolment does neither depend on the infrastructure nor geographical locations (Noe, 2005). As MoEVT (2011) argues that the HEIs should deploy e-learning for their day to day training activities, in order to minimize training cost and to remain competitive in the market. Furthermore, while MoCT (2003) articulates the need for harnessing ICT opportunities to meet the vision 2025 goals by blending strategic ICT leadership; ICT infrastructure; ICT Industry through Human Capital, MoEVT (2007) stipulates that Tanzania needs national e-learning sensitization by stressing the effort on applications such as distance education, e-learning, m-learning and blended learning.

E-LEARNING AT HEIs IN TANZANIA

Dr. Gajaraj Dhanarajan (2001:9), President of the Commonwealth of Learning, argued that:

One would be foolish to question the importance of the internet and www for education in this new decade; at worst it has the ability to connect communities of learners and teachers and at its best it could very well be the tool that education has been waiting for these past thousands of years; its promise is only limited by the imagination and capacity of the people who can apply and benefit from it.

This kind of vision of a future electronically driven and inclusive education has been a driving force for HEIs in Tanzania and has provided the spur to implement e-learning. As is the case with other African countries, the rate of implementation of e-learning platforms in Tanzania is still very slow despite the potential opportunities provided by open source technology and the conducive environments created by the respective governments. There have been some initiatives on the part of governments to develop ICT policies as a way forward in the implementation of e-learning. In addition, there have been different round table conferences and the formation of the Tanzania Commission of Universities (TCU) has fostered a debate on a common education delivery. For example, Tanzania has abolished all taxes related to computers and related equipment and reduced licence fees and royalties' payable by the telecommunication operators (Morrison & Khan, 2003 and McPherson & Nunes, 2008). The more established public and private HEIs have managed to implement e-learning platforms in Tanzania. They are implementing these using either open source or customized platforms such as WEBCT, Blackboard, Moodle, Joomla, etc. Other universities in the Tanzania have started the basic process of ICT infrastructure expansion to include local area network implementation, Internet, computer labs and other facilities, as a way forward to the establishment of e-learning (Sife, et al., 2007).

E-Learning Market and the Drivers of Change in Tanzania

While e-learning is not a new phenomenon in the developed world, it may be new to some developing countries. Its market is rapidly increasing globally. While Merrill Lynch (2003) argues that the e-learning is the fastest growing sector in the developed countries, many developing countries (including Tanzania) are striving to implement e-learning in HEIs. Doughty et al. (2001) and Saint (1999) have documented the rise of the virtual university in Africa (including Tanzania). There are many e- learning initiatives in progress in Tanzania, such as Schoolnet, e-learning centres, and African Virtual University (Ndume, et al., 2008; Sife et al., 2007). The increase in the demand for higher education is one of the driving forces for implementing e-learning. Higher population growth, lower education costs, increased access to education, and higher participation rates in higher education changes the way firms organize work and cost-effectiveness and are factors driving the implementing of e-learning in Tanzania (Ndume et al., 2008).

E-Learning Adoption in Higher Education Institutions

In Tanzania, the author conducted an empirical survey. Following the survey, which probed among others whether there is difference in the adoption of e-learning between the private and public HEIs, some factors facilitating the difference were revealed about e-learning adoption. When the respondents were questioned about e-learning, it was shown that there is a high level of ignorance among most of them about the use of Information and Communication Technologies (ICT) in teaching and learning. In the survey that was conducted, the research approach was qualitative. The aim was to explore a particular phenomenon in depth that could only be obtained through the collection and analysis of subjective data from the selected participants involved in the process under investigation (Shenton & Dixon 2004, 1). The study populations were e-learning providers (e-learning managers, lecturers) and informed users as well as corporate sponsors. What guided the pre-selection mechanism in this study was first to identify and interview e-learning providers broadly representative of university e-learning practices in Tanzania. The Universities of St Joseph, Makumira and Dodoma were chosen. The University of Dodoma (UDOM) represents a public owned institution, the Makumira University College (MUC) and St Joseph College of Engineering and Technology (SJCET) represents private institutions. Other selection criteria were those who possibly have been acknowledge by various stakeholders and informed user (e-learners); and those who are presumably knowledgeable with the e-learning practices (e-learning providers). The pre-selection approach enabled the researcher to select 190 respondents. The total number of respondents interviewed was 114 i.e. 60%. The other 76, i.e. 40%, were not interviewed because they were not interviewed for reasons such as lack of time, interest or absent during the visitation for the interview. Among the given questionnaire, respondents were 40 informed e-learners and 74 e-learning providers. The questionnaire was the main instrument used in the survey. The researcher administered the interviews by visiting the respondents in their respective places of work, as well as in conference venues to conduct the interviews. In order to obtain the differences of e-learning adoption, both e-learners and e-learning providers were asked to respond to the following question:

- *What are the current situations of e-learning adoptions in Tanzania higher learning institutions?*
- *What are the factors affecting the adoption of e-learning in higher learning institutions in Tanzania?*
- *What are the assessed similarities and differences in level of adoption between private and government institutions?*

RESULTS AND DISCUSSIONS

Experience with E-Learning Adoption in Higher Education Institutions

Following the survey, it emerged that several e-learning providers have little experience about e-learning adoption. The survey shows that only 72% of the lecturers neither taught of e-learning course nor use e-learning at their course delivery. Only 28% have ever taught e-learning course before or used it. One of the respondents noted that e-learning is not for teaching and learning is only for emails. As shown in Figure 1, the survey has revealed that the more lecturers never experienced e-learning. Figure 2 shows that more students have never took online courses. The cause could be the lack of awareness among the e-learning players (Sife et al., 2007; Omwenga et al., 2004; Ndume et al., 2008; Allen, 2007; MacFadzean & Newton, 2005; Bates, 2001; Manville, 2004). Several of them showed their ignorance in the survey when asked about the e-learning meaning. They were of option that e-learning is the act of the use of email only. One of the e-learning providers even put it bluntly (and incorrectly by saying that there is no such thing of learning and teaching online, because teaching or learning you need to see you students face to face. This long serving professor, added that for the past 30 years of service, never saw people teaching online. E-learning providers, especially lecturers were no better informed. One gave an ambiguous response, going some way toward confirming the suspicion that those who claim to study online are corrupted, and stealing degrees. Adding to their claims, noted that this has even made other accreditation bodies not accepting certificates earned online. On the base of asking the respondents if they have eve used e-learning, most e-learning providers responded that they ever taught e-learning

Figure 1. Lecturers e-learning experience survey

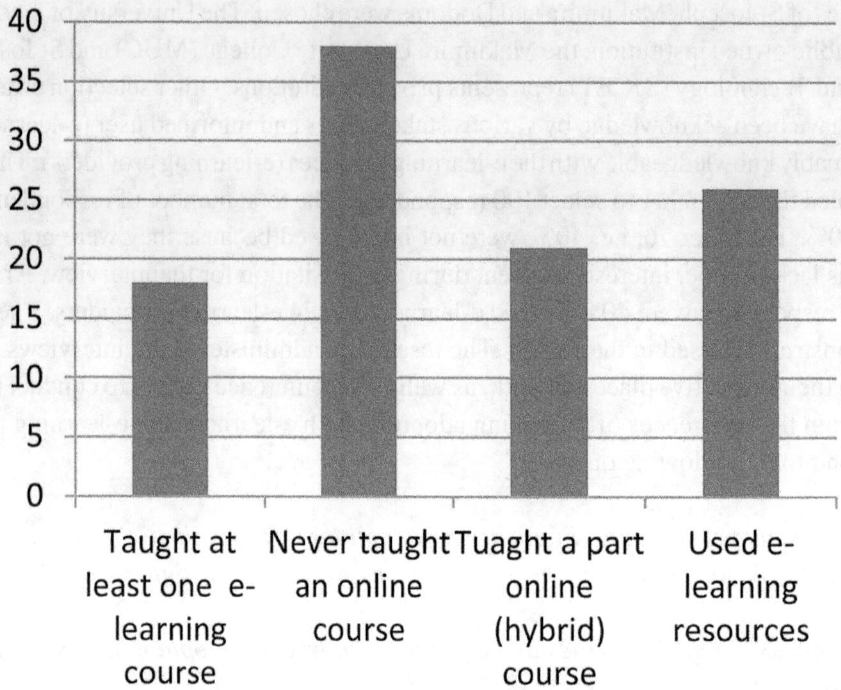

Figure 2. Student experience with e-learning

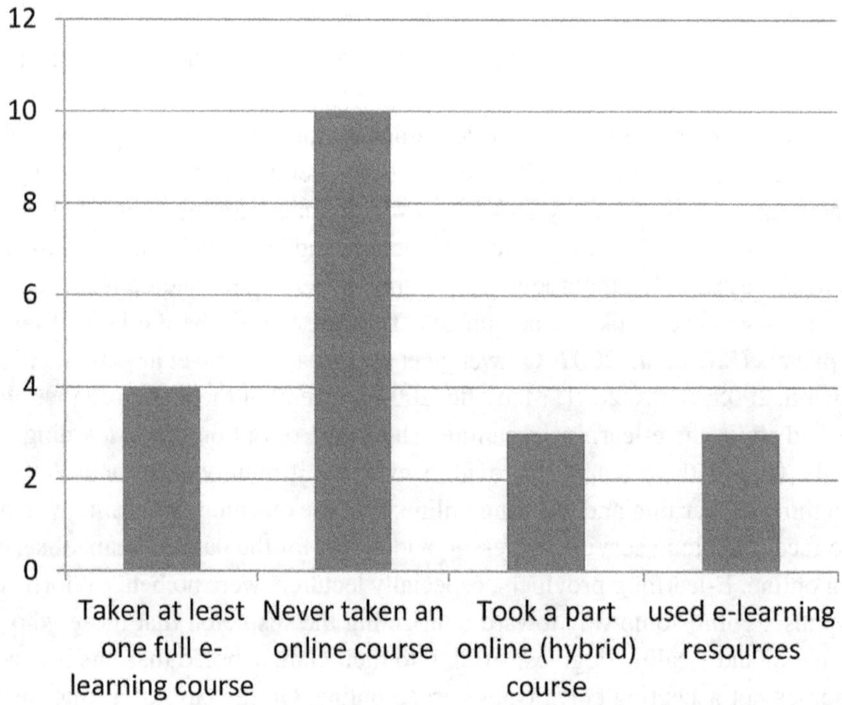

courses. On the same question, most e-learners said they ever been involved, and after all the e-learning providers (these lecturers, discourage the use of internet in learning).

Following the responses from the survey, it can be argued that the reason for the failure by some respondents to recognize e-learning adoption may be the idea that e-learning apply to only enjoyment/fun and not for learning and teaching. This is because, e-learning has been viewed as stealing others scholars work among the e-learning providers and e-learners in Tanzania. Twaakyondo (2008) corroborates this opinion with the statement most of our e-learners and e-learning providers, have little knowledge on the importance of e-learning. As result the e-learning experiences among these groups is slowing instead of increasing.

Furthermore, the lack of experiences may have been grounded in erroneous interpretations of the views expressed by some scholars (Twaakyondo, 2008) that e-learning raises new issues that cannot be addressed by current learning environment. E-learning raises issues of quality, copyright, as well as accreditation (Twaakyondo, 2008). According to Tossy (2012), contemporary advances in the use of ICTs in teaching and learning enabled the flexibility in education delivery. It can be argued that the views expressed by these scholars may have been interpreted erroneously. It can be argued that the survey showed that many of the respondents were unaware of the e-learning probably because they have had no formal tuition informing them of what copyright comprises. During the interview, a lecturer said" following the type of questions you are asking. I will strongly recommend that you come and give us a talk on these issues." Most of the lectures of e-learning are not versed in the importance of e-learning.

E-Learning User Preferences and Opinion

Study shows that over 60% of users prefer traditional courses as compared to e-learning courses (Figure 3).

Most users agree to take both online and offline courses as compared to pure online courses. Figure 4 shows that 79% of users potentially agreed on this while just one user strongly disagrees on taking combination of both online and offline course.

Most respondents answered the question on comparison of online and normal classroom courses believes negatively on e-learning courses. Figure 5 shows that 42% of the users believe on face-to-face courses as compared to 12% who relies on e-learning courses.

This result is discouraging on the level of e-learning adoption in Tanzania. It seems much has to be done to turn around the results.

Facilities Used for E-Learning Adoption

The respondents clearly articulate that over 33% of users have never used any e-learning facilities on their institutions while 26% have been using video and documents or text or power point on their training. See Figure 6.

This result is promising, over 66.66% have used e-learning resources at least once on their academic life. Although the rest 33.33% have not used e-learning before but with more knowledge concerning e-learning, this figure shall decrease, because user X says that "I didn't know if I was using e-learning resources".

Figure 3. Preferences on traditional courses over e-learning courses

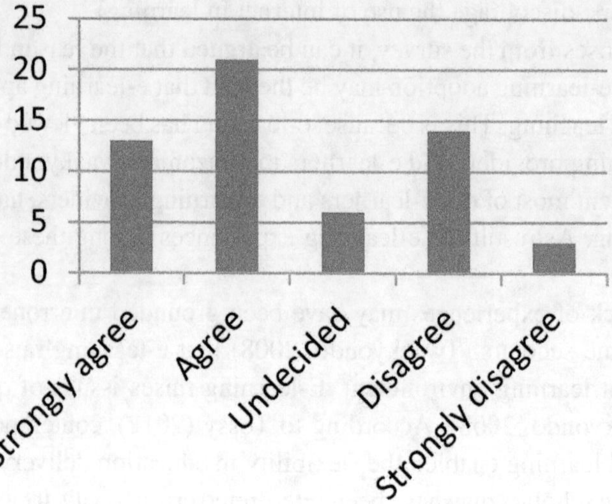

Figure 4. User preferences on e-learning courses

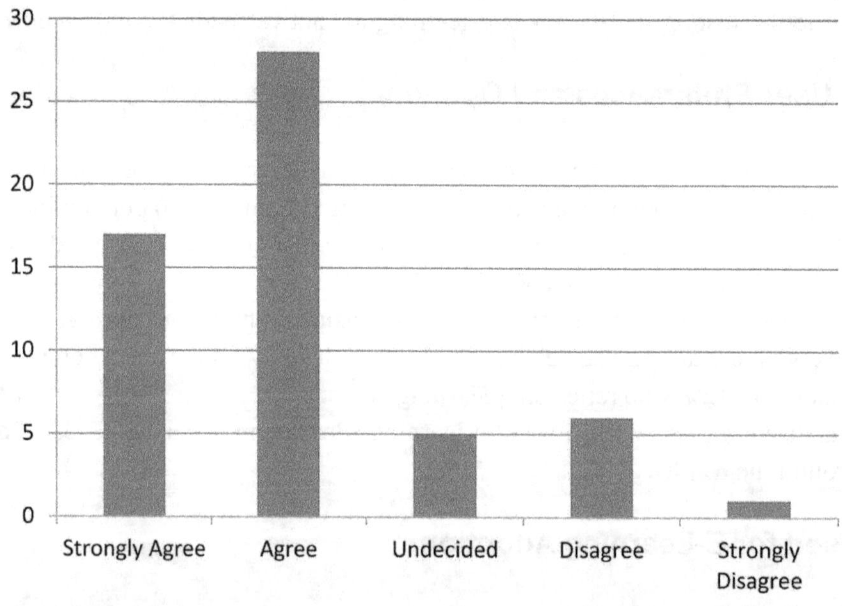

Factors Affecting the Adoption of E-Learning in Higher Learning Institutions in Tanzania

It seems that economic and technological factors are the ones hinder the e-learning deployment in Tanzania. Among 57 respondents only 7% has mentioned other factors that affects e-learning negatively as shown in Figure 7. During the study, author has observed that, although economically SJCET is sounding very good but it has very poor students' campus. I have never seen this kind of compact campus without open spaces, or car parking (see Figures 7 and 8).

Figure 5. Users` opinions on traditional courses vs. online courses

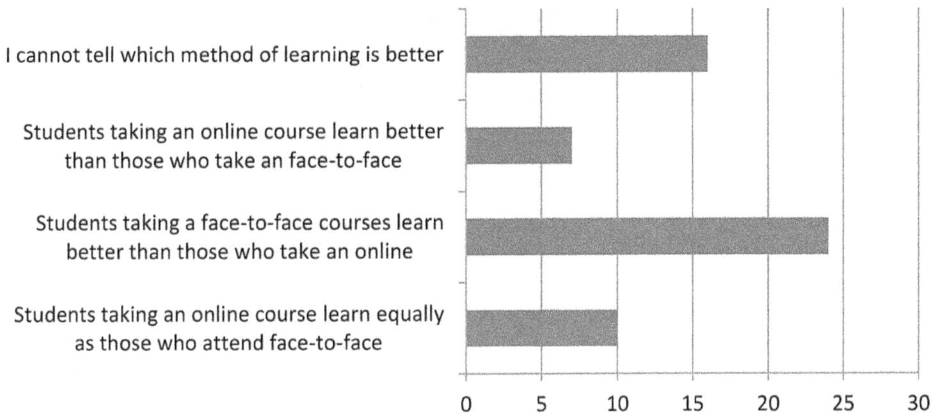

Figure 6. E-learning facilities used in higher education institution

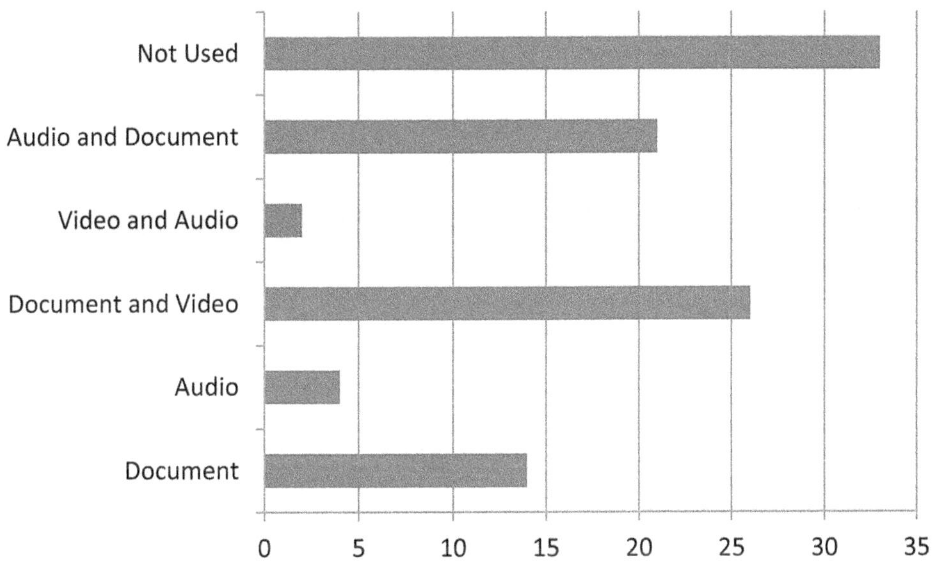

SIMILARITIES AND DIFFERENCES IN LEVEL OF ADOPTION BETWEEN PRIVATE AND GOVERNMENT INSTITUTIONS

The comparison between private and public institutions can be fetched from the resources they have, the way they use it, user`s acceptance, and the limitations they face on e-learning adoptions. This can be reflected from the user`s opinions offered on the questionnaire distributed. Generally, from question number four to the last question of the questionnaire addressing the issue of e-learning can be used as a comparative case study in this research.

Figure 7. Factors affecting e-learning positively

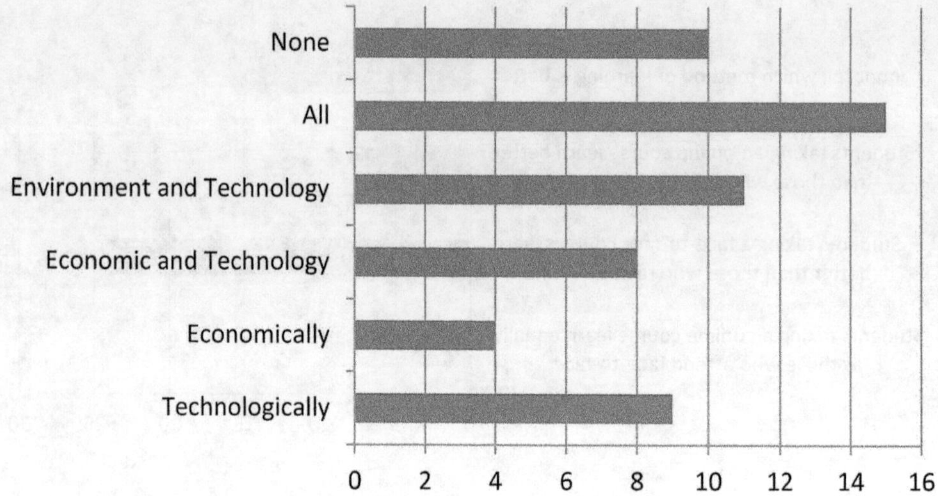

Figure 8. Factors affecting e-learning negatively

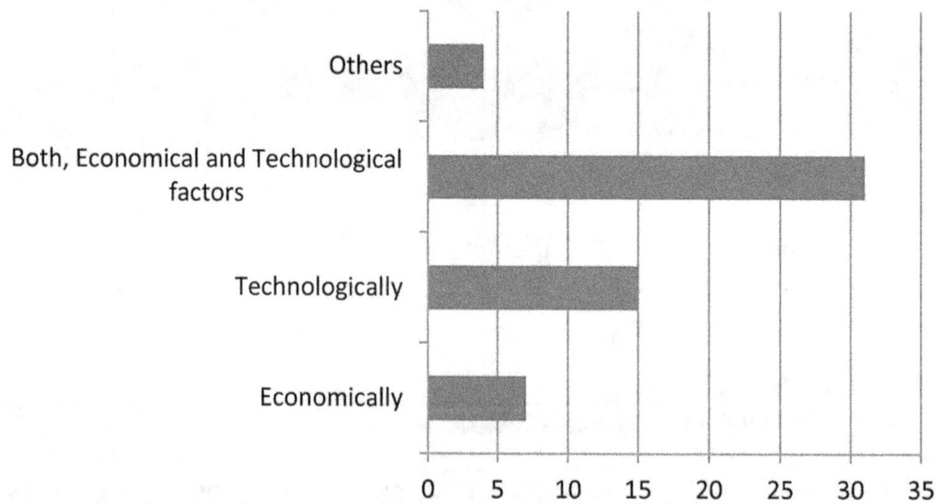

User`s Exposer to ICT Facilities and E-Learning

In this part, the respondent's opinions on level of computer skills, hours spent on computer, e-learning experience, level of internet connectivity, and overall satisfaction of e-learning resources will be reviewed.

Level of Computer Skills

Most of computer users in these institutions seem to tie at competent level at 33.1% however public college have very few beginners compared to private institutions which reflects on the number of available resources at these institutions see Figure 9.

Figure 9. Level of computer level between public and private HEIs

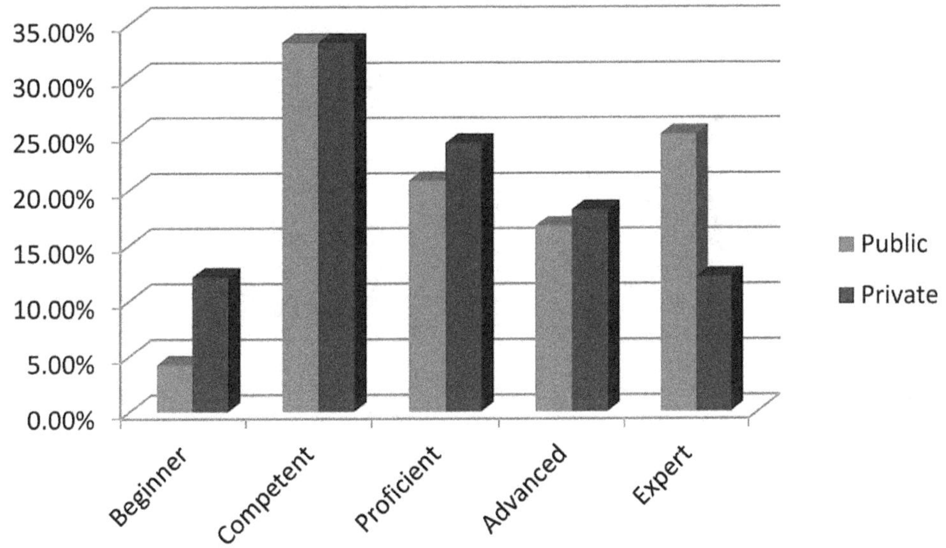

Hours Spent on Computer

The study shows that most of the public institutions` students and lecturers spend more hours per day compared to their private counterpart. Almost 25% of the respondents spend more than four hours in public institutions compared to 9% of the private institutions. Figure 10 shows further details on the number of hours spent on computers per day by both public and private institutions.

Experience in E-Learning

The users in public institutions are more experienced on e-learning at 75% compared to 48% of private institutions respondents. This can be depicted from the choice number two of the Figure 11, "*I have never teach/ taken an online course before*" its 25% for public and 51.5%. See Figure 11 for details.

Internet Connectivity in Public and Private Institutions

Despite of having differences on the percentages figures shown in Figure 12 but they seem to tie in all aspects for both private and public point of view. Almost 42% of public institutions claim to have very good and excellent internet connectivity which is the same figure for private institutions. See Figure 12 for further details.

Satisfaction with E-Learning Resources

In this question, the respondents of both public and private institutions share common mind, the resources at their institutions are not satisfactory for e-learning applications and overall implementations. However public institutions seem to have less sounding "NO" at 58% compared to 67% of private institutions as shown in Figure 13.

Figure 10. Hours spent on computers per day by respondents

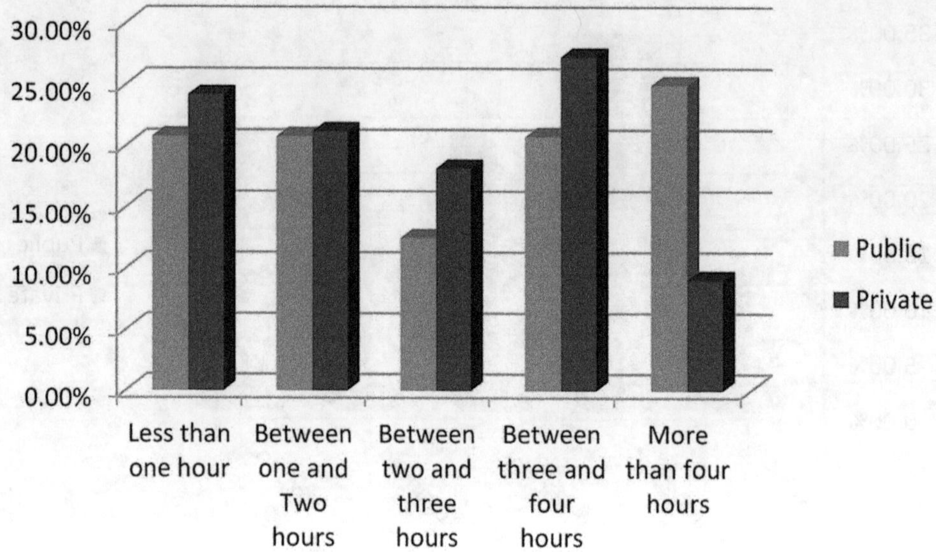

Figure 11. Experiences on the e-learning adoption

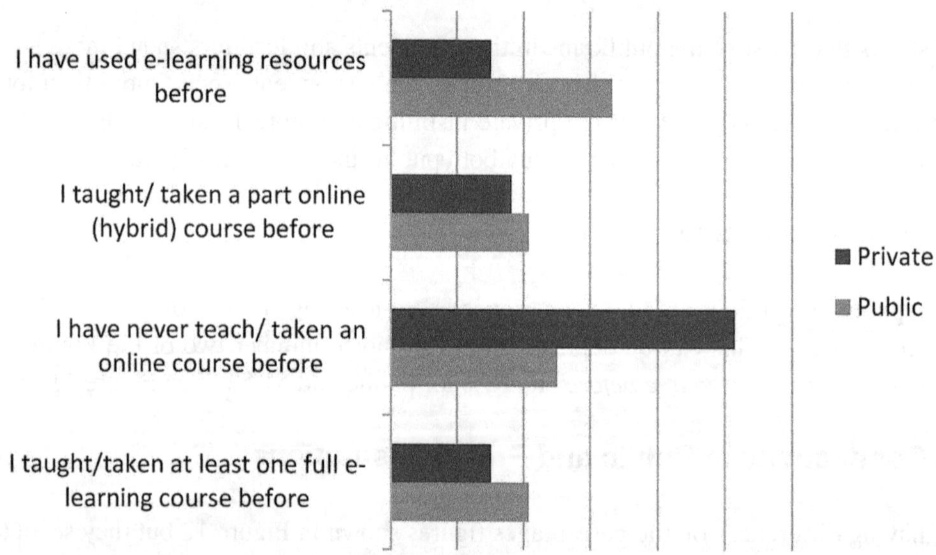

User`s Preference on E-Learning

The users from private and public institutions respond on several questions about e-learning preferences and the results are shown in Figures 14 and 18. Generally, from all figures mentioned above which shows; preference on traditional courses, preference on e-learning courses, and preferences on hybrid courses shows the similarities on responses for all answers from private and public respondents. The study shows the answers on "*strongly agree*" and "*agree*" to match on private and public, they are low, high or moderate for all questions.

Figure 12. Institution internet connectivity

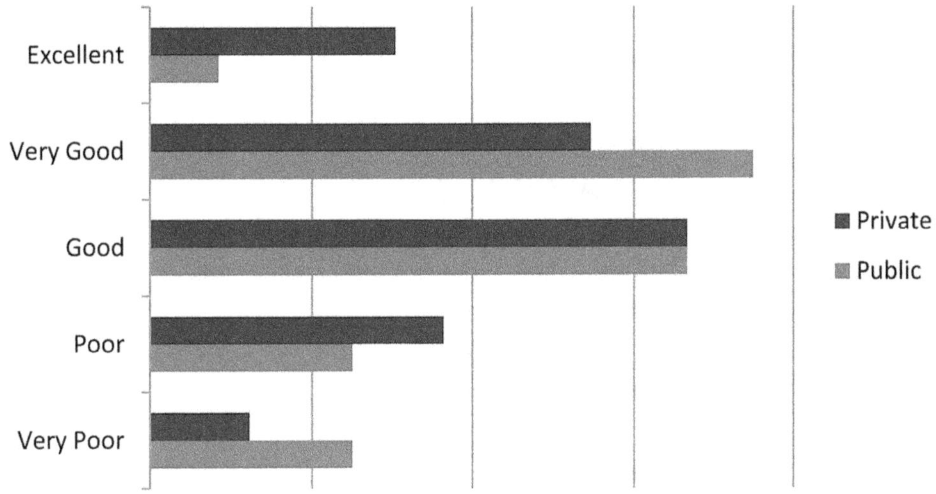

Figure 13. Satisfaction with e-learning

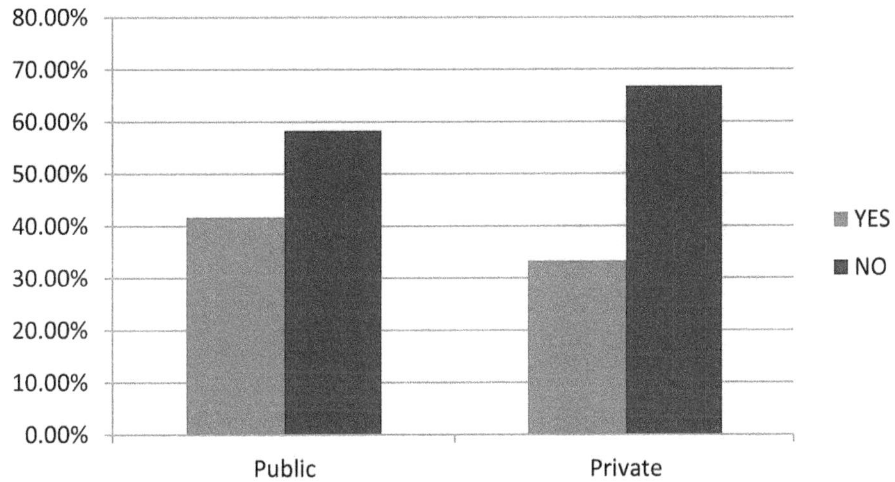

Furthermore, there are very few "*undecided*" respondents and very high "*disagree*" cases for both respondents. Also *"Strongly disagree"* answers are very few for both respondents and for all questions. The details on these are shown in Figures 14-17. In Figure 18 the user`s opinions were tested and the results indicate that face to face courses are more preferred.

Preferences on Traditional over E-Learning

Most of the respondents on this question seem to accept on traditional courses over online courses at over 50% overall; while there are very few strongly disagree in both cases private and public respondents (see Figure 14).

Figure 14. Preference on traditional (face to face)

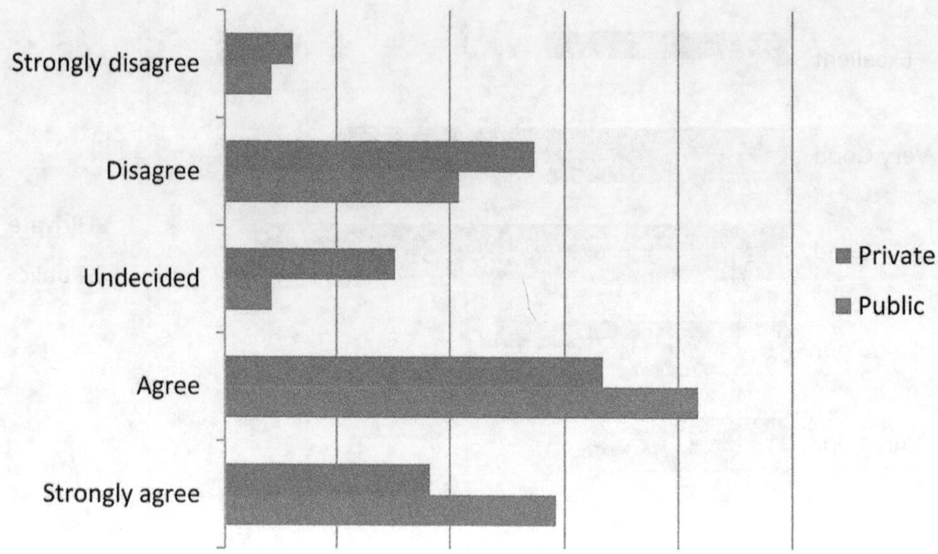

Figure 15. Preference of face to face than online

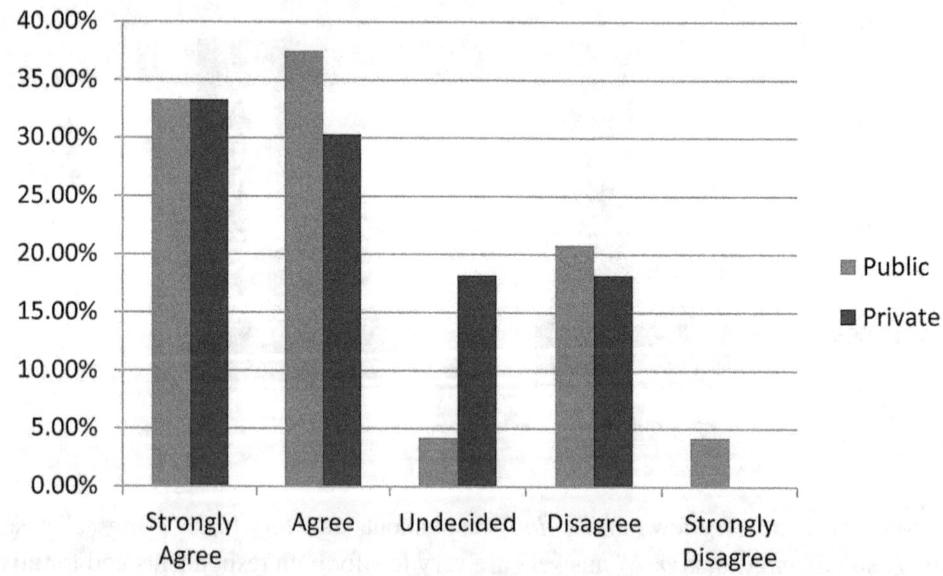

Face to Face Compared to Online Courses

This question tests the user`s acceptance on face-to-face compared to e-learning courses. Over 60% of the respondents contributed their views on "*Strongly agree*" and "*Agree*" for both public and private institution. Very few "*strongly disagree*" on this question at 4.2% for public and 0% for private respondents (see Figure 15).

Figure 16. Preference on hybrid course delivery

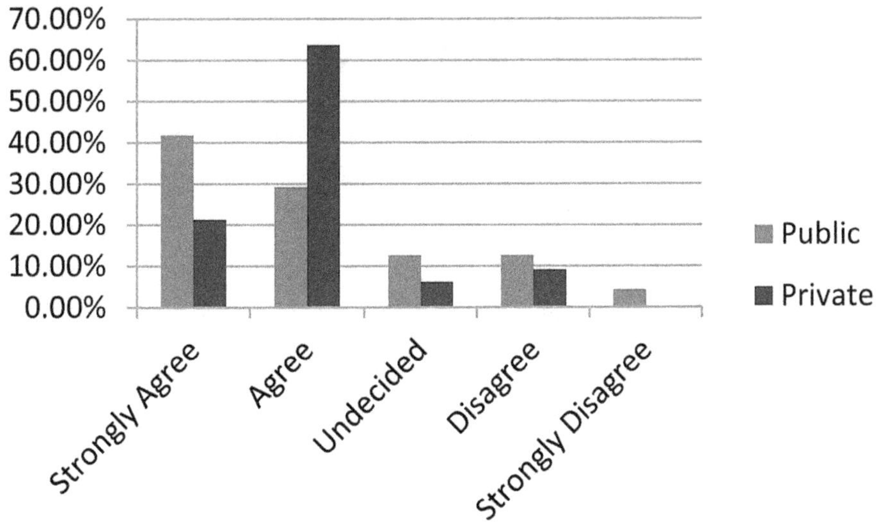

Figure 17. Future interest

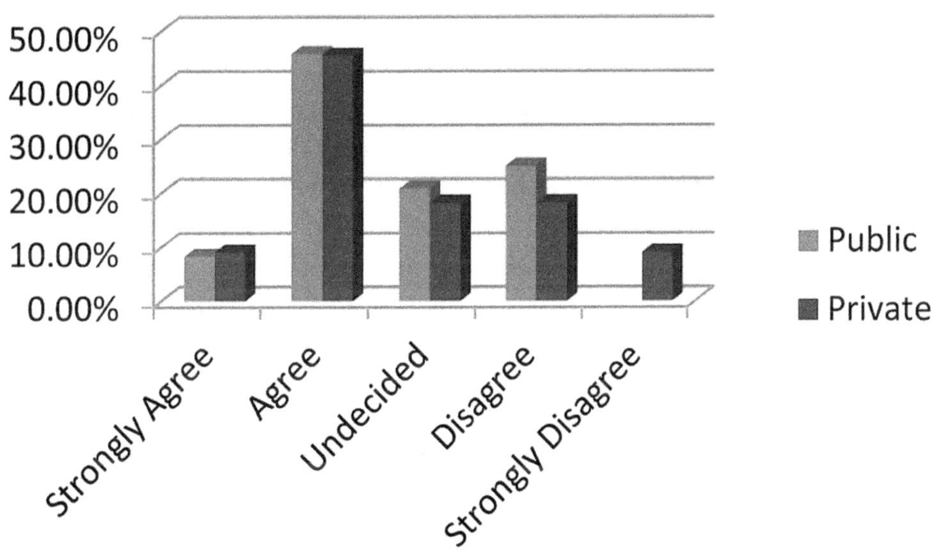

Preference on Hybrid Courses

The respondents exclusively agreed to have both online and offline courses than having just one form of education due to various reasons yet to be determined. As shown in Figure 16, more private HEIs prefer a blended e-learning compared to Public HEIs. While Public HEIs are more featured to strongly agree on blended learning, Private institution shows some dilemma on agreeing strongly, but they agree.

Figure 18. Comparison between e-learning and normal classroom

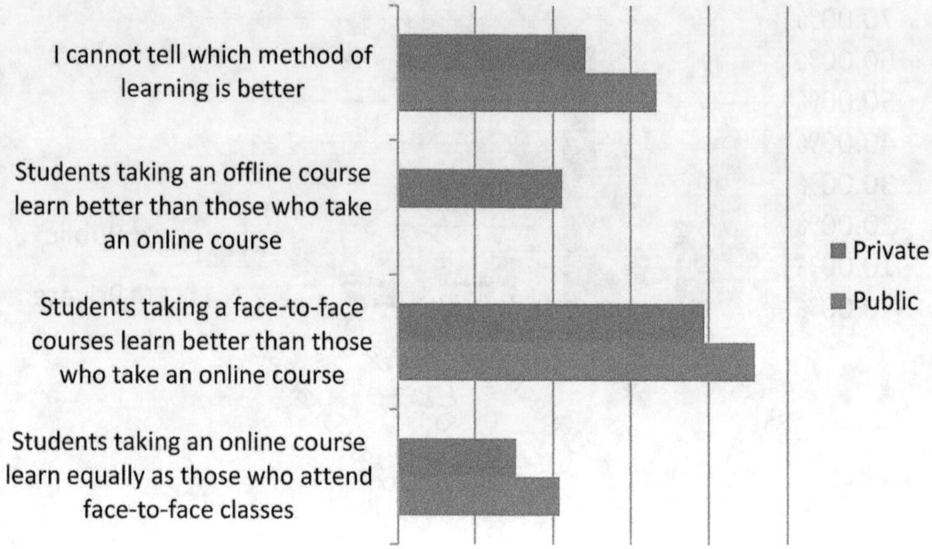

Future Prospects in Participating in Online Courses

The mindset has distributed between "*Agree, Undecided* to *Disagree*" and there are very few in "*Strongly disagree* and *strongly agree"* cases. However most of them seem to agree on teaching or taking online courses in the future. As shown in Figure 17, there is slightly difference between Private and Public sector. The results reveals that both private and public HEIs willingly to adopt e-learning practices.

Comparison between Online and Face to Face Courses

The study shows that face to face courses believed to be delivered much better than online counterpart at 45.8% for public and 39.4% for private participants of this study. The details are shown in Figure 18.

Factors Affecting E-Learning in Tanzania Higher Learning Institutions

In this part the respondents were asked on the various factors that might affect the adoption of e-learning in Tanzania higher learning institutions, despite of having various answers which ranges in all aspects of life; but it seems to end up on either economical problem or lack of technical knowhow. Generally, over 90% of both negatively and positively affected factors appear to be economical and technological factors. The details are shown in Figure 19.

The differences in opinion can be seen on the environmental and technological factors that affect e-learning adoption. The users believe that environmental or geographical problems causes e-learning to be adopted more easily because, it solves the problem of geographical location between trainers and trainees and thus no need of travelling to attain your academic qualifications with e-learning. Figure 20 shows further details on these factors.

Figure 19. Factors affecting e-learning negatively

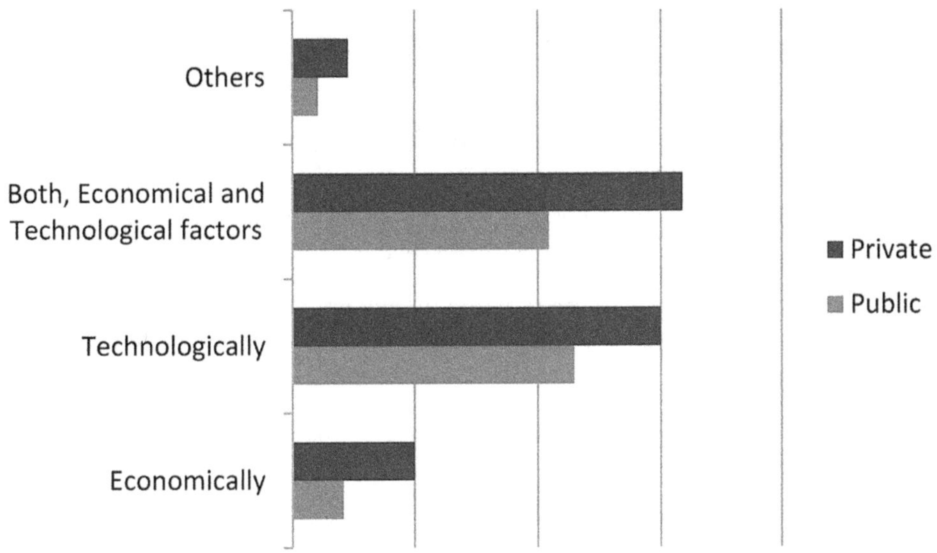

CURRENT SITUATIONS

In this category of questionnaire, the respondents were questioned on the e-learning facilities ever used on their institutions and delivery mode deployed on day to day basis. The e-learning facilities seem to ranges from audio, video, text to various forms of software and the answers for e-learning delivery mode indicate that both online and offline methods are deployed in these institutions as shown in Figure 21.

Figure 22 depicts that over 42% of private institution have never used any facilities on e-learning as compared to 21% for public institutions as shown in Figure 22.

HOW TO ENHANCE E-LEARNING.

In this part of questionnaire, over 84% of the respondents from public institutions believe that: enough electrical power, training on ICT and further investment on ICT is needed to boost the pace of e-learning adoption, while private institutions demand on more training on ICT shall boost the e-learning adoption on their institutions. Furthermore, none of the public institution respondents is satisfied with current situation there about 3% of the respondents are satisfied with what is delivered on their institutions (see Figure 23).

CONCLUSION

As per discussion above, this study shows that E-learning adoption with the HEIs is taking place. While the study shows that there is rapid adoption in public sector, there is low adoption in private institution. This is facilitated much due to lack of funding. The study has revealed that the government of Tanzania has been supporting e-learning initiative by introducing various polices to boost the level of adoption.

Figure 20. Factors affecting e-learning positively

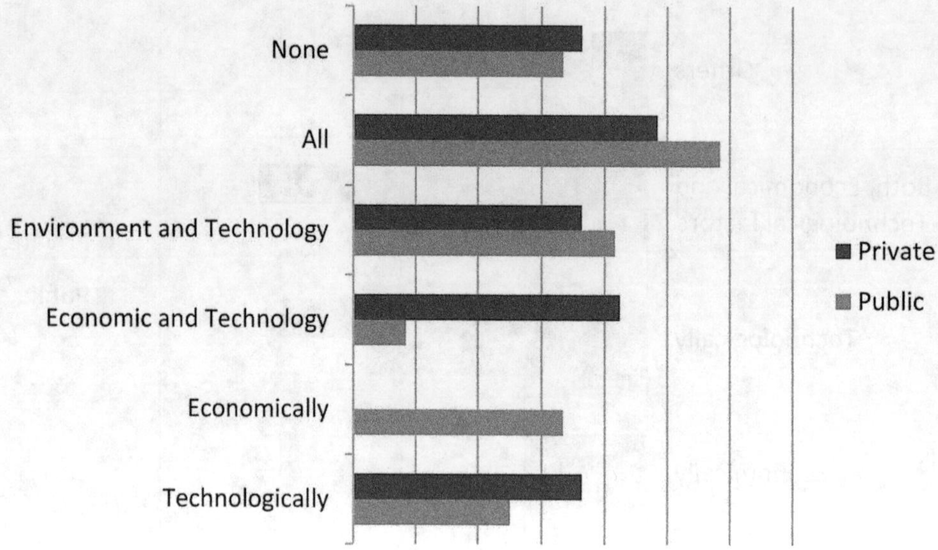

Figure 21. Delivery mode used

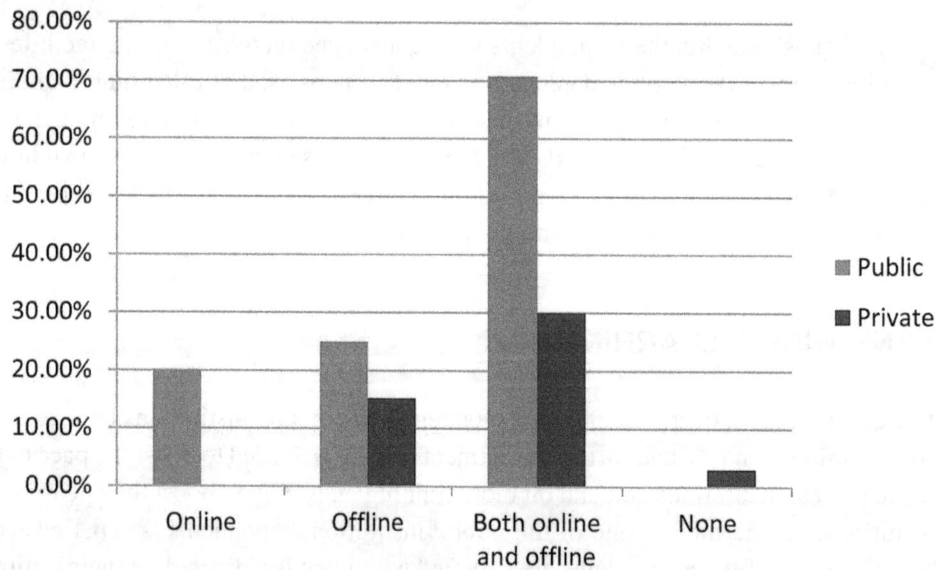

The current situation of e-leaning adoption depict that most of Tanzanian have not acquainted with e-learning of any form. In fact, very few have used e-learning resources at least once in their academic life. Furthermore, the study shows that, most of academicians do not believe on the potential benefits poised by e-learning. Despite of having tremendous opportunities in the market, the adoption of e-learning is at very low level in Tanzania due to the various barriers which includes, lack of management support, methodology, resource accessibility and availability, culture of education and learning styles, design of assistive tools, intellectual investment, and global business conceptual awareness of its benefits, lack

Figure 22. Facilities used for e-learning

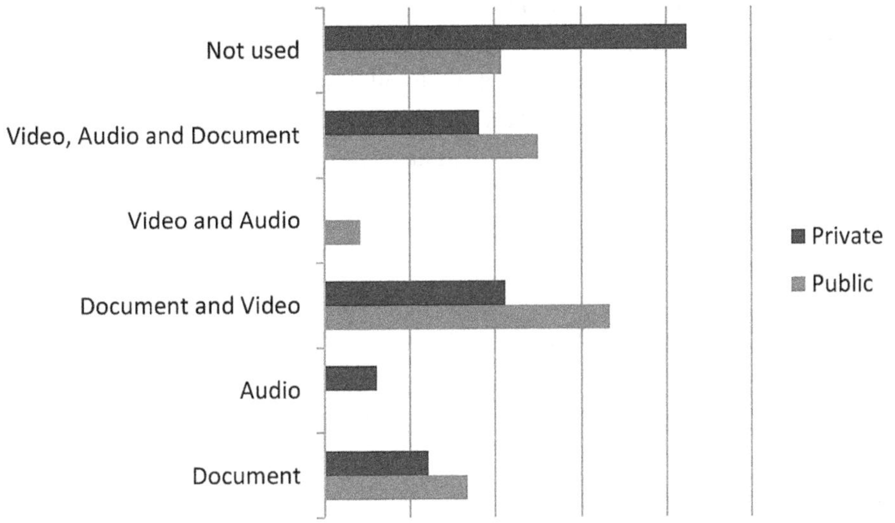

Figure 23. What should be done to enhance e-learning

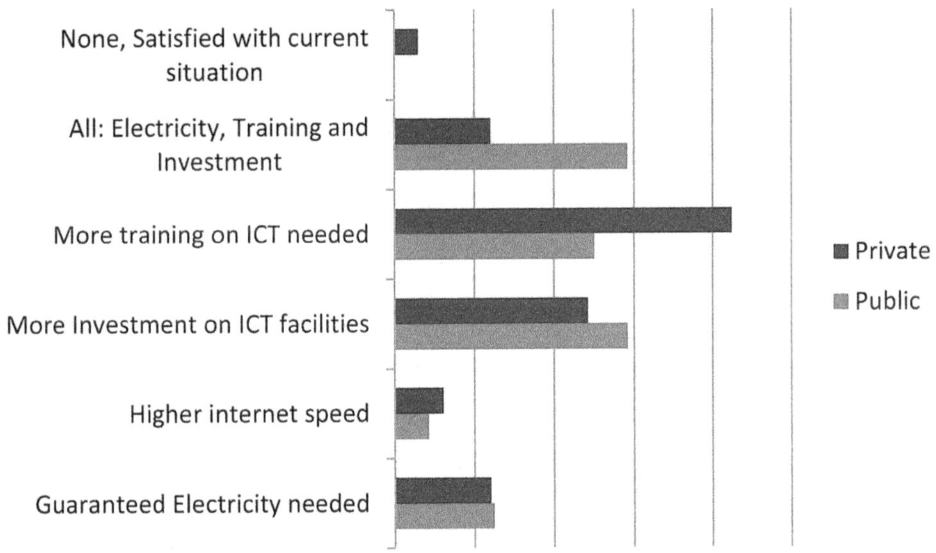

of technical knowhow, lack of enough funds, poor strategic change management plans, and lack of innovative ideas.

The study showed that the adoption level differs from private to public institutions, the private institutions suffers the most as compared to the public counterpart. They are characterized by: lack of e-learning resources, lack of experience on e-learning, lack of enough ICT infrastructures, poor economic background, fears of government intervention on innovations, and lack of technical expertise in the field.

The study recommends that the government of Tanzania should improve e-learning initiative support to higher learning institutions. It should provide some kind of scholarships for students doing online studies. There should be national e-learning programs on media to address various issues in this sector so as to bring the awareness among the society at large. It should also provide funds or loans to those institutions eagers to adopt e-learning. The study also recommends that; more funds should be injected to ICT facilities to improve its accessibility to users. Finally, should availability of electrical power and high speed internet be scarce, it might hinder the development and enhancement of e-learning adoption; and hence the government should guarantee its availability.

REFERENCES

Achterberg, J. S., Heng, M. S. H., & Gerrit, A. (1991). *Information systems research in the post-modern period*. Elsevier Science Publishers.

Alan, B. (2009). Handbook of Library Training Practice and Development. Abingdon, UK: Ashgate Publishing Group.

Angello, C. (2010). The awareness and use of electronic information sources among livestock researchers in Tanzania. *Journal of Information Literacy*, *4*(2), 6–22. doi:10.11645/4.2.697

Ball State University. (1997). *A Brief History of Distance Education*. Available at http://www.seniornet.org/edu/art/history.html

BLR-Employee Training Centre. (2012). *The Most Effective Training Techniques*. Available at http://trainingtoday.blr.com/employee-training-resources/How-to-Choose-the-Most-Effective-Training-Techniques

Boaden, R., & Lockett, G. (1991). Information technology, information systems and information management: Definitions and developments. *European Journal of Information Systems*, *1*(1), 23–32. doi:10.1057/ejis.1991.4

Brause, R. S. (2000). *Writing Your Doctoral Dissertation: Invisible Rules for Success*. London: Farmers Press.

Chiemelie, I. B. (2012). *Disadvantaged of traditional methods of training*. Available at http://bnetwork.us/showthread.php?337-Disadvantaged-of-traditional-methods-of-training

CIRTL Network. (2012). *Lecturing: Advantages and Disadvantages of the Traditional Lecture Method*. Available at http://www.cirtl.net/node/2570

Clark, R. C., & Mayer, R. E. (2011). *E-Learning and the Science of Instruction: Proven Guidelines for Consumers and Designers of Multimedia Learning* (3rd ed.). Hoboken, NJ: Pfeiffer. doi:10.1002/9781118255971

Coldeway, D. O. (1982). Recent Research in Distance Learning. In J. S. Daniel, M. A. Stroud, & J. R. Thompson (Eds.), *Learning at a Distance: A World Perspective* (pp. 29–37). Edmonton, Alberta, Canada: Alberta University.

Commons, C. (2012). *e-Learning Fundamentals*. Available at http://www.leerbeleving.nl/wbts/1/history_of_elearning.html

Creswell, J. (2003). *Research Design: Qualitative, Quantitative and Mixed Method Approaches* (2nd ed.). Thousand oaks, CA: SAGE publishers.

Cruthers, M. (2008). *Education technology gives teachers a wider reach.* Available from http://www.etni.org.il/etnirag/issue5/mark_cruthers.htm

Customer-Essay. (2012). *Dissertation Structures: Some Important Things to Consider.* Available at http://custom-essay.org/article/dissertation-structures

Davis, F. D. (1989). Perceived Usefulness, Perceived Ease of Use, and Users` Acceptance of Information Technology. *Management Information Systems Quarterly*, *13*(3), 318–340. doi:10.2307/249008

Davis, F. D. (1993). Users` acceptance of information technology: System characteristic, user perceptions and behavioral impacts. *International Journal of Man-Machine Studies*, *38*(3), 475–487. doi:10.1006/imms.1993.1022

E-learning Africa. (2012). *Perspectives on ICT and Education in Africa.* Available from http://www.elearning-africa.com/eLA_Newsportal/E298tanzania-beyond-tomorrow%E2%80%99-soon-ready/

Fulton, K. (1996). *Technology Infusion and School Change: Perspective and Practices.* Available at http://www.ncrel.org/sdrs/areas/issues/methods/technlgy/te600.pdf

Hawaii Public Schools. (2012). *How Can Technology Benefit Our Students?.* Available from http://www.k12.hi.us/~teono/teri/benefits_of_technology.htm

Howard, H. (2003). *Individual Preferences in E-Learning.* Gower Publishing Limited.

GBR Institute for Interactive Technologies. (2006). *E-Learning Concepts and Techniques.* Available from http://iit.bloomu.edu/Spring2006_eBook_files/chapter1.htm

Kaltan, G. (1983). *Introduction to Survey Sampling.* Beverly Hills, CA: Sage Publications.

Kizito, J. B., & Bijan, K. (2006). *An Empirical study on Education Strategies to E-leaning in Developing Countries.* Emerald Group Publishing Limited.

Kothari, C. R. (2002). *Research Methodology: Methods & Techniques* (2nd ed.). New Delhi, India: K.K Gupta.

Kruse, K. (2006). *The Benefits and Drawbacks of e-Learning.* Available at http://74.125.155.132/scholar?q=cache:DJhiHmX0drMJ:scholar.google.com/&hl=tr&assdt=2000

Kwan, F. R., & Joseph. (2005). Web-Based Learning. *Proceedings of the 4th International Conference, 'Technology and Pedagogy'.* River Edge, NJ: World Scientific Publishing Co.

MacKeogh, K., & Fox, S. (2009). Strategies for Embedding e-Learning in Traditional Universities: Drivers and Barriers. Electronic Journal of e-Learning, 7(2), *147 - 154.*

Mafu, S. (2004). From the Oral Tradition to the Information Era: The Case of Tanzania. *International Journal on Multicultural Societies*, *6*(1), 53–78.

Maliyamkono, T. L & Mason H. (2006). *The Promise.* TEMA Publishers Company Ltd and Siyaya Publishing (pty).

Matthews, D. (1999). The Origins of Distance Education.T.H.E. Journal, 27(2), 56-66.

Mgendi, M. F. (2010). *Introducing Web-Based E-learning Platforms at An African University*. Available http://www.isprs.org/proceedings/XXXVIII/part6/ papers/Mgendi/Mgendi_ISPRS_VI_symposium B1 D.pdf

Ministry of Education and Culture. (2001). *Basic Education Master Plan (BEMP): Medium Term Strategic and Programme Framework 2000-2005*. Available from www.mwanza.go.tz/kurasa/nyaraka_mbalimbali/BEMP.pdf

Ministry of Education and Vocational Training. (2007). *Information & Communication Technology (ICT) Policy For Basic Education*. Available from http://planipolis.iiep.unesco.org/upload/Tanzania/UR/Tanzania_ICT_Policy_for_BasicEducation_2007.pdf

Ministry of Education and Vocational Training. (2011). *Higher Education Baadhi ya Taarifa*. Available from http://www.moe.go.tz/pdf /HE_Organogram.pdf

Ministry of Education and Vocational Training. (2011). *Takwimu za Kielimu kwa Baadhi ya Miaka ya Uhuru wa Tanzanganyika*. Available from http://www.moe.go.tz/pdf/TAKWIMUTANGUUHURUHA-DIT2011.pdf

MOCT. (2003). *National Information and Communications Technologies Policy*. Available from http://www.tanzania.go.tz/pdf/ictpolicy.pdf

MUC. (2012a). *Academic Programmes*. Available at http://www.makumira.ac.tz/index.php/academic-programmes

MUC. (2012b). *Photo gallery*. Available at http://www.makumira.ac.tz/zindex.php/photo-gallery

Mungania, P. (2003). *The Seven E-learning Barriers Facing Employees*. Available at http://www.aerckenya.org/docs/ElearningReport.pdf

Mysite. (2012). *History of Distance Education*. Available at http://mysite.du.edu/~kkeairns/de/Text/Lessons/Lesson1.pdf

NACTE. (2012). *List of Accredited Institutions as of June 2011*. Available at http://www.nacte.go.tz/registered_institution.php?lang=kis

Naidu, S. (2006). *E-Learning: A Guidebook of Principles, Procedures and Practices*. Available from http://cemca.org/e-learning_guidebook.pdf

Ndume, V., Tilya, F.N., & Twaakyondo, H. (2006). Challenges of Adaptive E-learning at Higher Learning Institutions: A Case Study in Tanzania. *International Journal of Computing and ICT Research, 2*(1).

Newman. (2008). *Advantages and Disadvantages of E-learning*. Available at http://www.newman.ac.uk/Students-Websites/~m.m.friel/

Nichols, M. (2012). *International Forum of Educational Technology & Society: Formal Discussion Initiation-A theory for eLearning*. Available at http://ifets.ieee.org/discussions/discuss_march2003.html

Noe, R. A. (2005). *Traditional Training Methods*. Available from http://www.stamfordonline.com.my/courses/dhrm/dhr304/DHR%20304%20Lecture%207.pdf

O`Brien, J. A. (2004). *Management Information Systems: Managing Information Technology in the Business Enterprise* (6th ed.). New York: McGraw-Hill/Irwin.

Oomen-Early, J., & Murphy, L. (2008). Self-actualization and e-learning: A qualitative investigation of university faculty's perceived needs for effective online instruction. *International Journal on E-Learning*, *8*(2).

Open University of Tanzania. (2011). *PROSPETUS 2011/2012*. Retrieved from http://www.out.ac.tz/announcements/Prospectus 2011-2012 for printing3.pdf

OPPAPERS. (2012). *Traditional Training Methods*. Available from http://www.oppapers.com/essays/Traditional-Training-Methods/685211

Papalewis, R. (2003). *What are the Advantages /Disadvantages of Technology in the classroom?* Available at http://answers.yahoo.com/question/index?qid=20080712164252AA1iB7Y

Piskurich, G. M. (2002). *AMA Handbook of E-Learning: Effective Design, Implementation, and Technology Solutions*. AMACOM Books.

Ray, K., & Day, J. (1998). Students` attitudes towards electronic information resources. *Information Research, 4*(2). Available at http://informationr.net/ir/4-2/paper54.html

Resources.com. (2012). *Research Methodology - Key Concepts of the Scientific Method*. Available from www.experiment-resources.com/research-method

Rogers, E. M. (2003). *Diffusion of innovations* (5th ed.). New York: Free Press.

Romiszowski Alexander, J. (2004). How's the E-learning Baby? *Factors Leading to Success or Failure of an Education Technology Innovation*. Available at http://www.itslifejimbutnotasweknowit.org.uk/files/elearning_failure_study-romiszowsky.pdf

Scribd. (2005). *Traditional Training Methods*. Available at http://www.scribd.com/doc/35921013/Traditional-Training-Method

Simonson, M., Smaldino, S., Albright, M., & Zvacek, S. (2000). *Teaching and Learning at a Distance: Foundations of distance education*. Upper Saddle River, NJ: Prentice Hall.

SJCET. (2012). *School of Information Technology*. Available at http://www.sjcet.in/sit.html

Stockley, D. (2006, February). Implementing e-learning successfully: Building a successful e-learning strategy. *The EI Magazine*, *2*(7), 4–5.

TCU. (2010). *Statistics for Universities and University Colleges Enrollment of Students*. Available at http://www.tcu.go.tz/uploads/file/Statistics%20for%202009-2010.pdf

TRA. (2012). *Tax Exemption for deemed capital goods*. Available at www.tra.go.tz/index.php?option=com_content&task=view&id=238&Itemid=76

Tynjala, P., & Hakkinen, P. (2005). E-learning at Work: Theoretical Underpinnings and Pedagogical Challenges. *Journal of Workplace Learning*, *17*(5/6), 2005. doi:10.1108/13665620510606742

UDOM. (2012). *Call For Admission into Undergraduate Degree Programmes for Equivalent Applicants*. Available at http://udom.ac.tz/index.php/ admission/undergraduate-application

UDOM. (2012). *Organisation Structure*. Available at http://udom.ac.tz/index.php/admission/undergraduate-application

UDOM. (2012). *University History*. Available at http://udom.ac.tz/index.php/about-udom/history

University of Florida. (2012). *The History of Distance Education*. Available at http://iml.jou.ufl.edu/projects/spring01/declair/history.html

Wagner, E. (2000). *Strategies for leveraging learning objects: PowerPoint presentation delivered at the TechLearn2000 conference*. Available at http://www.techlearn.net/layout/default.cfm?page=sessiondetail&session=1236

Watson, J. F., et al. (2004). *Education evolution: The need to keep pace with development of K-12 online learning*. Available from www.learningpt.org/expertise/ schoolimprovement/pdf

Wiersma, W., & Jurs, S. G. (2009). *Research Methods in Education* (9th ed.). Pearson Education Inc.

Wiersma, W. (1995). *Research methods in education* (6th ed.). Boston: Allyn and Bacon.

Williams, P., & Nicholas, D. (2004). *What Literature Tell us About Distance Education*. Available from http://site.ebrary.com/lib/coventry/docDetail.actiondocID=10085710&p00=informatione-learning

Woodward, H., & Pilling, S. (1993). *The international serials industry: The International Serials Industry*. Aldershot, UK: Gower Publishing.

Woody, C. (1927). Research Methodology. *Journal of Social Sciences*.

World Business Education. (2012). *E-learning: Adoption Rates and Barriers*. Available at http://www.internettime.com/itimegroup/FrmRep1.pdf

Worldwidelearn. (2012). *E-learning Essentials: Benefits of E-learning*. Available from http://www.worldwidelearn.com/elearning-essentials/elearning-benefits.htm

Wright, J. (2008). *Primary ICT and E-Learning Co-Coordinator's Manual Kit*. SAGE Publications Inc.

Chapter 8
The Impact of Web 2.0 on E-Commerce Adoption and Use by Tourism Businesses - Can SMMEs Play the Trick?
A Case of the Eastern Cape Province

Pardon Blessings Maoneke
University of Fort Hare, South Africa

Naomi Isabirye
University of Fort Hare, South Africa

ABSTRACT

The subject of Information Technology (IT) adoption and use has been going on for some time (Jeyaraj & Sabherwal, 2008). In particular, to electronic commerce (e-Commerce) adoption and use, the invention of Web 2.0 presents new technological features for potential and current e-Commerce adopters as well as new challenges. With Web 2.0, customers' perceptions, preferences and decisions are not only based on information presented on e-Commerce websites, but are also influenced by content generated by people on social networks and interactive e-Commerce websites. This poses the following question: how can Small, Medium and Micro-Sized Enterprises in the tourism sector (tourism SMMEs) keep up with these technological advancements given their limited resources? Accordingly, this book chapter proposes a framework that shows challenges and incentives (critical success factors) of e-Commerce, identifies e-Commerce platforms tourism SMMEs should adopt in order to maximise benefits and outlines what tourism SMMEs should expect from their e-Commerce platforms.

DOI: 10.4018/978-1-5225-0539-6.ch008

BACKGROUND

Tourism Products and Customer Buying Behaviour

The tourism sector is not formally classified under the International Standard Industrial Classifications (Smith, 1998 in Rogerson, 2005). As such, the scopes and ranges of tourism service or products can range from very tangible products such as geographical areas and man-made facilities to more intangible specific attractions (Gilmore, 2003). In other words, a tourism product is a complex set of complementary tourism services produced by different economic agents, with completely different functions, resources and a different knowledge base (Wanhill, 2002; Werthner & Ricci, 2004). As a result, tourism services can be distinguished from manufactured goods in that they are intangible, inseparable, perishable and heterogeneous (Seabra, Abrantes & Lages, 2007; Gilmore, 2003; Palmer, 2005).

These characteristics of tourism services create uncertainty and perceived risk in customers' minds in such a way that before buying, consumers acquire a large amount of information and make procurement decisions based on the consequent images and expectations (Seabra at al., 2007). Palmer (2005) proposed a model of services purchase in which he argued that service customers recognise the need for a service first then search for information on potential services that could meet their needs, evaluate available services, decide and purchase the services, and then evaluate the services after use.

In line with Palmer's (2005) proposition, research by Ho, Lin and Chen (2012) and Litvin, Pan and Goldsmith (2008) suggest that the major sources of information that influence tourism customers' purchase behaviours include the Internet, adverts, movies, travel agents, word-of-mouth and interpersonal influence. In particular, to the Internet, Huang and Benyoucef (2013) recently noted that the invention of Web 2.0 has transformed e-Commerce from a product oriented environment to a social and customer oriented one. With Web 2.0, customers' perceptions, preferences and decisions are not only based on information presented on e-Commerce websites, but are also influenced by content generated by people on social networks and interactive e-Commerce websites (Constantinides & Fountain, 2008 in Huang & Benyoucef, 2013). A good example is that of consumers sharing their tourism opinions is the social media website tripadvisor.com — touted (by the company) as ''the largest site for unbiased travel reviews (which) gives you the real story about hotels, attractions, and restaurants around the world" (Litvin et al., 2008:458). It is therefore a primary task for tourism businesses' marketing and promotional effects to ensure relevant information is made visible and accessible to potential customers (Xiang & Pan, 2011).

The Impact of Web 2.0 on E-Commerce Platforms for Tourism Business

The invention of Web 2.0 brought about different e-Commerce platforms for tourism businesses to adopt. This book chapter uses the ICDT model to describe what is expected of e-Commerce platforms for tourism businesses. The ICDT model is generic and applies to various industries (Karanasios, 2008). Halawani, Abdullah, Rahman and Halawani (2013) argue that there is a dearth of research that uses the ICDT model to assess the characteristics of the use of the Internet in travel and tourism enterprises even though O'brien (1998 in Halawani et al., 2013) and Karanasios (2008) have used the model before to describe characteristics of e-Commerce platforms for tourism businesses. It is also important to reconsider the ICDT model due to the rapid growth of Internet use and Web technologies and its expansion to include many features (Halawani et al., 2013).

Figure 1. Tourism and the ICDT model (O'Brien, 1998 in Halawani et al., 2013, p. 263)

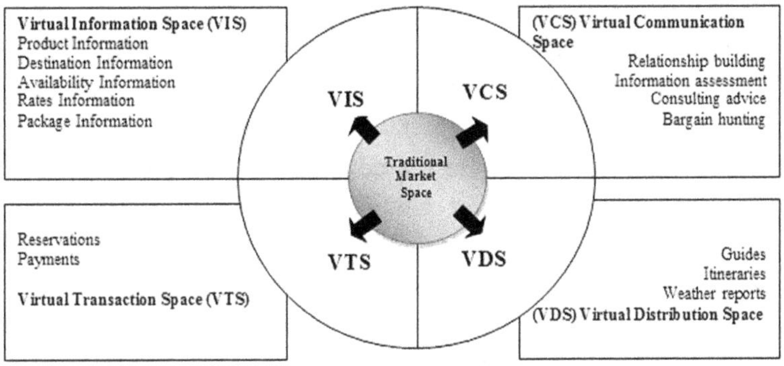

The ICDT model proposed that the Internet has extended the traditional market space by providing new spaces in which economic agents can interact by exchanging information, communicating, distributing different types of products and services and initiating formal business transactions (Angehrn, 1997). The ICDT model provides a basis for identifying how existing products and services could be extended and redesigned, as well as defining the characteristics of completely new services whose conception and development has been made possible by the Internet (Angehrn, 1997). The model categorises the opportunities presented to economic agents through the Internet into four different virtual spaces namely: the information, communication, distribution and transaction space (Halawani et al., 2013) as shown in Figure 1.

Guided by the ICTD model related studies by Abou-Shouk et al. (2012), Molla and Licker (2005), Maswera et al. (2008), Maswera et al. (2009) on e-Commerce, this book chapter proposes that e-Commerce platforms for tourism SMMEs should be characterised by the institutionalization of e-Commerce with SEM and/or listing services on intermediary websites, and engaging through social media websites. These characteristics are discussed below to demonstrate how tourism SMMEs can take advantage of the four virtual spaces presented by the Internet according to the ICDT mode.

1. **Institutionalization of E-Commerce:** Institutionalization of e-Commerce refers to organisations that have interactive, transactive or integrated e-Commerce capabilities (Molla & Licker, 2005). This level of e-Commerce adoption allows for the attainment of virtual information, distribution, transaction and to a certain extent, communication spaces as proposed by Angehrn's (1997) ICDT model. By so doing, tourism SMMEs can take advantage of the virtual information space provided by the Internet and display corporate information, information on products and services and Frequently Asked Questions (FAQs). Additionally, virtual distribution space can be met by making available information relating to destination weather or climate; how to get there; local transport information; other places to see or visit; where to stay; security; immigration procedures and customs (O'Brien, 1998 in Halawani et al., 2013; O'Brien, 1998 in Karanasios, 2008; Maswera et al., 2008). In addition, owning a transactive website implies that the virtual transaction space is catered for and will cover Internet-based channels through which economic agents can exchange formal business transactions such as payments and reservation facilities (Halawani et al., 2013; Maswera et al., 2008). Lastly, interactive attributes of the website also allow for parts of the vir-

tual communication space through the use of e-mails or comments on the website or instant chats (Maswera et al., 2008; Kshetri, 2007; Blyth, 2011).

a. **SEM:** In addition to the institutionalization of e-Commerce, tourism SMMEs needs to increase traffic to their websites. Online visibility allows for this by directing traffic towards the SMME's e-Commerce platform (Fiorini & Lipsky, 2012; Skiera & Nabout, 2012). Search engines can be seen as powerful ''gateway'' for online consumers to access travel-related information, as well as an important distribution channel for tourism destinations and businesses (Xiang, Wo¨ber, & Fesenmaier, 2008 in Xiang & Pan, 2011). Customers use different keywords and combinations of keywords to search for products through search engines (Skiera et al., 2012; Skiera & Nabout, 2012). When exposed to a list of search results, a number of factors influence the traveller's evaluation and selection of search results, particularly the ranking of search results link along with its relevancy to the search query (Xiang & Pan, 2011). Search engine providers like Google and Yahoo own and manage keywords specific to particular business sectors (Skiera, Eckert & Hinz, 2010). As such, SEM through Search Engine Optimisation (SEO), keyword advertising, and paid or sponsored search are effective forms of Internet marketing that tourism SMMEs can use to promote their visibility in Search Engine Result Pages (SERPs) (Moran & Hunt, 2005 in Xiang & Pan, 2011; Skiera et al., 2012; Skiera & Nabout, 2012). SEO activities can include the use of the right keywords in webpage titles and encouraging other websites to include links to a tourism business' site (Blyth, 2011).

2. **Listing Services on Intermediary Websites:** Tourism SMMEs often lack adequate resources that may hinder their ability to institutionalize e-Commerce (SEDA, 2012; Wanhill, 2002). In light of limited resources, some of the tourism SMMEs can take advantage of the business opportunities presented to them by the Internet through listing their services on intermediary websites like booking.com and travelground.com. These intermediaries provide interactive online platforms and can enhance tourism SMMEs' online visibility through inbuilt SEM. Previous studies show that SMMEs' reliance on ICT vendors for Web designing often result in them consulting vendors with unknown skills, or what Al-Qirim (2007a) refers to as "amateur" designers, who in turn design poorly structured websites that could negatively affect clients' trust as they contemplate on making online bookings. It is therefore assumed that by listing on intermediaries, tourism SMMEs stand a chance to benefit from using established and trustworthy payment systems run by established organisations. Today, it is claimed that thousands have conducted their bookings online through Travel Ground. Travel Ground has also been awarded winner of the Best Designed, Standards and Ease to use e-Commerce platform and was 2nd runner up in the Best e-Commerce Services Website in the 2012 South African e-Commerce Awards (TravelGround, 2013; e-Commerce Awards, 2013). Nevertheless, to engage an intermediary, tourism SMMEs have to pay a commission for every booking. Furthermore, they have no direct access to customers, and all customers that come through the intermediaries remain the intermediaries' customers. It can therefore be concluded that, just like with institutionalizing e-Commerce, tourism SMMEs can list their services on intermediaries and take advantage of the business opportunities presented by all the virtual spaces noted in the ICTD.
3. **Engaging Through Social Media Websites:** Through the virtual communication space, Angehrn (1997) proposes that economic agents (customers) meet to exchange ideas and experiences, influence opinions, negotiate potential collaborations, lobby, engage in relationships and create different types of virtual communities. Social media is expected to provide an alternative means of communication besides the one provided through interactive websites of tourism SMMEs and intermediary

websites on which services are listed on. As already noted by Xiang and Gretzel (2010), social media platforms represent various forms of Consumer-Generated Content (CGC) such as blogs, virtual communities, social networks, and collaborative tagging have gained substantial popularity in online travellers' use of the Internet. They assist consumers in posting and sharing their travel-related comments, opinions and personal experiences, which then serve as information for others (Xiang & Gretzel, 2010). Sparks and Browning (2011) found that the intentions to book and trust in a target hotel were higher following exposure to positive reviews on social media website. In addition, research by Xiang and Gretzel (2010) found that certain types of social media sites are grouped together, along with certain types of keywords, such that travellers are more likely to be exposed to social media websites than other tourism websites when using a search engine to find relevant travel information.

E-COMMERCE USES BY TOURISM SMMEs

The use of e-Commerce can be seen as the degree to which a firm is successful in linking with customers and converting its transaction documents into electronic form (Chong, 2006). The customer or supplier use is an important measure of success of e-Commerce systems because customer use is more often voluntary (Petter, DeLone & McLean, 2008). Research by Chong (2006), Hameed, Counsell and Swift (2012), Hong and Zhu (2006) proposed techniques for ascertaining e-Commerce use in an attempt to draw a line between adopting a technology and post adoption activities that defines the use of a technology for business purposes. Accordingly, this book chapter adopts and extends propositions by Chong (2006) and Hong and Zhu (2006) that depict the use of e-Commerce. These techniques were adjusted to accommodate tourism business activities to establish the following e-Commerce use activities:

1. **Online Sales:** Relates to the revenue generated through the website namely, the institutionalized e-Commerce platform and/or intermediary websites on which a tourism SMME's services are listed on. This technique is in accordance to previous studies (Chong, 2006; Hong & Zhu, 2006) and measures activities of the virtual transaction space.
2. **Reservations:** Tourism business involves customers booking for services such as reserving accommodation or any other particular service. Service reservations can be done manually and over the Internet through the institutionalized e-Commerce platforms and/or intermediary websites on which tourism SMMEs' services are listed on.
3. **Customer Service:** Relates to the variety of correspondence between tourism SMMEs and customers through the Web (Chong, 2006). Such documents include e-mails on enquiries, cancellation of reservations, sending of quotes or invoices or electronic marketing activities, and customer relationship management that includes sending out of promotions and special offers messages, use of customer loyalty systems and instant messaging (Maswera et al., 2008).

Critical Factors to E-Commerce Adoption and Use

Several theoretical propositions on understanding IT adoption by individuals or organizations have been made. This book chapter used the Unified Theory of Acceptance and Use of Technology by Venkatesh et al. (2003) and a Perceived e-Readiness Model proposed by Molla and Licker (2005) as sources of

Table 1. Critical success factors of e-commerce adoption and use

Factors of E-Commerce		Source Models or Studies	
		UTAUT	**PERM**
Internal Environmental Factors			
Organisational	ICT skills		✓
	Awareness		✓
	Commitment		✓
Knowledge	Ease of use or adoption	✓	✓
	Relative advantage	✓	✓
Technological	Cost		✓
External Environmental Factors			
Government support			✓
Market e-Readiness (Electronic Readiness)			✓
Information and Communication Technology (ICT) vendors		✓	✓
ICT infrastructure		✓	✓
Financial institutions		✓	✓
Social Influence		✓	

critical success factors of e-Commerce adoption and use. The UTAUT was developed through the consolidation of eight models making it the most complete model with the ability to explain 70 percent of variance in usage intentions (Venkatesh, Morris, Davis & Davis, 2003). Additionally, both the UTAUT and PERM have received a wide use and validation on different technologies in studies conducted in both developed and developing countries (Martins, Oliveira, & Popovi, 2013; Yang, Wu & Tao, 2008; Casey & Wilson-Evered, 2012; Im, Hong, & Kang, 2011).

To arrive at factors determining the adoption and use of e-Commerce by tourism SMMEs from the reviewed models, this research adopts Wymer and Regan's (2005) systematic qualitative process of charting, matching and consolidating variables. The identified factors were further categorised into internal environmental factors namely: organisational, technological and knowledge factors (Wymer & Regan, 2005) and external environmental factors (Chong, 2006) as shown in Table 1.

MAIN FOCUS OF THE CHAPTER

The need to propose a framework for this book chapter is motivated by the fact that most of the research on technology adoption has been focused on economically developed countries with little focus on the least economically developed countries (Molla & Licker, 2005; Tyler et al., 2007). Thus, businesses in developing countries face different challenges from those in developed countries which complicate the generalisability of such propositions or findings (Ghobakhloo, Arias-Aranda & Benitez-Amado, 2011; Kshetri, 2007; Molla & Licker, 2005; Tyler et al., 2007). Another reason for proposing a framework is that, commonly used and popular models such as the PERM, UTAUT, TAM and frameworks like the Technological-Organisational-Environmental (TOE) framework were proposed without looking much into the adopter's contextual characteristics such as the business size (whether the business under investigation is an SMME or large) and the business sector or industry the researched adopter is in. This limits the generalisability of subsequent models given that such issues can play a critical role in determining concerns around e-Commerce adoption such as the characteristics of respective e-Commerce platforms or websites. For instance, the major difference between SMMEs and large organizations is reflected by the scarce or restricted resources owned and controlled by SMMEs, which is commonly referred to as resource poverty (Ghobakhloo et al., 2011). In light of resource constraints, the owners or managers of SMMEs are often entrusted with the allocation of scarce resources for their businesses, including investing in e-Commerce adoption (Al-Qirim, 2007a).

In addition, it is generally accepted that e-Commerce has the capacity to transform not only internal practices but also the methods SMMEs use to interact with their trading partners, associates and customers (Chau, 2004 in Lin et al., 2005), hence the levels of sophistication of e-Commerce platforms can vary across SMMEs in different business sectors. Research by Hong and Zhu (2006) note that businesses in the service sector tend to have a higher e-Commerce usage as they deal with intangible products whose value chain processes can be easily digitized. Consequently, this difference in business sector of the researched SMMEs lead Ghobakhloo et al. (2011) into concluding that the generalisability of their study on manufacturing SMMEs to other business sectors or context can become problematic.

Proposed Framework of E-Commerce Adoption and Use

Using insights from the reviewed literature on e-Commerce platforms for tourism businesses, e-Commerce uses by tourism businesses and critical success factors of e-Commerce (discussed above), this book chapter proposes a framework of e-Commerce adoption and use by tourism SMMEs. The framework shows how the discussed components link along as Web 2.0 impact tourism SMMEs of the Eastern Cape's adoption and use of e-Commerce. Due to Web 2.0, tourism SMMEs are expected to institutionalize e-Commerce that includes SEM activities and/or listing services on intermediary websites and engaging through social media websites. Additionally, tourism SMMEs are expected to use these e-Commerce platforms to conduct online sales, reservations and customer services as shown in Figure 2.

Figure 2. A proposed framework of e-commerce adoption and use

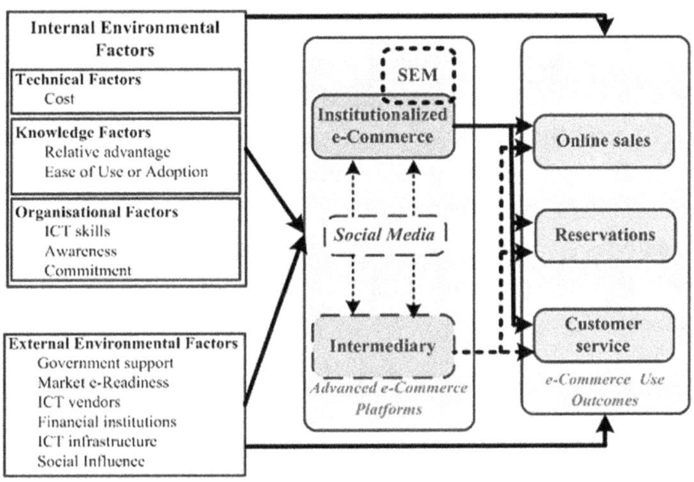

RESEARCH METHODOLOGY

The research onion proposed by Saunders, Lewis and Thornhill (2009) was adapted in this book chapter to provide a guideline on the core elements for this book chapter's research methodology. Accordingly, this book chapter subscribes to the interprevism philosophy. The, interpretivism philosophy argues that human beings are not mechanistic and they have multiple realities which need to be understood within their context (Roux, 2005). The book chapter uses multi-case studies within the interpretivism philosophy. Case studies are suitable for this book chapter as the researcher has no control over the phenomenon under study and the phenomenon cannot be studied outside the context in which it occurs (Dube & Pare, 2003; Roux, 2005).

Multiple Case Study Research Design

The book chapter defines tourism SMMEs in the Eastern Cape Province that have adopted e-Commerce as its unit of analysis. Four cases were used as sources of data. Eisenhardt (1989: 15) in Choi et al. (2011:331) motivated the notion that 4 to 10 cases "usually work well" and can help capture complexity of the real world. In addition, the fewer the number of cases, the greater the opportunity for in-depth observation, thereby increasing chances of better results (Voss et al. in Choi et al., 2011). The book chapter also used the snowballing sampling method to select its cases for data collection. Collis and Hussey (2009) state that snowball sampling is associated with interpretivist studies where cases with experience of the phenomenon being studied are selected. Triangulation (document analysis, questionnaire, observation and interviews) was then used for data collection. Hill and Wright (2001:436) in Karanasios, (2008: 110) argue that "when all of the SME factors and various competing orientations are taken into consideration, no one research approach is appropriate. What is needed, therefore, is an integrated approach to SME research."

A case study protocol containing the data collection instruments, procedures and general rules that were followed during data collection was used (Dube & Pare, 2003). Accordingly, general rules observed include labelling tourism SMMEs that participated in data collection as cases (Case 1, Case 2.) and no mention of their identity was done (Beverland, Ewing & Matanda, 2006; Irani, Alshawi & Missi, 2011). In addition to obtaining a University Ethical Clearance certificate prior to data collection, this book chapter was guided by four basic ethical principles, namely: autonomy and respect for the dignity of persons, beneficence, nonmaleficence and justice (Beauchamo and Childress 2001 in Wassernaar, 2006).

Data collected using a likert scale based questionnaire for evaluating the uses of e-Commerce were analysed using the view of the majority tourism SMMEs considered to establish their impact across cases considered. Data collected through observations, interviews and document analysis was collectively analysed as these approaches were used together during data collection to support thick descriptions (Choi et al., 2011; Dube & Pare, 2003; Iacono et al., 2011). Within and cross case analysis was used for data analysis. The analysis started with the transcribing of tape-recorded interviews to obtain a full record of the conversations. Coding was used for data reduction purposes during within-case analysis starting with open coding.

Data and Sample Profile

Tourism SMMEs engaged for data collection qualifies as SMMEs according to the criteria defined in the South African National Small Business Act of 1996. Based on the number of full-time employees, Case 1, Case 2 and Case 3 are classified as Micro enterprises, while Case 4 is classified as a Small business according to the South African National Small Business Act of 1996. Cases 1 and 2 have two full-time employees while Cases 3 and 4 have four and thirteen employees respectively. The cases considered for data collection have adopted considerable Web-based platforms that describe their e-Commerce adoption as prescribed in this book chapter's unit of analysis.

Tourism E-Commerce Platforms

Appendix A provides a summary of cases considered for this book chapter and their extent of online presence. As proposed in the framework, the majority of tourism SMMEs considered for this book chapter have shown attempts to institutionalize e-Commerce, list on at least one intermediary and engage through social media website. Arguably, tourism SMMEs have modified the way they extend their services through the utilization of virtual spaces proposed by Angehm (1997). For instance, two tourism SMMEs have institutionalized e-Commerce and the other two have interactive websites but not transactive. In addition, three tourism SMMEs have listed on at least two intermediaries and engage at least one social media website. It is interesting to note that, tourism SMMEs that have institutionalized e-Commerce are also listed on at least two intermediary websites and engage social media websites. The adoption of these platforms can be attributed to the invention of Web 2.0 that has seen social media websites (in addition to company websites) representing various forms of consumer-generated content (CGC) that influence customer buying behavior. In addition, tourism SMMEs have engaged measures to enhance their online visibility through meta tags, use of keywords, keyword purchasing and place links of their websites on other websites. Blyth (2011) reckon these measures as some of the steps one could consider when moving a website up the search engine ranks.

E-Commerce Uses

While tourism SMMEs are adopting e-Commerce platforms that commensurate with the demands of their industry or sector. The level and extent of use of these platforms remains low. This section analyses findings on how tourism SMMEs utilise e-Commerce to secure their business following the invention of Web 2.0. Three usage categories were assessed, namely online sales, reservations and customer service.

1. Online Sales

This attribute evaluated the extent to which tourism SMMEs rely or receive payments through their e-Commerce platforms as an indication of website use. Considered payments include those done through tourism SMMEs' transactive e-Commerce platforms and intermediaries. Collected data showed that there is a low usage of intermediaries and website to facilitate online sales by tourism SMMEs. This could be attributed to customers' desire to use traditional methods of payment over online payment system. For instance, one interviewee stated that: "I find very few, starting to become more, but still very very few, I say no more than 5 to 10 percent book directly on the website." In addition, some small business

Figure 3. Tourism SMMEs' use of online payments systems

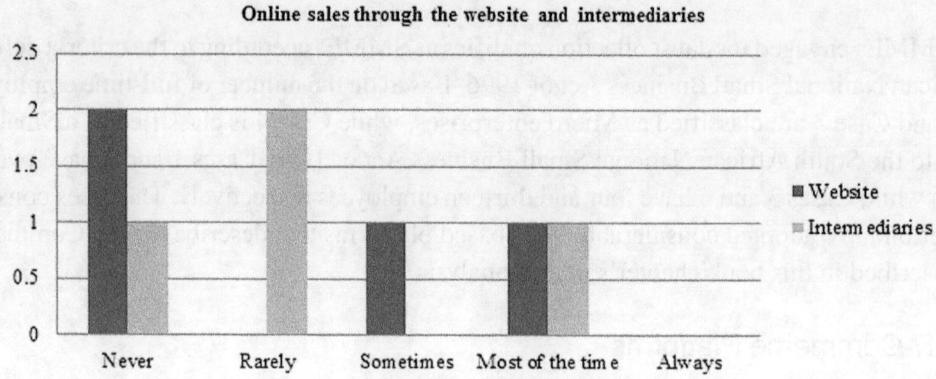

owners prefer cash over online payments as stated in the following statement by one participant: "…for me as a small business it's still for me better when I either take a credit card payment myself here or I actually get cash" (see Figure 3).

2. Reservation

Collected data reveals that tourism SMMEs makes use of their interactive and transactive website and intermediaries to facilitate reservations. As shown in the diagram below, most tourism SMMEs make use of their websites to facilitate reservations compared to intermediary websites. This could be attributed to the trust that customers gain after making phone calls to confirm what they would have seen on the website. One of the interviewees confirmed that clients still need that personal touch even when doing online business: "I still find most people go on the website and still prefer to call, they would even state to me I see you have available accommodation…" (see Figure 4).

3. Customer Service

This category assesses tourism SMMEs' use of e-mail for communication, sending out promotions, use of instant messaging for customer support, update website and intermediary content and use online loyalty system (see Figure 5).

All cases considered for data collection showed extensive use or reliance on e-mail when communicating with customers-two SMMEs indicated that they always use e-mail while the other two indicated that they use e-mail most of the time for communication. When it comes to sending promotions through e-mail only one SMME indicated that it uses e-mail for sending promotions while three other SMMEs indicated that they have never used email for sending out promotions. In addition, only one SMME confirmed that they use of instant messaging for online customer support. Despite Web 2.0's ability to allow for web content to be updated from customer or website owner end, tourism SMMEs considered indicated a very low rate of updating or regularly changing the contents of their website as shown in the diagram below.

Figure 4. Tourism SMMEs' use of online reservations

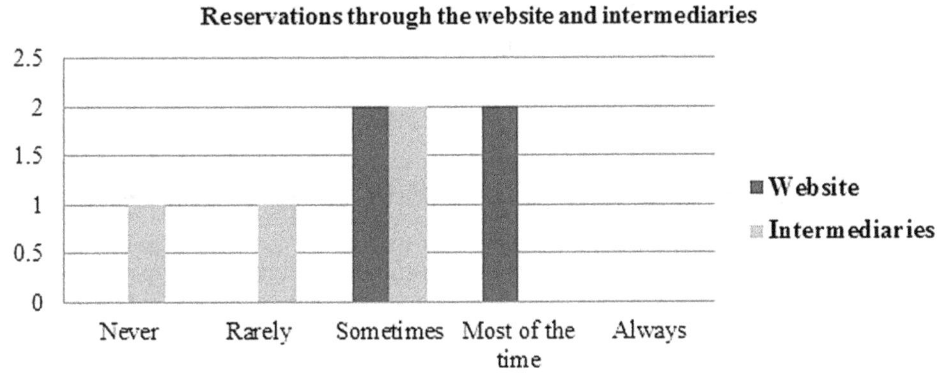

Figure 5. Web-content updating by tourism SMMEs

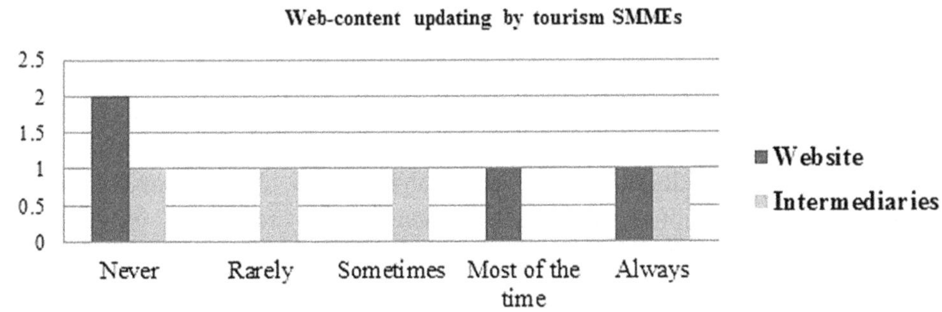

Figure 6. Tourism SMMEs' use of a loyalty system

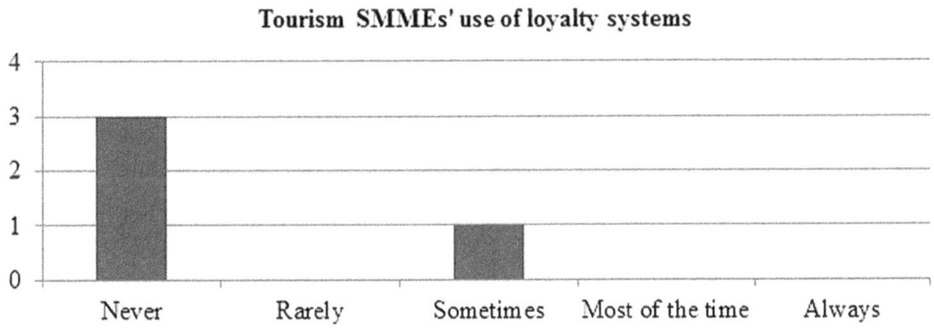

Use of Loyalty System

Only one case considered indicated that it sometimes uses a loyalty system on their website to promote repeat sales. Irrespective of the propositions by Jere, Mamello and Terzoli's (2010) that loyalty systems can enhance the competitiveness of small businesses in marginal areas, the majority of investigated tourism SMMEs indicated that they do not use such facilities on their website. The failure to use loyalty systems could be down to tourism SMMEs' failure to align the service with their e-Commerce platforms.

Blyth (2011) emphasizes the importance of strategically aligning services with e-Commerce platforms for successful online marketing. Accordingly, this attribute is dropped as a measure of customer service provided by tourism SMMEs. Figure 6 shows the tourism SMMEs' response on their use of a loyalty system.

CRITICAL SUCCESS FACTORS OF E-COMMERCE ADOPTION AND USE

This section discusses the critical success factors of e-Commerce adoption and use by tourism SMMEs.

1. Government Support

Empirical evidence shows that government support through the auspices of the Department of Environmental Affairs and Tourism has contributed to tourism SMMEs' knowledge on e-Commerce such as electronic marketing (e-Marketing) and online booking. Two cases engaged in data collection indicated that they regularly attend informative workshops organised by the local Tourism Board. In reference to the government's support through the Tourism Board's workshops, Case 1 and Case 3 stated: "The e-Marketing workshop was interesting"; "Through the tourism office, we get quite a lot of support from them…sort of classes that we can find out more about online booking and online marketing."

However, the other case considered for this research shows a lack of knowledge of the government's supportive initiatives and does not expect the government to assist them with issues pertaining to e-Commerce. Case 2 explains that: "I have never thought of the government helping me (laughs), it has never crossed my mind, and I must be brutally honest and say that."

Nevertheless, studies in developing countries show that good government policy and support can help e-Commerce growth (Li & Xie, 2012). Li and Xie (2012) and Tyler et al. (2007) note that on one hand; the Chinese government have some related policies that have encouraged e-Commerce adoption. But on the other hand, government control and restrictions, as well as the lack of legal regulation, is a source of considerable tension and may impede e-Commerce's further development within SMMEs (Li & Xie, 2012; Tyler et al., 2007). Similarly, Kabanda (2011) found that SMMEs in Tanzania lament lack of government support and low e-Readiness on their failure to institutionalize e-Commerce. All these findings affirm the finding that government support is playing an instrumental role in enhancing e-Commerce adoption and use by tourism SMMEs that have received the support. Government support is very critical considering that Web 2.0 is a bit technical and the provision of technical skills through related boards could enhance SMMEs' understanding and ability to operate their e-Commerce platforms.

2. Market E-Readiness

Early sections of this book chapter explains the nature of tourism products and customer buying behaviour in which, the role of the Internet to tourism products buying behaviour is explained. The tourism market is seen as using the Internet for searching information, purchasing services and leave comments on respective e-Commerce platforms. All cases agree that their customers use their e-Commerce platforms for bookings – something that certainly push tourism SMMEs into using their e-Commerce platforms-even though the number of users remain low with the young generations and most of European customers taking the lead. One interviewee stated that: "I find the younger generations do that (book directly on the website), the guests that have started have generally been the younger generation that trust that technology will work. I find the older generation may go on the website but they still prefer to phone." Interesting to note, customers were also posting comments with regards to their satisfaction or dissatisfaction over the services rendered to them. One of the tourism SMMEs considered showed, shown on online documents, that they respond to customer complains posted on their platforms.

These findings are in line with recent research findings by Abou-Shouk, Lim and Megicks (2012) on travel agents in Egypt, that the new generations as well as some current customers are more interested in using technology and the Internet. Customer pressure is thus perceived by the managers as one of the drivers of e-Commerce adoption (Abou-Shouk et al., 2012). On the other hand, research by Kabanda (2011) found that most Tanzanian SMMEs indicated that they would not be interested in institutionalizing e-Commerce because there was a lack of customer readiness. Accordingly, this finding implies that Market e-Readiness is pushing tourism SMMEs in the Eastern Cape into adopting and using e-Commerce.

3. ICT Vendors

The majority of cases considered in this research indicate they rely on ICT vendors' in Web designing, PC repairs and maintenance and Web hosting. The extent of reliance on ICT vendors suggest their availability something that can be explained by the fact that all cases considered are located in or near urban settings.

The importance and reliance on ICT vendors by tourism SMMEs is in agreement with previous research. For instance, Karanasios (2008) notes that tourism SMMEs in metropolitan and semi-rural areas of Ecuador suggest that there is easy access to all the necessary supporting services. Where the general support is not available locally, tourism SMMEs are forced to outsource the service from other towns. These findings affirm the finding that the availability of ICT vendors is very critical for tourism SMMEs in the Eastern Cape Province if they are to adopt and use e-Commerce with most Web 2.0 features included.

4. ICT Infrastructure

Data collected through observations, document analysis and interviews suggest tourism SMMEs located in urban setting have reliable options for connecting to the Internet as their only concern was that of slow Internet speed. However, the unavailability of ADSL in the outskirts of towns or urban settings (Case 1 and Case 4) suggests that the availability of Internet connection options suitable for business purposes remains scarce within the Eastern Cape Province. This could explain the reason why Case 1, located just outside East London, is still using an unreliable Internet connection – something that certainly affects their use of their e-Commerce platforms. Due to the invention of Web 2.0, e-Commerce platforms could come with heavy content that requires reliable internet connections with better bandwidth and broadband. Lack of reliable internet connection is therefore seen as a challenge to e-Commerce adoption and use.

In support of this finding, a full base line study by Pade, Plamer, Kavhai and Gumbo (2009) on the SLL in rural parts of the Wild Cost in the Eastern Cape found that teachers, students and the general public rely on mobile phones to access the Internet – a finding that points to the unavailability of alternative Internet connection in rural areas. This affirms the finding that ICT infrastructure is a limiting factor of e-Commerce adoption and use by tourism SMMEs in the Eastern Cape.

5. Financial Institutions

Empirical evidence shows that Case 2 and Case 3 acknowledge the availability of financial support-giving institutions that facilitate online payments. In addition, intermediaries also provide an alternative platform that could facilitate online payments. However, of particular concern to one of the cases is the time it takes for a financial transfer to be cleared and reflect in another account. Case 2 laments: "If I have a customer who pays on credit card its often four days before it actually appears in my bank…from that point of view is too long a delay before I actually see it in my bank. That for me as a small business it's still for me better when I either take a credit card payment myself here or I actually get cash." In addition, Case 2 cites challenges of Electronic Funds Transfer at international level given that tourism market spans across different countries.

Accordingly, such concerns are likely to negatively affect the adoption and use of online payments by tourism SMMEs. For instance, Kabanda (2011) state that SMMEs in Tanzania has not adopted transactive e-Commerce, instead they have adopted and use mobile payment systems like MPESA and Z-PESA that offer instant funds transfers without delay. MPESA and ZPESA are mobile phone based money transfer services designed to enable users to complete basic banking transactions without the need to visit a bank branch (Kabanda, 2011). While mobile payment systems can be of use to tourism SMMEs, their use in the tourism sector could be complicated by the fact that mobile payment use is currently limited to a particular country, whereas the tourism and hospitality market span across different countries requires that people abroad use different currencies to conduct commercial transactions. As such, the availability of financial institutions supporting instant online payments at local and international level is important for e-Commerce adoption and use.

6. Awareness

Empirical evidence shows that management or owner's awareness of changes in technological trends plays a critical role in the adoption and use of e-Commerce. It all depends on the absorptive capacity of the potential adopter, which refers to the knowledge and ability of an organisation to judge and process certain information in order to make as efficient as possible use of the information towards the adoption and use of technology within the organisation (Wang & Qualls, 2007). For instance, Case 1 stresses that: "It's just the way to go, everyone use the Internet, years ago we didn't have cell phones (but now) that's how we communicate, we have to move with time, that's what everybody is doing." In addition, Case 2's management and owner explain the specifics of their e-Commerce platforms clearly indicating how they expect to benefit from their investment which shows their awareness.

The importance of management or owner's awareness aligns with recent research findings by Abou-Shouk et al. (2012) in Egypt who noted that interviewed managers used words like 'revolution', 'trend' and 'fashion' to describe the age of Internet technologies. Some of the managers perceived e-Commerce as the fashion saying 'this is the trend in the world now' as they describe the need to adopt e-Commerce (Abou-Shouk et al., 2012). All these findings affirm the finding that management's awareness of e-Commerce can create an anticipation of its role and potential impact on the organisation, thereby enhancing chances of adoption and use.

7. Relative Advantage

The collected data shows relative advantage is an important factor that is common in all cases considered for this research. Participants cited similar and different e-Commerce advantages as the reasons behind its adoption and use. For instance, Case 1 seeks to benefit from advertising, communication and online bookings. Case 2 uses e-Commerce to market or publicise their business, display services, facilitate online bookings and offer immediacy in a highly competitive sector. Case 3 expects to get business around the world, while Case 4 seeks to generate new business and advertise.

This finding on relative advantage is constant with previous research findings. For instance, Chong (2006) found that e-Commerce adoption and use by Australian SMMEs is down to the anticipated relative advantage among other factors. Similarly, Abou-Shouk et al. (2012) found that relative advantage is a driver of e-Commerce adoption by travel agents. However, very often the lack of knowledge of the advantages that e-Commerce offer can cause impatient SMMEs to give up and discontinue the adoption and use of e-Commerce as noted by Chong (2006). As such, the knowledge of perceived benefits is seen as pushing SMMEs into adopting and using e-Commerce in their business.

8. Ease of Use or Adoption

Collected data suggest two cases see ease of use or adoption as an important factor to e-Commerce adoption and use. For instance, Case 2 indicated that they can change website content and communicate changes to their clients with ease. Accordingly, Case 3 cements the relevance of ease of use or adoption by explaining how they have made their website easy to use for them and their customers who mainly consist of an older age group. The presence of an instant message system on Case 3's website suggests their desire to make their website easy to use through the provision of instant customer support.

Arguably, research by Kabanda (2011) on the popularity of mobile payment systems among SMMEs in Tanzania can be attributed to the technology's ease of use as it is less complicated requiring very limited technical assistance. In addition, research a finding by El-Gohary (2012) show that e-Marketing's perceived ease of use is among the most important factors influencing its adoption by tourism SMMEs in Egypt. Martins et al.'s (2013) research shows that effort expectancy (ease of use or adoption) plays a critical role in influencing the behaviour intention to use technology. All these findings cement the relevance of ease of use or adoption of e-Commerce to tourism SMMEs in the Eastern Cape. e-Commerce's degree of ease to use or adopt is particularly important for tourism SMMEs considering their limited skills. On the other side, friendly and easy to use e-Commerce platforms mean that customers become more willing to use tourism SMMEs' online platforms for commercial purposes.

9. ICT Skills

Empirical evidence shows that tourism SMMEs rely on ICT vendors for technical issues like PC repairs and maintenance, and Web designing. Of all cases considered, none of them have hired an expert in IT or ICTs to provide in-house support except Case 3. This affirms Case 4's argument that ICT skills are not necessary in their organisation as they are not in line with their business. In addition, ICT skills are seen as not required on a daily basis, hence the decision to outsource. However, the major concern aligned to ICT skills lies on the availability of skills needed to effectively use e-Commerce especially now that Web 2.0 and latest versions accommodate many capabilities. For instance Case 2 feels that they are not effectively using social media for their B&B. Case 3 see Web marketing skills as necessary to keep their website visible on the Internet (to feature among the first search results on search engine) as an ongoing challenge.

In support of this finding, Abou-Shouk et al. (2012) found that many travel agents in Egypt do not have IT departments and most hire part-time technicians to solve their technical problems. In addition, Kabanda (2011) proposes that having a website requires SMMEs to acquire expertise which is difficult to acquire and retain given limited resources. Based on these arguments, this research concludes that hiring an IT specialist on a full-time basis within the context of tourism SMMEs can be a farfetched reality given SMMEs' limited resources. Whilst ICT skills remain a challenge, tourism SMMEs have shown that government support, researching and out-sourcing services of specialists could be effective measures that contribute to the alleviation of the magnitude of challenges resulting through ICT skills. As such, it is therefore concluded that the availability of ICT skills (through government support, research and out-sourcing services of specialists) within tourism SMMEs promote e-Commerce adoption and use, while their unavailability discourages its adoption and use.

10. Cost

The majority of cases considered show that costs associated with e-Commerce adoption and use are a major stumbling block. However, when viewed in light of the anticipated benefits, these costs are seen as having less impact on e-Commerce adoption and use. For instance, Case 2 and Case 4 concede that Web designing costs are high, but worth it.

In addition, Case 2, Case 3 and Case 4 all cite the cost for connecting to the Internet as very expensive. Furthermore, Case 2 laments the cost of commission charged by some intermediaries which is seen as crippling the industry.

Accordingly, previous research by Al-Qirim (2007b) found that cost is not an impending factor to both e-Commerce adopters and e-Commerce adopters due to the adoption of simple technologies and anticipated benefits respectively. In addition, Mapeshoane and Pather (2012) recently found that 45.2 percent of the respondents agree that the cost of computer equipment is too high, therefore e-Commerce is not an option in their business. However, the remaining 54.8 percent suggest cost in not an inhibiting factor among the tourism SMMEs that were considered (Mapeshoane & Pather, 2012). These findings suggest that without a clear understanding of e-Commerce to have a better anticipation of the benefits, Web and Internet related costs can impede e-Commerce adoption and use.

11. Commitment

Data collected shows that management or owner commitment positively influences e-Commerce adoption and use. Even though tourism SMMEs do not have policies directly supporting e-Commerce activities, commitment is reflecting through the amount of effort and extent of the management or owner's involvement. For instance, management of Case 2 was directly involved in the identification and selection of a suitable online payment system through research. In addition, the management for Case 2 is also committed to enrol for short courses where they learn how to use social media for their business. Case 3 management hired a Web Administrator who works extended hours dealing with issues that concerns their e-Commerce platforms. Accordingly, management or owners for Case 1 and Case 4 have shown their commitment towards e-Commerce adoption and use despite ICT infrastructural challenges within their locations.

These findings are constant with previous studies. For instance, Abou-Shouk et al. (2012) note the manager or owner's commitment to adopt technology as another driver of adoption. Similarly, Molla and Licker (2005) found management commitment as one of the important factors of e-Commerce adoption and institutionalization. On the other hand, Al-Qirim (2007b) identified that e-Commerce adoption in New Zealand SMMEs was hindered by a paucity of Chief Executive Officers' (CEOs) involvement. These proposals cement the research finding that management commitment push tourism SMMEs in the Eastern Cape into adopting and using e-Commerce.

12. Security

Empirical evidence shows two cases citing security concerns with regards to e-Commerce use. Case 1 and Case 2 often receive 'bogus' enquiries through their e-Commerce platforms. In addition, Case 2 indicated that they also get 'bogus' or fraudulent bookings through the intermediaries they are listed on, which they have to reject or reverse with their intermediaries to avoid being charged commission for a non-existing booking. On the other hand, Case 3 indicates that they have anti-virus hence security threats due to viruses is not seen as a concern: "We have no (security) concerns. We just make sure we have got the good security online; you have to have a good anti-virus and other security measures on the website and internet." Accordingly, Case 4 concede that "I don't allow people to use a memory stick on my computer" as one of the measures to cab the spread of security threats through viruses. In addition, Case 4's manager and owner is the only one who uses the company computer – something they see as improving security – as such, security is not seen as a threat. Thus, it is concluded that security issues present a threat to the use of e-Commerce platforms for business purposes.

13. Competitors' Pressure

Empirical evidence shows that only Case 2 see competitors' pressure as one of the factors driving them into adopting and using e-Commerce. It should be noted that Case 2 is a new entrant in the tourism business with only a year of existence, while other cases considered have been in business for over five years – a factor that can explain their view on competitors. In particular, Case 2 indicated that they designed their e-Commerce platforms looking at the reasons why customers should come to their B&B at the expense of competitors. In addition, the e-Commerce facilities that are linked to available rooms, prices, booking system and other services are used to improve immediacy or customer response that they instantly know what is and is not available, as Case 2 attempt to improve customer service in a competitive sector.

In line with these findings, Al-Qirim (2007b) notes that few SMMEs' owners indicated that they adopted different e-Commerce technologies to compete in the marketplace and that it was very important for them to adopt these technologies to maintain that market lead. On the other hand, Abou-Shouk et al. (2012) found competitive pressure as one of the strongest drivers behind e-Commerce adoption as tourism SMMEs compare themselves and fight for competitive positions, not wishing to be left behind by fellow competitors. These findings cement this research's argument that competitors' pressure can act as a driving force behind e-Commerce adoption and use by tourism SMMEs in the Eastern Cape.

Figure 7. The proposed framework of e-commerce adoption and use

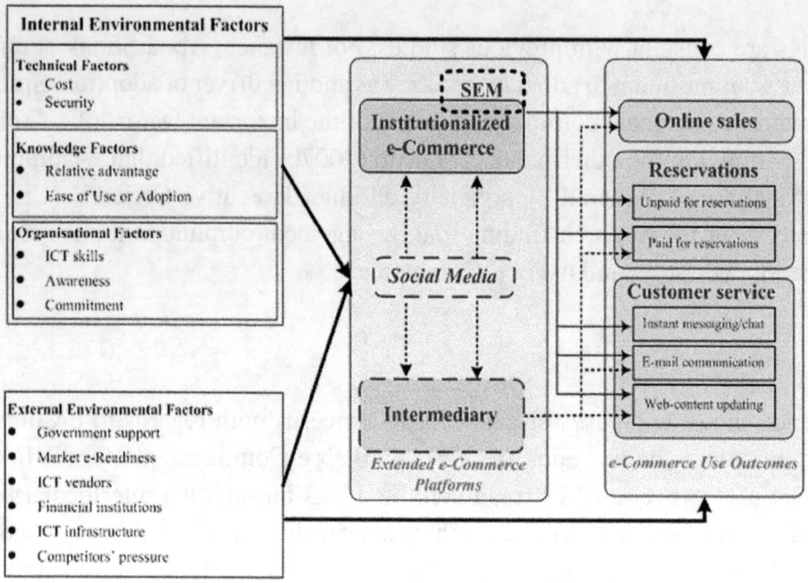

SOLUTIONS AND RECOMMENDATIONS

Based on the empirical evidence discussed in the previous section, this book chapter proposes a framework of e-Commerce adoption and use. As shown in Figure 7, tourism SMMEs that have institutionalized e-Commerce also implemented SEM activities for them to be visible online. The framework does not show SEM activities separately under intermediaries (see Figure 7-revised framework after data collection) because intermediaries offer SEM activities as part of the package of listing services with them. On the other hand, social media websites that provide for tourism customers' comments is also included and expected to provide useful information to current customers and potential customers who identify tourism SMMEs through the institutionalized e-Commerce and/or listings on intermediary websites. Such information can include comments on tourism SMMEs' quality of services and overall facilities that may influence a customer's buying decision-one of the major elements of Web 2.0.

Nevertheless, research on e-Commerce adoption and migration by Hong and Zhu (2006) suggests that quite often businesses with e-Commerce platforms do not conduct transactions online. Hence the need to capture the quantitative property that describes the amount of commercial activities conducted through e-Commerce platforms. Accordingly, research findings on tourism SMMEs in the Eastern Cape show that even though tourism SMMEs have shown a desire to adopt e-Commerce, the use of such platforms for business purposes remains low. For instance, there are cases where only a few customers are paying through online payment facilities provided by tourism SMMEs, instead they prefer to do direct transfer. Tourism SMMEs cite the lack of trust by customers (especially South African based customers) and the technology is still new to them as some of the reasons why online sales are low. On the other hand, tourism SMMEs investigated have shown a low use of their e-Commerce platforms to promote online sales. For instance, tourism SMMEs do not regularly send out periodic promotional e-mails and use loyalty systems to promote repeat purchase.

Empirical evidence shows that internal and external environmental factors play a pivotal role to the adoption and use of e-Commerce. Internal environmental factors found to be instrumental to e-Commerce adoption and use includes technical factors, namely cost and security. Costs of connecting to the Internet and Web designing costs are seen as high. Additionally, costs for subscribing to some intermediaries are also considered expensive. Nevertheless, quite often tourism SMMEs cite cost as a minor concern when compared to the anticipated relative advantage of adopting and using e-Commerce. On the other hand, unsolicited e-mails and fraudulent bookings through intermediaries are seen as major security concerns.

In addition, knowledge factors, namely relative advantage and ease of use or adoption, were seen as influencing e-Commerce adoption. Relative advantage shows to be the most influential factor pushing tourism SMMEs into adopting and using e-Commerce. Ease of use or adoption is seen as an important factor to the adoption and use of e-Commerce.

Organisational factors, namely ICT skills, awareness and commitment, show to have an influence on e-Commerce adoption and use. Due to tourism SMMEs' reliance on ICT vendors, only ICT skills to effectively use e-Commerce platforms are seen as a major concern, for example, the use of social media like Facebook for business purposes. Management awareness and commitment were also found to be beneficial in promoting e-Commerce adoption and use.

Within external environmental factors, government support, market e-Readiness, ICT vendors, financial institutions, ICT infrastructure and competitors' pressure were found to be influential to e-Commerce adoption and use. Government support is seen as contributing to tourism SMMEs' ICT and business skills, while market e-Readiness is seen as pushing tourism SMMEs into adopting and using e-Commerce platforms as most customers are seen as using the Internet on a daily basis. In addition, tourism SMMEs rely on ICT vendors for specialist services like Web designing and PC repairs and maintenance. One of the tourism SMMEs engaged through data collection argued that ICT skills are not their line of business and they do not need ICT skills on a daily basis, hence the reason behind outsourcing the service.

While tourism SMMEs acknowledge the availability of financial institutions, delay in clearing funds transfer is seen as a major concern to the use of e-Commerce platforms for business purposes. In addition, ICT infrastructure with regards to the availability of alternative Internet connection is seen as scarce. The Internet speed is also considered to be very low considering the Internet subscription costs. Lastly, competitors' pressure is seen as pushing tourism SMMEs into adopting and using e-Commerce. It should be noted that competitors' pressure was not in the initial proposed framework; instead it emerged through data collection.

FUTURE RESEARCH DIRECTIONS

This book chapter has successfully proposed a framework of e-Commerce adoption and use with the aims of accommodating Web 2.0 features. In terms of future research, it could be important to investigate the applicability of the proposed framework of e-Commerce adoption and use to tourism sectors of other countries to validate the framework. In addition, the proposed framework of e-Commerce adoption and use could be used for studying the adoption of e-Commerce in other service sectors. It may be worth examining whether the framework of e-Commerce adoption and use could be applied to the context of the service sector in general. Such empirical attempts may enable researchers to extend the generalisability of the proposed framework and develop a more incisive perspective on the relationship between business strategic orientations and e-Commerce adoption and use in other service sectors.

CONCLUSION

This book chapter researched and proposed a framework that shows how tourism SMMEs in the Eastern Cape Province could keep up with technological advancement in web. The framework of e-Commerce adoption and use shows a blue print on how tourism SMMEs could successfully adopt e-Commerce and benefit from features that come with Web 2.0. The framework shows challenges and incentives of e-Commerce adoption and use. It also shows the platforms tourism SMMEs could consider to keep up with technological advancements in the Web. Tourism SMMEs could use e-Commerce platforms to facilitate online payments, reservations and customer service though e-mail communication, Web-content updating with regards to new prices and or promotions and using instant messaging/chat for customer support.

REFERENCES

Abou-Shouk, M., Lim, W. M., & Megicks, P. (2012). e-Commerce and small tourism businesses in developing countries: Drivers versus boundaries of adoption. *Tourism Planning & Development.*

Al-Qirim, N. (2007a). The adoption of eCommerce communications and applications technologies in small businesses in New Zealand. *Electronic Commerce Research and Applications, 6*, 462–473.

Al-Qirim, N. (2007b). A Research Trilogy into e-Commerce Adoption in Small Businesses in New Zealand. *Electronic Markets, 4*, 263–285.

Angehrn, A. (1997). Designing mature Internet business strategies: The ICDT model. *European Management Journal, 15*, 361–369.

Beverland, M. B., Ewing, M. T., & Matanda, M. J. (2006). Driving-market or market-driven? A case study analysis of the new product development practices of Chinese business-to-business firms. *Industrial Marketing Management, 35*, 383–393.

Blyth, A. (2011). *Brilliant online marketing. How to use the Internet to market your business.* Pearson Education Limited.

Casey, T., & Wilson-Evered, E. (2012). Predicting uptake of technology innovations in online family dispute resolution services: An application and extension of the UTAUT. *Computers in Human Behavior, 28*, 2034–2045.

Choi, T. Y., Barratt, M., & Li, M. (2011). Qualitative case studies in operations management: Trends, research outcomes,and future research implications. *Journal of Operations Management, 29*, 329–342.

Chong, S. (2006). An empirical study of factors that influence the extent of deployment of electronic commerce for small-and medium sized enterprises in Australia. *Journal of Theoretical and Applied Electronic Commerce Research, 1*(2), 45–57.

Collis, J., & Hussey, R. (2009). *Business Research: A practical guide for undergraduate & postgraduate students* (3rd ed.). Palgrave Macmillan.

Dube, L., & Pare, G. (2003). Rigor in Information Systems Positivist Case Research: Current Practices, Trends and Recommendations. *Management Information Systems Quarterly, 27*(4), 597–636.

e-Commerce Awards. (2013). *e-Commerce Awards Source*. Retrieved May 2, 2013, from http://www.ecommerceawards.co.za/2012/nominations/travelground

El-Gohary, H. (2012). Factors affecting E-Marketing adoption and implementation in tourism firms: An empirical investigation of Egyptian small tourism organisations. *Tourism Management*, *33*, 1256–1269.

Fiorini, P. M., & Lipsky, L. R. (2012). Search marketing traffic and performance models. *Computer Standards & Interfaces*, *34*, 517–526.

Ghobakhloo, M., Arias-Aranda, D., & Benitez-Amado, J. (2011). Adoption of e-Commerce applications in SMEs. *Industrial Management & Data Systems*, *8*, 1238–1269.

Gilmore, A. (2003). *Service Marketing and Management*. London: SAGE Publications Ltd.

Halawani, F., Abdullah, M. M. B., Rahman, M. S., & Halawani, Y. (2013). A Proposed Framework for E-Commerce Usage and Competitive Advantage on Small and Medium Tourism Enterprises (SMTES) in Lebanon. *Journal of Social and Development Sciences*, *6*, 258–267.

Hameed, A. M., Counsell, S., & Swift, S. (2012). A Conceptual Model for the Process of IT Innovation Adoption in Organizations. *Journal of Engineering and Technology Management*, *29*, 358–390.

Ho, C., Lin, M., & Chen, H. (2012). Web users' behavioural patterns of tourism information search: From online to offline. *Tourism Management*, *33*, 1468–1482.

Hong, W., & Zhu, K. (2006). Migrating to internet-based e-commerce: Factors affecting e-commerce adoption and migration at the firm level. *Information & Management*, *43*, 204–221.

Huang, Z., & Benyoucef, M. (2013). From e-commerce to social commerce: A close look at design features. *Electronic Commerce Research and Applications*, *12*, 246–259.

Iacono, J. C., Brown, A., & Holtham, C. (2011). The use of the Case Study Method in Theory Testing: The Example of Steel eMarketplaces. *The Electronic Journal of Business Research Methods*, *9*(1), 57–66.

Im, I., Hong, S., & Kang, M. S. (2011). An international comparison of technology adoption Testing the UTAUT model. *Information & Management*, *48*, 1–8.

Irani, Z., Alshawi, S., & Missi, F. (2011). Organisational, technical and Data Quality Factors in CRM Adoption-SMEs Perspective. *Industrial Marketing Management*, *40*, 376–383.

Jere, N. R., Thinyane, M., & Terzoli, A. (2010). Enhancing e-commerce services for a rural community: online customer negotiation application. ZA_WWW 2010, Graduate School of Business (UKZN).

Jeyaraj, A., & Sabherwal, R. (2008). Adoption of information systems innovations by individuals: A study of processes involving contextual, adopter, and influencer actions. *Information and Organization*, *18*, 205–234.

Kabanda, S. (2011). "E-Commerce Institutionalization is not for us": SMEs perception of E-Commerce in Tanzania. *The African Journal of Information Systems*, *1*, 1–16.

Karanasios, S. S. (2008). *An e-Commerce framework for small tourism enterprises in developing countries.* (Unpublished Doctoral of Philosophy Thesis). Victoria University.

Kshetri, N. (2007). Barriers to e-Commerce and Competitive Business Models in Developing Countries: A Case Study. *e-Commerce Research and Applications, 6*, 443–452.

Litvin, S. W., Pan, B., & Goldsmith, R. E. (2008). Electronic word-of-mouth in hospitality and tourism management. *Tourism Management*, *29*, 458–468.

Mapeshoane, T. J., & Pather, S. (2012). Adoption of e-commerce in typical developing country context: Lesotho tourism industry.*Proceedings of the 14th Annual Conference on World Wide Web Applications.*

Martins, C., Oliveira, T., & Popovi, A. (2013). Understanding the Internet banking adoption: A unified theory of acceptance and use of technology and perceived risk application. *International Journal of Information Management.*

Maswera, T., Dawson, R., & Edwards, J. (2008). e-Commerce Adoption of Travel and Tourism Organisations in South Africa, Kenya, Zimbabwe and Uganda. *Telematics and Informatics*, *25*, 187–200.

Molla, A., & Licker, P. S. (2005). eCommerce adoption in developing countries: A model and instrument. *Information & Management*, *42*, 877–899.

Pade, C., Plamer, R., Kavhai, M., & Gumbo, S. (2009). *Siyakula Living Lab: Baseline study report.* (Unpublished report). Grahamstown: Rhodes University.

Palmer, A. (2005). *Principle of Services Marketing* (4th ed.). McGraw-Hill Education.

Petter, S., DeLone, W., & McLean, E. (2008). Measuring information systems success: Models, dimensions, measures, and interrelationships. *European Journal of Information Systems*, *17*, 236–263.

Rogerson, C. M. (2005). Unpacking Tourism SMMEs in South Africa: Structure, Support Needs and Policy Response. *Development Southern Africa*, *22*, 623–642.

Roux, L. l. (2005). *Qualitative research: method in the madness?* A Working paper, University of Stellenbosch Business school.

Saunders, M., Lewis, P., & Thornhill, A. (2009). *Research Methods for Business Students* (5th ed.). New York: FT/Prentice Hall.

Seabra, C., Abrantes, J. L., & Lages, F. L. (2007). The Impact of Using Non-media Information Sources on the Future Use of Mass Media Information Sources: The Mediating Role of Expectations Fulfilment. *Tourism Management*, *28*, 1541–1554.

Skiera, B., Eckert, J., & Hinz, O. (2010). An analysis of the importance of the long tail in search engine marketing. *Electronic Commerce Research and Applications*, *9*, 488–494.

Skiera, B., & Nabout, N. A. (2012). Return on Quality Improvements in Search Engine Marketing. *Journal of Interactive Marketing*, *26*, 141–154.

Skiera, B., Nabout, N. A., Stepanchuk, T., & Gerstmeier, E. (2012). An analysis of the profitability of fee-based compensation plans for search engine marketing. *International Journal of Research in Marketing*, *29*, 68–80.

Sparks, B. A., & Browning, V. (2011). The impact of online reviews on hotel booking intentions and perception of trust. *Tourism Management*, *32*, 1310–1323.

TravelGround. (2013). *The easiest way to find and book accommodation across South Africa*. Retrieved May 2, 2013, from: http://www.travelground.com

Tyler, K., Tan, J., & Manica, A. (2007). Business-to-business adoption of e-Commerce in China. *Information & Management*, *44*, 332–351.

Venkatesh, V., Morris, M. G., Davis, G. B., & Davis, D. F. (2003). User acceptance of information technology: Toward a unified view. *Management Information Systems Quarterly*, *2*, 425–478.

Wang, Y., & Qualls, W. (2007). Towards a Theoretical Model of Technology Adoption in Hospitality Organizations. *Hospital Management*, *26*, 560–573.

Wanhill, S. (2002). *Sustaining Tourism SMEs.VII Congreso Internacional del CLAD sobre la Reforma del Estado y de la Administración Pública*, Lisboa, Portugal.

Wassenaar, D. R. (2006). Ethical issues in social science research. In *Research in practice: Applied methods for the social sciences*. Cape Town: UCT Press.

Werthner, H., & Ricci, F. (2004). e-Commerce and Tourism. *Communications of the ACM*, *47*, 101–105.

Wymer, S. A., & Regan, E. A. (2005). Factors Influencing e-commerce Adoption and Use by Small and Medium Businesses. *Electronic Markets*, *15*(4), 438–453.

Xiang, Z., & Gretzel, U. (2010). Role of social media in online travel information search. *Tourism Management*, *31*, 179–188.

Xiang, Z., & Pan, B. (2011). Travel queries on cities in the United States: Implications for search engine marketing for tourist destinations. *Tourism Management*, *32*, 88–97.

Yang, P., Wu, Y., & Tao, Y. (2008). The use of unified theory of acceptance and use of technology to confer the behavioral model of 3G mobile telecommunication users. *Journal of Statistics & Management Systems*, *11*, 919–949.

APPENDIX

Table 2. A description of cases considered for this book chapter

Case	Location and Business Description	Online Presence	Interviewees Engaged	Position of Participant
Case 1	Self-catering bed and breakfast and weddings venue. Located 20 kilometres away from East London off the N2 road to Butterworth in a location surrounded by farming plots.	• Interactive but not transactive website, • Not listed on any intermediary or social media website.	1	Business owner and manager
Case 2	Bed and breakfast provider. Located in the low density area of Port Elizabeth.	• Interactive, transactive and integrated website • Listed on intermediaries namely: www.roomsforafrica.com; www.safarinow.com and on social media www.tripadvisor.com. • Uses SEO, keyword purchasing and places links of their websites on other websites to enhance online visibility.	1	Business owner and manager
Case 3	Game reserve and safari lodge located between Grahamstown and Port Elizabeth road. Administration offices are located in Port Elizabeth.	• Transactive and interactive website, • Listed on intermediaries: www.safarinow.com; www.booking.com; www.nature-reserve.co.za and social media website: www.tripadvisor.com. • Uses SEO, meta tags and places links of their website on other websites to enhance online visibility.	2	Web administrator and Receptionist
Case 4	A B&B located 5 kms south of Queenstown surrounded by farming plots.	• Interactive website • Listed on intermediaries such as www.sa-venues.com; www.booking.com ; www.safarinow.com and social media www.tripadvisor.com	1	Business owner and manager

Chapter 9
Cultivating Recognition:
A Classic Grounded Theory of E-Learning Providers Working in East Africa

Titus Tossy
Mzumbe University, Tanzania

Irwin Brown
University of Cape Town, South Africa

ABSTRACT

The E-learning provision in East Africa including Tanzania, Kenya and Uganda is rapid gaining its way. E-learning providers are all groups tasked with e-learning activities including design, delivery, implementation and others. The e-learning providers are surrounded by issues and challenges (main concerns). The purposed of this study was to identify and examine the main concern amongst e-learning providers in East Africa (Tanzania, Kenya and Uganda). In this study the Classic Grounded Theory Research Methodology was used in order to identify the main concern of e-learning providers. In the course of repeated encounters with e-learning providers, this study found the concept of Cultivating Recognition to emerge as the main concern or core variable amongst these e-learning providers. The core variable within the Classic Grounded Theory research methodology is the main entity that accounts for most of the variation in the data. The core variable of the theory, the basic social psychological process of Cultivating Recognition is characterized by Legitimizing and Credentializing. The process of legitimizing involves convincing the e-learning stakeholders that the e-learning programs and projects will be delivered in a timely, valid and sustainable manner. Legitimizing process is achieved through Collaborating (reciprocal and asymmetric), Referral Networking and Strategic alliancing. Credentializing aims to enhance the stakeholders' belief in the e-learning providers' competence prior to the provision of the e-learning programs and projects. While Credentializing is achieved through Endorsementizing, Result Orienting and Prioritizing Duties, Result Orienting itself is achieved through Visualizing, Professionalizing, Focalizing and Role Delineating. This study will help e-learning providers and other groups to cultivate recognition and pave a way for their success in implementation of e-learning.

DOI: 10.4018/978-1-5225-0539-6.ch009

INTRODUCTION

In a contemporary setting, in the 21st century, it is unlikely that there is a sector in East Africa, including the education sector, which functions without integrating Information and Communication Technologies (ICTs) into its operations. There is increased use of ICT in every aspect of the economy in East Africa (Sife, et al. 2007). It does not take an ICT specialist to see that there is on-going ICT use and development in East Africa. ICT use is concentrated more in urban areas than in rural areas due to the availability of supporting infrastructure, such as electricity and ICT expertise (Ndume et al. 2008). Sife et al. (2007) argue that, due to lack of infrastructure (electricity, internet access, bandwidth and others), the use of ICTs in rural areas is yet to reach a level which compares with that in urban areas. There are on-going debates about the use of ICTs in various sectors, not excluding the education sector, and including conferences and other forums. The education sector has witnessed rapid growth in the use of ICTs in comparison with other sectors such as agriculture (Sife et al. 2007). The use of ICTs in teaching and learning opens new vistas to education and its accessibility (Galagan, 2000). Netland (2007) argues that ICTs have become a potent force in transforming social, economic, educational, and political life both globally and in East Africa and other developing regions. ICTs provide an opportunity for educational institutions to harness and use technology to complement and support the teaching and learning process. E-learning is a popular known term used to refer the use of these ICTs to support teaching and learning (Wentling, et al, 2000; Khan, 2005; Galagan, 2000). Wentling et al. (2000:5) defined e-learning as:

The acquisition and use of knowledge distributed and facilitated primarily by electronic means. This form of learning currently depends on networks and computers but will likely evolve into systems consisting of a variety of channels (e.g. Wireless, satellite), and technologies (e.g. Cellular phones,etc.) as they are developed and adopted. E-learning can take the form of courses as well as modules and smaller learning objects. E-learning may incorporate synchronous or asynchronous access and may be distributed geographically with varied limits of time. (Wentling et al., 2000:5).

E-learning facilitates both learner engagement and the engaging of experiences (Uys, 2004; Meyen, 2000; 2002). Meyen (2002) demonstrate how e-learning helps to overcome the traditional barriers to education delivery. These barriers include lack of physical infrastructure, lack of qualified teaching staff, absence of adequate education budgets, and the failure of traditional pedagogy and curricula. East African countries are characterised by these barriers (Ndume et al, 2008). The failure of the government's efforts in building physical classrooms has created an opportunity for innovative education delivery via e-learning (Yieke, 2005).The implementation of e-learning in East African Higher Education Institutions (HEIs) is taking place despite the various constraints. Chacha (2009:4) argues for the importance of ICTs in education:

There is a need to tap the potential of ICTs to enhance data collection and analysis, and to strengthen management systems in educational institutions; to improve access to education by remote and disadvantaged communities; to support initial and continuing professional development of teachers; and to provide opportunities to communicate across classrooms and cultures.

Alavi and Leidner (2001) argues that e-learning's importance will grow right across the educational spectrum from primary to higher education institutions (HEIs). This prediction has been borne out in the United States, where in 2007 the market for 'wholly online' degrees was estimated to be worth 17.2 billion US$ compared to 5 billion US$ in 2004, with an estimate of 4 billion US$ online tutoring at the growth rate of 10 -15% annually. However, in the UK in 2004 the Higher Education Funding Council for Education (HEFCE) announced the effective demise of the £50 million UKeU online university project. Various individuals and bodies, blamed the failure of UKeU on its largely supply-driven, rather than demand-led, approach, arguing that:

There was no systematic evaluation of the markets, no thorough and robust market research and no understanding of consumer demand. This was typical of UkeU's supply-driven rather than demand-led approach. (Clarke, 2003:17)

The failure of UKeU and other e-learning initiatives (see Appendix 1) indicates that e-learning implementation should be demand-led and carefully planned from a provider's point of view. Li (2009) stated that a clear understanding of both e-learning providers (University e-learning teaching and training professionals) and customers (e-learners) was required. There are various e-learning projects in East Africa which have failed, but have not been documented. Unfortunately governments have learned little from these failures. It is failures such as these which motivated the researcher to undertake this study. Paulsen (2003) argues that a successful e-learning project should be both robust and sustainable. E-learning education offered so far has been transient, unsuccessful and far from sustainable. Most of the e-learning projects in East Africa face challenges due to inadequate knowledge or understanding of ICTs (Ndume et al., 2008; Khan, 2005; Bebee, 2004; Allen, 2002 and Alexander, 2001). Aranda (2007) and Allen (2007) have shown how conflicting agendas in terms of the appropriate use of technology, the application of pedagogy and delivery of the curriculum, lack of resources, lack of well-trained knowledgeable staff, lack of management support and poor infrastructure have exacerbated hindrances to e-learning implementation. Issues related to institutional, management, technological, pedagogical, ethical, interface design, resources and design have been cited by Khan (2005) as issues which need to be addressed at all stages of e-learning implementation, operation and evaluation processes. According to Clarke (2003) the shortcomings of e-learning implementation are due to the lack of both systematic evaluation of the market and thorough and robust market research. Macpherson et al. (2005) suggest that understanding of such shortcomings can be enhanced through developing the *"navigation skills"* to evaluate specific contexts before determining the best course of action to take in implementing e-learning. Navigation skills include understanding the main concerns of e-learning recipients as noted by Letch (2006:32):

Lack of a clear understanding of both e-learning providers' (e-learning providers that include lecturers) and customers' (e-learners) main concerns to the implementation of e-learning was required.

Scott (2007) researched the main concerns from the e-learner's perspective, and discovered "The Temporal Integration of Connected Study into a Structured Life" to be the main concern of online learners. While there are many well documented research studies, using various research methodologies on various ICT topics, there is no documented research where the main concerns from the perspective of e-learning providers have been investigated. While there is work being done by leading researchers and practitioners in the field, most of these researchers have used reductionist research methodology. In view of the recent history of the ineffective introduction of e-learning in EA, a strong case can be made for employing a new research approach. This study has focused on the perspectives of the e-learning providers, rather than on the various e-learning modes and IT involved in its implementation. It aims to fill the research gap by means of developing an understanding of the main concern of the e-learning providers working in East Africa, and how that concern is addressed, in order to deliver more effective and sustainable e-learning provision in practice through the use of an inductive research methodology.

THE STUDY

Motivation

In 2007 the Government of Tanzania appointed the researcher as the just Associate Director of e-learning based at the Open University of Tanzania. One of the contributions made by the researcher to the Government of Tanzania in e-learning was to write a successful e-learning funding proposal, which was accepted by the Swedish Program for ICTs in Developing countries (SPIDER) for funding. The researcher was able to witness how the universities have been and are struggling to implement e-learning projects. Some of these projects fail, some are approved, and of these only a few have managed to survive. The researcher was keen to discover the reasons for this failure in the context of the government's attempts to encourage the universities to implement e-learning (Ndume et al., 2008) in spite of the insistence of Alvi and Leidner (2001) that the importance of e-learning will grow right across the education spectrum from primary to higher learning institutions (HEIs) in developing countries (East African being one of them). The researcher come progressively to begin to understand the main concern of e-learning providers working in East Africa, and this embryonic understanding was the beginning of this study.

Research Objective

Considering the detailed identified gaps in the previous sections, the study's research objectives were therefore as follows *(1) To identify the main concerns of e-learning providers in East Africa; (2) To discover the most important and challenging issues that e-learning providers face in East Africa; and (3) To understand how their main concerns are processed and resolved in order to inform better e-learning provision in practice.* The research objectives were found to be of most important to uncover and recommend ways for successful e-learning implementations in East Africa. As supported by Leitch (2006), who argues for the need to know the main concerns and issues of e-learning providers and how these are addressed and resolved by them.

Research Method

To discover the main concern of respondents as regards providing e-learning required a thorough investigation of "real-life" situations relevant to the specific problem (Gasson, 1998). The research methodology was chosen on the basis of its allowing the investigation of the real-life situations relevant to the main concern of the e-learning providers working in East Africa. Due to lack of inductive studies in this research area, it was decided that there exists a need to do a pure inductive study. Among the inductive research methodologies available, it was decided that the Grounded Theory (GT) research methodology was the most appropriate for the purposes of this study. Due to the existence of many GT research methodologies, the question that needed to be answered was "Which Grounded Theory research methodology approach should be used and why?" This question is discussed in detail in next section.

Which Grounded Theory

Matavire and Brown (2011) have demonstrated the existence of different GTM approaches in Information Systems (IS) research. Such approaches include the "Classic" GT (Glaser & Strauss, 1967; Glaser, 1978; 1992; 1996; 1998; 1999a; 1999b; 2001; 2003; 2005; 2006; 2008; 2009; 2011), "Evolved" GT,(Strauss & Corbin, 1990; 1989), "Analytical" GT (Strauss & Corbin, 1990), and "Mixed" GT Method (Mingers, 2001; Eisenhardt, 1989; Baskerville and Pries-Heje, 1999.

Classic Grounded Theory Methodology (CGTM) approach is accepted as being faithful to the original formulation and follows the original tenets of Glaser & Strauss (1967). It has been further elaborated and refined by Glaser (1978; 1992; 1996; 1998; 1999a; 1999b; 2001; 2003; 2005; 2006; 2008; 2009; 2011). CGTM is a general inductive research method designed to reveal deep seated latent patterns of human behavior and how the main concerns are being continually being resolved (Glaser, 1978).

The "Evolved" Grounded Theory (EGT) approach originated primarily from Strauss & Corbin (1990), and involved some modification of and departure from the CGTM. EGT approach is also labelled as "Straussian" (Strauss & Corbin (1990) grounded theory method) (Matavire & Brown, 2008; Van Niekerk & Roode, 2009). Hernandez (2006), amongst others, points out that the Strauss and Corbin (1990) method (evolved) differs fundamentally from the CGTM of Glaser and Strauss (1967). The two fundamental differences between CGTM and EGT methodology concern the difference between the ideas of 'emergence' versus 'forcing', and whether grounded theory is a qualitative or general research method. Glaser (1978, 1992, 2008, and 2009) argues that the Strauss & Corbin (1990) method, the EGT method, changes the research method from one of emergence to one of forcing. Glaser (1982) asserts that it cannot deliver high level conceptual explanations of latent patterns which transcend both time and location.

The "Mixed" method approach attempts to combine the GT research method with other research methods (Mingers, 2001; Eisenhardt, 1989; Baskerville & Pries-Heje, 1999). Among the most popular combinations are grounded action research (Baskerville & Pries-Heje, 1999; DeLuca et al, 2008) and the grounded case study approach (Eisenhardt, 1989). Glaser & Holton (2004) are against combining classic GTM with other research methods. Such combinations, they believe, may erode the ability to generate conceptual theory (Glaser, 1992, 1998, 2001; Glaser & Holton, 2004).

When using the "Analytical" GT approach, the researcher does not necessarily follow either the evolved or the Classic GT approach. The researcher only uses the GTM techniques such as open and/or axial and/or selective coding (Matavire & Brown, 2011). When using the analytical GT approach, there is no evidence for following principles of CGT, rather than simply using GT for the purpose of analysis (Matavire & Brown, 2011). It merely follows the tools of CGT but does not utilise its tenets. As a result, the "Mixed" approach does not develop and augment theory. Glaser (1978, 1992) argues that mixing CGTM with other research methods neither delivers theory nor discovers the main concern of the substantive area of research.

Due to the existence of these many GT approaches, Van Niekerk and Roode (2009) argue that a choice of the GT research approach must be approached with care and done within the specific context of the research. The choice of research methodology should have a rationale, and such rationale should be justified in terms of the researcher's preference of methodology over other research methodologies. Heath & Cowley (2004:1) argue that:

... rather than debate the relative merits of the 'GT approaches', it is suggested that novice researchers need to select the method that best suits their cognitive style and develop analytic skills through doing research.

Firstly, being fully aware of what has been suggested by Heath & Cowley (2004), it was decided that the Classic Grounded Theory (CGT) Research Methodology (Glaser & Strauss, 1967; Glaser, 1978, 1992, 1996, 1998, 1999a,1999b, 2001, 2003, 2005, 2006, 2008, 2009, 2011) was the most appropriate for this study.

Secondly, the CGTM was chosen because it claims to deliver the main concerns of e-learning providers in East Africa, the key research objective in this study. Prior research in the field of e-learning implementation has been more concerned with comparing received wisdom of existing theories with practitioner behaviour (Scott, 2007). Most research tends to view the data from the perspective of a prior theory rather than from those subjects being researched. CGTM is an approach which ensures that the contributions to knowledge in and of the area are not generated from existing theories but are grounded in the data collected from one or more empirical study (Gasson 2004; Adolph et al., 2008; Glaser, 1978). Thus, CGTM fits both the nature of the phenomenon being researched (E-e-learning providers) as well as the key research question (the main concern) and follows Lowe's (1996:1) description of CGT as being designed to:

Develop and integrate a set of ideas and hypotheses in an integrated theory that accounts for behaviour in any substantive area. (Lowe, 1996:1)

Thirdly, the choice of CGTM over other theories, methods and approaches was due to two aspects of the uniqueness of the Classic grounded theory (Glaser, 1978, 1992) approach related to this specific study: (1) The theory is based upon patterns found in empirical data, not on inferences, prejudices, or the association of ideas, (2) There is constant ongoing comparison between emergent theory (codes and constructs) and new data. Constant comparison confirms that theoretical constructs are found across and between data samples, thus driving the collection of additional data until the researcher feels that "theoretical saturation" (the point of diminishing returns from any new analysis) has been reached.

A fourth reason for the choice of CGTM was because it provides a flexible set of inductive strategies for collecting and analysing data, is a comparative and interactive method, starts with gathering focused data and stays close to the data-developing concepts that synthesise and conceptualise the collected data. Glaser (1992) and Glaser & Strauss (1967) have outlined the key distinguishing characteristics of the CGTM research methodology as follows: *(1)simultaneous involvement in the data collection and analysis phases of research; (2)developing analytic codes and categories from data, not from preconceived hypotheses; (3) constructing middle-range theories to understand and explain behaviour and processes; (4) Memo-writing, that is, analytic notes to explicate and fill out categories; (5) Making comparisons between sets of data, data and concept, and between different concepts; (6) Theoretical sampling, that is sampling for theory construction to check and refine conceptual categories, not for representativeness of a given population;* and *(7) Delaying the literature review until after the emergence of a core variable.* Thus the distinguishing characteristics of CGTM have made it the most appropriate research methodology for a study of the main concerns of e-learning providers in East Africa.

CLASSIC GROUNDED THEORY

Ontology and Epistemology

Since the discovery of CGTM in 1967, many authors have explained and discussed ontological and epistemological stances, labelling such research paradigms positivism, interpretivism and criticalism (Burton-Jones 2009; Campbell & Russo 2001; Cupchik 2001; Mingers 2001; Burrel & Morgan 1979; Fitzgerald & Howcroft, 1998b; Adolph, et al., 2008)). Ontology is the term for an assumption or group of assumptions about 'perceived reality' (Burrell & Morgan 1979; Fitzgerald & Howcroft, 1998). This means that Ontology is the study of being - essentially studying questions of what kinds of entities exist. Ontology is viewed as a formal representation of 'knowledge' by a set of concepts within a domain, and the relationships between those concepts (Fitzgerald & Howcroft, 1998). The ontological question is whether 'reality' is objective, single and external to the individual, or in other words, a given, or taken-for-granted 'out there' in the world, or whether it is subjective, with multiple realities that exist as subjective constructions of the mind (Fitzgerald & Howcroft, 1998). Epistemology is the branch of philosophy that is concerned with the "nature of knowledge its possibility scope and general bias"(Honderich 1995, p.242). Epistemological questions are grounded in knowledge and on communicating this reality that exists out there in the world. Positivism as an epistemology and a research paradigm, builds on the ontological realist assumptions of existence, where the subjective reality and the subjective techniques of enquiry about the truths are dismissed. Instead an objective reality is assumed (Burke, 2007). Positivism is premised on the supposed 'existence of the existence of a prior relationships within phenomena' *(*Orlikowski & Baroudi, 1991:5). Positivism is aligned with experimental and quasi-experimental design, using the rules of formal logic and the rules of deductive logic (Howcroft & Trauth, 2005). It often follows hypothesis formulation and the testing of theories, focusing on enabling predictions and generalisations about a phenomenon (Myers, 1997). Fitzgerald & Howcroft (1998) argue that the realism, positivism, and the nomothetic beliefs of a researcher position the researcher as a positivist, while nominalism, antipositivism, and ideographic beliefs constitute an interpretivist view of reality or the world. While logical positivism and post-positivism are two versions of the positivist school, critical theorists and constructivists belong to the interpretivist school, constructing four major research paradigms with their respective purists (Fitzgerald & Howcroft, 1998).

CGT is essentially ontologically and epistemologically neutral (Glaser and Strauss, 1967; Glaser, 1978, 1992). As such, Glaser (2005) has argued that discussions of ontology (what we believe about the world) and epistemology (how we can come to know what we know) are moot within classic grounded theory. Christiansen (2005:81-84) argues that classic GT is an "ontology free and epistemology free methodology". Glaser (1978) furthermore reiterates that ontological assumptions have no place in the use of Classic GT since GT is grounded in data. Glaser (2004:76) argues that:

It is only a question of applying a rigorous and systematic method for discovering and explaining these patterns. Thus, just do it.

Glaser (1978:45) explains further that in the CGTM methodology all ontological and epistemological suppositions are irrelevant and are not useful:

... logical conjectures or preconceptions that pre-frame the research, and pre-framing has to be avoided in any CGTM work, in order to keep close to what the data has to say, and to discover and name emergent latent patterns in the data.

Criteria for Evaluating Classic Grounded Theory Methodology Research

The grounded theory research method is a general inductive methodology which can be applied to all types of data, both quantitative and qualitative, or the combination thereof as has been outlined by Glaser (1978, 1998). GT is not a sub set of Qualitative Data Analysis (QDA). Therefore QDA criteria are inappropriate for the evaluation of research done using the grounded theory research method (Glaser, 1978). Glaser (1978, 1998, 2000, and 2001) gives explicit examples of the four main criteria with which to evaluate grounded theory study: fit, workability, relevance and modifiability.

DATA COLLECTION AND ANALYSIS

Research Population

The data in this Classic Grounded Theory study was drawn from e-learning providers working in East Africa. Following the Classic Grounded Theory Methodology (as detailed in chapter 2), the number, scope and range of those being interviewed were determined in an ongoing process based on the emergent data as shown in Appendix 2.

Data Collection

A total of 24 e-learning providers respondents from Tanzania, Kenya and Uganda were interviewed on several occasions. This meant that the researcher had a total of 101 respondent encounters. All the respondents were e-learning providers working in East Africa with successful career track records, ranging from directors and e-learning stakeholders (consultants etc.) to e-learning lecturers and tutors. The meetings with respondents were scheduled after the researcher submitted a request for permission letter to the respective authorities to allow the researcher to meet the e-learning providers within the organization. Permission to meet the staff was granted verbally from the authorities. The appointment to meet the staff was made, scheduling date, time and location. This was accomplished either through a personal secretary or a phone call to the staff data collection department. Three types of data collection techniques were used: interviews, observation and secondary data as discussed in the next section below.

Interviews/Meetings/Respondent Encounters

Data was collected through oral conversations, defined by Payne and Payne (2004) as "data collection in face-to-face settings". The most important aspect of these interviews was that they were of short duration, conducted without tape recordings, and were followed immediately by detailed memo writing (Glaser & Strauss, 1967; Glaser, 1978). The round of interviews was conducted with 24 people and each interview started from a respondent's perspective. Respondents were e-learning providers at various institutions in East Africa, including Higher Learning Institutions.

As per Glaser (1978), no notes or tape recordings were used in any of the respondent encounters. There were no semi-structured interview questions. Instead the researcher concentrated in the initial conversations on establishing rapport and making the respondents feel comfortable and unthreatened. These were deliberate strategies recommended by Glaser (1978). The rationale for this was that the respondent encounters should be more like normal conversations where both parties felt sufficiently relaxed to participate in a candid exchange of ideas. Immediately following each respondent encounter the researcher made very detailed notes that were later converted into theoretical memos. The Grounded Theory research method is known as a "delayed action phenomenon" (Glaser, 1978). This is because it takes some time for both the researcher and the respondent to consider what is and has been said. For this reason it is essential not to attempt to collect all the data first and then proceed to attempt to analyse it. It is essential in CGTM that the researcher follow wherever the data leads. This is because the GT researcher has to be able to move from the low level realm of contextual description to the higher level of theoretical explanation.

Once the researcher had completed at least six initial interviews with different respondents, appointments for subsequent interviews were made with respondents. Prior to the next round of respondent encounters more theoretical memos were written. The focus of the respondent encounters shifted and followed the issues emerging from the most recent iteration of theoretical memos. Although there is no standard template which all theoretical memos must adhere to, there are some basic issues which all theoretical memos must address. If these are not present there will be no transparency in terms of how the researcher progressed from raw data to an emerged core variable. In order to meet transparency requirements all memos included the following (see Appendix 3 for memos sample): *(1) Memo title; (2) Summary of any substantive codes, categories and properties; (3) Conceptual indicators (These are important links that form the bridge from substantive description to conceptual explanation); (4) Emergent theoretical codes, categories and properties; (5) A statement as to the type of data the researcher believes has been given to him or her by the respondent (Baseline, properlined vague, Zero, or interpreted data)*(see Appendix 3). It actually does not matter when doing CGT what type of data the researcher has because everything is considered to be data. What is of the utmost importance is that the researcher fully understands what type of data s/he has, rather than worry about accuracy.

Each of the interviewees was re-interviewed several times until latent patterns of behaviour began to emerge. Once these patterns emerged the second stage of interviewing took place with more university professionals as well as other constituencies identified by the interviewees as being important people to speak with concerning this issue. As key concept indicators emerged theoretical sampling was the feature of the second stage of interviewing. The interviewing and memo writing and sorting continued until no more new patterns emerged. When this happens the data is said to be 'saturated': no more data is required because predictable patterns have begun to emerge.

The meetings with respondents were scheduled at least two weeks before meeting them to ensure a respondent was free. Each meeting was introduced with an explanation. It was made clear to the respondent that the research focus was to reveal, understand and explain what the main issues were which confronted these professionals in the execution of their role as e-learning providers in the implementation of e- learning projects. Respondents were encouraged to let the researcher know if they felt uncomfortable about any line of enquiry. The absence of any tape recording or note taking during these encounters helped to make the respondents feel more relaxed. This made it possible to transform the nature of these encounters from an interview to something akin to a normal conversation during which both interviewer and respondent were able to pose questions to each other. All encounters were kept deliberately brief so that the researcher was able to make detailed notes immediately after the conclusion of each meeting. The purpose of the series of meetings was to discover the respondents' main concerns regarding online learning including the challenges, how they solve problems and what they consider to be achievements and successes in the course of their daily working lives. It was considered vital to schedule several meetings with each respondent as this would give each respondent the time and space to reflect on subsequent encounters and enabled the researcher to encourage a respondent to think more deeply about what their real concerns were.

Following this incremental process of having several interviews with each respondent ensured that in each case the respondent's agenda emerged rather than that of the researcher. The choice of the locations was an important consideration. The idea was to try and ensure that the respondents felt relaxed. Thus a variety of locations, other than their own offices was selected. These included restaurants, university canteens, meeting rooms, public bars, walking in the street and driving in the car.

Observation

E-learning providers were observed during their daily activities. These observations were systematically documented in the course of writing theoretical memos. The researcher observed e-learning patterns first by using substantive coding and then by theoretical coding until the core variable emerged.

Secondary Data Sources

A review was done of written existing documents in the higher education institutions such as those dealing with e-learning contexts and trends, performance ratings, program logs, tally sheets, and other existing indicators. This contributed to an analysis of the main concerns relating to e-learning by university professionals. Much of this data was sourced from the Ministry of Higher Education, universities and other governmental and non-governmental sources.

Data Analysis

The Grounded Theory research method is a general research method which treats everything as data, whether quantitative or qualitative (Glaser, 1992). Grounded Theory primarily uses an inductive research approach in which the researcher has to be subordinate to the data. Grounded Theory is not a sub-set of Qualitative Data Analysis (QDA). This is because QDA values accuracy above all within a narrow research context whereas Grounded Theory is mainly concerned with the discovery of latent patterns of human behaviour which are transcendent of the original context of the data (Glaser, 1978).

According to Figure 1, the following Classic Grounded Theory Research process was followed as outlined by Glaser (1978, 1992): *(1)Theoretical sampling and substantive coding; (2)Theoretical coding; (3) Theoretical memo writing; (4) Constant comparison between and within memos; (5) Sorting codes; (6)Data saturation; and (7)Emergence of the core variable.* Although is represented in a linear format, in practice the exact sequence varied according to what patterns seemed to be emerging from conversations with respondents. The reason for this is that GT is a process, not a unit based approach (Glaser, 1992). The data was simultaneously, subsequently, and sequentially collected, analysed and synthesized (See Figure 1). Once most of the variations in the data were explained in terms of a key concept, the

Figure 1. Doing classic grounded theory

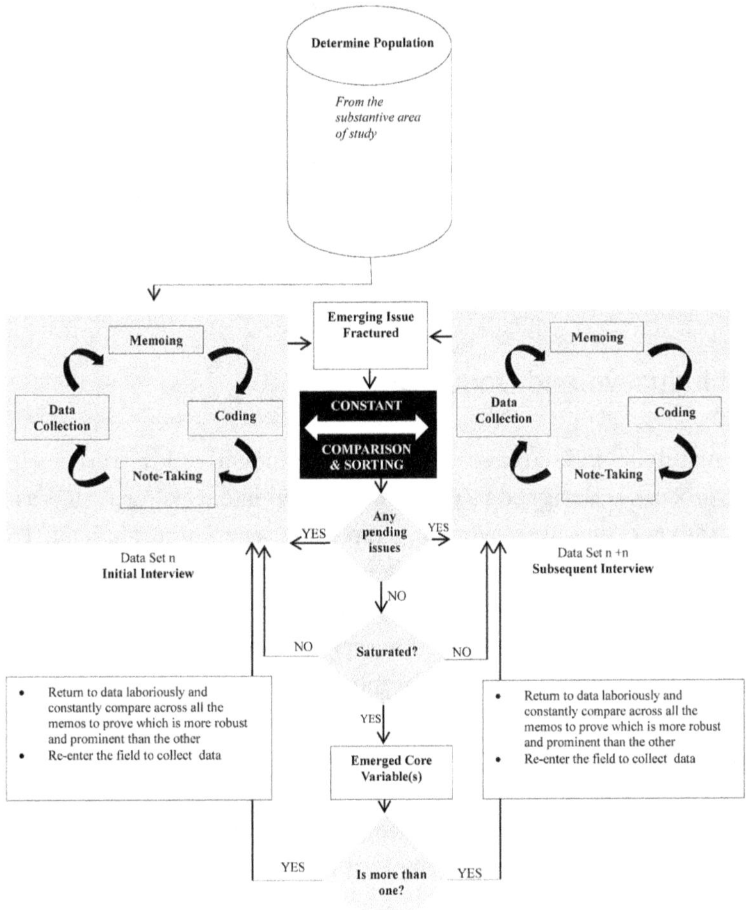

data was said to be saturated and therefore there was no further need to collect more data since the core variable could be said to have emerged.

How Principles of Theoretical Sampling Have Been Used in This Study to Aid the Researcher's Progressions

Glaser (1978:36) defines the theoretical sampling process in the following terms:

Theoretical sampling is the process of data collection for generating theory where by the analyst jointly collects, codes and analyzes his data and decides what data to collect next and where to find them, in order to develop his theory as it emerges. (p. 36)

Theoretical sampling is exclusively controlled and guided by the emergent theory as it arises. As an inductive approach, it is not constrained from the outset and is able to follow the data to the existing latent patterns embedded within it. In this study, the four most important principles involved in the use of theoretical sampling are as follows; it is a non-linear process, it is a combination of induction and deduction, the data is always in control of the direction of the research and the essential relationship between the data and the theory is the conceptual code (Glaser, 1978).

A Non-Linear Process

Throughout the research process, the researcher was always mindful that whenever insights arose the researcher would immediately have to write a theoretical memo about it. Frequently these insights would arise when the researcher was doing a quite unrelated tasks and at inconvenient times. Theoretical sampling enabled the researcher to examine right across all the data so that the emergent theory could be discovered.

A Combination of Inductive and Deductive Logic

The CGT method is mainly an inductive research method. Inductive logic comes into play because the theoretical sampling process is designed to reveal embedded latent patterns. It was not possible for the researcher to know in advance precisely what to sample and where it might lead. This is why the use of induction was essential. This allowed the researcher to discover the connections between the emergent substantive codes. These latent patterns in which the substantive codes are configured are called theoretical codes. Once the theoretical codes started to emerge the researcher was able to use deduction to figure out where to go next in order to sample for more data to generate the theory.

The Data Is Always in Control

As theoretical codes emerged the researcher kept on returning to the theoretical memos and any other data to see how prevalent they were. This also meant that once theoretical codes had emerged I went to respondents with a much more focussed approach. I was only sampling for the emergent theoretical codes. Those codes which were robust were found in all respondent encounters. Sometimes this meant returning to a respondent and asking them to give ever more detailed examples of issues which were of most concern to them in their work. Obviously the researcher never coached the respondents for specific responses instead the encounters were always open and we would often discuss many other things. Once the respondents became more relaxed they also became more candid about their work situation. At this point the researcher had to concentrate very hard because throughout this process notes were taken within 24 hours after the respondent encounters.

The Essential Relationship between the Data and the Theory Is the Conceptual Code

After following the principles of theoretical sampling two distinct theoretical codes were persistently present. These were "asymmetrical collaboration" and "cultivating recognition". At this point the researcher was very confused. To clear confusion the researcher was advised to re-read the section entitled core categories in Glaser (1978:94-100). Glaser (1978:94-100) explains in detail how and why this possibility will often occur. He also provides an 11 point explanation (see Appendix 4) as to how the researcher can discern which of the completing theoretical codes is the core variable. After re-reading this chapter I returned to my data and applied the 11 point guide lines. In addition I also arranged more respondent encounters and only one of the two theoretical codes finally emerged as the core variable. It did so because it was able to explain most of the variation in the data. The Figure 1 demonstrates how the theory of "cultivating recognition" emerged from the data.

Using Figure 1, this study began with determining the population through the researcher's networks and IT providers' networks. Following this, e-learning providers working in east Africa were chosen. As Glaser (1978) argues after determining the population, the theoretical sampling process will involve initial data collection and analysis. This study did an initial data collection using 6 respondents. After fracturing the data collected, detailed memos were written. In initial stages of analysis, codes were elicited rapidly through a joint process of theoretical sampling and memo-writing. Through this process, codes were corrected, trimmed, and continually fitted to the data (Glaser, 1978). The memo writing enabled the researcher to conceptualise the boundaries and properties of each category and illuminated gaps in the emerging theory, thus highlighting where to sample next and for what theoretical purpose (Glaser, 1978). The memo writing included all basic issues detailed in section 3.3.1 (see sample memos in Appendix 2).

The constant comparison of codes yielded a provisional set of conceptual categories, from which point new categories emerged and new incidents were fitted and re-fitted into existing categories. The researcher sampled for both theoretical similarity and difference in order to expound the properties of each category, and attempted to saturate all categories until the emergence of a core category (Glaser and Strauss, 1967). Once latent patterns were discovered, the researcher had to return to the data and see how wide spread across all the data the emerging phenomena actually was. At this point there were no more categories emerging- it was saturated. As soon the categories were saturated, the study realized that two competing core variable candidates did emerge. As has been explained earlier after applying the 11 criteria which distinguished between competing core theories, the core variable emerged.

RESULTS

Main Concepts of the Theory of "Cultivating Recognition"

The study discovered "Cultivating Recognition" to be an emergent core variable (see Figure 2), amongst e-learning providers who participated in this research, who were based in East Africa and had continually to resolve the issue of "Cultivating Recognition" in order that e-learning projects could be successfully implemented. The concept of "Cultivating Recognition" emerged as the core variable because it was able to explain most of the variations in the data. The core variable of "Cultivating Recognition" is defined by the researcher as follows:

The recurrent creation and recreation of gaining acknowledgement of achievement in order to both commission and implement e-learning initiatives.

This definition of "Cultivating Recognition" explains in conceptual terms the perpetual process in which e-learning providers have to engage in order to secure the necessary financial support for their e-learning projects to be commissioned and implemented. "Cultivating Recognition" was found to emerge in a variety of different contexts and business relationships in which e-learning providers were working, including providers[1]-to-government, government-to-providers, community-to-providers, providers-to-community, providers-to-funding bodies, funding bodies-to-providers, manager-to-employees, employee-

Figure 2. Cultivating recognition

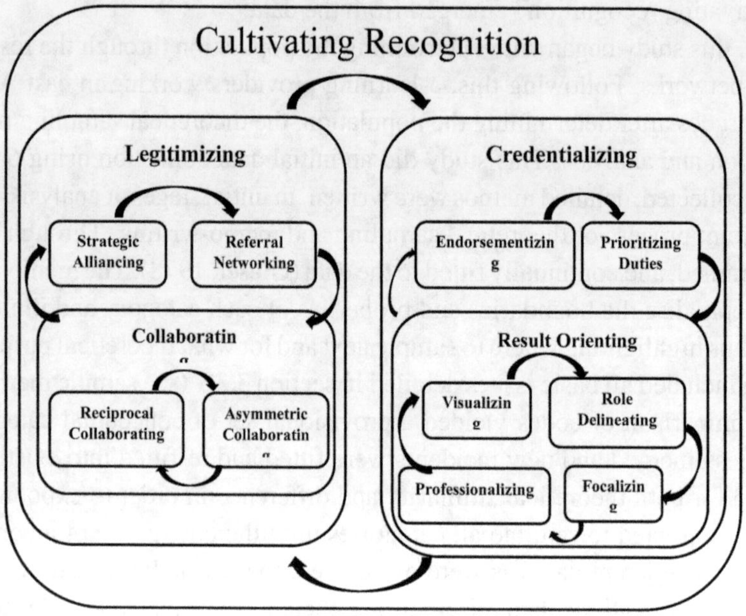

to-manager, peer-to-peer, and providers-to-partners, partners-to-providers. The two sub core variables ("Legitimizing" and "Credentializing") of "Cultivating Recognition" are discussed in the next section

The Two (2) Primary Sub Core Variables of "Cultivating Recognition"

The e-learning providers do cultivate recognition through "Legitimized" and "Credentializing". Both these concepts, "Legitimizing" and "Credentializing", are sub core variables of cultivating recognition. If either of these variables is absent then the process of "cultivation of recognition" cannot take place. Both concepts emanate from the external endorsement by key professionals. They constitute a theoretical code that connects them. This indicates the two primary sub core variables to constitute the process necessary to the emergence of "Cultivating Recognition".

"Legitimizing" is defined in this context by e-learning providers as the process of ensuring the project is delivered, valid and sustainable. This is achieved by the accumulation of social credit bestowed on the e-learning providers by their peers and clients. "Legitimizing" is the process of ensuring the e-learning activities (e.g. designing, delivery, evaluation) are delivered, and such deliverables are valid and sustainable. The validity and sustainability of the project is ensured and maintained due to the "Legitimizing" process, which is an on-going and perpetual process. Thus for their projects to be continuously sought after, E-learning providers need to ensure a project is legitimized by a client or funder and is accorded respect throughout the process. The main symptom of and reason for a lack of legitimization is the lack of respect shown to e-learning providers by some clients and funders. A "Legitimizing" process is undertaken by e-learning providers as they need to gain respect from their clients. The process of "Legitimizing" involves three main activities or processes: "Collaborating", "Referral Networking" and "Strategic Alliancing" At the start of this legitimizing process e-learning providers collaborate with other experienced e-learning partners prior to a formal funding submission for their e-learning projects. This collaboration can be either with partners who have a considerable degree of influence and power (asymmetric collaboration) or with partners with a similar or equal level of power and influence (reciprocal collaboration). Secondly, they make full use of their existing referral networks so that the potential client or funder is more likely to have confidence in the proposal. Finally, e-learning providers are more successful if they have already entered into formal strategic alliances with other respected partners and these inform, and are integrated into, the proposal document. Successful e-learning providers are able gain more traction in their projects when they are able to simultaneously synchronise several different projects. This is because of the potential for sustainability from one project to another that synchronicity could deliver.

"Collaborating" is built by "asymmetrical Collaborating" and "Reciprocal Collaborating" as its sub-core variables. "Collaborating" is defined by a researcher as" "the process of creating business opportunities through the advantages of partnerships". "Collaborating" partners are of two types: (1) those who are "Asymmetrically Collaborating" and (2) those "Reciprocally Collaborating". "Asymmetrical collaborating" is a concept which explains how the weaker party in the collaboration is able to obtain the benefits that collaboration brings (e.g. funding opportunities, scholarships) whilst skilfully avoiding being dominated by the more powerful partner. "Reciprocal Collaborating" is the process of generating mutual benefit for the parties by sharing power and status. The power comes from funding opportunities and decision making. The collaborating parties in reciprocal collaboration tend to share power equally regarding the implementation of the e-learning project. The benefits from the e-learning project are shared equally between the "Reciprocal Collaborating" partners.

"Referral Networking" is a process of stimulating existing, and creating new, contacts. These connections and networks can be useful for endorsing e-learning projects, thereby increasing their legitimacy. "Referral Networking" can be of value to the e-learning providers in terms of capturing more opportunities for funding, collaboration, and for increasing the popularity of the online programs. "Strategic Alliancing" is defined as the mutually beneficial process of working together for the purpose of delivery of online courses or securing funding. "Strategic Alliancing" is a medium to long term process of connecting with other parties in a particular field. This includes matching the strategic objectives of one institution to the other.

"Credentializing" refers to the creation of increased client belief in the e-learning providers' competence to execute a project on time and within budget, and according to the required specifications. This is achieved by the endorsement both by members of the e-learning providers' peers and by their client. "Credentializing" involves the creation of increased client belief in the e-learning provider's competence to execute an e-learning project on time and within budget". Three elements or criteria necessary for the "Credentializing" process to ensure the belief of the client in the e-learning providers' competence. Firstly all claims and statements of accomplishments and achievements made by the e-learning providers are only seriously considered when endorsed by respected external third parties; secondly, the e-learning providers are able to both understand and execute project priorities by the skilful accomplishment of the process of prioritizing. Finally, the e-learning providers are able to develop a results-orientated mind set by cultivating the skills involved in result-orienting.

"Endorsementizing" is defined as a process whereby the e-learning provider presents his professional work experience and academic qualifications to the client in a format which the client can easily comprehend, and which are perceived by the client to be relevant. Such a process makes the e-learning professional tailor each curriculum vitae presented specifically to meet the needs of a particular client. "Result Orienting" can be defined as the process of ensuring concrete results and achievements ensue from the planned e-learning project so that clients can recognise such efforts. "Result Orienting" is achieved by means of the process of "Visualizing", "Professionalizing", "Role Delineating" (ensuring the role of each e-learning provider is clear and understood), and "Focalizing", or being obsessive about details and the use of professionalism and competence so that unforeseen crises such as conflict gaps can be dealt with effectively.

"Prioritizing duties" is defined as the process of ensuring that all tasks are sequenced in order of importance to ensure the project's successful completion (i.e. the process helps to increase competence and recognition).

The Six Secondary Sub Core Variables of "Cultivating Recognition"

As noted before, the two primary sub core variables of "Cultivating Recognition" are "Legitimizing" and "Credentializing" each emerged and were found to contain three sub core variables. As Figure 2 shows, the "Legitimizing" process can emerge by means of "Collaborating" with partners (either asymmetrical collaborating or reciprocal collaborating), "Referral Networking" and "Strategic Alliancing".

"Credentializing" arises from the joint link of "Endorsementizing", "Result Orienting" and "Prioritizing Duties". The six secondary sub core variables have their distinct categories and sub-processes .

'Collaborating Partners" involves creating a working together agreement in such a way as to ensure each partner gains from the agreement. The process of collaborating with partners can be either asymmetrical or reciprocal. These two labels were derived by researcher due to his deep understanding of Theoretical Sensitivity by Glaser (1978). "Asymmetrical Collaborating" is a form of collaboration in which the weaker party in the collaboration is able to obtain the benefits that the collaboration brings, whilst skilfully avoiding being dominated by the more powerful partner. However the weaker partner is potentially always more vulnerable. The more leverage the powerful partner has, the more this partner will strive to ensure that the project is implemented. In "Asymmetric Collaborations" the aim of the weaker partner is to manage to skilfully extract resources and finances from the dominant party without being overwhelmed. The e-learning providers ideally should attempt to transform the "Asymmetric Collaboration" to one characterised by a reciprocal relationship wherever possible. The reason for this is that asymmetric relationships tend to be limited or reduced to being contractual relationships whereas reciprocal relationships always have a greater potential for developing beyond a single project. Glaser (1978) and Glaser and Strauss (1967) states several types of temporal properties including cyclical, spasmodic, perpetual. Cyclical refers to the event occurring in defined interval repeating. Spasmodic refers to occurring once within the process and perpetual refer to the properties occurring frequently (Glaser and Strauss, 1967; Glaser, 1978). The temporal properties of "Asymmetrical Collaborating" and other forms of collaborating are detailed in Appendix 5 (see Figure 3).

"Reciprocal Collaborating" involves two parties coming together to achieve agreed objectives by contributing different and unique skills and investment to the partnership. For e-learning providers to share the right of ownership of e-learning projects equally, they have to manoeuvre themselves into a position of parity. To ensure this, e-learning providers have been designing and signing various agreements and memoranda of understanding. All parties are obliged to sign such agreements to enter into the collaboration. The parties coming into a reciprocal collaboration expect to have equal power and status. However, if one partner gains a higher status and greater power than the other in the collaboration, the collaborating process will change to an asymmetrical collaboration, thus ceasing to be a reciprocal collaboration. The collaboration tends to be entered into by the partners in order to secure funds, the strengthening of the programs and recognition.

Figure 3. An overview of the conceptual hierarchy of levels (vertically) and some of the characteristics of the concepts (horizontally)

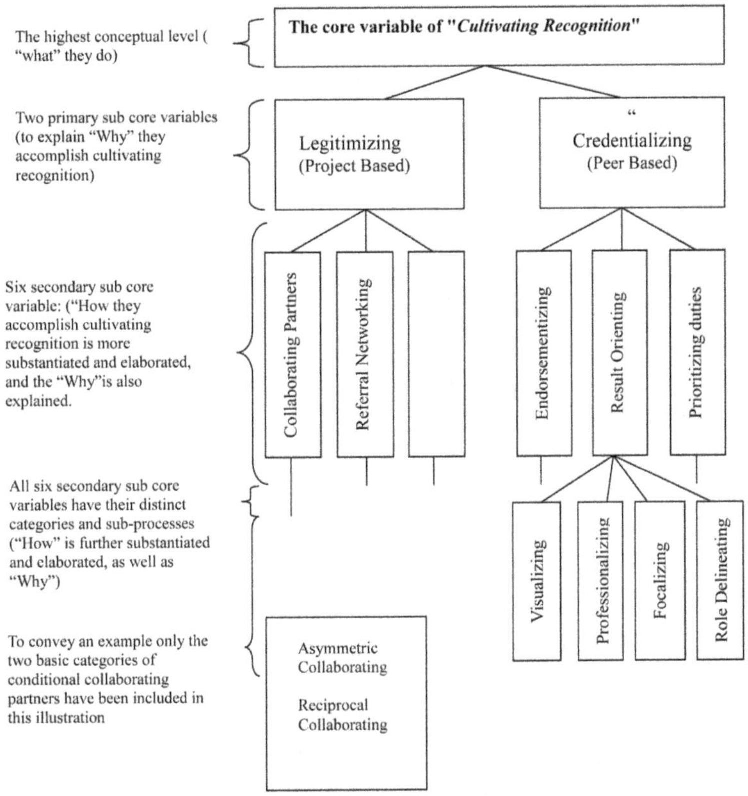

Referral networking has been defined as the process of both stimulating existing, and creating new, contacts. These connections and networks can be used for endorsing online learning projects, thereby increasing their credibility. This networking process is undertaken in order to capture more opportunities for e-learning providers in terms of funding and further collaboration, and for increasing the popularity of the online programs. Clients tend to believe in networking as soon as they see more e-learning providers' networks for the same project and they tend to commission the program more willingly. One of the pre-requisites for securing funding from donors is to be part of and activate referral networks. Such networks should be strong and well known in their host country. Successful e-learning providers work hard to capture and build referral networking to ensure that they get recognized and in turn capture more funding. The referral networking process is ongoing and perpetual. E-learning providers are looking for new networks all the time to ensure a project is recognised and sustainable. The need for referral networking is more urgent when the party seeking resources or influence is lacking in credibility and experience. The process of referral networking is dynamic and fluid and changes as a member's credibility and influence increases or diminishes. Referral networking may begin with the weaker party being vulnerable but at later stages, as this party demonstrated more competence, the receiver of credibility and influence can become the bestower of resources and influence. It is a continuous and reciprocal rather than a one way process.

"Strategic Alliancing" occurs when a party or parties enter into a medium to long term process of connecting with other parties in order to secure the completion of a project with mutually beneficial outcomes. It is a process of working together for the purpose of delivery of online courses or to secure funding. Such processes include matching of the strategic objectives of one institution with the other. When matches are found, the parties whose objectives 'match' each other form such alliances to ensure that all benefit from the alliance. Strategic alliances can increase the rate of program delivery, recognition, secure credibility, popularity, recognition, project quality, and funding opportunities. A strategic alliance helps both institutions grow and design sustainable programs. Most funders tend to require a strategic alliance configuration (e.g. a university-university partnership) as a precondition of funding. Such alliances can be made and cemented through exchange of programs and staff. The 'alliancing' institutions have reciprocal alliancing characteristics in that they have equal power and status in the management of the project. They tend to sign agreements to ensure smooth delivery of the program. Strategic Alliancing usually takes place prior to the commencement of a project and is usually triggered by the client's wish to reduce uncertainty and risk which can take the form of lack of secure funding, loss of students and of marketing access. The greater the perceived uncertainty and risk of the online learning project on the part of the client, the greater the need for strategic alliancing.

"Endorsing" is the process of presenting professional work experience and academic qualifications to the client in a format which the client can easily comprehend and perceive to be relevant to the particular needs of the client or clients. Endorsing's purpose is to create trust on the part of the client. The presentation of the professional's qualifications, experience and accumulated social credit is aimed at reducing the perceived risk to potential strategic partners and as a form of reassurance. E-learning providers use this process for proving their capability, competence and ability to ensure that the implementation and use of the product will go smoothly and the project will be effective. In addition e-learning providers use this professional endorsement strategy with the aim of creating confidence in their clients in terms of delivering projects on time and of the desired quality. The e-learning providers should first believe in themselves besides demonstrating a high level of knowledge and expertise in e-learning. If e-learning providers are able to do this they are more likely to gain the approval and confidence of their clients (e.g. funders, donors). In order to demonstrate this they should be able to provide proof of their ability to solve any and all e-learning implementation problems as well as displaying high levels of self confidence in their decision making. All of these are made more possible through the kind of endorsement derived from strategic alliances This endorsement by respected third parties in turn increases their credibility in the eyes of senior managers.

"Result orienting" is the "process of ensuring there are concrete results and achievements in the planned project so that clients can clearly recognize and acknowledge such efforts. It involves perpetual concentration on all key issues relating to the project, including synchronizing e-learning with conventional learning. This is achieved by doing four things: visualizing, role delineating to avoid conflicts, focalizing, or being obsessive about details, and professionalising. The skilled e-learning professional tends to be very aware of what the successfully delivered project will look like long before any project has begun or is in its planning stage. In a few cases the process of result orienting happens cyclically. As soon one project is completed, the e-learning professional needs to look for another thus are making this a continuous and cyclical process. To ensure more funding, support, and promotion, e-learning providers try to ensure positive results from the previous projects. In order to activate recognition they tend to be result oriented.

"Prioritizing Duties" involves a process of ensuring that all tasks are sequenced in order of importance to ensure the project's successful completion. The process helps to increase both competence on the part of those involved, as well as recognition. This process, like that result orienting, is usually cyclical and dynamic. It starts when a tendency of a failure to recognize the e-learning professional starts to become apparent. Sometimes prioritizing happens spasmodically (i.e at intervals). The prioritizing process is usually triggered by the obvious lack of respect or recognition on the part of the top managers of respective institutions. When an e-learning provider does not receive recognition from top managers for what he/she is doing, s/he tends to prioritize his/her duties in order to increase results and reclaim recognition.

These six secondary sub core variables are of vital importance in the process of resolving the main concern of e-learning providers, that of cultivating recognition. Embedded within the six secondary sub core variables is a number of important concepts which help to explain the dynamics of the process of cultivating recognition and are discussed here below.

The Lesser Concepts of Cultivating Recognition

The lesser, or lower level, concepts in the cluster of concepts around cultivating recognition are so called because, although important to the overall process of recognition, they form part of the third tier of the hierarchy of the respondents' main concern .

Asymmetrical Collaborating was defined earlier as a concept that explains how the weaker party in the collaboration is able to obtain the benefits that collaboration with a stronger partner brings. Asymmetrical Collaborating is achieved by the weaker partner skillfully avoiding being dominated by the more powerful partner. In all collaborations each partner will have a slightly differing agenda from the other partner. The weaker partner is potentially always more vulnerable. The more leverage the powerful partner has, the more s/he will strive to ensure that the project is implemented. In asymmetric collaborations the weaker partners often skillfully manage to extract resources and finances from the dominant party without being overwhelmed.

Reciprocal Collaborating was defined in section 5.5 as the process of achieving mutual benefit for the parties in the sharing of power and status in a collaborative partnership by contributing their different and unique skills and investments to the project. E-learning providers will attempt to transform an asymmetric collaboration to a reciprocal relationship or collaboration wherever possible. The reason for this, as has been pointed out, is because asymmetric relationships tend to be limited to being contractual relationship (mainly involving signing of agreements), whereas reciprocal relationships always have a greater potential for development beyond a single e-learning project.

Result Orienting has been defined as the process of ensuring there are results and achievements in the planned project so that clients can clearly recognize such efforts and its four elements: visualizing with clarity and in detail prior to the commencement of the project what its final outcome will be. . . Role Delineating was described as the process of defining the specific duties and responsibilities of each partner in the project in order to avoid conflicts, achieved by instituting and policing role delineating regimes to avoid conflicts.

Focalizing was defined in terms of e-learning providers of staying attentive and having an obsessive attention to detail in designing and implementing a project. Professionalizing, the process of e-learning provider showing professionalism, and experience order to boost the e-learning provider's credibility.

DISCUSSION

The Core Variable of "Cultivating Recognition"

The "Cultivating Recognition" process involves the recurrent creation and recreation of attempts on the part of professionals to gain public acknowledgement of achievement in order to both commission and implement ongoing e-learning implementation. In the research process "Cultivating Recognition" represents a latent pattern found everywhere in the data and which emerges during constant comparison of the properties of theoretical memos. "Cultivating recognition" represents: (1) A commonality within the wide variation in behaviour of the e-learning providers. It integrates practically all the e-learning providers' practices that have been compared and coded; and (2) A huge variation in behaviour that can be explained by a few distinctions within the same underlying meaning.

"Cultivating Recognition" as a core variable is highly stable in its core meaning and highly variable due to its representation of, and its accounting for, high variation in behaviour. "Cultivating Recognition" is the main concern of the e-learning providers, representing the essence of relevance reflected in the data as seen from the perspective of the participants. "Cultivating Recognition" shows how the main concern is resolved. "Cultivating Recognition" is basically what these e-learning providers engage in on a continual basis. They gain "public acknowledgment" by "Legitimizing" projects and by "Credentializing" their professionalism. Both substantive concepts and theoretical codes are the building blocks of "Cultivating Recognition" theory.

"Cultivation of Recognition" can be explained in terms of "basic social processes" (BSP). This BSP connects substantive concepts (the six secondary sub core variables), and the theoretical code into two inseparable three-in-one structure (Glaser 1978). This is briefly explained with reference to the two primary sub core variables of "Cultivating Recognition". The attachment of theoretical codes (e.g. processes, sub-processes, loops) to substantive concepts (variables, categories) within the theory of "Cultivating Recognition" is explained in the next paragraph.

Without the gaining of external acknowledgement recurrently taking place, the very survival of both e-learning providers and e-learning projects may be at stake. Thus people in e-learning instinctively have a pivotal sensitivity and alertness to the importance of the emergence of "gaining external acknowledgement". The "achievement of external acknowledgement", of many kinds and types is their pivotal main concern. Thus, "Cultivating Recognition" as a main concern issue occurs in many guises and in many facets and sub-facets. "Gaining of acknowledgement" does not occur by accident. It emerges from a series of deliberate activities on the part of those seeking to gain acknowledgement for their services. Accordingly, people involved in e-learning business deliberately seek and sustain public acknowledgement. This process is ongoing even during a period which is unfavourable for "gaining public acknowledgement". Opportunities are continually being re-created or remodelled by e-learning providers to gain an edge over competitors in e-learning practice and delivery. Essential to this process is "Cultivating Recognition" which is also the recurrent, pivotal and prime solution to the problematic nature of the main concern. Thus, "Cultivating Recognition" drives and directs behaviour. In this sense "Cultivating Recognition" as a concept partly explains itself and its own variation. It explains certain problems occurring in achieving e-learning objectives and gives the reasons why e-learning providers pursue a certain course of action or non-action and how they do this. As has been described, Cultivating Recognition's theoretical and substantive codes signify the two main emerging features of Cultivating

Recognition. One feature of "Cultivating Recognition" is the problematic nature of the main concern of the participants being studied, together with the problematic nature of its potential resolving. This feature is represented by its variables, categories and properties.

The other feature of "Cultivating Recognition" is that of accounting for the actual resolving of the main concern. This feature is modelled by the emerging theoretical codes or theoretical concepts. On the primary sub core level an "inseparable two-in-one structure" is emergent. On the secondary sub core level and below, the principal theoretical codes emerge as varieties of a basic social process that forms a loop. The secondary sub core process-loop has been called the core process-loop of "Cultivating Recognition". This core process loop has its sub-processes which may in turn have sub-processes. "Cultivating Recognition" emerges as a concept in many guises in the data and can be conceived of as a multivariate variable with many dimensions, being highly variable and dependent. Conditions and context can easily cause variations (Glaser, 1978), yet is unchangeably omnipresent in the data. . The main dimensions can be conceived of as the two primary sub core variables of "Cultivating Recognition" as well as being conceived of as the six secondary sub core variables of "Cultivating Recognition". Compared to "Cultivating Recognition" as the core variable, the six secondary sub core variables represent the concept of "Cultivating Recognition" as more specific and less general (at a lesser conceptual level) patterns of it. These six secondary sub core variables represent distinct, stable and exhaustive stereotypes of variability. Yet, these six secondary sub core variables can also themselves be conceived of as multivariate variables with many dimensions. These dimensions constitute the categories, or sub-categories, or sub-sub-categories of "Cultivating Recognition".

The main dimensions of "Cultivating Recognition" can be conceived of in terms of its main sub-processes which can also themselves be conceived as multivariate processes with many dimensions thus making "Cultivating Recognition" a highly complex multivariate theory. Much of the process of the fitting of the multivariate concepts or substantive and theoretical codes in a Classic GT can be conceived of as the practising of a kind of "qualitative mathematics" (Glaser 1978). By a process of grounding in the data collected from the e-learning providers, being studied, these qualitative variables or theoretical concepts are made specific. Their fit, their use in the process of explaining the main concern, and their relevancy to e-learning implementation have been verified repeatedly by properties and interchangeable indices in the data during the process of grounding. The two features of "Cultivating Recognition" are semantically identical in the definition of "Cultivating Recognition". Their actual linking is seen recurrently in the data. The emergence of convenient and accessible practices to users of e-learning practices is the pivotal main concern and these also emerged from a series of deliberate activities on the part of e-learning providers. "Cultivating Recognition" as a main concern recurrently attaches itself, for example, to the requirements inherent in a two-in-one structure, or a basic social process for the resolving of the main concern (Glaser, 1978). The common properties of the substantive variables, and the variety of emerging theoretical codes, indicate how this processing of "Cultivating Recognition" is being triggered. Some of these common properties are, for example, "Credentializing" and "Legitimizing". Because "Cultivating Recognition" possesses these two features, both the problem and its actual resolution, it is not so much a theory describing "what" e-learning providers do, as a theory that conceptually accounts for "why" they accomplish, or do not accomplish, the task or project. Its variables and categories as well as its processes and sub-processes, conceptually explain "how" "Cultivating Recognition" is carried out, basically "how" e-learning providers manage to gain recognition on a continuous basis.

"Legitimizing": A Project-Based Variable of "Cultivating Recognition"

The process of "Legitimizing" is a fairly tricky one for e-learning providers to achieve. "Legitimizing" can only happen with the approval of respected third parties. These third parties sometimes have difficulty prior to the project's commencement understanding the full implications of e-learning in terms of increasing educational effectiveness and quality. One of the main barriers experienced by these third parties is the inability to understand clearly the technical jargon employed by e-learning providers working in this field. E-learning providers also often find it problematic to communicate clearly and precisely to their collaborative partners what it is they are doing. In order to reduce the potential for misunderstanding they have to legitimize their actions. As soon as the "legitimizing" has been achieved, the project gains greater acceptance, which in turn leverages future funding. As was described in previous sections of chapter 5, "legitimizing" is a perpetual process throughout which the e-learning providers have to ensure the project is legitimized and is respected (See Appendix A.7). The main symptom of the lack of legitimization is the lack of respect shown to e-learning providers by some clients and funders. The instances of increased respect from top managers has been shown to be more frequent when e-learning providers employ three strategies in the course of working with partners: "Collaborating" "Referral Networking" and "Strategic Alliancing". These processes need to be present if legitimizing is to be successful.

Firstly e-learning providers need to collaborate asymmetrically or reciprocally with other experienced partners prior to the formal submission of funding submissions, as well as initiating the process (e.g. proposal writing, feasibility study,). Secondly, they need to make full use of their existing referral networks so that the potential client or funder has more confidence in their proposal. Finally, e-learning providers are more successful when, prior to and as part of the proposal document, they have already entered into formal strategic alliances with partners respected in the field.

It has also been found that successful e-learning providers are able to gain more acceptance in their e-learning projects when they are able to simultaneously synchronize several different e-learning projects due to the potential for sustainability from one project to another that synchronicity could deliver.

"Credentializing": A Peer-Based Variable of "Cultivating Recognition"

In order to successfully accomplish the "Credentializing" process, e-learning providers need to skilfully articulate three key processes. Firstly, all the claims and statements of accomplishments and achievements, made by the e-learning providers must be endorsed by respected external third parties. Secondly, the e-learning providers must be able to both understand and execute project priorities by the skilful accomplishment of the process of "Prioritizing". Finally, the e-learning providers must develop a results orientated mindset by cultivating the skills involved in "Result Orienting".

The Six Secondary Sub Core Variables of "Cultivating Recognition"

The data collected in this research focused on the main concern of e-learning providers working in East Africa, namely that of "Cultivating Recognition". "Cultivating Recognition" was achieved by e-learning providers by means of the systematic integration of several different variables. The primary sub core variable of "Legitimizing" is created by the interaction of three secondary sub core variables: "Collaborating, "Referral Networking" and "Strategic Alliancing". The primary sub core variable of

"Credentializing" occurs when the following three secondary sub core variables are present: "Results Orienting", "Prioritizing Duties" and "Endorsing".

"Collaborating"

The research has discovered that "Collaborating" with partners can happen in two different ways, either asymmetrically or reciprocally. This is because in all collaborations each partner will have a slightly differing agenda. Where there is a power imbalance between the parties "Asymmetrical Collaborating" will manifest. Where there is a greater degree of power equity between the parties, "Reciprocal Collaborating" will occur. It always requires more skill on the part of the weaker party in an asymmetrical collaboration. E-learning providers who are part of collaboration tend to attempt to transform the collaboration from an asymmetric to a reciprocal relationship wherever possible. The reason for this is that asymmetric relationships tend to be limited to being contractual relationships. "Asymmetric Collaborating" is likely to have several different temporal characteristics. The temporal characterises can be cyclical, spasmodic and also perpetual. "Asymmetric Collaborating" occurs because, once a project is completed, the collaborating partners start again on a new e-learning project. Thus the process becomes cyclical as the need for change arises. For example, when it is necessary to convert direct learning to e-learning, this becomes the trigger for "Asymmetric Collaborating". Such a change is facilitated or triggered by lack of physical infrastructure and resources in the direct learning or face to face learning situation. During the implementation process the need for more partners will often arise. This in turn will trigger more collaboration within the e-learning project. Other factors which facilitate the refining of such collaboration can be the rapid changes in technology and the availability of funding and of e-learning experts. A change involving one or more of these factors will lead to the start of a new e-learning collaboration or agreement, or to a process of refining the current e-learning collaborations. This in turn rejuvenates the e-learning project. As soon as the junior e-learning partner is able to stand, the e-learning provider tends to start e-learning projects running in tandem with partnering projects and with new e-learning partners. "Asymmetrical Collaborating tends to be spasmodic when certain relationships create unanticipated problems and necessitate the putting of strategies in place to deal with the unforeseen problems. Asymmetric relationships can also be perpetual. Even in harmonious relationships conflict of interest can exist, or agenda priorities differ. To avoid and deal with these issues it is also quite feasible that "Asymmetric Collaborating" will also have to be perpetual.

As has been described, "Reciprocal Collaborating" involves mutual benefit accruing to the e-learning provider parties in the sharing of power and status, both contributing different and unique skills and investment and having a greater potential for development beyond a single e-learning project. For e-learning providers to share equally the right of ownership of e-learning projects, they have to manoeuvre themselves into a position of parity. To ensure this, e-learning providers design and sign various agreements and memoranda of understanding and so enter into the collaboration. While the parties in such an arrangement are accorded equal power and status, should one e-learning partner acquire higher status and power than any or all of the others, the collaborating relationship changes to an "Asymmetrical Collaboration" and ceases to be reciprocal. "Reciprocal Collaborating" is spasmodic, can happen at any stage of the project and initially tends to be for the purposes of securing funds and strengthening the programs and their recognition. As soon as such goals are achieved, the process ends. At this point the relationship tends to evolve into new e-learning collaborations as the situation changes. On the one hand, while rapid changes in technology result in an increase in the numbers of students attending universities resulting in increased pressure on the government to enrol as many students as possible, thus creating a situation in which reciprocal relationships flourish, on the other hand, the lack of qualified staff and expertise triggers the process of the universities looking for partners in order to start staff exchange programs to develop the staff and secure sufficient numbers of experts. One of the criteria for a university to receive funding for a particular project, is to have a collaborating partner. However the completion of the project will signify the end of the Reciprocal Collaboration. This research has revealed the creation of more referral networks, and an increase in the degree of mutual benefits derived in the process, to be the properties of "Reciprocal Collaborating".

"Referral Networking"

One of the pre-requisites for securing recognition and success (e.g. funding from donors) is for the e-learning provider to possess and to activate referral networks. Such networks should be strong and be recognised in their respective countries. Successful e-learning providers work hard to capture and build "Referral Networking" in order to ensure that they get recognized and are able to capture more funding. "Referral Networking" processes are ongoing and perpetual. E-learning providers continually search for networks to ensure their projects are sustainable. The need for "Referral Networking" is more urgent when the party seeking resources or influence is short of credibility and experience.

The process of "Referral Networking" is dynamic and changes as credibility and influence fluctuates. A "Referral Networking" process may begin with the weaker party being vulnerable but at a later stage, as more competence is demonstrated and recognition gained, the receiver of credibility and influence can become the bestower of resources and influence. It is not a one way process. This research has discovered that the creation of more connections can be the catalyst for receiving more project funds. It also ensures a higher level of client confidence through the process of legitimization.

"Strategic Alliancing"

It is a process in the course of which parties work together for the purpose of delivering online courses, or for securing funding. Such processes include the matching of the strategic objectives of one institution to the other. When two parties find a 'match' they form an alliance in such a way as to ensure that all parties benefit from the alliance. Most e-learning providers in East Africa tend to form strategic alliances with the purpose of increasing program delivery, recognition, securing credibility, popularity, project quality, and funding opportunities with other e-learning providers (e.g. universities, consultants,) in Africa or abroad.

"Strategic Alliancing" helps both institutions to develop and ensure sustainable programs and most potential funders tend to require a "Strategic Alliancing" configuration as a precondition of funding. Such alliances can be made and cemented through exchange of programs and staff. The alliancing institutions have "Reciprocal Alliancing" characteristics in that they have equal power and status in the management of the projects. They tend to sign agreements to ensure smooth delivery of programs. Strategic alliancing usually happens prior to the commencement of a project and is triggered by the desire of the clients to reduce uncertainty and risk. The greater the perceived uncertainty and risk of the e-learning project by the client, the greater the need for strategic alliancing. This research discovered that increased collaboration and funding tend to lead to increased legitimization.

This research has revealed that "Legitimizing" must take place in order for e-learning providers to "Cultivate Recognition" and cannot happen unless the e-learning providers are also able to credentialize themselves. The three secondary sub core variables involved in the achievement of "Credentializing": "Endorsementizing", "Results Orienting" and "Prioritizing Duties" have been mentioned.

"Endorsementizing"

Endorsement by respected third parties bestows credibility on the professional e-learning provider. E-learning providers are required to prove to their clients, their communities and their donors that they are capable of implementing and sustaining a project before obtaining their trust and support. Endorsement helps e-learning providers find ways to create trust and confidence by means of demonstrating their professional competence to their clients through a combination of their qualifications and prior experience.

"Result Orienting"

"Result orienting" allows the clients to recognize whether these agreed stages in the project have been fulfilled. The skilled and successful e-learning provider will be aware of what the successfully delivered end project will look like long before it has begun. The issue for the e-learning provider is how to communicate this clearly to the client and partners prior to the start of any project. This can be achieved by the e-learning provider doing four things. Firstly, s/he should be skilled in the process of "Visualizing" involving clearly envisaging the successfully completed project prior to its commencement. Secondly, s/he should institute and police or monitor "Role Delineating" regimes in order to avoid conflicts. Thirdly s/he should not hesitate to be obsessive about details within each stage of the project. This ensures the ongoing concentration of all involved in the project on all key issues relating to the e-learning project, including synchronizing e- learning with conventional learning. Finally s/he should demonstrate a high degree of professionalism and competence throughout so that unforeseen crises such conflict hiatus can be dealt with timeously and effectively.

The more skilled the e-learning professional is in this process the more likely staff will work together to ensure the intended results are achieved, which is the necessary prerequisite to "Cultivating Recognition". For example, some teachers are reluctant to introduce or incorporate e-learning in their teaching. This reluctance can be as a result of a lack of result orienting on the part of these teachers who then to fail to be recognized. Those who tend to be result oriented, and are prepared to sacrifice their time and energy and "volunteer" their services, manage to achieve their teaching goals. They do not only volunteer to participate in new ventures, but they also are also strongly attuned to "Result Orienting" and are recognized for their achievements.

"Result Orienting" is a perpetual process. E-learning providers get recognition benefits from the leverage which result orienting brings. To attract more partners and clients one needs endorsement from trusted partners. That means that, as soon as one project is completed, the e-learning professional needs to start searching for another, thus engaging in a continuous cyclical process. E-learning providers and e-learning projects tend to lack recognition, while the demand for recognition in this field is growing, thus triggering result orienting as a way of gaining recognition. To ensure more funding, support, and promotion, both those involved in e-learning projects and e-learning providers need to ensure positive results from their previous projects. Thus being more result oriented initiates and ensures recognition.

"Prioritizing Duties"

"Prioritizing Duties" means ensuring that all tasks in the delivery of a project are sequenced in order of importance to ensure the project's successful completion, as well as increasing the competence of staff and securing recognition for the provider. E-learning providers make use of prioritizing as a way of "Credentializing" their work. This process starts when there is a failure on the part of potential clients to recognize a particular e-learning provider. Sometimes "Prioritizing" happens spasmodically. A lack or absence of a prioritizing process in the design and implementation of a project can trigger various negative responses from an existing or potential client. These include lack of respect for or recognition of the provider. The acceptance and implementation of the innovative ideas of e-learning providers by their clients and by respective institutions, underpins and reinforces their professionalism. Thus all of the three secondary sub core variables have to be present in order to ensure that the E-learning providers are able to gain effective "Credentializing".

COMPARATIVE LITERATURE REVIEW

The purpose of this comparative review is to situate the generated CGT theory of "Cultivating Recognition" within the body of extant knowledge and to assess its position and place within the main body of relevant literature. Such comparison should be conceptual in nature (i.e. the focus should be on the comparison of concepts), rather than contextual (i.e. should not be based on the background to or origin of the data), thus suggests a selective comparison (Glaser, 1978).

"Cultivating Recognition" is an example of the theoretical code of "the inseparable two-in one structure" that connects the two primary sub core variables. The existence of this theoretical code would suggest that much existing theory cannot be compared with the model of "Cultivating Recognition", at least at the present conceptual level of the model of "cultivating recognition". Since the comparison has nothing to do with the question of which is better, it would be quite unreasonable to require that this literature review should encompass all the main literature within the Information Systems (IS) area. It would, on the other hand, be reasonable to expect that some comparison would be made with literature that has never, or rarely, been cited within the literature of Information Systems. In any case, to compare every single piece of literature that has been written on some particular concept to establish if it is potentially related to some concept within the theory of "Cultivating Recognition" would be impossible.

It is also impossible and unnecessary to make comparisons with the literature dealing with each emergent concept or every indicated conceptual relationship found within the theory of "Cultivating Recognition". As was described in the previous chapter, the theory of "Cultivating Recognition" is an integrated theory, based on its core variables, which are on three different conceptual levels (core, primary sub core and secondary sub core), and the conceptual relationships among these core variables, as well as their lesser-level categories. Each concept and each conceptual relationship is accorded significance according to its place in the integrated theory, that is according to its relationship to each of the other concepts and each conceptual relationship. For example, if one of the six secondary sub core variables of "Cultivating Recognition" is omitted, the five remaining secondary sub core variables will lose most of their meaning. The reason for this is the level of importance of the indirect and direct relationships between these variables in terms of explaining the meaning of "Cultivating Recognition" in all its facets. Whatever is to be compared within the literature will have a definite place and significance in an integrated theory, although the comparing of one concept with another will to some degree involve other related concepts. In order to fully understand the conceptual delimiting of this literature review, it is necessary to fully understand the delimiting of this study. Thus some points explained in previous sections will be briefly repeated.

Due to the choice of research methodology, this research has been delimited to discovering respondents' main concern and continuously resolving this concern for the subjects of this research. In generating the theory of "Cultivating Recognition", the researcher has taken the approach of delimiting the findings in terms of what is of particular importance to, and problematic for, these e-learning providers working in East Africa. The agenda and substantive interests of those being studied set the agenda for and ground the research outcomes in this agenda. Use of this particular research methodology has up to this point been rare, and in a sense "contrary" to 'mainline' research methodology in its avoidance of many of the elements of traditional research methodology, such as a pre-framed "professional interest" perspective, favouring the "a priori" over the "a posteriori" regarding fit of concepts, and delivering research that is grounded in the agendas of the established professional research communities rather than in the agenda of those being studied (Glaser, 1992). Thus when a comparison is drawn, the standard "professional interest" approach for delimiting a research project is substantially different to the GTM. The standard approach essentially delimits the research to what is seen as professionally important, and hence appropriately professionally problematic, for the researcher and the research community involved. This may be due to a long standing attachment to a particular research methodology, or to an adherence to a particular research program. Thus in standard methodology, the agenda of the researcher and/or his/her research community sets the agenda for the research by preframing it from the perspective of their own research community, and results in researchers delivering research outcomes that are grounded in a particular research agenda. While it cannot be denied that all research is in some way 'grounded', Classic GTM as it is used in this study is based on a different concept of grounding to that used in standard research methodology, one that has little or nothing to do with the standard meaning of the concept of grounding (Glaser, 1978; 1998). CGTM cannot be said to be informed by a pre-framed grounding in pre-existent literature, or by a pre-determined theoretical perspective and in terms of predetermined conceptual usage(Glaser, 2001). In standard research methodology there is a significant amount of "a priori" in place before the start of research and the "a posteriori" requirements are fulfilled by statistical testing or data description. This can in turn result in the criteria for a literature review of such research themselves becoming standardized.

These particular standards for a literature review cannot apply to a Classic GT study (Glaser, 1978). The particular research approach of Classic GT methodology means that the outcomes of the research may be conceptually very different from those that are pervasive in the literature. This also means that the scope for discursive and meaningful comparison with other literature may become restricted due to some degree of incommensurability (Glaser, 1998). It is unlikely that I will find in standard research theory and methodology literature, within the context from which my research data has been derived, exactly the same concepts as those in the literature on emergent theories of "Cultivating Recognition". This delimiting also makes it more possible to achieve 'saturation' in the literature comparison process. Saturation point is reached when the addition of new pieces of literature to the literature review does not provide new or noteworthy conceptual properties, insights or perspectives in terms of the specifics of one's research. Literature reviews of classic GT studies tend to be much shorter than those of more traditional studies (Glaser, 1978). Firstly, this GT literature is delimited to emergent concepts. Secondly, through saturation the comparison delimits itself. Besides being conceptual, a comparative

Classic Grounded Theory literature review should be discursive in nature; it should not follow a 'passive', uncritical, descriptive or listing process (Glaser, 1978; 1992; 1998). Discursive comparing implies a comparison that is characterised by some form of analysis, argument and critical evaluation (Glaser, 1992). It may question pre-existent literature according to ground indications, and it may offer directions for new research, including evidentiary research (e.g. statistically testing, qualitatively corroborating, or hypothetical-deductive research) (Glaser, 1978). A discursive comparison of the literature also involves finding indications of fits to concepts in the pre-existent literature that may indicate usability in terms of "Culitivating Recognition" theory . It also means finding indications of how an emergent concept of the theory of "cultivating recognition" could be modified or not modified in the light of a particular concept, hypothesis or theory found in the literature. As mentioned before, "fit" is another term for validity, but according to CGT it means fit 'in action' and usage, as part of an ongoing process, not simply fit by testing. In a discursive comparison it may become necessary whenever possible to synthesize much of the literature, and thus, in a sense, to 'transcend' it. This synthesizing may be carried out in different ways. It may for example, be carried out by delimiting a comparison to a group of paradigms or research programs. It may also occasionally also be carried out by comparing just one particular piece of representative literature (article or book) that could be said to be representative of a much larger body of literature. A discursive and conceptually delimited comparison of the literature also implies a process that is coherent from topic to topic (Glaser, 1992). It will become unavoidable to exclude some issues that some researchers may find relevant, and to include some issues that others may find less relevant, depending on the theoretical or methodological perspectives of those evaluating the research. This also means that some literature will be reviewed without being included or referred to in the research. The bypassing of this literature does not mean that it is considered of less merit or as having less relevance to the research. It simply means that this literature is not considered important in the given context of conceptual comparison, a comparison which follows the chosen methodology (Glaser, 1992). Thus, the literature review and comparison will be conceptually and not contextually delimited. Conceptual delimiting means comparing emergent concepts - substantive codes, theoretical codes, conceptual hypotheses - to pre-existent concepts and hypotheses such as those in existing literature (Glaser, 1992; 1998; 2001). No comparison will be made with literature where conceptual relatedness cannot be found. The potential literature to be compared is not restricted to the contextual literature on Information Systems and related subjects; it includes subject areas and disciplines as contextually distant as the natural sciences. In a sense, the compared literature can be seen as new "data" to be constantly compared with the emergent theory. This new "data" may modify or refine the theory, or provide new perspectives on the emergent theory and its potential role in the literature, as well as requiring other views from other positions in the research community. The structure of this chapter is as follows: First, some examples and brief explanations are given to illustrate and highlight the difference between a conceptual comparison and a contextual comparison. Second, a comparison is made with other Classic GT studies within business and related fields. A justification for such a comparison is also the relative uniformity of methodological approach as well as similarity of research fields. Third, a conceptual comparison is made with other literature that has been selected by applying two main selection procedures: (1) a variety of search criteria have been used to find conceptually related literature and a comparison has been made with whatever emerged from this search, (2) comparison is made with other more or less known and recognised literature that was selected because it had some obvious conceptual relatedness. Fourth, the whole model representation of Cultivating Recognition is conceptually compared with other models within and outside the realms of the social sciences.

Comparison with Other Classic Grounded Theory Research in Related Fields

The Classic Grounded Theory studies within the area of Information Systems or business and management are few in number. It may be useful to start the essential literature review with a comparison of the majority of these studies. The focus of these studies is on special issues within business and management, while the theory of "*Cultivating Recognition*" focuses on general issues to do with Information Systems, management and business.

"Cultivated Relationship" as Core Variable

In 1972, Odis Bigus carried out a study of a home delivery milk business (Bigus, 1972: 131-165). The title of this study, "The Milkman and His Customers: Cultivated Relationships", dates it. Today most milk is sold in supermarkets. Although the context of Bigus's study is the US over 40 years ago, it is concerned with understanding how people in a service environment use 'Cultivating Techniques' for their own personal benefit. Bigus (ibid) argued that, for the balance of power to be one of high power asymmetry because of the low demand for the service and high supply of the commodity, different cultivating techniques should be employed. Bigus (ibid) describes this process as follows:

Cultivating techniques are employed with the intent of either directly or indirectly gaining a reward (usually monetary). Cultivated relationships are relationships which are carried out with the primary intent of gaining such a reward...They are usually asymmetrical, with the less powerful party utilizing cultivating tactics to bring relationships closer to a state of symmetry

However, Bigus (ibid) who did not use Classic Grounded theory (he used social Anthropological Methodology), discovered the core variable to be "Cultivated Relationships" and that it processes through a number of stages or cycles. Bigus (1972:137-142) highlights three stages in the acquiring of new customers:

1. Detecting (canvassing amongst friends and associates)
2. Soliciting (personalizing and dealing)
3. Trust-inducing tactics (the sincerity act, contrived disclosure, situational mitigation, accentuated honesty and predicting credit worthiness)

Bigus (1972: 142-145) reveals some more subtle ways in which trust can be further cultivated: *Nurturing pseudo friending; Effecting obligation and Favouring.*Bigus (1972: 144-145) then goes on explain how it proved to be a wise strategy to deliberately avoid certain types of customers by decultivating relationships by means of *(1) Inhibiting conduct and (2) Terminating tactics.* In the final section of Bigus's (1972:147) paper he observes that "High-expertise service occupations, particularly the professions, seem to involve a lower amount of cultivating activity than low expertise occupations. Cultivating in such relationships seems to be directed mainly at trust maintenance". A critique or evaluation of Bigus's (1972) paper reveals:

1. The quality of his line of argument is compelling.
2. The development and synthesis of concepts are very clear

3. The analysis of his data is difficult to assess because he has not been very transparent with his data sources, but the discussion of his research is clear and convincing.
4. The conclusions of his research are very plausible and widely applicable
5. It has both content and construct validity
6. Its reliability is not possible to assess because he has not been totally transparent with his data
7. Its main strength is at the conceptual level

Within the theory of "Cultivating Recognition", E-learning providers in this study specifically, were also in an inferior status and position vis-a-vis their clients (the universities and the government sponsors). Expressed in Bigus (1972:136) terms, the twin "Coping Strategies" were "Legitimizing" and "Credentializing". Even though, in this study, in a "Cultivating Recognition" situation specific factors were very different from those of Bigus's research, respondents in both studies shared the need to cultivate relationships. The people in Bigus's research had to cultivate relationships in order to gain monetary reward. According to the theory of "Cultivating Recognition" in this study, e-learning providers had to "Cultivate Recognition" before they could obtain any monetary rewards. In both cases the process of cultivation took place.

All the trust-inducing techniques for Cultivating Relationships that Bigus (1972) explains have their correspondence to "Credentializing" within the theory of "Cultivating Recognition". "Credentializing" involves the creation of trust by means of "Endorsementizing", resulting in the "Orienting" and "Prioritizing of Duties". Bigus (1972) describes one of the trust inducing techniques as the "sincerity act" ("Result Orienting"), "contrived disclosure" (transparency), "situational mitigation" ("endorsementizing"), and "accentuated honesty" ("Prioritizing Duties").

As noted briefly in previous paragraphs, this researcher would agree with Bigus (1972) concerning the importance of trust cultivation in professional relationships. However Bigus (1972) only indicates "WHAT" is likely to be of importance in the gaining of trust and not "HOW" that trust can be both gained and sustained. It could be argued that this study builds on and goes beyond Bigus (1972) because it was able to demonstrate not only WHAT professionals have to do to gain trust, but also HOW they are able to do this by means of "Cultivation of Recognition". The study also shows how and why cultivation is just as important in high expertise occupations or enterprises (East African E-learning providers and their clients) as in low expertise occupations or enterprises, the main difference being the greater emphasis on the maintenance of trust. The main weakness of the Bigus (1972) research (which uses Social Anthropological Methodology) is that it was published before sufficient knowledge and understanding of CGT had emerged and thus all he was able to do was to create an insightful description and explanation of Cultivating Relationships without being able to state which, of all the variables he cited, was the core variable. This study was able to identify the core variable by facilitating the emergence of the "Cultivation of Recognition" core variable.

It can be stated with certainty that the concept and the meaning of the pattern of "Cultivating Recognition", as it has been defined and explained in this study, has not been applied before as a concept of any significance in any theory building process within Information Systems, Business, Management, Economics or Social Sciences in general. When one searches for this concept or for synonyms or keywords, or terms in titles, or under frequently used words in electronically available texts, it is not possible to find directly relatable terms or literature. The same applies to the set of two primary sub core variables and the set of six secondary sub core variables. This means that there is little or virtually nothing in the literature to which it can be directly or usefully compared.

This means that an indirect literature review comparison has to be attempted. Such an indirect review can lead to indications or explanations in terms of how this particular study can make a contribution to the literature on e-learning providers, CGTM, and "Cultivating Recognition". In the search for such literature in order to make comparisons, the search criteria or key words used have been terms such as "Cultivating Recognition", "gaining recognition", and "public acknowledgement", or "public acknowledgement creation". A small body of literature within Information Systems, Business, Management and related fields has been found. Besides being sparse, this literature is characterized by the frequent use of the terms "acknowledgment" or "recognition". The next section details the outcomes of the review.

Comparing Literature That May Have Relatedness Due to Similarity in Conceptual Vocabulary

A search for literature dealing with "Cultivating Recognition" using the keywords, "Cultivating recognition", "Gaining Recognition", and "Public Acknowledgement" or reverse wording, such as "Public Acknowledgement Creation", yielded a few journal articles and a couple of books. This literature mostly uses terms such as "recognition" or "gaining recognition" or "public acknowledgement" as part of a cluster of auxiliary descriptive terms and not as part of a basic substantive or theoretical concept. For example, within the study of entrepreneurship, the term "Opportunity Recognition" has often been used for contextual descriptive or focusing purposes and not as a pivotal concept in a theory (e.g SWOT analysis). In spite of the remote and occasion absence of conceptual relatedness, a comparison will be attempted.

One Classic Grounded Theory thesis showed "Connected Learning" (Scott, 2007) as focusing on the main concern of the recipients (receivers) of e-learning. Although Scott's (2007) thesis helped to clarify the main concerns of e-learning recipients), it never broached the main concerns of those charged with delivering those services. Thus, unlike Scott's 2007 study, this study has revealed the main concern of the e-learning providers who were the respondents in the study.

Another model pertaining to campus based learners is Tinto's (1993) model of institutional departure. The institutional departure study looked at how campus based learners struggle to gain knowledge. This study has gone beyond Tinto's (1993) work in looking at online and e-learning providers and challenges Tinto's (1993) contention of the centrality of the role itself of the online learning providers, and explains that "Cultivation of Recognition" is of greater significance to online learning providers than simply fulfilling a role. This study has enriched Kember's (1999) model for student persistence, replacing it with this Classic Grounded Theory. Kember's (1999) model is based on Tinto's (1993) model but is criticised in this study for misunderstanding the role of e-learning providers in successful implementation of e-learning. Further Kember's (1999) claim to have conducted data analysis according to the grounded theory research method is disputed since the explanation offered falls short of the coherent explanation of the relationships between variables, a relationship which is an integral part of Classic Grounded Theory. Woodley et al. (2001) dispute Kember's (1999) model from a quantitative perspective.

The professional literature on attrition and retention in e-learning is extended by a conceptual framework with which to understand the relationships between those variables which impact on online and e-learning providers. This study offers a framework with which to organise future research efforts. In distinguishing between the temporal nature of persistence and that of academic success it becomes clear that academic success is dependent upon the success and strength of the providers. This study goes beyond Simpson's (2004) notion of a 'maximum possible increase in retention', in the case of learners withdrawing from a course because of a change in their personal commitment structures. It adds that retention could be facilitated from the providers' side and be due to a lack of "Cultivating Recognition". Lastly, in postulating that reducing the cost of integration and increasing the value of the learning opportunity will increase a learner's propensity to study, this study goes beyond Yorke's (2001) contention that improving conditions for students will increase persistence rates, by arguing that "Cultivating Recognition" is of more significance to the success of online and e-learning projects than the upgrading of conditions.

The most exciting contributions of the research of this study are those opportunities it provides to inform practice. Different aspects of the theory are of value and relevance to different stakeholders. Providers of connected learning will be the ones most interested in the concept of 'Cultivating Recognition'. The development of e-learning programs depends on the providers, such as higher learning institutions and other stakeholders. The success of such institutions will depend on how they cultivate recognition.

The Leitch (2006) report on skills called for institutions of higher education to provide a flexible response to the needs of employers in the provision of demand led vocational learning. This study supports Leitch's (2006) call for providers to change from a supply-driven approach to a demand-led approach to the provision of vocational learning. From the point of view of the vocational learners of this study, the providers need to offer learning opportunities which are relevant and useful for the learners they are serving and to design for persistence. In a fast changing, technology driven world, and to the extent that the needs of employers and employees coincide, providers of connected learning will need to remain up to date and to be flexible in the sense of providing programs which are flexible for learners rather than for the providers.

Research on "Legitimizing"

Only a few IS studies which involve and discuss legitimization have so far been carried out (Brown, 1995; Keable et al., 1998). While Elsbach & Sutton (1992) saw "Legitimatizing" as focussing on organization objectives, artefacts, and activities, Klein and Hirschheim (1989) related legitimatization directly to IS success or failure and defined legitimization as "a social process by which institutions, practices and ideas gain social acceptance" (Klein & Hirschheim 1989: 7). All of these studies included discussion on how IS projects in organizations are unlikely to succeed unless they have gained legitimization from their users. According to Banville (1991) these authors have only mentioned legitimization as a general concept and have not systematically discussed legitimization strategies or types of legitimization. Brown (1998), and Kohl & Kettinger (2004) argue that little is known about the process of legitimization-seeking in IS, and the strategies that can help achieve the required legitimization from users are discussed by Flyn & Zalid (2003). This dearth of detailed studies on "Legitimizing" makes this study an important contributor to literature on "Legitimizing".

Flynn & Hussain (2004) used structuration theory and activity theory to come up with a Legitimization Activity Model (LAM). LAM consists of a number of actions, each of which has a goal. Figure 4 shows the LAM and explains the process of gaining legitimization. The LAM process involves various stages which include: constructing the target, identifying the learning process, closing the gap, learning the norms, granting legitimization, comparing targets and monitoring legitimization.

Flyn & Du (2004) discussed the various activities involved in the process of seeking and maintaining legitimization. Figure 4 shows three activities: gaining legitimization, evaluating and monitoring legitimization and maintaining and/or repairing legitimization. Oliver (1991) argues that IS managers can apply many tactics to gain legitimization for a new IS and many of these tactics fall into two basic categories: conformity and manipulation. Conformity means that, when introducing a new IS, manag-

Figure 4. Legitimization activity model
(Source: Flynn & Hussain, 2004)

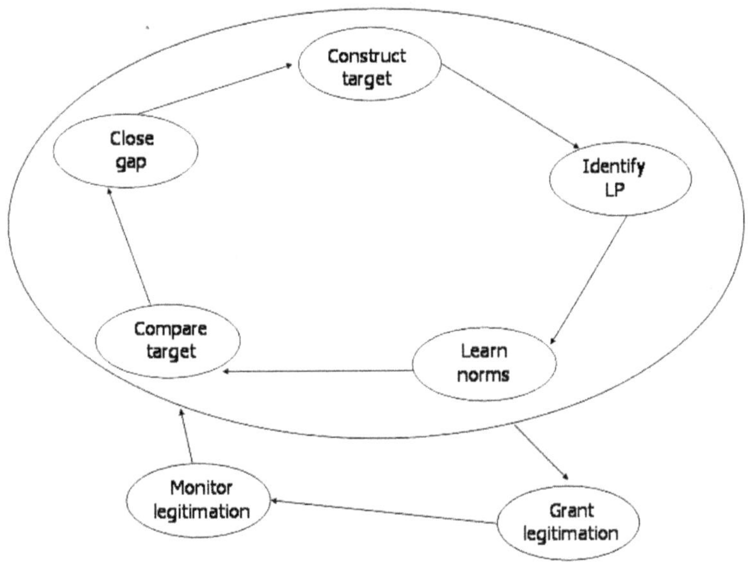

ers conform to the dictates of organizational stakeholders and, if necessary, and, according to Suchman (1995), manipulate only the characteristics of the IS necessary to achieve such conformity. With the manipulation approach, managers take a more active role in influencing organizational stakeholders and in creating new legitimating criteria.

None of the authors outlined above is able provide an answer to the question: "How does the process of legitimization occur?"

This is in stark contrast to the Classic Grounded Theory of "Cultivating Recognition" which has emerged in this thesis. "Cultivating Recognition" explains in detail how "Legitimizing" happens through a combination of "Partner Collaborating", "Referral Networking" and "Strategic Alliancing" (see Figure 5).

Research on "Partner Collaborating"

While public agency directors define collaboration as the act or process of "shared creation" or discovery which involves the creation of new value by doing something new or different, Thomson (2001:23) defines it as:

... a process in which autonomous actors interact through formal and informal negotiation, jointly creating rules and structures governing their relationships and ways to act or decide on the issues that brought them together; it is a process involving shared norms and mutually beneficial interaction.

This definition suggests a higher-order level of collective action than cooperation or coordination. The extensive literature on collaboration does not agree on the terms included in the process, drawing as it does from a wide variety of perspectives, including inter-organizational relations: Alexander (1995); Ring and Van de Ven (1994); Warren et al. (1975); Alter and Hage (1993); O'Toole (1997); O'Toole,

Figure 5. Integrated legitimation activity model
(Source: Flynn & Hussain, 2004)

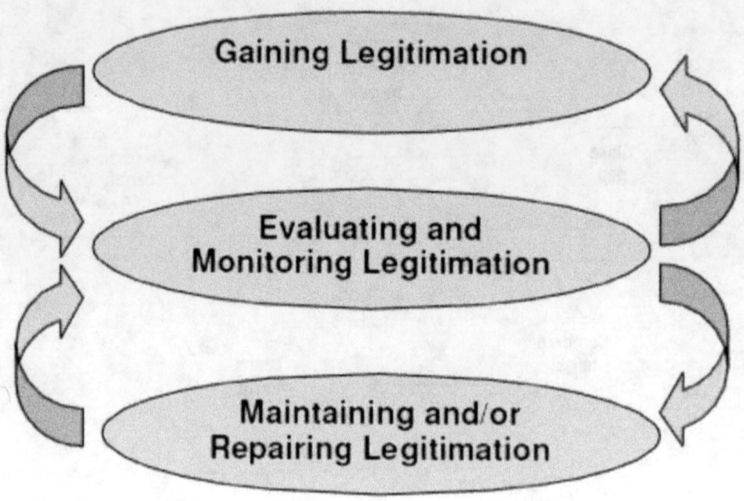

Figure 6. A process framework of collaboration
(Source: Thomas & Perry, 2006)

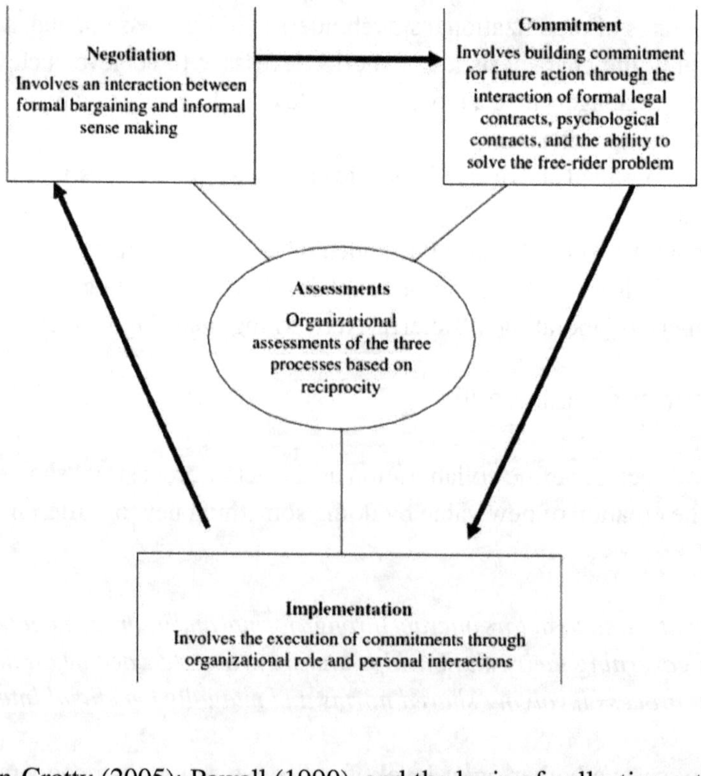

Meier, and Nicholson-Crotty (2005); Powell (1990), and the logic of collective action (Ostrom, 1990). However Thomas and Perry (2006) outlined the Process Framework of Collaboration (Figure 6).

Thomas and Perry (2006) deliver an interesting context based descriptive framework although they are unable to explain how the process of partner collaborating actually happens. This is in contrast to the grounded theory of "Cultivating Recognition" in this thesis, which explains how the process of collaboration actually occurs, demonstrating how the process of collaboration occurs in both "Asymmetrical Collaborating" and "Reciprocal Collaborating" forms.

Research on "Strategic Alliancing"

Effective strategic alliances can have important performance implications for higher learning institutions. Such strategic alliances require recognition that they are embedded in the strategies of each of the partners in the alliance. The success of the alliance increases when the symmetry in the strategic exploitation/exploration intent of the partners is present at the start and is re-calibrated and maintained over time (Koza & Lewin 2000).

Koza & Lewin (2000) in a paper titled "Managing Partnerships and Strategic Alliances: Raising the Odds of Success", outlined a framework for considering the strategic decisions for entering into an alliance and some of the key issues involving the management process of alliances. In Table 1, various reasons are shown for business entering into alliances. Such reasons include gaining access to a restricted market, or overcoming barriers to entry, gaining market power, maintaining market stability, acquiring new technology products and/or new skills, pooling resources, reducing uncertainty, sharing risky research and development projects, speeding up entry into new markets, and deriving new incremental sources of revenue from combining complementary assets (Hitt et al., 1997).

Koza and Lewin (2000) argue for three types of alliances, see Figure 7.

These alliances have as their primary strategic intent the reduction in ignorance of the partners (Balakrishnan & Koza, 1993; Dierickx & Koza, 1991). Such alliances can reveal new information and insights about (1) markets, including local competition, regulations, customer tastes and habits, marketing infrastructure, and the like; (2) new core competencies such as just-in-time processes, negative working capital, one-on-one marketing, and mass customization, and (3) new technologies, such as competency innovations, new complementary technologies, as well as franchising capabilities such as the Pizza Hut brand. The critical success factor in alliances is the ability of the partners to design, manage and continuously adapt organizational processes and informal linkages which keep the alliance on track (Koza & Lewin 2000).

Each of the strategic alliance types has different management characteristics. See Table 2.

Various Doz (1996) and Smith and Van de Ven (1994) outline the stages in the alliance process (Figure 8).

From the above explanations and graphic representations, one can see that creating recognition theory withstands and reveals the important concept of strategic alliancing. The grounded theory of creating recognition, described and discussed in this thesis, explains that strategic alliancing is part of the process of legitimizing the project.

Research on "Referral Networking"

In the context of SME and internationalizing businesses, Dana (2001), using a multi-case descriptive study, analyses the role of networks in creating business opportunities. Networking among small companies and large businesses involving smaller businesses in networks, has apparently become very important, especially given the increasing necessity of operating across state borders (Dana, 2001). Within the theory of cultivating recognition, networking is a property of the concept of Legitimizing, possibly

Table 1. Reasons for strategic alliances by market type

Market	Reasons
Slow Cycle	Gain access to a restricted market Establish franchise in a new market Maintain market stability (e.g. establishing standards)
Standard Cycle	Gain Market power (Reduce industry overcapacity) Gain access to complementary resources Overcome trade barriers Meet competitive challenge by other competitors Pool Resources for very large capital projects Learn New Business techniques
Fast Cycle	Seep up new goods or service entery Speed up new market entry Maintain market leadership Form and Industry technology standard Share Risk R&D expenses Overcome Uncertainty

(Source: Hitt et al., 1997)

Figure 7. Three types of strategic alliances
(Source: Koza & Lewin, 2000)

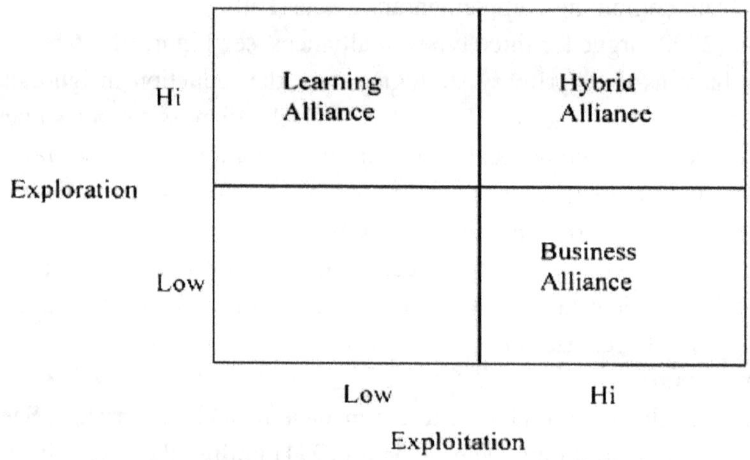

Table 2. Managerial characteristics of strategic alliances

	Learning Alliance	Business Alliance	Hybrid Alliance
Loyalty Control Mechanism	Parent Behaviour and Process	Child Out	Parent (Transformed) Behaviour and Process and output
Ability to absorb	Hi	Lowest	Highest
Time Horizon	Limited term (Learning Cycle)	Open-ended (Industry Cycle)	Multiple time horizons
Success criteria	Pacing the partner	Performance	transformation

(Source: Koza & Lewin, 2000)

Figure 8. The alliance process
(Source: Mackenzie & Callahan, 1999)

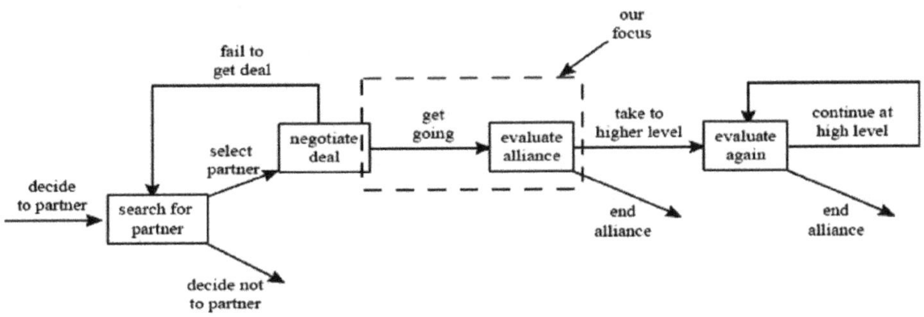

a property of some importance since a network may involve many different patterns, and each of these patterns may offer very different potentials for the creation and sustaining of business opportunities.

In the 21st century networking has played an important role in capturing more businesses and opportunities. Given the rapid growth of technology, for the business to remain competitive in the midst of change, networking has become more important. The increase in 'knowledge workers' (i.e. those involved in 'knowledge management') in turn increases the importance of networking among businesses. Rapid technological changes, and the increasing speed of change, have increased the pressure on businesses and providers to restructure networks, and for these networks to transcend globally. This could lead to some concrete operative solutions to business problems instead of researchers and consultants delivering theoretical outcomes.

HOW THE CORE VARIABLE (CULTIVATING RECOGNITION) CAN ADVANCE E-LEARNING IN EAST AFRICA

As noted earlier, various e-learning implementation efforts failed in East Africa. The emergence of Cultivating recognition will help avoid such failures. It will help to advance the implementation of e-learning in East Africa through creating collaborating partners, strategic alliances, etc.

HOW CGT ANALYSIS CAN ADVANCE E-LEARNING IN EAST AFRICA

Firstly, CGT is a low tech and low cost research method (Glaser, 1978). This means in places such as East Africa, where economic resources for research are very limited, it has enormous potential to be used more widely. E-learning failures in East Africa, have caused lack of funding in this region (Ndume, et al 2008). As a result there is little done towards advancing e-learning research due to lack of funding. Use of CGT will help more researchers to embark on research. Although the GT research method involves a low level of costs prior investment in GT research seminars and workshops are a vital prerequisite before any future major GT research should take place in this region. This is because it is very difficult to understand the GT research by only referring to the published works of of its co-founder Dr Glaser. Dr Glaser's publications are very densely written and they need to be interpreted by other experienced and skilled GT practitioners. Secondly, the e-learning researchers in East Africa have concentrated more on the technical part of e-learning technologies (Lwoga et al 2007), forgetting the social concerns of the people involved. The CGT research method is a powerful tool towards discovering the social concerns (main concerns) (Glaser, 1978). As a result it is thought to be a potential tool towards filling the gap. Thirdly, CGTM has made e-learning qualitative inquiry legitimate. CGTM provides the means to conduct a systematic and detailed analysis the data, and thus gives researchers ample evidence to back up their claims. Considered particularly useful in this respect are the techniques of constant comparison and theoretical sampling, encouraging a constant interplay between data collection and analysis. Overall, the detailed guidance may provide a certain level of comfort that the data are being analysed in a systematic and rigorous manner - especially for novice researchers (Myers, 2009; Fendt & Sachs, 2008)

POTENTIAL CONTRIBUTION OF THE STUDY TO E-LEARNING PRACTICES IN EAST AFRICA

There are three main reasons why this study has the potential to contribute towards e-learning implementation practices in East Africa. Firstly, the research is sought as a way forward for successful implementation of e-learning. E-learning providers can make use of cultivating recognition theory to advance e-learning practices in East Africa. Through this research, the e-learning providers are shown as able to create partners, collaborations, referral networks and credentialize their operations. Secondly, the research has activated the use of CGT in research in East Africa. As noted in the previous section, CGT is low a cost effective research method. This means that it will help East Africa to conduct make more research in various disciplines. The third main reason why this research is of importance to the introduction of e-learning to East Africa is to do with those who are part of the delivery system; the IT experts and the funders of the e-learning process. This thesis demonstrates that it is not sufficient for the IT professional just to have technical competence in e-learning implementation. It is of equal importance that that these professionals invest time and energy to understand the decision making processes involved by those with the power to authorize funding. They can do this by cultivating recognition. Even if the IT professional manages to cultivate recognition sufficiently to ensure the funding of e-learning projects it should be understood that if further CGT research is to be carried out by others more investment in CGT seminars is required. In chapter 7 comments have been made of the complexities of the CGT method and the rather densely written books on the CGT method by Dr Glaser. Novices to using the CGT method are often confused. Doing the CGT method cannot ever be simply restricted to reading the research methodology books by Glaser. There are many nuances and experiences which can only be learnt by experience. This is why the introduction of a series of CGT trouble of seminars is an essential prerequisite to extend the use and understanding of the CGT method in East Africa.

ASPECTS OF THE CLASSIC GROUNDED THEORY OF "CULTIVATING RECOGNITION" WHICH HAVE THE POTENTIAL FOR FUTURE RESEARCH DEVELOPMENT

More research is needed in both of the two sub core variables of cultivating recognition: "Legitimizing" and "Credentializing". The complex processes involved in "Legitimizing", as it has been described in this study, need further investigation so that people who need to be more skilled in this process can be formally trained. The processes of "Collaborating partners", both reciprocally and asymmetrically, in particular need to be better understood. This could be achieved by researchers doing a more extensive cutting edge analysis of new data to find out how existing negative and ineffective asymmetrical collaborative relationships could be transformed into more positive reciprocal relationships. Secondly, within the process of "Credentializing" more data and more analyses are needed in order to better understand the interactions between the four elements of "Results Orienting", namely "Visualizing", "Professionalizing", "Focalizing" and "Role Delineating". Amongst these four main aspects of "Credentializing" this researcher believes that more research into the intricacies of "Visualizing" holds the most promising potential for further development. This kind of research has the potential to make the process of visualizing accessible and of value to a wider range of constituencies other than those at present involved in the delivery of e-learning implementation in East Africa. This would also be of value to those tasked with making complex decisions whose outcomes are uncertain: they could benefit from honing and strengthening their visualizing skills.

The ultimate goal and ideal of any grounded theorist is to be able to discover a formal classic Grounded Theory which is completely transcendent of the time, location and context from which the data originates.

A formal Grounded Theory is a theory of a SGT (Substantive Grounded Theory) core category's general implications generated from, as wide as possible, other data and studies in the same substantive area and in other substantive areas. (Glaser, 2006:34)

A formal theory of "Cultivating Recognition" could be developed using more and various kinds of data which would address not only how online and e-learning providers can successfully implement their e-learning projects, but also how career academics could develop their professional careers.

ACKNOWLEDGMENT

The authors wish to express their thanks to the National Research Foundation (NRF) for funding this research. Our further thanks go to the co-founder of Classic Grounded theory Research, Dr. Barney Glaser. And finally we thank all Classic Grounded Theory Fellows including Dr. Andy Lowe for continuous mentoring process during this research.

REFERENCES

Abouchedid, K., & Eid, G. (2004). E-learning Challenges in the Arab World: Revelation from a Case Study Profile. *Quality Assurance in Education*, *12*(1), 75–81. doi:10.1108/09684880410517405

Adolph, S., Hall, W., & Kruchten, P. (2008). A methodological leg to stand on: lessons learned using grounded theory to study software development. In *Proceedings of the 2008 Conference of the Center For Advanced Studies on Collaborative Research: Meeting of Minds*. ACM. http://doi.acm.org/10.1145/1463788.1463806

Adrich, C. (2004). *Simulations and the Future of Learning*. London: Pfeiffer.

Ajzen, I. (1998). *Attitudes, Personality and Behaviour*. Chicago, IL: Dorsey Press.

Ajzen, I. (2008). *Theory of Planned Behaviour*. Retrieved on January 10, 2008, from http://hdppa.files.wordpress.com/2008/04/azjen.jpg

Akabar, S. (2005). E-learning in developing countries: Challenges and opportunities Bangladesh perspectives. *Proceedings of the2nd International conference on e-learning for Knowledge-based society*.

Akar, E., Ozturk, E., Tuncer, B., & Wiethoff, M. (2004). Evaluation of a collaborative virtual learning environment. *Education + Training*, *46*(6/7), 343-352.

Alavi, M., & Leidner, K. (2001). Knowledge Management and Knowledge management systems: Conceptual Foundations and Research Issues. *Management Information Systems Quarterly*, *25*(1), 107–136. doi:10.2307/3250961

Albert, D., & Mori, T. (2001). Contributions of Cognitive psychology to the future of e-learning.*Bulleting of Graduate School of Education, Hiroshima University*, *50*(1), 25–34.

Alexander, S. (2001). E-learning development and experience. *Education + Training*, *43*(4), 240-248.

Allen, M. (2002). Michael Allen's Guide to E-Learning. Addison-Wesley.

Allens, M. (2007). *Designing successful e-learning*. John Wiley & Sons, Inc.

Andrews, R., & Haythornthwaite, C. (2007). E-learning Journey. In R. Andrews & C. Haythornthwaite (Eds.), The Sage Handbook of E-learning research. Sage Publication.

Andrews, T. (2007, November). Reflections on 'The Discovery of Grounded Theory'. *The Grounded Theory Review*, 56.

Angelina, P. (2002a). Local initiatives in e-learning. *The Star Tech Plus*, *29*(Aug), 11.

Angelina, P. (2002b). E-learning: The alternative leaning process. *The Star Tech Plus*, *29*(Aug), 9–12.

Anolina, A. (2005). Analysis of the terminologies used in the field of virtual learning. *Educational Technology and Society Journal*, *8*(3), 91-102.

Appleton, J. V., & King, L. (2002). Journeying from the philosophical contemplation of constructivism to the methodological pragmatics of health services research. *Journal of Advanced Nursing*, *40*(6), 641–648. doi:10.1046/j.1365-2648.2002.02424.x PMID:12473042

Aranda, N. (2007). *A Brief History of E-learning and Distance Education*. Retrieved March 23, 2009, from http://ezinearticles.com/?A-Brief-History-of-E-learning-and-Distance-Education&id=496460

Aroyo, L., & Dicheva, D. (2004). The New Challenges for E-learning: The Educational Semantic Web. *Journal of Educational Technology & Society*, *7*(4), 59–69.

Baker, R. K. (2003). A Framework for Design and Evaluation of Internet-Based Distance Learning Courses Phase One – Framework Justification, Design and Evaluation. *Online Journal of Distance Learning Administration*, *6*(2), 43–51.

Banathy, B. H. (1991). *Systems designs of education: A journey to create the future*. Englewood Cliffs, NJ: Educational Technology Publications.

Barbara, A. (2002). *E-learning and teaching in Library and information services*. Neal-schuman publishers.

Barbara, A. (2004). *Project Management: Tools and Technique for Today's LIS professionals*. Neal-schuman publishers.

Barron, J. (2006). Top Ten Secretes of Effective E-learning. *Industrial and Commercial Training*, *38*(7), 360–364. doi:10.1108/00197850610704552

Barry, B. (2002). *ISD and the e-learning framework*. Available online at http://www.wit.ie/library/webct/isd.html

Baskerville, R., & Pries-Heje, J. (1999). Grounded action research: A method for understanding IT in practice. *Accounting. Management and Information Technology*, *9*(1), 1–23. doi:10.1016/S0959-8022(98)00017-4

Baskerville, R., & Myers, M. (2002). Information Systems as a Reference Discipline. *Management Information Systems Quarterly*, *26*(1), 1–14. doi:10.2307/4132338

Baskerville, R., & Pries-Heje, J. (2001). Racing the E-Bomb: How the Internet Is Redefining Information Systems Development Methodology. In B. FitzGerald, N. Russo, & J. DeGross (Eds.), Realigning Research and Practice in Is Development: The Social and Organisational Perspective (pp. 49-68). New York: Kluwer.

Baskerville, R., & Pries-Heje, J. (1996). Information Technology Diffusion: Building Positive Barriers. *Proceedings of the 4thEuropean Conference on Information Systems*. AIS.

Baskerville, R., & Pries-Heje, J. (1999). Grounded Action Research: A Method for Understanding IT in Practice. *Accounting, Management and Information Technology*, *9*(1), 1–23. doi:10.1016/S0959-8022(98)00017-4

Bates, A. (2005). *Technology, e-learning and Distance Education*. London: Routledge. doi:10.4324/9780203463772

Bates, A. (2009). *Is e-learning failing in higher education?* Available at www.tonybates.ca/2009/06/is-e-learning-failing-in-higher-education/

Bates, T. (2000). *Managing Technological Change: Strategies for College and University Leaders*. Jossey Bass.

Bates, T. (2001). *National Strategies for e-Learning in Post-Secondary Education and Training*. UNESCO.

Bates, T. (2007). *Strategic Planning for E-learning in Polytechnic*. Tony Bates Associates Ltd. doi:10.4018/978-1-59140-950-2.ch004

Becker, H. J. (2000). Findings for the teaching, learning and computing survey: Is Larry Cuban right? School Technology Leadership conference of the Council of Chief State Schools officers, Washington, DC.

Bednarek-Michalska, B., & Wolodko, A. (2007). E-learning model for Polish libraries: BIBWEB. *The Electronic Library*, *25*(1), 80–89. doi:10.1108/02640470710729137

Beebe, M. (2003). *Partnerships, Alliances and Networks for eLearning*. New Delhi: Tata McGraw-Hill Publishing Company, Ltd.

Beebe, M. (2004). *Impact of ICT Revolution on the African Academic Landscape*. CODESRIA Conference on Electronic Publishing and Dissemination, Dakar, Senegal. Retrieved from: http://www.codesria.org/Links/conferences/el-publ/beebe.pdf

Beebe, M. et al. (2003). *AfricaDotGov: IT Opportunities and Higher Education in Africa*. New Delhi: Tata McGraw-Hill Publishing Company, Ltd.

Bennett, C. (2004). E-learning in Scotland: The North Lanarkshire Experience. *New Library World*, *105*(11/12), 410–416. doi:10.1108/03074800410568743

Bigus, O. E. (1972). The Milkman and his Customer: A Cultivated Relationship. *Journal of Contemporary Ethnography*, *1*(2), 131–165. doi:10.1177/089124167200100201

Bird, J., & Morgan, C. (2003). Adults Contemplating University Study at a Distance: Issues, themes and concerns. *International Review of Research in Open and Distance Learning*, *4*(1), 24–36.

Bleimann, U. (2004). Atlantis University: A new pedagogical approach beyond e-learning. *Campus-Wide Information Systems*, *21*(5), 191–195. doi:10.1108/10650740410567536

Block, J. (1971). *Mastery Learning: Theory and Practice*. New York: Holt, Rinehart and Winston.

Bloom, B. (1968). Learning for Mastery. *Evaluation Comment*, *1*(2), 1–5.

Bonk, C. J., & Reynolds, T. H. (1997), Learner-centered Web instruction for higher-order thinking, teamwork and apprenticeship. In B. H. Khan (Ed.), Web-based instruction (pp. 167-178). Englewood Cliffs, NJ: Educational Technology Publications.

Bose, K. (2003). An E-learning Experience: A written Analysis Based on my Experience with Primary School Teachers in an E-learning Pilot Project. *International Review of Research in Open and Distance Learning*, *4*(2).

Bourne, J. R., Harris, D. A., & Mayadas, A. F. (2005). Online Engineering Education: Learning Anywhere, *Anytime. The Journal of Engineering Education*, *94*(1), 131–146. doi:10.1002/j.2168-9830.2005.tb00834.x

Breckenridge, J., & Jones, D. (2009). Demystifying theoretical sampling in grounded theory research. *The Grounded Theory Review*, *8*, 113–126.

Broadbent, B. (2003). *Championing e-learning*. Retrieved from http://www.elearninghub. com/articles/chamioning.htm#Pro%20and%20cons%20of%20%20e-learning

Brusilovsky, P. (2003). Developing adaptive educational hypermedia systems: From design models to authoring tools. In *Authoring Tools for Advanced Technology Learning Environment* (pp. 377–409). Dordrecht: Kluwer Academic Publishers. doi:10.1007/978-94-017-0819-7_13

Bryant, A., & Charmaz, K. (2007). Grounded theory research: Methods and practices. In A.Bryant & K. Charmaz (Eds.), The Sage handbook of grounded theory (pp. 1-29). London: Sage Publications Limited.

Bryant, A., & Charmaz, K. (Eds.). (2007). *The Sage Handbook of Grounded Theory*. London: Sage Publications.

Bryant, A. (2002). Re-grounding Grounded Theory. *Journal of Information Technology Theory and Application*, *4*(1), 25–42.

Bryant, A. (2009). Grounded theory and pragmatism: The curious case of Anselm Strauss. *Forum Qualitative Sozial Forschung*, 10.

Burawoy, M. (1991). Reconstructing social theories. In M. Burawoy (Ed.), Ethnography unbound. Power and resistance in the modern metropolis (pp. 8-27). Berkeley, CA: University of California Press.

Burke, M. E. (2007). Making Choices: Research Paradigms and Information Management (IM): Practical Application of Philosophy in IM Research. *Library Review*, *58*(6).

Burrell, G., & Morgan, G. (1979). *Sociological Paradigms and Organisational Analysis*. London: Heinemann.

Burney, S. M. A., & Mahmood, N. (2006). A Brief History of Mathematical Logic and Applications of Logic in CS/IT. *Karachi University Journal of Science*, *34*(1), 61–75.

Burrell, G., & Morgan, G. (1979). *Sociological Paradigms and Organizational Analysis: Elements of the Sociology of Corporate Life*. London: Heinemann.

Burton-Jones, A. (2009). Minimizing Method Bias through Programmatic Research. *Management Information Systems Quarterly*, *33*(3), 445–471.

Calder, J., & McCollum, A. (1998). *Open and Flexible Learning in Vocational Education and Training*. London: Kogan Page.

Calvert, J. (2005). Distance Education at the Crossroads. *Distance Education*, *26*(2), 227–238. doi:10.1080/01587910500168876

Calvert, J. (2001). Deakin University: Going online at a dual mode university. *International Review of Research in Open and Distance Learning*, *1*(2), 12.

Campbell, D. T., & Russo, M. J. (2001). *Social Measurement*. Thousand Oaks, CA: Sage Publications. doi:10.4135/9781412985857

Carr, J. (1999). The Role of Higher Education in the Effective Delivery of Multimedia Management Training to Small and Medium Enterprises. *Journal of Educational Technology & Society*, *2*(2), 31–38.

Carr, N. G. (2003, May). IT Doesn't Matter. *Harvard Business Review,* 41-49.

Carrier, C. A., Davidson, G. V., Williams, M. D., & Kalweit, C. M. (1986). Instructional options and encouragement effects in a microcomputer-delivered concept lesson. *The Journal of Educational Research*, *79*(4), 222–229. doi:10.1080/00220671.1986.10885681

Castells, M. (1996). *The Rise of the Network Society* (Vol. 1). Malden, MA: Blackwell.

Chacha, E. (2009). *Understanding the Tools of the Social Learning Landscape*. Available online at http://www.elearninglearning.com/social-media/delicious/2009/?query=chacha

Chadha, G., & Kumail, S. M. N. (2002). *E-Learning: An Expression of the Knowledge Economy*. New Delhi: Tata McGraw-Hill Publishing Company Limited.

Charmaz, K. (1983). The grounded theory method: An explication and interpretation. In R. Emerson (Ed.), *Contemporary field research: A collection of readings* (pp. 109–126). Boston, MA: Little Brown Company.

Charmaz, K. (1990). 'Discovering' chronic illness: Using grounded theory. *Social Science & Medicine*, *30*(11), 1161–1172. doi:10.1016/0277-9536(90)90256-R PMID:2360052

Charmaz, K. (2003). Grounded theory - Objectivist and constructivist methods. In N. K. Denzin & Y. S. Lincoln (Eds.), Strategies of Qualitative Inquiry (pp. 249-291). London: Sage.

Charmaz, K. (2006). *Constructing grounded theory: A practical guide through qualitative analysis*. London: Sage Publications Limited.

Charmaz, K. (2008a). Grounded theory. In J. A. Smith (Ed.), Qualitative psychology: A practical guide to research methods (pp. 81-110). Los Angeles, CA: SAGE.

Charmaz, K. (2008b) Constructionism and the grounded theory method. In Handbook of Constructionist Research. Guilford Press.

CHEMS (Ed.). (2000). *Cases, Issues and Themes in Higher Education Distance e-Learning: a Study on Market Issues for the Proposed e-University*. Higher Education Academy.

Chin, K. L., & Kon, P. N. (2003), Key Factors for a Fully Online E-learning Mode: a delphi study. In Crisp G., D. Thiele, I. Scholten, S. Barker, & J. Baron (Eds.), *Interact, Integrate, Impact:Proceedings of the 20th Annual Conference of the Australasian Society for Computers in Learning in Tertiary Education*.

Christiansen, Ó. (2005). The theory of "opportunizing" and the sub-process of "conditional befriending". *Journal of Business & Economic Research Journal*, *3*(4), 73–88.

Christiansen, Ó. (2006). Opportunizing: A classic grounded theory study on business and management. *The Grounded Theory Review*, *6*(1), 109–133.

Christensen, Ó. (2007). A simpler Understanding of Classic GT: How it is a fundamentally different methodology. *The Grounded Theory Review*, *6*(3), 39–61.

Clarke, C. (2003). Towards a Unified E-learning Strategy, A consultant Document submitted to the. Department for Education and Skills. Available online at http://www.education.gov.uk/consultations/downloadableDocs/towards%20a%20unified%20e-learning%20strategy.pdf

COL. (2002). *The Commonwealth of Learning, "An Introduction to Open and Distance Learning"*. Available online at http://www.col.org/ODLIntro/introODL.htm

Collins, C., Buhalis, D. & Peters, M., A. (2003). Enhancing SMTEs' Business Performance through the Internet and Elearning platforms. *Education + Training*, *45*(8/9), 483-494.

Collins, G. R. (2002). Case Study: A Satellite-based Internet Learning System for the Hospitality Industry. *Online Journal of Distance Learning Administration*, *5*(4), 23–34.

Collins, J. C., & Jerry, I. P. (2000). *Built to Last – Successful Habits of Visionary Companies*. London: Random House.

Collis, B. (1996). *Tele-learning in a digital world-the future of Distance Learning*. London: International Thomson Computer Press.

Collis, B. (1997). *Implementing ICT in the Faculty: Letting a 1000 Flowers bloom or managing change?* Keynote presentation for the national conference, Studying in Digital Learning Environments, University of Utrecht. Retrieved from http://education2.edte.utwente.nl/teletophomepage.nsf/PapersNLViewForm?readform

Collis, B., & Moonen, J. (2001). *Flexible Learning in a Digital World: Experiences and Expectations*. London: Kogan Page.

Cramer, S., Krasinki, S., Crutchfield, M. D., Sackmary, B., & Scalia, L. (2000). Using Collaboration and the Web to Implement the new CEC Standards. *Teaching Exceptional Children*, *32*(5), 12–19. doi:10.1177/004005990003200502

Cross, J. (2004). An Informal History of eLearning. *On the Horizon*, *12*(3), 103–110. doi:10.1108/10748120410555340

Crotty, M. (1998). *The foundations of social research: Meaning and perspective in the research process*. London: Sage Publications Limited.

CTL. (2001). *A vision of e-learning for America's workforce*. Available online at http://www.nga.org/cda/files/ELEARNINGREPORT.pdf

Cuban L. (1993). Computers Meet Classroom: Classroom wins Teachers College Record. *Volunteer*, *95*(2), TCR ID 82.

Cupchik, G. (2001). Constructivist Realism: An Ontology That Encompasses Positivist and Constructivist Approaches to the Social Sciences. *Forum: Qualitative Social Science*. Available online at http://www.qualitative-research.net/index.php/fqs/article/view/968/2112

Cutcliffe, J. R. (2004). Adapt or adopt: Developing and transgressing the methodological boundaries of grounded theory. *Journal of Advanced Nursing*, *51*(4), 421–428. doi:10.1111/j.1365-2648.2005.03514.x PMID:16086811

Dabbagh, N. H., Bannan-Ritland, B., & Silc, K. (2000). Pedagogy and Web-based Course Authoring Tools: Issues and implications. In B. H. Khan (Ed.), Web-based training (pp. 343-354). Englewood Cliffs, NJ: Educational Technology Publications.

Dam, N. (2004). *The E-learning Fieldbook. Implementation Lessons and Case Studies from Companies That Are Making E-learning Work*. New York, NY: McGraw-Hill.

Dana, L. P. (2001). Networks, Internationalization & Policy. *Small Business Economics*, *16*(2), 57–62. doi:10.1023/A:1011199116576

Dearnley, C. (2003). Student Support in Open Learning: Sustaining the Process. *International Review of Research in Open and Distance Learning*, *4*(1), 67–74.

Dhanarajan, G. (2001). *Open and Flexible Learning: Commonwealth Experience and Indian's Challenge.* Inaugural Remarks made at the 76th Annual Meeting of the Association of Indian Universities, Central University of Chandigarh, Chandigarh, India. Available at http://www.col.org/resources/speeches/2001presentations/pages/2001-12-08.asp

Dobson, P. J. (2002). Critical Realism and Information Systems Research: Why Bother with Philosophy?. *Information Research*, *7*(2). Available at http://InformationR.net/ir/7-2/paper124.html

Doughty, P. L., Spector, M., & Yonai, B. A. (2003). Time, Efficacy and Cost Considerations of e-Collaboration in Online University Courses. *Brazilian Review of Open and Distance Learning.* Available at http://www.abed.org.br/publique/cgi/cgilua.exe/sys/start.htm?UserActiveTemplate=1por&infoid=746&sid=70

Dreyfus, H. L. (1999). Anonymity versus Commitment: The Dangers of Education on the Internet. *Ethics and Information Technology*, *1*(1), 15–21. doi:10.1023/A:1010010325208

Dringus, L. P. (2003). From both sides now: On being an Online Learner and Online Instructor. *E-Learn Magazine.* Available at URL:http://www.elearnmag.org/subpage/sub_page.cfm?section=3&list_item=1&page=1

Driscoll, M. (2002). *Web-based Training: Designing E-learning Experiences.* Jossey-Bass.

Drucker, P. (1994, November). The Age of Social Transformation. *Atlantic Monthly*, 68.

Drucker, P. (2000). *Need to Know: Integrating e- Learning with High Velocity Value Chains.* A Delphi Group White Paper.

Duchaste, A. (1996). Design for web-based learning.*Proceedings of the WebNet-96 World Conference of the Web Society.*

Duncan, E. A. S., & Nicol, M. M. (2004). Subtle realism and occupational therapy: An alternative approach to knowledge generation and evaluation. *British Journal of Occupational Therapy*, *67*(10), 453–456. doi:10.1177/030802260406701006

EAC. (2001). *Retreat to establish East African Community*. East African Community. Available at http://www.eac.int/treaty/

Edland, M. (2011). *ELearning and the Benefits That It Offers.* Ezine@Articles. Available at http://ezinearticles.com/?ELearning-and-the-Benefits-That-It-Offers&id=5715586

Edmundson, A. (2007). *Globalized e-learning cultural challenges*. London: Idea Group. doi:10.4018/978-1-59904-301-2

Eisenhardt, K. M. (1989). Building theories from case study research. *Academy of Management Review*, *14*(4), 532–550.

E-learning in India. (2001). *E-learning in India.* Presented in the E-Learning in India Conference. Available at http://www.slideshare.net/missan/e-l-india-finl-presentation

Ellington, H. (1995). Flexible learning, your flexible friend. In C. Bell, M. Bowden & A. Trott (Eds.), Implementing flexible learning (pp. 3-13). London: Kogan Page.

El-Tigi, M. A., & Khan, B. H. (2001). Web-based Learning Resources. In B. H. Khan (Ed.), Web-Based Training (pp. 59-72). Englewood Cliffs, NJ: Educational Technology Publications.

Ennew, C., & Fernandez-Young, A. (2006). Weapons of Mass Instruction? The rhetoric and reality of online learning. *Marketing Intelligence & Planning*, *24*(2), 148–157. doi:10.1108/02634500610654008

Ettinger, A., & Holton, V. (2005). E-learning lessons from the sharp end: The BBC. *Training & Management Development Methods*, *19*(2), 7–13.

Ettinger, A., Holton, V., & Blass, E. (2006b). E-learner experiences: What is the future for e-learning? *Industrial and Commercial Training*, *38*(4), 208–212. doi:10.1108/00197850610671991

Ettinger, A., Holton, V., & Blass, E. (2006a). E-learner experiences: Key questions to ask when considering implementing e-learning. *Industrial and Commercial Training*, *38*(3), 143–147. doi:10.1108/00197850610659409

Evans, C., & Fan, J. P. (2002). Lifelong Learning through the Virtual University. *Campus-Wide Information Systems*, *19*(4), 127–134. doi:10.1108/10650740210438810

Evans, J.R. & Haase, I.M. (2001). Online Business Education in the Twenty-First Century: An Analysis of Potential Target Markets. *Internet Research: Networking.*

Fendt, J., & Sachs, W. (2008). Grounded Theory Method in Management Research. *Organizational Research Methods*, *11*(3), 430–455. doi:10.1177/1094428106297812

Fitzgerald, B., & Howcroft, D. (1998). Towards the dissolution of the IS debate. From polarisation to polarity. *Journal of Information Technology*, *13*(4), 313–326. doi:10.1057/jit.1998.9

Fitzgerald, B. & Howcroft, D. (1998b). Competing Dichotomies in IS Research and Possible Strategies for Resolution. *ICIS 1998 Proceedings.*

Flinn K. & Lawrence, M. (2003). *E-learning Development in Sub-Saharan Africa: Issues and Challenges.* Paper Presented to National ICT policy round table.

Flynn, D., & Hussain, Z. (2001). *Using Structuration Theory To Explain Information Systems Development And Use In A Public Health Organization.* Paper presented to the 9th European Conference on Information Systems, Bled, Slovenia. Available at http://sdaw.info/asp/aspecis/20010034.pdf

Flynn, D., & Hussain, Z. (2004). Seeking Legitimation for an Information System: A Preliminary Process Model. *ECIS 2004 Proceedings.* Available online at http://aisel.aisnet.org/ecis2004/27

Fresen, J. W., & Boyd, L. G. (2005). Caught in the Web of Quality. *International Journal of Educational Development*, *25*(3), 317–331. doi:10.1016/j.ijedudev.2004.12.002

Fry, K. (2000). *Forum Focus and Overview. The Business of E-learning: Bringing your Organization in the Knowledge Economy, Telcam Group.* Sydney: University of Technology.

Galagan, P. (2000). Getting Started With E-learning. *Journal of Training and Development*, *54*(4), 62–64.

Garrison, D. R., & Shale, D. (1987). Mapping the Boundaries of Distance Education: Problems in Defining the Field. *American Journal of Distance Education*, *1*(1), 7–13. doi:10.1080/08923648709526567

Gasson, S. (1998). Framing Design: A Social Process View of Information System Development. In *Proceedings of the Nineteenth International Conference on Information Systems* (pp. 224 – 236). Atlanta, GA: Association for Information Systems (AIS).

George, T., & Mcgee, M. K. (2003). Educational Advantage. *Information Week*, 57–58.

Gilbert, P. K. (2002). *The Virtual University an Analysis of Three Advanced Distributed Leaning Systems*. Available at http://gseacademic.harvard.edu/~gilberpa/homepage/portfolio/research/pdf/edit611.pdf

Glaser, B. (1964). *Organizational Scientists: Their Professional Careers*. Chicago: Aldine.

Glaser, B. (1965). The *Constant Comparative Method of Qualitative Analysis*. *Social Problems*, *12*(4), 436–445. doi:10.2307/798843

Glaser, B. (1978). *Theoretical Sensitivity: Advances in the Methodology of Grounded Theory*. Mill Valley, CA: Sociology Press.

Glaser, B. (1992). *Emergence vs Forcing: Basics of Grounded Theory Analysis*. Mill.

Glaser, B. (1992). *Basics of grounded theory analysis*. Mill Valley, CA: Sociology Press.

Glaser, B. (1993). *Examples of Grounded Theory: A Reader*. Sociology Press.

Glaser, B. (1995a). *Grounded Theory 1984-1994, Volume One*. Mill Valley, CA: Sociology Press.

Glaser, B. (1995b). *Grounded Theory 1984-1994, Volume Two*. Mill Valley, CA: Sociology Press.

Glaser, B. (1996). *Gerund Grounded Theory: The Basic Social Process Dissertation*. Sociology Press.

Glaser, B. (1998). *Doing Grounded Theory: Issues and Discussions*. Mill Valley, CA: Sociology Press.

Glaser, B. (1999a). *The Discovery Of Grounded Theory: Strategies for Qualitative Research*. Mill Valley, CA: Sociology Press.

Glaser, B. (1999b). *The Future of Grounded Theory, Qualitative Health Research. Sage Publications*, *9*(6), 836–845.

Glaser, B. (1999c). The Future of Grounded Theory, Keynote address at the 4th annual Qualitative Health Research Conference. *The Grounded Theory Review*, *1*(1), 1–9.

Glaser, B. (2001). *The Grounded Theory Perspective: Conceptualization Contrasted with Description*. Mill Valley, CA: Sociology Press.

Glaser, B. (2002). Constructivist grounded theory? *Forum Qualitative Sozial Forschung*, 3.

Glaser, B. (2003). *The Grounded Theory Perspective II: Description's Remodeling of Grounded Theory*. Mill Valley, CA: Sociology Press.

Glaser, B. (2004a). *Glaser's explanations on ontology and epistemology*. Academic Press.

Glaser, B. (2004b). *Staying open*. Unpublished paper.

Glaser, B. (2004c). Naturalist inquiry and grounded theory. *Forum Qualitative Sozial Forschung*, 5.

Glaser, B. (2005a). The Impact of Symbolic interactionism on GT. *GT Review*, *4*(2), 1-23.

Glaser, B. (2005b). *The grounded theory perspective III: Theoretical coding*. Mill Valley, CA: Sociology Press.

Glaser, B. (2005c). *The Roots of Grounded Theory*. Keynote presentation delivered to the 3rd International Qualitative Research Convention, Johor Bahru, Malaysia.

Glaser, B. (2006). *Doing Formal Grounded Theory: A proposal*. Mill Valley, CA: Sociology Press.

Glaser, B. (2008). *Doing Quantitative Grounded Theory*. Mill Valley, CA: Sociology Press.

Glaser, B. (2009). *Jargonizing: Using the grounded theory vocabulary*. Mill Valley, CA: Sociology Press.

Glaser, B. (2011). *Getting out of the data: Grounded theory conceptualisation*. Mill Valley, CA: Sociology Press.

Glaser, B., & Holton, J. (2004). Remodeling grounded theory. *Forum: Qualitative Social Research, 5*(2), Article 4. Retrieved from http:// www.qualitative-research.net/fqstexte/2-04/2-04glaser-e.htm

Glaser, B., & Holton, J. (2005). Basic Social Processes. *The Grounded Theory Review*, *4*(3), 1–21.

Glaser, B., & Kaplan, D. W. (Eds.). (1996). *Gerund Grounded Theory: The Basic Social Process Dissertation*. Mill Valley, CA: Sociology Press.

Glaser, B., & Strauss, A. (1965). *Awareness of Dying*. Chicago: Aldine Publishing Company.

Glaser, B., & Strauss, A. (1967). *The Discovery of Grounded Theory: Strategies for Qualitative Research*. Hawthorne, NY: Aldine de Gruyter.

Goodear, L. (2001). *Cultural Diversity and Flexible Learning*. Presentation of Findings 2001 Flexible Learning Leaders Professional Development Activity. South West Institute of TAFE. Australia. Available at http://www.flexiblelearning.net.au/leaders/events/pastevents/2001/statepres/papers/lyn-handout.pdf

Gotschall, M. (2000). E-learning Strategies for Executive Education and Corporate Training. *Fortune*, *141*(10), S5–S59.

Goulding, C. (1998). Grounded theory: The missing methodology on the interpretivist agenda. Qualitative Market Research. *International Journal (Toronto, Ont.)*, *1*(1), 50–57.

Goulding, C. (2001). Grounded theory: A magical formula or a potential nightmare. *The Marketing Review*, *2*(1), 21–33. doi:10.1362/1469347012569409

Grange, L. (2004). E-learning: Some Critical Thought, South Africa. *The Journal of Higher Education*, *18*(1), 87–97.

Grant, D., & Cliff, O. (1996). *Metaphors and Organization*. London: Sage Publications.

Grant, L. K., & Spencer, R. E. (2003). The Personalized System of Instruction: Review and Applications to Distance Education. *International Review of Research in Open and Distance Learning*, *4*(2), 28–37.

Grix, J. (2002). Introducing students to the generic terminology of social research. *Politics*, *22*, 175–186.

Grooms, L. D. (2003). Computer-Mediated Communication: A Vehicle for Learning. *International Review of Research in Open and Distance Learning*, *4*(2), 70–76.

Grix, J. (2002). Introducing students to the generic terminology of social research. *Politics*, *22*(3), 175–186. doi:10.1111/1467-9256.00173

Gunga, S., & Ricketts, I. (2007). Facing the Challenges of E-learning Initiatives in African Universities. *British Journal of Educational Technology*, *38*(5), 896–906. doi:10.1111/j.1467-8535.2006.00677.x

Haigh, J. (2004). Information Technology in Health Professional Education: Why IT Matters. *Nurse Education Today*, *24*(7), 547–552. doi:10.1016/j.nedt.2004.07.008 PMID:15465170

Hall, B. (1997). *Web-based training cookbook*. New York: Wiley.

Hall, B. (2001). E-Learning: Building Competitive Advantage Through People and Technology, A Special Section on Elearning. *Forbes Magazine*. Available at http://www.forbes.com/specialsections/elearning/

Halse, M.L., & Mallinson, B.J. (2009). Investigating popular Internet applications as supporting e-learning technologies for teaching and learning with Generation Y. *International Journal of Education and Development using Information and Communication Technology*, *5*(5), 58-71.

Hamid, A. A. (2002). E-Learning - Is it the "e" or the Learning that Matters. *The Internet and Higher Education*, *4*(1), 311–316.

Hartman, H. J. (2001). *Metacognition in learning and instruction*. Academic Press.

Harvey, B., & Beards, D. (2004). E-learning in Scottish Further and Higher Education. *Education + Training, 46*(6/7), 353-360.

Heath, H., & Cowley, S. (2004). Developing a Grounded Theory Approach: A Comparison of Glaser and Strauss. *International Journal of Nursing Studies*, *41*(2), 141–150. doi:10.1016/S0020-7489(03)00113-5 PMID:14725778

Hernandez, C. A. (2006). *Are there two methods of grounded theory?: Demystifying the debate*. Paper presented at the Grounded Theory Seminar, New York, NY.

Hernandez, T. (2006). Next Generation E-Learning: A Buyer's Market. *Certification Magazine*. Available at http://www.certmag.com/issues/jul00/feature_hernandez.cfm

Herselman, M. (2005). An Investigation into E-learning Practices Implication for Providers of Education and Training. *South African Journal of Higher Education*, *19*(2), 393–410.

Hiltz, S. R. (1997). Impacts of College Level Courses via Asynchronous Learning Networks: Some Preliminary Results. *Journal of Asynchronous Learning Networks*, *1*(2), 1–18.

Hinson, R. (2006). The Internet for Academics: Towards a Holistic Adoption Model. *Online Information Review*, *30*(5), 542–554. doi:10.1108/14684520610706415

Hirschheim, R., & Klein, H. (1989). Four paradigms of information systems development. *Communications of the ACM*, *32*(10), 1199–1216. doi:10.1145/67933.67937

Hirschheim, R. (1985). Information systems epistemology: An historical perspective. In E. Mumford, R. Hirschheim, & R. Fitzberald et al. (Eds.), *Research Methods in Information Sysfems* (pp. 13–38). Amsterdam: North-Holland.

Hitt, M. A., Hoskisson, R. E., & Kim, H. (1997). International Diversification: Effects on Innovation and Firm Performance in Product-Diversified Firms. *Academy of Management Journal*, *40*(4), 767–798. doi:10.2307/256948

Hofmann, J. (2004). *The Synchronous Trainer's Survival Guide*. Pfeiffer.

Holton, J. A. (2007). The Coding Process and Its Challenges. In The Sage Handbook of Grounded Theory. Sage Publications.

Holton, J. A. (2009). Qualitative tussles in undertaking a grounded theory study. *The Grounded Theory Review*, *8*, 37–50.

Holton, W. K. (2000). *Designing Web-based Training: How to Teach Anyone Anything Anywhere Anytime*. John Wiley & Sons.

Homan, G., & Macpherson, A. (2005). E-learning in Corporate Universities. *Journal of European Industrial Training*, *29*(1), 75–90. doi:10.1108/03090590510576226

Honey, P. (2001). E-learning: A performance appraisal and some suggestions for improvement. *The Learning Organization*, *8*(5), 200–202. doi:10.1108/EUM0000000005913

Horton, W. (2001). *Leading e-Learning: American Society for Training and Development*. Available at http://www.elearninggurus.com/articles.html

Hotspot Service Area. (2004). *Telekom Malaysia*. Available at http://isp.tm.net.my/hotspot/service-klangv.shtml

Howcroft, D., & Trauth, E. M. (2005). Choosing Critical IS Research. In Handbook of Information Systems Research: Critical Perspectives on Information Systems Design, Development and Implementation. Edward Elgar Publishing.

Howell, S. L., Williams, P. B., & Lindsay, N. K. (2003). Thirty-two Trends Affecting Distance Education: An Informed Foundation for Strategic Planning. *Online Journal of Distance Learning Administration*, *6*(3).

Hsbollah, H., & Idrisk, K. (2009). E-learning Adoption: The Role of Relative Advantages, Trialability and Academic Specialization. *Campus-Wide Information Systems*, *26*(1), 23–37.

Hussain, Z., Taylor, A., & Flynn, D. (2004). A case study of the process of achieving legitimation in information systems development. *Journal of Information Science*, *30*(5), 408–417. doi:10.1177/0165551504046725

Iivari, J., Hirschheim, R., & Klein, H. K. (1998). A Paradigmatic Analysis Contrasting Information Systems Development Approaches and Methodologies. *Information Systems Research*, *9*(June), 164–193. doi:10.1287/isre.9.2.164

Locke, J. (2001). *An Essay Concerning Human Understanding* (A. C. Fraser, Ed., Vol. 1). Oxford, UK: Clarendon.

INACOL. (2011). *Use of ICT in Learning and Teaching*. The International Association for K-12 Online Learning. Available at http://www.inacol.org

InfoDev. (2005). Knowledge Map: ICT in Education. Author.

IUCEA. (2001). ICT for Learning and Training. *Proceeding during roundtable meetings of The Inter-University Council of East Africa (IUCEA)*. Available at http://www.iucea.org/resources/ict_for_learning_and_training/

Ivergard, T., & Hunt, B. (2005). Towards a Learning Networked Organization: Human Capital, Compatibility and Usability in E-Learning Systems. *Applied Ergonomics*, *36*(2), 157–167. doi:10.1016/j.apergo.2004.09.006 PMID:15694069

James-Gordon, Y., Young, A., & Bal, J. (2003). External Environment Forces Affecting E-learning Provider'. *Marketing Intelligence & Planning*, *21*(3), 168–172. doi:10.1108/02634500310474984

Jennings, M. M. (1994, November 28). Business Schools' Formula for Irrelevance. *Wall Street Journal, Manager's Journal*.

Johnston, J. (1987). *Electronic Learning: From Audio Tape to Videodisc*. London: Lawrence Erlbaum Associates.

Jones, M. G., & Farquhar, J. D. (1997). User Interface Design for Web-Based Instruction. In B. H. Khan (Ed.), Web-based instruction (pp. 239-244). Englewood Cliffs, NJ: Educational Technology Publications.

Juhary, J. (2005). A Step Towards E-learning: Some Pedagogical Issues. *International Journal of Pedagogies and Learning*, *1*(1), 48–58. doi:10.5172/ijpl.1.1.48

Juma, M. (2001). *The Establishment of a Higher Education Open and Distance Learning Knowledge Base for Decision Makers in Kenya*. Available online at http://www.UNESCO.org/education/studyingabroad/highligt/od/kb/kenya.doc

Juma, M. (2002). *The Impact of Implication of Open Courseware for Higher Education in Developing Countries*. Available online at http://www.wcet.info/unesco/juma_kenya.pdf

Kao, D., Tousignant, W., & Wiebe, N. (2000). A Paradigm for Selecting an Institutional Software. In *Proceedings ISECON 2000*. AITP Foundation for Information Technology Education.

Karon, R. L. (2000). Bankers go Online: Illinois Banking Company Learns Benefits of E-training. *E-learning*, *1*(1), 38–40.

Kathawala, Y., Abdou, K., & Elmulti, D. S. (2003). The Global MBA: A Comparative Assessment for its Future. *Journal of European Industrial Training*, *26*(1), 14–23. doi:10.1108/03090590210415867

Kavulya, J. (2007). Digital Libraries and Development in Sub-Saharan Africa: A Review of Challenges and Strategies. *The Electronic Library*, *25*(3), 299–315. doi:10.1108/02640470710754814

Kearsley, G., & Shneiderman, B. (1999). *Engagement Theory: A Framework for Technology-Based Teaching and Learning*. Available online at http://home.sprynet.com/~gkearsley/engage.htm

Kearsley, G. (1996). The World Wide Web: Global Access to Education. *Educational Technology Review*, (5), 26-30.

Kearsley, G. (2000). *Teaching & Learning in Cyberspace*. Available at http://home.sprynet.com/~gkearsley/chapts.htm

Keegan, D., et al. (2007), E-learning Initiatives that did not Reach Targeted Goals. Megatrends Project 2007.

Keegan, D. (1986). *The Foundation of Distance Education*. Routledge, Kegan & Paul.

Kelle, U. (2005). Emergence vs forcing of empirical data? A crucial theory of grounded theory reconsidered. *Forum: Qualitative Social Research, 6*(2), Article 27. Retrieved from http://www.qualitativeresearch.net/fqs-texte/2-05/05-2-27-e.htm

Kelly, K. B., & Bonner, K. (2005). Digital Text, Distance Education and Academic Dishonesty: Faculty and Administrator Perceptions and Responses. *Journal of Asynchronous Learning Networks, 9*(1), 43–52.

Kember, D. (1995). *Open Learning Courses for Adults. A model of Student Progress*. Englewood Cliffs, NJ: Educational Technology Publications.

Kember, D. (1999). Integrating Part-time Study with Family, Work and Social Obligations. *Studies in Higher Education, 24*(1), 109–124. doi:10.1080/03075079912331380178

Kember, D., Armour, R., Jenkins, W., Lee, K., Leung, D. Y. P., & Li, N. et al. (2001). Orientation to Enrollment of Part-time Students: A Classification System Based upon their Perceived Lifelong Learning Needs. *Higher Education Research & Development, 20*(3), 265–280. doi:10.1080/07294360120108359

Kember, D., Ying, C. K., Wan, C. S., & Yung, C. S. (2005). How students cope with part time study: An analysis of coping mechanisms through an on-line forum. *Active Learning in Higher Education, 6*(3), 230–242. doi:10.1177/1469787405057662

Khan, B. (2005). *Managing E-learning Strategies: Design, delivery, implementation and Evaluation*. London: Idea group. doi:10.4018/978-1-59140-634-1

Khan, B. H. (1997a). Web-based Instruction: What is it and why is it? In B. H. Khan (Ed.), Web-based instruction (pp. 5-18). Englewood Cliffs, NJ: Educational Technology Publications.

Khan, B. H. (2001a). Virtual U: A Hub for Excellence in Education, Training and Learning Resources. In B. H. Khan (Ed.), Web-Based Training (pp. 491-506). Englewood Cliffs, NJ: Educational Technology Publications.

Khan, B. H. (2001b). Web-Based Training: An Introduction. In B. H. Khan (Ed.), *Web based Training*. Englewood Cliffs, NJ: Educational Technology Publications.

Khan, B. H. (2005a). *E-Learning QUICK Checklist*. Hershey, PA: Information Science Publishing. doi:10.4018/978-1-59140-812-3

Khan, B. H. (2007a). Introduction to open, flexible and distributed learning. In B. H. Khan (Ed.), *Flexible Learning*. Englewood Cliffs, NJ: Educational Technology Publications.

Khan, B. H. (2007b). Obstacles Encountered During Stages Elearning Process. In B. H. Khan (Ed.), *Flexible Learning*. Englewood Cliffs, NJ: Educational Technology Publications.

Khan, B. H., & Ealy, D. (2001). A Framework for Web-Based Authoring Systems. In B. H. Khan (Ed.), Web-based Training (pp. 355-364). Englewood Cliffs, NJ: Educational Technology Publications.

Khan, B. H., & Smith, H. L. (2007c). Student Evaluation Instrument for Online Programs. In B. H. Khan (Ed.), *Flexible Learning*. Englewood Cliffs, NJ: Educational Technology Publications.

Khan, B. H., Waddill, D., & McDonald, J. (2001a). Review of Web-Based Training Sites. In B. H. Khan (Ed.), Web-Based Training (pp. 367-374). Englewood Cliffs, NJ: Educational Technology Publications.

Klein, D. E., Mallory, C. A., & Safstrom, D. W. (1997). Analysis, Design, And Implementation of a web-Based Training System for Multi-Criteria decision Support, Integrating Hypertext. In Multimedia-Based Case Studies and Training Software. Monterey, CA: Naval Postgraduate School.

Klein, H. K., & Myers, M. D. (1999). A Set of Principles for Conducting and Evaluating Interpretive Field Studies in Information Systems. *MIS Quarterly*, *23*(1), 21–29.

Klein, H. K., Nissen, H. E., & Hirschheim, R. (1991). *A Pluralist Perspective of the Information Systems Research Arena. In Information Systems Research: Contemporary Approaches & Emergent Traditions*. Amsterdam: North-Holland.

Kohli, R., & Kettinger, W. J. (2004). Informating the Clan: Controlling Physicians' Costs and Outcomes. *Management Information Systems Quarterly*, *28*(3), 363–394.

Koza, M., & Lewin, A. (2000). Managing partnerships and strategic alliances: Raising the odds of success. *European Management Journal*, *18*(2), 146–151. doi:10.1016/S0263-2373(99)00086-9

Krauth, B. (1998). Distance Learning: The Instructional Strategy of the Decade. In G. P. Connick (Ed.), The Distance Learner's Guide. Upper Saddle River, NJ: Prentice Hall.

Kuchi, R., Gardner, R., & Tipton, R. (2003). *A Learning Framework for Information Literacy and Library Instruction Programs at Rutgers University Libraries, Recommendations of the Learning Framework Study Group*. Rutgers University Libraries.

Kypreos, T. (2003). Building a Business Case for Elearning. *E-Learn Magazine*. URL:http://www.elearnmag.org/subpage/sub_page.cfm?section=3&list_item=13&page=1

Lahn, L.L. (2004). Dilemmas in the Development of Elearning at Work. *Journal of Workplace Learning*, *16*(8), 466-478.

Laird, D. (1985). *Approaches To Training and Development* (2nd ed.). Reading, MA: Addison-Wesley.

Land, S. M., & Hannafin, M. J. (1997). Patterns of Understanding with Open-Ended Learning Environments: A Qualitative Study. *Educational Technology Research and Development*, *45*(2), 47–73. doi:10.1007/BF02299524

Larson, A., Teisberg, E. O., & Johnson, R. R. (2000). Sustainable Business: Opportunity and Value Creation. *Interfaces*, *30*(3), 1–12. doi:10.1287/inte.30.3.1.11658

Lau, M. P. (2002). *Online Teaching & Learning*. Nanyang Siang Pau's New Century.

Laurillard, D. (1995). Multimedia and the Changing Experience of the Learner. *British Journal of Educational Technology*, *26*(3), 179–189. doi:10.1111/j.1467-8535.1995.tb00340.x

Lave, J., & Wenger, E. (1991). *Situated Learning: Legitimate Peripheral Participation (Learning in Doing: Social, Cognitive and Computational Perspectives)*. Cambridge, UK: Cambridge University Press. doi:10.1017/CBO9780511815355

Laws, R. D., Howell, S. L., & Lindsay, N. K. (2003). Scalability in Distance Education: "Can we Have Our Cake and Eat it Too?". *Online Journal of Distance Learning Administration*, *6*(4), 80–89.

Lee, B. (1997). *The Power Principle – Influence with Honor*. New York: Simon & Schuster.

Lee, C. S., & Al-Hawamdeh, S. (2001). New Mode of Course Delivery for Virtual Classroom. *Aslib Proceedings*, *53*(6), 238–242. doi:10.1108/EUM0000000007057

Lee, F. (2002a). *Post Keynesian Price Theory*. Cambridge, UK: Cambridge University Press.

Lee, F. (2002b). Theory Creation and the Methodological Foundation of Post Keynesian Economics. *Cambridge Journal of Economics*, 26789–26804.

Lee, F. (2005). Grounded Theory and Heterodox Economics. *The Grounded Theory Review*, *3*(1), 70–86.

Lehaney, B. (2004). *Beyond Knowledge Management*. Idea Group publishing. doi:10.4018/978-1-59140-180-3

Leitch, S. (2006). *Prosperity for all in the global economy - world class skills*. London: HM Treasury.

Levy, S. (2003). Six Factors to Consider when Planning Online Distance Learning Programs in Higher Education. *Online Journal of Distance Learning Administration*, *6*(1), 35–43.

Li, B. (2009). The Use of E-learning in Pre-Service Teacher Education. *Campus-Wide Information Systems*, *26*(2), 132–136. doi:10.1108/10650740910946855

Litto, F. M. (2002). The Hybridization of Distance Learning in Brazil – An Approach Imposed by Culture. *International Review of Research in Open and Distance Learning*, *2*(2), 54–64.

Locke, K. (2001). *Grounded Theory in Management Research*. London: Sage Publications Ltd.

Lowe, A. (1998). Managing the Post-Merge Aftermath by Default Remodeling. *Management Decision*, *36*(2), 102–110. doi:10.1108/00251749810204188

Lowe, A. (1995). The Basic Social Processes of Entrepreneurial Innovation. *International Journal of Entrepreneurial Behaviour and Research*, *1*(2), 54–76. doi:10.1108/13552559510090622

Lowe, A. (1996). *An Explanation of Grounded Theory* (Working paper). Glasgow, UK: University of Stratchlyde, Department of Marketing.

Lowe, A. (2005). *Trust in Emergence*. Keynote presentation delivered to the 3rd International Qualitative Research Convention, Johor Bahru, Malaysia.

Luambano, I., & Nawe, J. (2004). Internet Use by Students of the University of Dar es salaam. *Library Hi Tech News*, *2*(10), 13–17. doi:10.1108/07419050410577550

Lujara, S. K. (2008). *Development of e-Learning Content and Delivery for Self Learning Environment: Case of Selected Rural Secondary Schools in Tanzania*. Blekinge Institute of Technology. Available online at http://www.bth.se/tks/teknovet.nsf/bilagor/Suzan%20Lujara%20Lic%20manuscript_pdf/$file/Suzan%20Lujara%20Lic%20manuscript.pdf

Lytras, M. D., Pouloudi, A., & Poulymenakou, A. (2002). Knowledge Management Convergence – Expanding Learning Frontiers. *Journal of Knowledge Management*, *6*(1), 40–51. doi:10.1108/13673270210417682

MacFadzean, S., & Newton, B. (2005). Models of ELearning: Technology Promise vs Learner Needs Literature Review. *The International Journal of Management Education*, *3*(1), 43–55.

Macpherson, A., Homan, G., & Wilkinson, K. (2005). The Implementation and Use of E-learning in the Corporate University. *Journal of Workplace Learning*, *17*(1/2), 33–48. doi:10.1108/13665620510574441

Malopinsky, L., Kirkley, J., Stein, R., & Duffy, T. (2000). *An Instructional Design Model for Online Problem Based Learning (PBL) Environments: The Learning to Teach with Technology Studio*. Paper Presented at Association for Educational Communications and Technology Conference (AECT), Denver, CO.

Manville, B. (2004). *Organizing Enterprise-Wide eLearning and Human Capital Management*. Chief Learning Officer.

Martin, V. B. (2006). The postmodern turn: Shall classic grounded theory take that detour? *The Grounded Theory Review*, *5*, 119–128.

Matavire, R., & Brown, I. (2008). Investigating the use of 'Grounded Theory' in Information Systems research. *Proceedings of the SAICSIT 2008 Conference, Wilderness*.

Matavire, R., & Brown, I. (2011, August). Profiling grounded theory approaches in information systems research. *European Journal of Information Systems*, *30*. doi:10.1057/ejis.2011.35

Materi, R. R., & Fahy, P. J. (2004). Interim Report: A Case Study of Internet-Based Distance Education Program Development in Vietnam. *International Review of Research in Open and Distance Learning*, *5*(1), 7–16.

McAdam, D. (2003). *E-Learning for Management and Marketing in Libraries*. IFLA Publications Series No. 115.

McCraken, H. (2004). Extending Virtual Access: Promoting Engagement and Retention through Integrated Support Systems. *Online Journal of Distance Learning Administration*, *6*(1), 44–52.

McKnight, M. (2000). Distance Education: Expressing emotions in video-based Classes. In *Conference on College Composition and Communication*, Minneapolis, MN.

McPherson, M., & Nunes, J. (2005). Critical Issues for E-learning; What May Seem Obvious is not Always Put Into Practices. *Journal of Computer Assisted Learning*, *5*(2), 34–42.

Mello, R. (2002). 100 Pounds of Potatoes in a 25-Pound Sack: Stress, Frustration, and Learning in the Virtual Classroom. Teaching *With Technology Today*, *8*(9). Available at http://www.elearningmag.com/elearning/article/articleDetail.jsp?id=2031

Meredith, S., & Newton, B. (2003). Models of e-Learning: Technology promise vs. learner needs–literature review. *The International Journal of Management Education*, *4*(3), 43–56. doi:10.3794/ijme.33.73

Mersham, G., & Viviers, D. (2007). Looking at E-learning Through the Lens of a Communication Model. *SOLSTICE 2007 Conference*. Edge Hill University.

Meyen, E. (2002). E-learning: A Programmatic Research Construct for the Future. *Journal of Special Education Technology*, *17*(3), 37–46.

Meyen, E. L. (2000). *Using Technology to Move Research to Practice: The Online Academy*. New York: National Center for Learning Disabilities.

Meyer, K. A. (2003). Face-To-Face Versus Threaded Discussions: The Role of Time and Higher-Order Thinking. *Journal of Asynchronous Learning Networks*, *7*(3), 55–65.

Mihhailova, G. (2006). E-learning as Internationalization Strategy in Higher Education: Lecturer's and student's perspectives. *Baltic Journal of Management*, *1*(3), 270–284. doi:10.1108/17465260610690926

Miles, M., & Huberman, M. (1994). *Qualitative Data Analysis*. Thousand Oaks, CA: Sage.

Miller, M. D., & Corley, K. (2001). The Effect of E-Mail Messages on Student Participation in the Asynchronous On-Line Course: A Research Note. *Online Journal of Distance Learning Administration*, *4*(3), 17–24.

Mingers, J. (2001). Combining IS research methods: Towards a pluralist research methodology. *Information Systems Research*, *12*(3), 240–259. doi:10.1287/isre.12.3.240.9709

Mitchell, A., & Honore, S. (2007). Criteria for Successful Blended Learning. *Industrial and Commercial Training*, *39*(3), 143–149. doi:10.1108/00197850710742243

Moeng, B. (2004). *IBM Tackles Learning in the Workplace, IBM*. Management Development Solutions.

Moore, M. G. (1998). Introduction. In C. C. Gibson (Ed.), *Distance learners in higher education*. Madison, WI: Atwood Publishing.

Morgan, G. (2001). Thirteen "Must Ask" Questions About Elearning Products and Services. *The Learning Organization*, *8*(5), 203–210. doi:10.1108/EUM0000000005914

Morrison, J. L., & Khan, B. H. (2003). *The Global E-learning Framework: An Interview with Badrul Khan. The Technology Source.* A Publication of the Michigan Virtual University. Available at http://ts.mivu.org/default.asp?show=article&id=1019#options

Morse, J. M., Stern, P. N., Corbin, J., Bowers, B., Charmaz, K., & Clarke, A. E. (2009). *Developing Grounded Theory: The Second Generation*. Walnut Creek, CA: Left Coast Press.

MoST. (2003). *National Information and Communication Technologies Policy*. Tanzania: Ministry of Science and Technology of Tanzania.

MoST. (2006). *The Use of ICT in National Development, an Opening Remark given during the Inaugural of "Tume ya Sayansi na Tekinoloji", (2-4 May) by Minister of Ministry of Science and Technology*. Available at www.most.gov.tz/resources/speeches/2006/tume_ya_sayansi_na_tekinolojia/

Mulkeen, A. (2003). What Can We do to Encourage ICT Integration? Evidence from the Irish School System. Education Department, NNI Maynooth, Ireland.

Muyinda, P. (2007). M-learning: Pedagogical, Technical and Organisational Hypes and Realities. *Campus-Wide Information Systems*, *24*(2), 97–104. doi:10.1108/10650740710742709

Myers, M. D. (1997). Qualitative research in information systems. *Management Information Systems Quarterly*, *21*(2), 241–242. doi:10.2307/249422

Myers, M. (1997b). Interpretive Research in Information Systems. In J. Mingers & F. Stowell (Eds.), *Information Systems: An Emerging Discipline?* (pp. 239–266). London: McGraw-Hill.

Myers, M. D., & Newman, M. (2007). The Qualitative Interview in IS Research: Examining the Craft. *Information and Organization*, *17*(1), 2–26. doi:10.1016/j.infoandorg.2006.11.001

Myers, M. D., & Walsham, G. (1998). Exemplifying Interpretive Research in Information Systems: An Overview. *Journal of Information Technology*, *13*(4), 233–234. doi:10.1057/jit.1998.2

Myers, M. D. (2009). *Qualitative Research in Business & Management (3rd ed.)*. London: Sage Publications.

Nanas, N. (2001). *Literature Review: Information Filtering for Knowledge Management*. The Open University.

Nasirin, S., Birks, D. F., & Jones, B. (2003). Re-examining fundamental GIS implementation constructs through the grounded theory approach. *Telematics and Informatics*, *20*(4), 331–347. doi:10.1016/S0736-5853(03)00012-1

Ndume, V., Tilya, F. N., & Twaakyondo, H. (2008). Challenges of Adaptive eLearning at Higher Learning Institutions: A Case Study in Tanzania, International. *Journal of Computing and ICT Research*, *2*(1), 47–59. Available online at http:www.ijcir.org/volume1-number2/article6.pdf

Nettland, G. (2007). E-learning in a Large Organisation: A Study of the Critical Role of Information Sharing. *Journal of Workplace Learning*, *19*(6), 392–411. doi:10.1108/13665620710777129

Neuman, W. L. (1997). *Social Research Methods* (3rd ed.). Allyn and Bacon.

Newton, R. (2003). Staff Attitudes to the Development and Delivery of E-learning. *New Library World*, *104*(1193), 412–425. doi:10.1108/03074800310504357

Ngugi, C. (2006). Report on the AVU (African Virtual University). *ODeL Seminar: Supporting Open, Distance and e-Learning in African Universities.*

Nicholson-Crotty, S. (2005). Bureaucratic Competition in the Policy Process. *Policy Studies Journal: the Journal of the Policy Studies Organization*, *33*(3), 341–361. doi:10.1111/j.1541-0072.2005.00119.x

Njenga, J., & Fourie, L. (2008). The Myths About E-learning in Higher Education. *British Journal of Educational Technology*, *3*(1), 1–14.

Nunan, T. (1996). Flexible Delivery - What is it and Why is it a Part of Current Educational Debate? Different Approaches: Theory and Practice in Higher Education. *Proceedings HERDSA Conference 1996*. Available at http://www.herdsa.org.au/confs/1996/nunan.html

O'Malley, J., & McGraw, H. (1999). Students Perceptions of Distance Learning, Online Learning and the Traditional Classroom. *Online Journal of Distance Learning Administration*, *2*(4), 23–36.

O'Regan, K. (2003). Emotion and E-learning. *Journal of Asynchronous Learning Networks*, *7*(3), 78–92.

Olaniran, B. (2007). Challenges to Implementing E-learning in Lesser-Developed Countries. In A. Edmundson (Eds.), Globalized E-learning Cultural Challenges. Idea group.

Omwenga, E. I., Waema, T. M., & Wagacha, P. W. (2004). A Model for Introducing and Implementing E-learning for Delivery of Educational Content Within the African Context, *African Journal of Science and Technology. Science and Engineering Series*, *5*(1), 34–46.

Orlikowski, W. J., & Baroudi, J. J. (1991). Studying Information Technology in Organizations: Research Approaches and Assumptions. *Information Systems Research*, *2*(1), 1–28. doi:10.1287/isre.2.1.1

Orlikowski, W. J. (1993). CASE Tools as Organizational Change: Investigating Incremental and Radical Changes in Systems Development. *Management Information Systems Quarterly*, *17*(3), 37–45. doi:10.2307/249774

Pachnowski, L. M., & Jurczyk, J. P. (2003). Perceptions of Faculty on the Effect of Distance Learning Technology on Faculty Preparation Time. *Online Journal of Distance Learning Administration*, *6*(3), 89–102.

Palloff, R., & Pratt, K. (1999). *Making the Transition: Helping Teachers to Teach Online*. In the 15th Annual Conference on Distance Teaching and Learning, Madison, WI.

Patalong, S. (2003). Using the Virtual Learning Environment WebCT to Enhance Information Skills Teaching at Coventry University. *Library Review*, *52*(3), 103–110. doi:10.1108/00242530310465906

Paulsen, A. (2003). *Online Education Obituaries*. Available online at www.studymentor.com/studymentor/Obituaries.pdf

Pergler, C., & Littlejohn, A. (2007). *Preparing for Blended E-learning*. New York: Routledge.

Phang, J. (2004). Impact of ICT Innovations: The Islamic Context. *The Star in Tech*, *8*(April), 18.

Pig, L. C. (2003). Exploring Critical Aspects of Information Technologies Integration in Singapore Schools. *Australasian Journal of Educational Technology*, *19*(1), 1–24.

Porter, L. R. (1997). *Creating Virtual Classroom: Distance Learning with the Internet*. New York: J. Wiley & Sons.

Powell, W. W. (1990). Neither market nor hierarchy: Network forms of organization. In B. M. Staw & L. L. Cummings (Eds.), *Research in Organizational Behavior* (pp. 295–336). JAI.

Puteh, M. (2008). E-learning Implementation in Malaysian Universities: The Universiti Teknologi Malaysia Experience. *Proceedings of the 3rd International Conference on E-learning*. University of Cape Town.

Reigeluth, C. M., & Khan, B. H. (1994). *Do Instructional Systems Design (ISD) and Educational Systems Design (ESD) Really Need each other?* Paper presented at the Annual Meeting of the Association for Educational Communications and Technology (AECT), Nashville, TN.

Relan, A., & Gillani, B. B. (1997). Web-Based Instruction and Traditional Classroom: Similarities and Differences. In B. H. Khan (Ed.), Web-based instruction (pp. 41-46). Englewood Cliffs, NJ: Educational Technology Publications.

Richards, T. (1998). The Emergence of Open Standards for Learning Technology. Macromedia Interactive Learning Division.

Ritchie, D. C., & Hoffman, B. (1997). Incorporating Instructional Design Principles with the World Wide Web-Stride Handbook. In B. H. Khan (Ed.), Web-based instruction (pp. 135-138). Englewood Cliffs, NJ: Educational Technology Publications.

Ritchie, J., & Lewis, J. (Eds.). (2004). *Qualitative Research Practice*. London: Sage.

Rivera, J. C., & Rice, M. L. (2002). A Comparison of Student Outcomes & Satisfaction between Traditional & Web Based Course Offerings. *Online Journal of Distance Learning Administration*, *5*(3), 75–83.

Rodriguez, J. (2008), Development of an ICT Training Model for In-service Teachers in Nicaragucen Schools. *23rd International Technical Conference on Circuits/system, Computers and Communication* (ITC-CSCC 2008).

Roffe, I. (2002). E-learning: Engagement, Enhancement and Execution. *Quality Assurance in Education*, *10*(1), 40–50. doi:10.1108/09684880210416102

Rogers, D. (2000). A Paradigm Shift: Technology Integration for Higher Education in the New Millennium. *Educational Technology Review*, *6*(3), 19–27.

Romiszowski, A. J. (2004). How's the E-learning Baby? Factors Leading to Success or Failure of an Educational Technology Innovation. *Educational Technology*, *44*(1), 5–27.

Rosenberg, M. J. (2001). *E-Learning: Strategies for Delivering Knowledge in the Digital Age*. New York: McGraw-Hill.

Rossett, A., & Sheldon, K. (2001). *Beyond The Podium: Delivering Training and Performance to a Digital World*. San Francisco: Jossey-Bass/Pfeiffer.

Rourke, L., & Anderson, T. (2002). Using Web-based, Group Communication Systems to Support Case Study Learning at a Distance. *International Review of Research in Open and Distance Learning*, *3*(2), 101–112.

Rovai, A. P., & Jordan, H. M. (2004). Blended Learning And Sense Of Community: A Comparative Analysis With Traditional And Fully Online Graduate Courses. *International Review of Research in Open and Distance Learning*, *5*(2), 65–71.

Roy, M.H. (1996). Collaborating Across Campuses: The Virtual University Experience. *Campus Wide Information Systems*, *13*(4), 6-10.

Rumble, G. (2000). The Globalisation of Open and Flexible Learning: Considerations for Planners and Managers. *Online Journal of Distance Learning Administration*, *3*(3), 1–15.

Rutter, D. J. (2005). *From Diversity to Convergence: British Computer Networks and the Internet, 1970-1995*. (Unpublished Doctoral thesis). University of Warwick, Warwick.

Salmon, G. (2000). *E-Moderating The Key to Teaching and Learning Online*. Sterling, VA: Kogan Page.

Salmon, G. (2002). Pedagogic Requirements of VLEs: Pets and Planets. *UCISA TLIG-SDG User Support Conference*. Available at http://sstweb.open.ac.uk:8282/oubs/gilly

Salomon, G. (1993). No Distribution without Individuals' Cognition. In G. Salomon (Ed.), *Distributed Cognitions: Psychological and education consideration* (pp. 111–138). New York: Cambridge University Press.

Saltzbert, S., & Polyson, S. (1995). Distributed Learning on the World Wide Web. *Syllabus*, *9*(1), 10–12.

Sambrook, S. (2003). E-learning in Small Organizations. *Education + Training*, *45*(8/9), 506-516.

Schank, R. C. (2001). *Designing World-class E-learning*. McGraw-Hill.

Schifter, C. C. (2004). Compensation Models in Distance Education: National Survey Questionnaire Revisite'. *Online Journal of Distance Learning Administration*, *7*(1), 54–63.

Schott, M., Chernish, W., Dooley, K. E., & Linder, J. R. (2003). Innovations in Distance Learning Program Development and Delivery. Online Journal of Distance Learning Administration, 6(2), 81-92.

Schreiber, D. A., & Berge, Z. L. (1998). *Distance Training: How Innovative Organizations are Using Technology to Maximize Learning and Meet Business Objectives*. San Francisco: Jossey-Bass.

Scott, H. (2004). *Design for Collaborative Learning Online: The Integrated Pedagogical and Community Model*. Paper presented at the Seamless Learning: Integrating Teaching, Research & Technology, Stony Brook State University of New York.

Scott, H. (2007). *The Temporal Integration of Connected Study into a Structured Life: a Grounded Theory*. (PhD Thesis). Portsmouth University.

Sife, A., Lwoga, E., & Sanga, C. (2007). New Technologies for Teaching and Learning: Challenges for Higher Learning Institutions in Developing Countries. *International Journal of Education and Development using ICT, 3*(2). Available: http://ijedict.dec.uwi.edu/viewarticle.php?id=246

Sifuna, N. D. (1998). *Crisis in the Public Universities in Kenya*. Available online at http://ahero.uwc.ac.za/index.php?module=cshe&action=downloadfile&fileid=81806115511732720564415

Simpson, O. (2003). *Student Retention in Online, Open and Distance Learning*. Kogan Page.

Simpson, O. (2004). The impact on retention of interventions to support distance learning students. *Open Learning*, *19*(1), 79–96. doi:10.1080/0268051042000177863

Singh, H. (2003). Building Effective Blended Learning Programs. *Educational Technology*, *44*(1), 5–27.

Smith, A.D., & Rupp, W.T. (2004). Managerial Implications of Computer-Based Online/Face-to-Face Business Educational Case study. *Online Information Review*, *28*(2), 100-109.

Smulders, D. (2003). Designing for Learners, Designing for Users. *E-Learn Magazine*. Retrieved from http://www.elearnmag.org/subpage/sub_page.cfm?section=3&list_item=11&page=1

Sorokin, P. A. (2006). *Social and cultural dynamics: a study of change in major systems of art, truth, ethics, law, and social relationships.* London: Transaction Publishers.

Sorokin, P. A. (1937). *Social and Cultural Dynamics*. New York: American Book Company.

Stockley, D. (2003). *E-learning definition and Explanation*. ECampus.com.au.

Stockley, D. (2006). *E-learning Definition and Explanation*. Retrieved from http://derekstockley.com.au/e-learning-definition.html

Strauss, H. (2003). Getting all of our Courses Online: A Euphoric State Case Study. *E-Learn Magazine*. Retrieved from http://www.elearnmag.org/subpage/sub_page.cfm?section=3&list_item=4&page=3

Strauss, A., & Corbin, J. (1989). Grounded Theory's Applicability to Nursing Diagnostic Research. In Monograph of the Invitational Conference of Research Methods for Validating Nursing Diagnoses, Palm Springs, CA.

Strauss, A. (1987). *Qualitative Analysis for Social Scientists*. Cambridge, UK: Cambridge University Press. doi:10.1017/CBO9780511557842

Strauss, A., & Corbin, J. (1990). *Basics of Qualitative Research*. Sage.

Strauss, A., & Corbin, J. (1994). Grounded Theory Methodology – An Overview. In N. K. Denzin & Y. S. Lincoln (Eds.), *Handbook of Qualitative Research*. Sage.

Strauss, A. L., & Corbin, J. (1990). *Basics of qualitative research: Grounded theory procedures and techniques*. Thousand Oaks, CA: Sage.

Strauss, A. L., & Corbin, J. (1998). *Basics of qualitative research: Grounded theory procedures and techniques* (2nd ed.). Thousand Oaks, CA: Sage.

Strauss, A., & Glaser, B. (1970). *Anguish*. Mill Valley, CA: Sociology Press.

Suddaby, G., & Milne, J. (2008). Coordinated Collaborative and Coherent: Developing and Implementing E-learning Guidelines Within a National Tertiary Education Systems. *Campus-Wide Information Systems*, *25*(2), 114–122. doi:10.1108/10650740810866602

Telematica Instituut E-learninggroep. (2001). E-learning, voordelen en kansen op Organisatie-niveau. *M&I, 2001*(3), 5.

The eLearning Guild. (2002). *The E-Learning Development Time Ratio Survey*. Available online at http://www.elearningguild.com/pdf/1/time%20to%20develop%20Survey.pdf

Thomas, D. R. (2003). A General Inductive Approach for Qualitative Data Analysis. University of Auckland. Available online at www.health.auckland.ac.nz/hrmas/resources/qualdatanalysis.html

Thomson, A. M., & Perry, J. L. (2006). Collaboration Processes: Inside the Black Box. *Public Administration Review*, *66*(s1), 20–32. doi:10.1111/j.1540-6210.2006.00663.x

Thomson, W. (2001). *A Guide for the Young Economist: Writing and Speaking Effectively*. Available at www.leaonline.com/doi/abs/10.1207/S15327736ME1501_7

Tinto, V. (1975). Dropout from Higher Education: A Theoretical Synthesis of Recent Research. *Review of Educational Research*, *45*(1), 89–125. doi:10.3102/00346543045001089

Tinto, V. (1993). *Leaving College: Rethinking the Causes and Cures of Student*. Academic Press.

Tinto, V. (1997). *Classrooms as Communities: Exploring the Educational Character*. Academic Press.

Trochim, W. M. K. (2006). *Philosophy of Research: Induction and Deduction, Research Methods Knowledge Base*. Available at http://www.socialresearchmethods.net/kb/dedind.php

Urdan, T. A., & Weggen, C. C. (2000). *Corporate E-learning: Exploring a new Frontier*. WR Hambrecht + Co. Available at http://www.wrhambrecht.com/research/coverage/elearning/ir/ir_explore.html

Urquhart, C. (1997). Exploring Analyst-Client Communication: Using Grounded Theory Techniques to Investigate Interaction in Informal Requirements Gathering. In Information Systems and Qualitative Research. London: Chapman & Hall.

Urquhart, C. (1998a). Analysts And Clients In Conversation: Cases In Early Requirements Gathering. *Proceedings of theNineteenth International Conference on Information Systems.*

Urquhart, C. (1998b). Strategies for Conversation and Systems Analysis in Requirements Gathering: A Qualitative View of Analyst-Client Communication. *Qualitative Report*, *4*(1/2). Available at http://www.nova.edu./ssss/QR/QR4-1/urquhart.html

Urquhart, C. (1999a). Reflections On Early Requirements Gathering – Themes From Analyst- Client Conversation. In T. J. Larsen, L. Levine, & J. I. DeGross (Eds.), *Information Systems: Current Issues and Future Changes*. Laxenburg, Austria: IFIP.

Urquhart, C. (1999b). Themes In Early Requirements Gathering – The Case Of The Analyst. *The Client And The Student Assistance Scheme. Information Technology & People*, *3*(1), 25–34.

Urquhart, C., & Fernandez, W. (2006). Grounded theory method: the researcher as blank slate and other myths. *ICIS 2006 Proceedings*. Retrieved from http://aisel.aisnet.org/icis2006/31

Uys, P. M. (2004). A Syntagm of Networked Educational Management: Case Study – University of Botswana. *Campus-Wide Information Systems*, *21*(1), 22–28. doi:10.1108/10650740410512293

Uys, P. M. (2003). Critical Success Factors in the Infusion of Instructional Technologies for Open Learning in Development Settings: The Case of the University of Botswana. *International Review of Research in Open and Distance Learning*, *4*(2), 33–39.

Van Dam, N. (2004). *The E-Learning Fieldbook*. New York: McGraw-Hill.

Van Niekerk, J. C., & Roode, J. D. (2009). Glaserian and Straussian grounded theory: similar or completely different? *Proceeding SAICSIT '09 Proceedings of the 2009 Annual Research Conference of the South African Institute of Computer Scientists and Information Technologists*. Available at http://dl.acm.org/ft_gateway.cfm?id=1632163&ftid=694294&dwn=1&CFID=81873259&CFTOKEN=70405415

VanBrakel, P., & Chisenga, J. (2003). Impact of ICT-based distance learning: The African Story. *The Electronic Library*, *21*(5), 476–486. doi:10.1108/02640470310499867

Vernon, M. (2002). Understanding E-learning. *Financial Times*. Available at http://specials.ft.com/elearning/FT35PSCM2ZC.html

Viviers, D., & Mersham, G. (2007). Looking at E-learning through the lens of a Communication Model. *SOLSTICE 2007 Conference*. Edge Hill University.

Wang, M. Y. (2003). The Strategic Role of Digital Libraries: Issues in E-Learning Environments. *Library Review*, *52*(3), 111–116. doi:10.1108/00242530310465915

WB. (2002). *World Bank Population growth Report*. New York: WB.

Wentling, T. L., Waight, C., Gallaher, J., La Fleur, J., Wang, C., & Kanfer, A. (2000). *Elearning -A Review of Literature*. Available at http://learning.ncsa.uiuc.edu/papers/elearnlit.pdf

Wesley, D. (2002). A Critical Analysis on the Evaluation of E-learning. *International Journal on E-Learning*, *2*(1), 45–56.

White, C. (2005). Contribution of Distance Education to the Development of Individual Learners. *Distance Education*, *26*(2), 165–181. doi:10.1080/01587910500168835

White, K. W., & Weight, B. H. (2000). *The Online Teaching Guide*. Canada: Tony Bates Associates Ltd.

Wild, R.H., Griggs, K.A., & Downing, T. (2002). A Framework for E-learning as a Tool for Knowledge Management. *Industrial Management & Data Systems*, *102*(7), 371-380.

Willging, P. A., & Johnson, S. D. (2004). Factors that influence students' decisions to drop out of online courses. *Journal of Asynchronous Learning Networks*, *8*(4), 105–118.

Willis, B. (Ed.). (1994). *Distance Education: Strategies and Tools*. Englewood Cliffs, NJ: Educational Technology Publications.

Wolodko, A., & Bednarek-Michalska, J. (2007). E-learning Models for Polish Libraries. *BIBWEB, the ElectronicLlibrary*, *25*(1), 80-89.

Woodley, A. (2004). Conceptualizing student dropout in part-time distance education: Pathologizing the normal? *Open Learning*, *19*(1), 47–64. doi:10.1080/0268051042000177845

Woodley, A., de-Lange, P., & Tanewski, G. (2001). Student Progress in Distance Education: Kember's model re-visited. *Open Learning*, *16*(2), 63–71. doi:10.1080/02680510123105

Yieke, F.A. (2005). Towards Alternatives IN Higher education: the benefits and challenges of E-learning in Africa. *CODESRIA Bulletin*, (3-4), 73.

Yong, A. (2003). Success Factors in E-learning Implementation. *The Star*, *19*, 93.

Yorke, M. (2004). Retention, persistence and success in on-campus higher education, and their enhancement in open and distance learning. *Open Learning*, *19*(1), 19–32. doi:10.1080/0268051042000177827

Young-Ju, J., Bong, M., & Choi, H. J. (2000). Self-Efficacy for Self-Regulated Learning, Academic Self-Efficacy, and Internet Self-Efficacy in Web-Based Instruction. *Educational Technology Research and Development*, *48*(2), 5–13. doi:10.1007/BF02313398

Yum, J. C. K., Kember, D., & Siaw, I. (2001). Study Examines Coping Methods in Part-time Students. *The Newsletter of the National Resource Center for the First Year Experience and Students in Transition*, *14*(1), 4–5.

Zahm, S. (2000). No Question About it – E-learning is here to Stay: A Quick History of the E-learning Evolution. *E-learning*, *1*(1), 44–47.

Zhang, J., Khan, B. H., Gibbons, A. S., & Ni, Y. (2001). Review of Web-Based Assessment Tools. In B. H. Khan (Ed.), Web-Based Training (pp. 137-146). Englewood Cliffs, NJ: Educational Technology Publications.

ENDNOTE

1 "Providers' here refers to people engaged/participating in e-learning provision- e.g. universities, consultant bodies, e-learning companies, university professionals, etc.

APPENDIX 1

Table 3. List of failed institution sorted by money spent (source: keegan et al., 2007)

Institution	Country	Years in Operations	Number of Years in Operation	Total Amount Spent	Type of Initiative	Education Level
UK E-University	UK	2000-2004	4	£50M	Political	University
United States Open University	USA& UK	1998-2002	4	$27M	Institutional	University
Alliance for Lifelong Learning	USA & UK	2000- 2006	6	$12M	Consortium	Further Education
IT Fornebu	Norway	1998-2001	3	€9M	Political	Further Education
The Scottish Interactive University	UK	2002-2007	5	£2.3M	Political	University

APPENDIX 2: TYPES OF CLASSIC GROUNDED THEORY DATA

1. **Base Line Data:** The respondent is being as truthful as possible, as far as the interviewer can gauge.
2. **Interpreted Data:** The respondent is filtering his/her message either to deliberately complicate or simplify her/his responses.
3. **Properlined Data:** The respondent deliberately edits the data so that it is not in conflict with the "official" line of argument from his employing authority.
4. **Vague Data:** The respondent deliberately gives an indistinct version of reality to try and confuse the researcher, obfuscate the issues and throw the researcher 'off the scent'.
5. **Zero Data:** The respondent refuses to say anything at all in order to avoid saying anything which he might later regret.

APPENDIX 3: CLASSIC GROUNDED THEORY METHODOLOGY SELECTIVE CRITERIA FOR CORE CATEGORY

Source: Glaser, 1978, pp. 93-100.

- Central.
- Reoccur frequently.
- More time to saturate.
- Connections not be-forced.
- Clear and grabbing implication for formal theory.
- Carry through.

- Completely variable.
- Is also a dimension of the problem.
- Prevent to other sources of establishing a core.
- See the core category in all relations.
- It can be a kind of theoretical code.

APPENDIX 4

Table 4. Temporal characteristics of the secondary sub core variables of "cultivating recognition"

Secondary Sub Core Variables	Temporal Characteristics (Dynamics)
Asymmetric Collaborating Partners	Several different temporal characteristics: cyclical, spasmodic and perpetual
Reciprocal Collaborating Partners	spasmodic
Referral Networking	perpetual
Strategic Alliancing	perpetual
Endorsing	perpetual
Result Orienting	perpetual and cyclical
Prioritizing duties	cyclical

Chapter 10
Assessment of Contribution of ICT for Sustainable Livelihoods in Kilosa District

C. Shirima
Sokoine University of Agriculture, Tanzania

Camilius Sanga
Sokoine University of Agriculture, Tanzania

ABSTRACT

ICT plays a key role in socio-economic development and ensuring sustainable livelihoods. The purpose of this chapter was to assess the contribution of ICT for sustainable livelihood in Kilosa district. Data were obtained from both primary and secondary sources of data. Structured questionnaires were used in collection of primary data. The collected data was analyzed using SPSS. A sample size of 60 respondents was interviewed. The study determined various contributions resulting from using ICT by rural people to enhance their livelihoods. In addition, this study highlighted several constraints which rural people face in accessing and using ICT. The results from this chapter revealed that majority of respondents are using ICT tools in their daily activities but the socio-economic development through ICT are yet to be realized fully.

BACKGROUND INFORMATION

The advancement of Information and Communication Technology (ICT) since the end of 20th century has affected all sectors (NICTP, 2003). Pigato (2001) argues that the current wave of globalization towards integration of markets is spurred by the development of ICT. The most noticeable ICT are the Internet and mobile phones (Chilimo, 2008; Sife et al., 2010). ICT can enable the realization of social development objectives to the extent that there is appropriateness and long lasting solution (Mallaliev *et al.*, 2010). Thus, the use of ICT can take into accounts not only the technologies itself but also the ultimate development objective and other factors which have an impact on sustainability.

DOI: 10.4018/978-1-5225-0539-6.ch010

In Africa, there is insufficient evidence to suggest a direct link between ICTs and development. Ngwenyama *et al.,* (2006:3) argues that

Recent studies have found a positive correlation between investment in ICTs and economic growth in development countries, but evidence for developing countries is not as extensive.

Many if not most countries in Africa lag behind in basic requirements for ICTs to play a meaningful and sustained role in improving people's life (Bernard & Vonk, 2003).

The development gaps in Africa are compounded by the problem of the digital divide, which exacerbated not only by lack of access to ICTs but also the challenges of inadequate pool of skilled persons and the use, maintenance and rapid obsolesces of the ICTs innovations and development (Vukanikinds-DTI, 2005).

Various studies have been done to access the use of ICTs to improve delivery of services and well being of rural people. For example, in Ghana, most users access the Internet through Internet cafes (Frempong, 2005). In Uganda few people use e-mail due to the lack of access to ICTs (Tusubira *et al.,* 2005). Access to email is through internet cafes which are only available in large towns. The reverse side of this coin as highlighted by Barnard and Vonk (2003) is that around 70% of the African continent population lives in rural areas, with many Africans lacking the basic needs. In Kenya there is generally no incentive to develop the ICT sector and this problem is compounded by inadequate ICT skills and lack of Research and development in rural areas in ICT (Weema, 2005).

In Tanzania for example, effective adoption and use of ICT requires basic infrastructure like electricity, which is badly missing in many rural areas. It is also important to note that unless there is a critical mass of ICT adopters, its value for business is greatly compromised. Thus, many more people should be facilitated to access and use ICT to make it commercially viable in the country.

However, a lot more needs to be done to ensure reliability, affordability, availability and usability of the ICTs devices. One of the challenges in Tanzania is that key sectors of the National economy do not perceive ICTs as a critical business imperative. However, the Government of Tanzania recognizes that ICTs is indispensable for development and is creating a suitable ICT environment starting with a National ICT policy framework[1].

Chapman and Slaymaker (2012) argue that in order to further understand how ICT can contribute to wider development objectives of sustainable economic growth which equitably benefits those living in developing countries it is necessary to consider the existing role of information in people's livelihoods, the current communication context and how recent developments might influence them.

As Fleming (2012) reported that ICT and development are linked both in terms of "information and communication technologies as the engine of development".

Therefore, it is against this background that this study was designed to assess the contribution of ICT on sustainable livelihoods of rural people in Kilosa District.

ICT is regarded as one among the sources of development, Tanzania UNDP reports of 2011 states that ICT is one of the key factors in social-economic development (UNDP, 2011). Access to relevant information and knowledge improves efficiency and productivity, enhance social services delivery, increase access to market opportunities, and improves government performance among others (UNDP, 2011). For these reasons, ICT has been considered as vital in most developing countries including Tanzania. It has been incorporated in the poverty alleviation and other socio-economic development strategies.

Like in other developing countries, in Tanzania ICT is regarded as an important tool for accelerating poverty reduction through its role in raising productivity, generating economic growth, creating jobs, facilitating learning, knowledge sharing and global information flows (ESRF, 2007). According to Tanzania Development Vision (2025) Tanzania is expected to become a knowledge based society, with a vision to have a universally access broadband infrastructure in ICT as well as expertise that will enhance sustainable socio-economic development and accelerate poverty reduction, and to become the ICT development hub regionally. To date there is remarkable progress towards this goal (ESRF, 2007).

Therefore, there is an increasing evidence that access to ICT's can have a direct impact on raising living standard and quality of life of the poor people in different parts of the world.

However, very little is known with respect to the contribution of ICT to sustainable livelihoods in Kilosa district[2] and hence the information remains widely fragmented (Chilimo, 2008). Therefore, this study was aimed to assess the contribution of ICT for sustainable livelihoods (ICT4SL) of people in Kilosa District.

Through understanding the role of ICT4SL of the people, the Government of Tanzania through the Ministry of Communication and Transport developed the National ICT policy of 2016 to guide the uses of ICT in the country. The number of objectives has been set to ensure that ICT is used in different aspects (i.e. in learning, commerce, tourism and other sectors).

As written in statement No 3.4.4 of the policy that, "The Government in collaboration with the private sector will develop and put in place appropriately designed schemes of service for different cadres of ICT personnel in order to secure their retention and encourage innovative behavior" (National ICT Policy, 2016).

This study is important to generate the empirical evidences on the contribution of ICT4SL of people and the strategies which should be taken by different stakeholders, including the Government on the usage of ICT to enhance sustainable livelihoods of the people.

The study will be useful to the Ministry of Communication, Science and Technology to understand the real situation with respect to the use and adoption of ICT tools by local people. Also planning Department in Kilosa District will understand the potential of ICT to sustainable livelihoods.

Lastly, the study will also be useful to policy makers, and other development stakeholders, Ministry of Employment and Youth Development to understand how the use of ICT can lead to sustainable livelihoods of people and hence economic development.

The general objective of this study was to examine the contribution of ICT4SL of people in Kilosa District. The specific objectives were:

1. To assess how rural people use ICT to enhance their sustainable livelihoods.
2. To identify constraints to effective use of ICT to sustainable livelihoods.
3. To investigate the impact of ICT on people's income generating activities.

The chapter is organised as follows: the first section presents the background which outlines the introduction, problem statement and justification of the study. The next section presents the literature review as per specific objectives. Thereafter, the research methodology is present. Following this, the results and discussion is presented. Finally, the conclusion and recommendations and future research directions especially on what should be done regarding the assessment of contribution of ICT4SL in Tanzania is presented.

LITERATURE REVIEW

Overview

This section presents the literature review for the study in terms of the specific objectives.

Definition of Key Concepts

Information and Communication Technology (ICT)

ICT refers to a wide range of computerized technologies. ICT is any technology that enables communication and the electronic capturing, processing and transmission of information (Ashrafi & Murtaza, 2008). These technologies include products and services such as desktop computers, laptops, handheld devices, wired or wireless intranet, business productivity software such as text editor and spreadsheet, enterprise software, data storage and security, network security mobile phones and so on (Ashrafi & Murtaza, 2008).

Livelihood

A livelihood comprises the capabilities, assets (stores, resources, claims and access) and activities required for a means of living. A livelihood is sustainable when it can cope with and recover from stress and shocks, maintain or enhance its capabilities and assets, and provide sustainable livelihood opportunities for the next generation and which contributes net benefits to other livelihoods at the local and global levels and in the short and long term (Chamber & Conway, 1992).

Sustainable Livelihood

A livelihood is sustainable when it can cope with and recover from stresses and shocks and maintain or enhance its capabilities and asserts both now and in the future, while not undermining the natural resources base (Chamber and Conway, 1992).

ICT and Sustainable Livelihood Approach (ODI, n.d.)

Sustainable Livelihoods Approaches (SLA)

The livelihood approach is a way of thinking about the objectives, scope and priorities of development. The SLA developed to put people at the centre of development, thereby increasing the effectiveness of development assistance. Sustainable livelihood approach and their application have been subject to considerable debate in recent years. As noted by Krantz (2001), donor interpretations of a Sustainable Livelihoods Approach commonly incorporate the various set of principles. A set of principles which are specified in developmental activities to be as done using a SLA are:

- **People Centered:** This means beginning with people's own views of their priorities, opportunities and needs, the approach works out technically and financially feasible responses. In this way, it seeks to be responsive and participatory.
- **Differentiated:** This means it recognize that the characteristics of poverty and appropriate policy responses, differ among different groups of the poor.
- **Multi-Level:** This means it recognized that poverty cannot be addressed by local action alone. Approaches are needed which link the local level perspectives obtained by sustainable livelihood into higher level processes of designing and implementing policies which impinge on the poor.
- **Conducted in Partnership:** This can be joint venture between public and private sectors. Thus both NGOs and private commercial agencies have roles to play which complement those of government.
- **Sustainable:** This need to be in several dimensions-economics, institutional, social and environmental but does not imply set patterns of livelihood which must be sustained indefinitely on the contrary.
- **Dynamic:** This means that the poor manage complex 'portfolios' of a number of (usually) part-time activities, changes the balance among them with changes in the opportunities and constraints they face.

SLA as an analytical framework (provides a broad and systematic understanding of the various factors that constraints or enhance livelihood opportunities and how they related to each other). The framework is a useful means of highlighting key aspects of the approach and the way it relates to each other but is not intended to capture all of sustainable thinking. Rather it is one of many tools which can be employed when implementing a sustainable approach (Krantz, 2001).

A developmental objective is to enhance the overall level and sustainability of livelihood to reduce poverty. A sustainable livelihood approach therefore aims to bring together the lessons of 'best practice' in a set of guiding principles. It also provides a common framework and language for analyst and policy makers from different sectors by emphasizing cross-sectoral collaboration and providing space for different discipline to work together towards a common goal (Ashley and Camey, 1999) (see Figure 1).

Figure 1. Sustainable livelihoods framework
Source: DFID (1991)

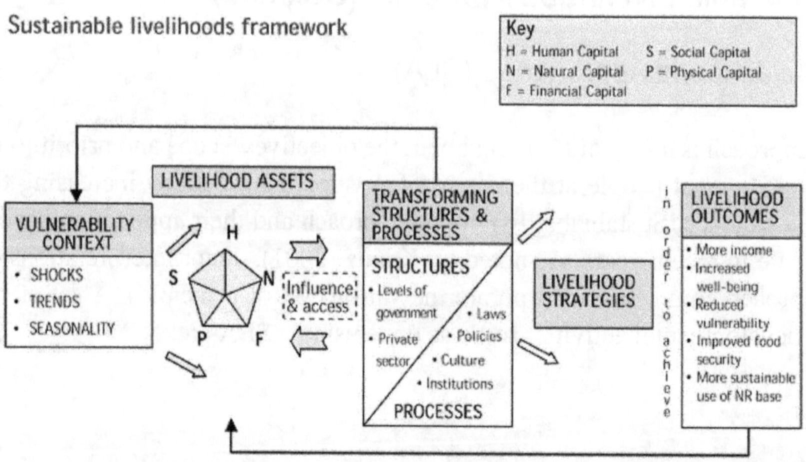

Sustainable livelihoods framework (SLF) focuses on the relationship between people's assets and their resilience in the face of external shocks, in particular the relationship between poverty and vulnerability. This can be usefully informing a differentiated approach to information needs assessment by highlighting the potential role of information in enhancing the different livelihood assets.

As Michiels and Van Crowder (2007) show the relation between ICT and livelihood assets. They argue that "ICT impact on livelihood assets in a number of way's depending on the local context in which they are introduced". Also, they explain livelihood assets in relation with ICT as follows:

- **Human Capital:** This improves access to education and training through distance learning programmes and the education tools in a wide range of different formats (Sanga, et al., 2007). The potential to transfer digital content to remote locations easily in the form of text, images, video, and radio combined with the vast storage capacity of PC's, CD's and DVD's reduce many of the cost associated with barriers to broad based information access. The impact of increased information flow on human capital development will depend equally on the effective translation of material into different languages and appropriate formats for the intended users and their local cultural context.
- **Natural Capital:** This improves access to institutional dealing with different aspects of natural resources management, including administrative and legal information such as land records. Communication channels can be enhanced with appropriate authorities, landowners, Government ministries and local Government officials. The experiences of other individuals and communication can also be shared and the information used to compare strategies and develop local solutions to problems and conflict situations.
- **Financial Capital:** This support and strengthen local financial institutions including micro-credit organizations to improve information provision of services and facilities such as loans and saving schemes. Extended access to financial information can also improve transparency and more equitable services provision such as through highlighting excessive rates of interest charged by money lenders. Community based financial management such as saving schemes can be introduced together with extended communication among a wider community of financial institution.
- **Social Capital:** This improves 'networking' both at the community with the existing networks and potentially amongst a much wider community. The ability to build new social networks at a regional and national level can help to bring benefits to existing networks and institutions at a local level, such as community based organizations, farmer's organizations etc. The reduction in the cost and time taken to travel to pursue social networking goals can also have a positive impact at a household level with family members spending less time away and less money on transport. Expanded social networks may also result in increased opportunities for employment both local and further afield.
- **Physical Capital:** This deals with access to markets and market information which help to improve choices for the sale of goods on local markets to enhance information on prices and comparative supply and demand for products. In the longer term, new markets, techniques and processes for production, processing and marketing of products both farm and non-farm can be improved.

Overview: Contribution of ICT4SL

The ICT play a greater role to enhance the sustainable livelihoods of many people in different parts of the world. According to World Bank report (2007) on using ICT to support rural livelihood, it shows the contribution of ICT to rural livelihoods in four aspects (i.e. agricultural extension, access to market information, rural access to finance and access to weather information, disaster early warning and country preparedness).

1. Agricultural Extension

Extension organizations serve two vital roles in helping rural communities access and use ICT as tools to improve their livelihoods. First, these organizations often serve as a crucial catalyst for awareness raising and advocacy, educating and mobilizing rural stakeholders to demand strategies, policies and investments that will increase their access to affordable ICT infrastructure and services. By aggregating and educating the demand side, they help both to express the need for and increase the livelihood of success of such initiatives by giving Government and the private sector active and eager local partners. Second, these organizations help the rural poor to understand, adapt and effectively use ICT tools and services once they are available, in ways that will have positive impact on their incomes and livelihoods (World Bank report 2007).

2. Access to Market Information

Among the wide range of factors shaping the incomes of rural poor farmers, issues related to market information and marketing effectiveness are crucial. Improving timely access to information about market prices, transport arrangements and costs, the locations and desires of buyers, and related issues such as weather has a tangible positive impact on farmer incomes. Traditionally, farmers have accessed this information largely through social networks and broadcast media (particularly radio) (Sanga et al., 2013).

However, the spread of the Internet and particularly mobile phones in rural areas is substantially expanding affordable and timely access to relevant and accurate market information, with positive effects both on farmers' incomes and on the performance of rural agricultural markets and hence improve rural livelihood (World Bank report 2007).

3. Rural Access to Finance

The majority of the rural poor in developing countries are "un-banked". They do not have access to, or do not use, the services of formal financial institutions. They exist in a predominantly cash economy, with low incomes and little or no savings. Furthermore, even when they do have savings or enough regular income to provide some liquidity, they often are unfamiliar with or suspicious of financial institutions.

Most traditional banks, at the same time, consider the poor to be unattractive customers, because of the small size of their liquid assets relative to the transaction costs entailed in servicing them (which are compounded by distance.) This lack of access to financial services or their high cost when the poor occasionally use them, for example to receive remittances is another key impediment that the rural poor face in making the most of the assets that they do possess, including their labor.

There are many ways that ICT can improve the reach and affordability of financial services for the rural poor. ICT can help traditional financial institutions to address some of the costs and inefficiencies entailed with serving remote rural populations hence enable poor people to access financial services such as loan also through the use of mobile services such as M-pesa, Tigo-pesa, Airtel Money and Z-pesa help poor people in rural areas to access financial services as saving and receiving money thus improve their livelihood (World Bank report 2007).

4. Access to Weather Information, Disaster Early Warning, and Country Preparedness

The poor, especially the rural poor, are particularly vulnerable to the negative effects of extreme weather and natural disasters. Yet accurate forecasting and timely warning can mitigate the effects of natural disasters such as floods, and improved weather forecasting can improve crop yields and lessen the effects of severe weather or drought. ICT has a crucial role to play in all links of the chain, from detection to modeling and forecasting to advance warning and localization.

Yet the vast majority of the poor in developing countries still have very poor access to such information and very little advance warning of adverse events. Accurate forecasting and the increasingly sophisticated computer models depend on a vast array of data at a global scale, and national meteorological organizations. These institutions play a key role both as suppliers of data for global forecasting and as consumers of information and forecasting that they localize and share (World Bank report 2007).

Theoretical Approaches towards the Use and Adoption of ICT

Understanding how ideas and technologies diffuse or spread among people has been studied in many fields. To explain the factors that promote or hinder the acceptance of a technology, several models have been proposed, such as the Technology Acceptance Model (Venkatesh & Bala, 2008) and the Lazy User Model (Tetard & Collan, 2009). Perhaps the leading and most influential model is the Everett Rogers's Diffusion of Innovations (2003). Although several researchers preceded him, Rogers (2003) is viewed as the pioneer of technology adoption research. Studying rural and agricultural sociology, his doctoral dissertation in 1957 was on the usage patterns of a new weed spray among Iowan farmers. He reviewed other studies of how groups adopted a new technology or idea. This section provides a description of his framework.

Everett Rogers's Diffusion of Innovations

One of Rogers's key insights was in not just focusing on technology or commercial products. Instead, he developed the concept of innovation, which he defined an as any object, idea, technology, or practice that is new. An innovation can include tangible, physical objects such as a new device or medicine. An innovation may also be intangible, such as a new design methodology or pedagogical technique. Furthermore, the notion of an innovation's newness can be relative to both place and population. An innovation may be cutting edge communication technology (Rogers, 2003).

By defining innovation in this way, Rogers effectively dissolved the barriers between disciplines and could openly consider adoption studies from multiple fields. With such a broad scope, the commonalities in findings from various studies are much more potent. Rogers's model thus readily generalizes and has wide applicability.

The Innovation-Decision Process

One of the general findings of Rogers's literature review was what he termed the innovation decision process (Rogers, 2003). The innovation decision process describes the steps an entity goes through in deciding whether to adopt an innovation. The entity involved may be a solitary individual or a group such as a community or company.

Roger's explain five stages where by an individual relay before adopting and use a certain technology. This stage includes Knowledge-One where an adopter cannot begin the adoption process without knowing about the innovation. In this stage, a person first becomes aware of the technology. Secondly is a persuasion which begins to show interest in the technology and seeks out information about the technology, costs, features, user reviews, etc. Thirdly, decision-at the decision stage where a person makes the choice to reject or adopt the technology. This personal process involves the weighing of advantages, disadvantages, costs, benefits, and trade-offs. The decision to not adopt, rejection, is an active choice to not acquire the technology or ever use it. Forth is an implementation whereby during this stage, re-invention may occur. Re-invention refers to the process by which a person adapts or modifies a technology to better meet his/ her needs and improve its overall compatibility. Fifth is the confirmation, whereby at this point the person finalizes their decision regarding the adoption of the technology. Lastly is discontinuance after the adoption of a technology the person does not always continue to use the technology.

Also Rogers explain the factors which influence people to adopt a certain technology this include: relative advantage, compatibility, complexity, trialability and observability. Therefore, Rogers theory can be linked with the adoption and use of ICT by rural people to enhance their livelihood. It is from this end that, this study reveals the relevance of Roger's theory with respect to adoption and use of ICT tools by rural people in Kilosa District.

Generally, both theoretical approach and previous empirical studies shows that the adoption and uses of a certain technology can have a positive and negative outcome to people's livelihood.

Major lesson learnt from literature review are:

- ICT have the greater role in people's daily lives
- ICT enhance the agriculture development in different parts of the world
- ICT improve the income of rural poor in developing countries
- ICT has an influence in the country's economy

Conceptual Framework

The conceptual framework for this study which is presented in Figure 2 comprises a set of background variables, independent and dependent variables. The background variable shows the socio-economic characteristics i.e. age, sex, education, income, and marital status. Independent variables show the communication devices i.e. television, radio, internet, mobile phone, computers, and satellite system and dependent variables shows the livelihood outcome (i.e. more income, increased well being, improved food security, reduce vulnerability, and sustainable use of resources).

Figure 2. Conceptual framework

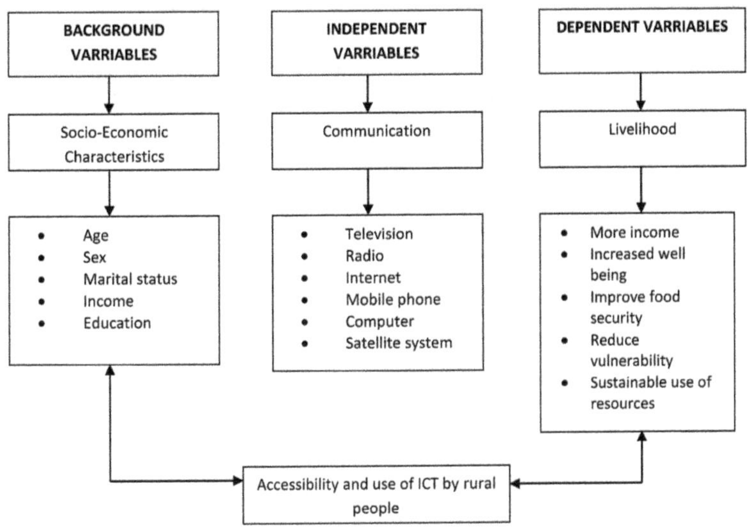

RESEARCH METHODOLOGY

Overview

This section describes methodological aspects related to the study.

Description of the Study Area

The study was conducted in Kilosa district in Morogoro region. Kilosa district is surrounded by Tanga region and Arusha region in North, Morogoro rural district in East, Kilombero district and Iringa region in South and to the west is surrounded by Dodoma region. Kilosa district has a land of 14,918 Km Square. Kilosa district administratively has 9 divisions, 36 wards and 132 villages, the main economic activities of people in Kilosa district is farming and livestock keeping,

Research Design

This study was conducted in a one round survey using cross-section research design. The cross section research allows data collection at a single point of the study in one time. The design was precise and favorable because of limitation of resource in terms of time and funds.

Sampling Techniques and Sample Size

Simple random sampling method was employed to get a sample from both male and female respondents to fill in questionnaires. Proportionate stratified sampling was applied to get a proportionate balance of male and female respondents. Three wards were chosen then from each 20 respondents male and female

Table 1. Data collection research methods

Specific Objectives	Data Collection	Data Analysis
To assess how rural people use ICT tools to enhance their living	Questionnaires were used as a tool for data collection.	Statistical Package for Social Science (SPSS) was employed during the analysis of collected data.
To identify constraints to effective use of ICT tools.	Questionnaires were used as a tool for data collection.	SPSS was employed during the analysis of data.
To investigate the impact of ICT on peoples income generating activities.	Questionnaires were used as a tool for data collection	SPSS was employed during the analysis of data.

were chosen randomly to fill the questionnaires. The reason behind choosing this sample is the population size of people in this wards and heterogenity of this population.

Data Collection and Analysis

Table 1 shows the data collection from the research methods.

Primary Data

The data was collected using structured questionnaires. The questionnaires consist of both open and closed-ended questions prepared in English and translated in Swahili to allow easy understand to the respondents.

Secondary Data

In order to obtain the readily available information with regard to the stated objectives the data was obtained from documents obtained from study area.

Data Analysis

Primary data was summarized and coded before being entered into SPSS for analysis. Descriptive statistics were done by computing frequencies, percentage, measures of central tendency and measure dispersion. The pilot study which was done before the actual study ensured the viability and reliability of data collection and analysis instruments. In addition, in order to strengthen the reliability and validity of the data, data were coded before analyzed through SPSS.

RESULTS AND DISCUSSION

Overview

The study was conducted in Kilosa district. The selected wards were Mkwatani, Mbumi and Kasiki. Various techniques were used in data collection; primary data was collected through questionnaire.

Questionnaires were used to collect data from community members of selected wards. The targeted respondents for this study were residents from the selected places. Residents both who accessing ICT tools and not accessing ICT tools were chosen randomly taking into account different age groups and gender parity. A total of 60 individuals from selected wards were interviewed. Documentary review of some previous studies on ICTs and peoples livelihood in Tanzania was done for collecting of secondary data. Quantitative data were analyzed using SPSS.

Respondents' Socio-Economic Characteristics

Socio-economic and demographic characteristics of the study area were based on the sample of respondents who were studied (n=60). The findings with respect to socio-economic characteristics are presented in Table 2.

Table 2. Respondents' socio-economic characteristics

	Socio-Economic Characteristics	Frequency (n=60)	Percentage (%)
Age of Respondent	20-25	5	8.3
	26-30	16	26.7
	31-35	7	11.7
	36-40	10	16.7
	41-45	6	10.0
	46-50	4	6.7
	51-55	5	8.3
	56-60	4	6.7
Sex of Respondent	Male	38	63.3
	Female	22	36.7
Level of Education of Respondent	Primary education	14	23.3
	Secondary education	25	41.7
	Diploma	8	13.3
	Certificate	5	8.3
	Degree	8	13.3
Main Occupation of Respondent	Peasant	23	38.3
	Business person	19	31.7
	Public servant	18	30.0
Marital Status of the Respondents	Single	15	25.0
	Married	40	66.7
	Divorced	5	8.3

Sources: Field data (2013)

Age of Respondents

Age is a very important variable determining the access and use of ICT tools. The respondents interviewed were with mean age of 20 years; the maximum age recorded in the sample was 70 years while minimum was 19 years. The majority of the respondents were aged 26-30 years forming 26.7% of the whole respondents. Therefore, the majority of respondents interviewed were young who are passionate of ICT tools.

Sex of Respondents

As it has been explained earlier, the study interviewed a total of 60 respondents of whom 63.3% of respondents were male and 36.7% of respondents were female.

Respondents' Level of Education

Education is always valued as a means of liberation from ignorance. It is the only principal mechanism for developing human skills (URT, 2002). Respondents' educational status may have an implication towards the use of ICT tools. Results show that 23.3% of the respondents having primary school education, 41.7% having secondary school education, 13.3% having diploma education, 8.3% having certificate education, and 13.3% were having a university degree education.

Respondents' Main Occupation

Main occupation examines the labour force of a given locality engaged in. Majority of respondents admitted to be conducting single income generating activities. Findings reveal that 38.3% of the respondents were peasants, 31.7% of the respondents were business persons, and 30.0% of respondent were public servants.

Respondents' Marital Status

Since the purpose of this research was to assess the contribution of ICT4SL, it was imperative to investigate marital status of respondents so as to examine who use ICT tools much than the other between male-headed household and female-headed household. Findings show that 25.0% of respondents were single, 66.7% were married and 8.3 were divorced.

Income Level of Respondents

Income in this study is defined as the earnings a person get from various activities. It was observed that 65.0% of respondents earn below 500,000 Tanzania Shilling (Tsh) per month, 20% of respondents reported to earn above 1,000,000 Tsh per month and 15% of the respondents reported to earn below 5,000,000 Tsh per month. Income level of respondents influences positively accessibility of ICT tools.

Results in Table 2 show clearly that adoption and usage of ICT tools are pre-determined by socio-economic status of an individual or community.

Accessibility and Use of ICT by Local People

Accessibility is the degree to which a product, device, servicers, or environment is available to as many people as possible (Lynn, 2009). While radio and television are considered to be conventional ICTs, unfortunately these technologies are still not yet fully accessible in rural areas. 98.3% of the respondents admitted to access radio and only 1.7% do not access ICT tool (Radio). For those who access radio they prefer to listen news especially national news on business and weather condition. 88.3% of respondents' use and access television and only 11.7% are not having television.

Also results show that 96.7% of the respondents access and use mobile phone to communicate with buyers and sellers, to access e-mail address and to look for price of different commodities in the market, while 3.3% of the respondents are not using and accessing via mobile phones. Mobile phones are seen as important ICT tools to many people in rural areas. Results in Table 3 are in line with what Samuel et al. (2005) found in his study on socio-economic impacts of mobile communications on households, rural communities and small businesses in South Africa, Tanzania and Egypt. Samuel et al. (2005) found that mobile phones were reducing travel needs, assisting job searching, improving access to business information, and contacts with families and friends.

For those respondents who access and use mobile phones were asked to state the type of phones they use. The results show that 46.7% of respondents accessing Internet enabled mobile phones, 35.0% accessing not Internet enabled mobile phones and 18.5% of the respondents they do not know whether their phones are Internet enabled or not.

However, for those who are currently not owning any mobile phone, when they were asked if they had ever own one, they said that they had mobile phones but due to various reasons they do not have anymore. Among these reasons include being stolen, damaged and others admitted to have sold them due to economic hardship.

Results also show that 40.0% of respondent use and access computer, 60.0% are not using computer and 48.3% of respondents are having an e-mail address while 51.7% of the respondents are not having an e-mail address.

Table 3. Accessibility and use of ICT by local people

	ICT Tool Accessed	Frequency (n=60)	Percentage (%)
Radio Accessibility	Access radio	59	98.3
	Not access radio	1	1.7
Television Accessibility	Access television	53	88.3
	Not access television	7	11.7
Mobile Phones Accessibility	Access mobile phone	58	96.7
	Not access mobile phones	2	3.3
Kind of Phone Used	Internet enable	28	46.7
	Not internet enable	21	35.0
	I do not know	11	18.3
Computer Accessibility	Access computer	24	40.0
	Not accessing computer	36	60.0
E-Mail Usage	Use e-mail address	29	48.3
	Do not use e-mail address	31	51.7

Source: Field Data (2013)

Table 4. Constraints to effective use of ICT

Constraints	Frequency (n=60)	Percent (%)
Very expensive	4	6.7
Low usage of ICT tools	10	16.7
Government does not recognize ICT	2	3.3
Low recognition and appreciation of ICT	4	6.7
Implementation of ICT policy is very low	6	10.0
Precise of public private partnership	8	13.3
Cost, low usage, low recognition and appreciation	17	28.3
Unspecified	9	15.0
Total	**60**	**100.0**

Source: Field data (2013)

Results from Table 3 shows that there is a variation when it comes to accessibility and usage of ICT tools by rural people. These variations are caused by inadequate fund, lack of knowledge and awareness on the importance of ICT tools to people's activities.

Constraints to Effective Use of ICT

Findings reveal several constraints facing rural people on using ICT tools as identified by respondents. The respondents who identified the constraints are: 6.7% said very expensive, while 16.7% said low usage of ICT tools, 3.3% said Government does not recognize ICT, 6.7% said low recognition and appreciation of ICT, 10.0% said implementation of ICT policy is very low, 13.3% said precise of public private partnership, and 28.3% highlighted cost, low usage, low recognition and appreciation being the challenges facing rural poor in using and accessing ICT tools and 15.0% are unspecified because they do not use any ICT tools.

Results in Table 4 are in line with what Kaino (2004) reported in his study on Impact of ICT to peoples' livelihood. Kano identified many constraints which are common in developing countries. The constraints include: access to computers (email and internet), affordability of computers and connectivity, telephone and electricity infrastructure, computer literacy, expertise, etc. While such problems have been acknowledged, the main reasons behind the slow pace of adoption have been identified as lack of effective policies on ICTs (Kaino, 2004). McNamara (2008) stated that inadequate infrastructure, such as road, electricity and landline telephones are the main impediment towards ICT-for-livelihood interventions in Tanzania.

Impact of ICT to Income Generating Activities

Pigato (2001) in his study on ICT Impact on Development and Economic Transformation explained that ICTs has impact on poverty alleviation. ICT has impact on productivity and income generation activities as it improves different sectors whereby the poor are likely to be most actively engaged. For instance, productivity increase in the SMEs can have poverty reducing effect to the extent SMEs gain access to

Figure 3. ICT tools used for sharing information on income generating activities
Source: Field Data (2013)

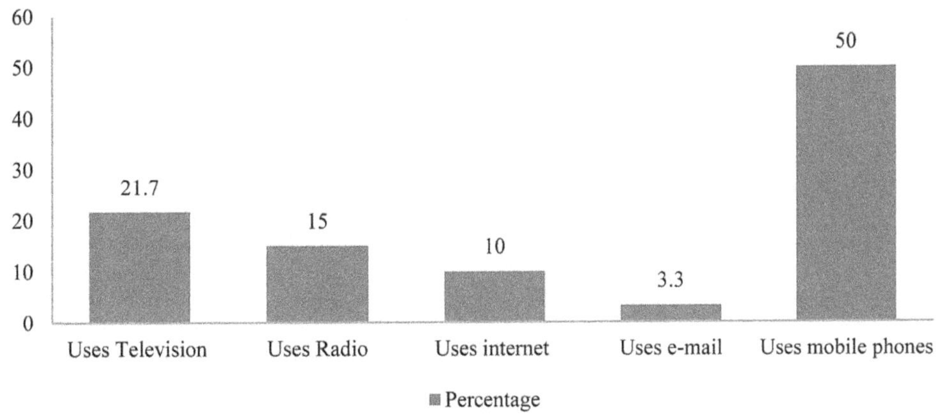

market information (faster and more cheaply), access to information on input prices and output markets and to the extent they strengthen forward linkages to the market and backward linkages to the domestic suppliers of inputs.

With respect to what kind of ICT people use to enhance their income generating activities: 21.7% of respondent uses TV's to access information on their daily activities, 15.0% of respondents uses radio to access information on their daily activities, 10.0% of respondents uses internet, 3.3% of respondents uses E-mail address, and 50.0% of respondents uses mobile phones as their way to access information on their daily activities.

Results in Figure 3 are in line with what FAO (2013) explains in their report on uses of ICT in fishers sector. They explained that the use of ICT for development goes beyond direct support for income-generating activities. ICT for pro-poor development can be a powerful means of reducing people's vulnerability through fostering equity and social inclusion and in mobilizing communities to take charge of their own development. In conjunction with traditional communications activities such as meetings and theatre, community radio, video/television, mobile phones, tele-centers and print publications can be used to share information, especially marketing information, and knowledge, as well as raise awareness and stimulate discussion of issues such as gender, health, education, local development and diversification of income-generation activities.

Also, respondents were asked to state if ICT tools facilitate innovation in their daily activities, 56.7% of respondents acknowledge that the use of ICT tools facilitate innovation in their daily income activities, also 73.3% of the respondent use ICT tools to access business information and communicate on social related issues. 68.3% of respondents uses ICT tools to access market information on the issues concerning prices of commodity, and the demand of the products in the market, apart from that 48.7% of respondents uses ICT tools like Televisions and Radio to be informed on the weather information and disasters if might happen so that they can overcome earlier and not affect their income generating activities. 66.7% of the respondents agree that the uses of ICT tools improve their income level and ensure income growth, 55.0% of respondents acknowledged that the uses of ICT tools ensure sustainable development.

With respect to improvement in production process, 66.7% of the respondents said that the use of ICT tools help to increase the production process since farmers will be in a position to access market and be

Figure 4. Contribution of ICT to income generating activities
Source: Field Data (2013)

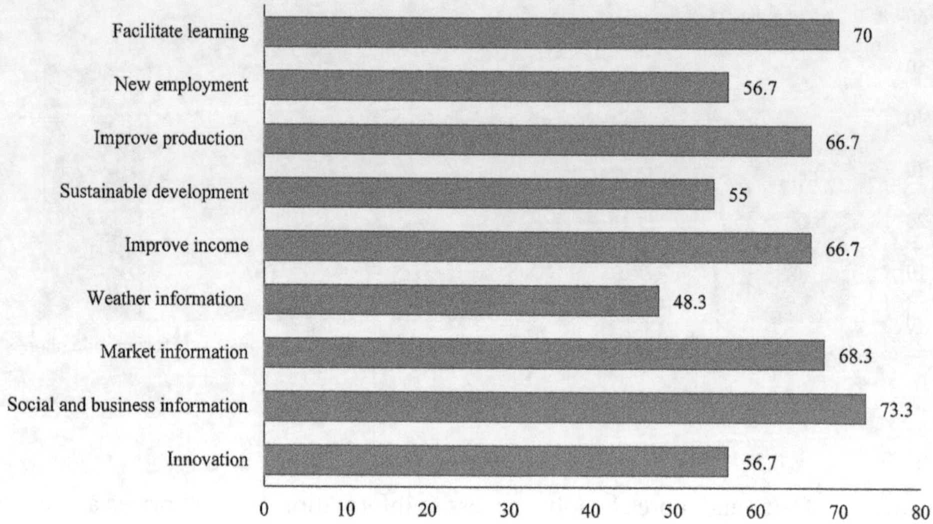

able to increase their production process. 56.7% of the respondents highlighted that ICT tools gives a room for new employment and 70.0% of respondents agree that the use of ICT facilitate learning process and thus can ensure sustainable livelihood. Note that the percentage obtained shows that respondents mentioned more than one item that is why the total percent don't make 100 percent (see Figure 4).

Contribution of ICT4SL

Another way in which ICT can improve livelihoods of people is increasing their access to potential markets. ICT used to facilitate contact between sellers and buyers, promote agricultural exports, facilitate online trading, and make producers aware of potential market opportunities including consumer and price trends in markets. By increasing awareness among producers on consumer trends and new production techniques, ICT can contribute to the diversification of production. This research found that ICT ensure sustainable livelihoods by giving people chance to access market and improve market system. ICTs improve the bargain power of farmers and hence eliminated some of the middlemen who normally benefit much in comparison with farmers.

From the sample (n=60) 40% of respondents agree that ICT improve market system while 26.7% of respondents indicated that ICT give information alerts earlier, furthermore 23.3% of respondents indicated that ICT strengthening social network and hence gives room to meet new people especially in business and last 10% of respondents acknowledge that ICT help to enhance development process and thus make sustainable livelihoods.

Results from Figure 5 are in line with Hassan (2009) who explains that ICT has the potential to reduce poverty and improve rural livelihoods through several channels including: (i) improved market information access for better decision making on enterprise choices, combinations and commodity marketing and helping to mitigate negative impacts of the weather (ii) creating awareness and increasing general knowledge of the people regarding production technologies as well as their access to micro-finance information and packages and family health matters (iii) raising education levels on general and specific

Figure 5. Contribution of ICT4SL
Source: Field Data (2013)

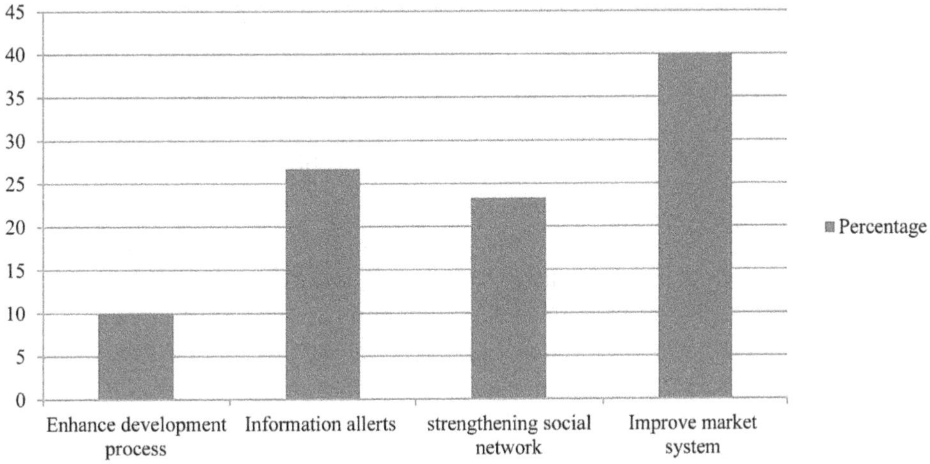

life skills; (iv) investments in rural infrastructure, especially electricity and roads. Thus, ICT tool as a means to enhance sustainable livelihoods can contribute to total income in terms of expanded employment opportunities and result to sustainable livelihoods.

Relevance of Sustainable Livelihoods Approach in ICT for People's Livelihood

One of the key supporters of a SLA argues that the way development professionals conceptualise development and poverty is very different from how poor people themselves view these (Chambers, 1995). Poor people perceive poverty in a much more complex manner than do development professionals and they employ a range of strategies, not only to maximize income, but also to minimise risk and to protect or increase other things that they value. Poor people's priorities are often different from those imputed to them by development experts, and their strategies are often more complex, both in terms of activity and motivation. Thus it is argued, the sustainable livelihoods framework (Figure 1) provides a conceptualisation that is more appropriate to the perspectives and realities of poor people (Chambers, 1995).

The focus of "livelihood" in sustainable livelihoods (SL) frameworks is an attempt to move away from narrow definitions of poverty, and as such reframes the broad aim of development as an effort to improve people's livelihood options. "Livelihood" refers broadly to a means of making a living, and includes the assets, access to institutions and processes, and strategies that a person utilizes to achieve livelihood outcomes (Ashley and Carney, 1999). The term "sustainable" refers both to the characteristic of a livelihood to endure the various shocks and uncertainties likely to be encountered in the environment, and to avoid contributing to long-term depletion of natural resources (Chambers, 1987).

The quality of the SL framework that is immediately apparent is that it requires interpretation and adaptation to fit any particular context. As such, the SL framework is not expected to be used in a fixed prescriptive way: "The framework is centred on people. It does not work in a linear manner and does not try to present a model of reality. Its aim is to help stakeholders with different perspectives to engage in structured and coherent debate about the many factors that affect livelihoods, their relative importance and the way in which they interact" (DFID, 2001). Of particular relevance to ICTs, the framework embraces multiple dimensions that are interrelated in a dynamic manner. In this sense, the SL framework is a systemic representation, and one that appears most promising in the context of impact assessment of ICT (Bryden, 1994).

The rationale for using the SL framework for ICT-related issues is that it is comprehensive and it helps us to think about ICTs in a more "bottom-up" way. For example, in what ways, positive, negative or neutral, are ICTs in the context of a specific project likely to interact with different livelihoods strategies? Which demographic groups within the population are most likely to use these strategies? And thus, who is most likely to be impacted by ICTs, and in what ways?

The application of the SL framework creates a kind of "big picture" context that moves away from the linear cause-effect thinking that is so clearly ineffective in ICT-related development projects (Heeks, 2002). ICTs are introduced into an existing and already complex web of mutual causality. People as actors who are seeking to make their livelihood and to maintain increase or minimize loss to their existing asset base are likely to use ICTs when they can do so in a way that provides a net benefit to these goals. Existing behaviour and resources thus become the foundation for building new behaviour that incorporates the use of ICTs.

With respect to this study, SL Framework is mostly applicable in understanding the use of ICT by rural poor to enhance their living because the livelihoods approach is concerned first and foremost with people. It seeks to gain an accurate and realistic understanding of people's strengths (assets or capital endowments) and how they endeavor to convert these into positive livelihood outcomes. The approach is founded on a belief that people require a range of assets (i.e. ICT tools such as computer, radio, mobile phones, television etc) to achieve positive livelihood outcomes; no single category of assets on its own is sufficient to yield all the many and varied livelihood outcomes that people seek.

Generally, the major lesson learnt from this study is whenever the development planners need to think on establishing an ICT intervention program or project they suppose to apply the SL framework from the design stage because SLA is people centered and may change to cope with shocks and trends and help poor people come out from vulnerability and improve their livelihood.

Applicability of Rogers' Theory Based on ICT and Sustainable Livelihood (Theoretical Review)

Diffusion is the process by which an innovation is adopted by members of a certain community. There are four factors that influence adoption of an innovation. These include a) the innovation itself, b) the communication channels used to spread information about the innovation, c) time, and d) the nature of the society to whom it is introduced (Rogers, 1995). The results from this study show that 98.3% of respondents' access radio which they use it as a communication channel to access information on different issues pertaining their livelihood, also results show that majority of interviewed respondents were accessing and using ICT tools (i.e. computers, mobile phones, television, fax, radio etc.).

Thus, these findings indicate clearly that most of rural people in Kilosa District are adopting and using ICT tools to enhance their livelihood. Therefore, Rogers' diffusion of innovation theory is at infancy stage in Kilosa District because majority of people does not follow systematic the steps identified by Rogers'. The five stages where by an individual relay before adopting and use a certain technology are namely: Knowledge, Persuasion, Decision, Implementation and Confirmation (Rogers, 1995). This study reveals that for an individual to adopt and use a certain ICT tool needs to have purchasing power –ability to buy something at any valuing price.

Applicability of Rogers' theory in this study is based on "Knowledge" as one of the stage which Rogers' indicate that before individual or community adopting any technology there must be knowledge on such innovation or technology before adopting. Findings from this study reveal that 40% of interviewed respondents were using computer while 60% were not using computer the reason being they do not have a knowledge base on using computers. Also the study shows that 48.3% of respondents use e-mail address to communicate while 51.7% of respondents were not exposed to e-mail address due to lack of knowledge on the use of e-mail services. Thus, knowledge can hinder or contribute to the adoption and use of ICT tools by poor people in rural areas.

Consequently, many more people should be facilitated to access and use ICT to make it agent of attainment of sustainable development to citizens in Kilosa District. However, a lot more needs to be done to ensure reliability, affordability, availability and usability of the ICT tools to rural population.

Relevance of Conceptual Framework

Conceptual framework for this study was used to measure the impact of ICT tools to people's livelihoods. This framework conceptualizes a clear and direct relationship between the use of ICT tools and changes in people's livelihoods. Mainly, it was undertaken using a positivist, survey-based approach that requires identifying people both using and not-using the ICT tools and to analyze their living standards, and the levels of their livelihoods after using the ICT tools.

This framework analyzes the cause- effect link whereby the dependent variable is the livelihood outcome, independent variable is communication tools and background variable are socio-economic status of respondents. The impact assessment therefore involves figuring out how the use of ICT tools impacts livelihood in terms of increased or decrease in knowledge over socio-economic aspects important for improved livelihoods. This knowledge is expected to influence people' productivity hence changes their livelihoods in a positive way.

This study also regards the issue of sustainable livelihoods as important. It conceptualized as a standard of living that is sustained and not shaken by any force either internal or external, including the onset of the ICT phenomenon. The study measured impact of ICT tools to people's livelihood. These impact include the improved access to information; empowerment through information about choices that affect people and decreased vulnerability to risk due to the possibility to send and receive information via radio and television. The conceptual framework for this study is presented above in Figure 2.

Tanzania National ICT Policy (2003) and Its Impact on Sustainable Rural Livelihood

Tanzania adopted a National ICT Policy in 2003, following a consultative process which involved both Government and an influential group of ICT professionals organized as eThinkTankTZ. This National Policy established 10 focus areas for action to improve the availability and developmental use of ICTs, as set out in the Table 5 (URT, 2003). However, this National Policy has not been followed up with an agreed national implementation strategy. As a result, no guidance or responsibility has been assigned to any Government body, and follow-up at a sectoral level has also been unsatisfactory (Yonazi, 2010).

ICTs are believed to contribute to improving development outcomes in two main ways: (a) the production of ICT-based knowledge and products contributes directly to wealth creation (b) the use of ICTs contributes indirectly to national development through its impact in social and economic sectors such as agriculture, health and education. Individuals also benefit from the availability and use of ICTs

Table 5. ICT policy 10 areas of focus

Focus Area	Description
Strategic leadership	To create a focused ownership and visionary leadership in favor of ICT
ICT infrastructure	To foster for supportive and interoperable ICT infrastructure
ICT industry	To build and promote and regulate a conducive ICT environment
Human capital	To bolster the human capital in favor and exploitation of ICT potentials
Legal & regulatory framework	To create an ICT enabling legal, legislative and regulatory framework
Productive Sectors	To encourage and support the utilization of ICT in all productive sectors
Service Sectors	To promote and encourage effective and efficient ICT application in service sectors
Public service	To support and become a role model in utilizing ICT in the public service
Local content	To allow, encourage, and promote the development of indigenous knowledge and traditional cultures
Universal access	To address the digital divide for better and affordable universal access

Source: United Republic of Tanzania (2003)

in many ways – for example, by substituting phone calls for travel, which costs time and money, and by using information on prices, which ICTs can make available, to sell their own produce and to make purchases. In these various ways, ICTs can have a significant impact on a country's ability to achieve the Millennium Development Goals (MDGs) (Yonazi, 2010).

Findings reveal several constraints facing rural poor in Kilosa district to access and use ICT tools i.e. very expensive was argued by 6.7% of respondents, low usage of ICT tools was argued by 16.7% of respondents, government does not recognize ICT 3.3% of respondents, low recognition and appreciation of ICT was argued by 6.7% of respondents, implementation of ICT policy is very low was argued by 10.0% of respondents, precise of public private partnership was argued by 13.3% of respondents, and 28.3% of respondents highlighted cost, low usage, low recognition and appreciation being the challenges facing rural poor in using and accessing ICT tools and 15.0% of respondents are unspecified because they do not use any ICT tools. These constraints may hinder an effective utilization of ICT policy to enhancing sustainable livelihoods.

There are, however, also constraints on the potential impact of ICTs in many developing countries. These constraints include inadequate technical infrastructure, limited human skills to use available networks and services, the relatively high cost of communications equipment, and poor policy and regulatory environments. These factors reduce the scope for communities in development countries to use ICTs for development (ICT4D), and may even increase exclusion and marginalization.

Conclusion, the Government of Tanzania should exploit opportunities through developing the ICT solutions as per different sectors, encouraging the development of local ICT hardware, software and applications. It must support implementation of ICTs through establishment of a more strategic approach to ICT4D, building awareness of this in the community, and establishing better organizational arrangements for implementation. Promote affordability and access through tax relief to ICTs and related services, as well as hardware and tackling challenges to enhance connectivity. This will result into sustainable livelihoods of people in rural areas through the use of ICT.

CONCLUSION AND RECOMMENDATIONS

The findings of the study had determined various contribution of ICT to people's livelihoods. These factors include: facilitate innovation, ensure market accessibility and market information, helps people to access weather and disaster information earlier, improve income growth, ensure sustainable development, create a room for new employment, facilitate learning, strengthening social network among different people in the community, and enhance the development process through saving time and money.

Findings also reveal several constraints facing rural poor in Kilosa District to access and use ICT tools. These constraints include: ICT tools are very expensive to access, low usage of ICT tools, government does not recognize ICT, low recognition and appreciation of ICT sector, implementation of ICT policy is very low and low public private partnership.

Lastly, results show that ICT tools help to improve the livelihoods of poor individuals, their families and communities in rural areas. Also ICT tools help to increase their income opportunities by improving their chances of escaping from persistent poverty.

Generally, the use of ICT tools contributes much to the sustainable livelihoods of people in Kilosa District. Thus, the Government of Tanzania should address the challenges facing rural poor in accessing and using ICT tools so as to create an enabling environment for the growth of ICT for development.

Recommendations

Although the uses of ICT tools have a greater impact on people's livelihoods but still the ICT sector in Tanzania faces a lot of challenges such as poor infrastructure system, the ICT tools are not affordable to rural people (i.e. very expensive), low usage of ICT tools especially those in rural areas, low recognition and appreciation of ICT by the Government, and the implementation of ICT policy is very low. Due to these challenges the following should be done so as to enable rural poor to access ICT tools and hence ensure sustainable livelihoods.

- The government should invest in ICT and appropriate ICT enabling infrastructural facilities such as wind and solar power generation in rural areas to make services available and affordable. It should also take deliberate initiatives to encourage investment in ICT in the rural areas (i.e. implementations of Last miles project and Rural Electrification project).
- Government should create awareness among leaders and the public, and championing of ICT to be used in all sectors from national level to household level.
- The stakeholders like non-governmental organizations, Government and other agencies should put emphasis on the uses of ICT in fulfilling their functions. This means they should put more efforts in provision of education with regard to importance of ICT to people in rural areas. The efforts could be in terms of building awareness on the benefits of ICT tools to their sustainable livelihoods.
- Since most of the people living in rural areas are poor, the Government should ensure the cost for accessing ICT is lower so as every individual could be able to use and access ICT tools.

Finally, ICTs offer special opportunities to stimulate growth and increase innovation in local setting. Thus through enabling individuals and institutions to interact more productively with the global knowledge economy, local people are advisedly to use ICT tools to enhance efficiency, effectiveness and efficacy in their daily activities towards sustainable livelihoods.

REFERENCES

Ashley, C., & Carney, D. (1999). *Sustainable Livelihoods: Lessons from early experience*. London: Department for International Development.

Ashrafi, R., & Murtaza, M. (2008). Use and Impact of ICT on SMEs in Oman. *Electronic Journal Information Systems Evaluation, 11*(3), 125–138.

Banard, D., & Vonk, T. (2003). *Information and Communication Technologies in African Local Governments*. Freiburg, Germany: EDITOSIA – Electronic Distance Training on Sustainability in African Local Governments. ICLEI European Secretariat GmbH.

Chambers, R., & Conway, G. R. (1992). *Sustainable Rural Livelihoods: Practical Concepts for the 21st century, Institute of Development Studies*. Sussex, UK: IDS.

Chilimo, W. (2008). Information and communication technologies and sustainable livelihoods: a case of selected rural areas of Tanzania (Doctoral dissertation, University of KwaZulu-Natal, Pietermaritzburg, South Africa).

ESRF. (2007). *Tanzania Country Report: Enhancing Livelihoods of the Rural Poor: The Role of Information and Communication Technologies (ICT) – The Final Report*. ESRF, Dar-es-Salaam.

Farrington, J., Chapman, R., & Slaymaker, T. (2001). *Sustainable Livelihoods approaches in practice: Potentials and constraints*. Paper prepared for the SIDA Poverty Workshop, London, UK.

Fleming, S. (2012). Information and Communications Technologies (ICTs) and Democracy Development in the South: Potential and Current Reality. *EJISCD, 3*(10), 1–10.

Frempong, G., & Stork, C. (2005). Towards an African e-Index: Household and individual ICT access and usage across 10 African countries. In *Research ICT Africa*. The LINK Centre, Wits University School of Public and Development Management. Available: www.researchictafrica.net

Krantz, L. (2001). *The Sustainable Livelihood Approach to Poverty Reduction: An introduction*. Stockholm: Division for Policy and Socio-Economic Analysis.

Mallaliev, K. I. (2010). *Selecting Sustainable ICT Solutions for Pro-Poor intervention, Department of Electrical and Computer Engineering, The University of west indies. St.* Augustine, Trindad and Tobago.

Michiels, S. I., & Van Crowder, L. (2007). Discovering the 'Magic Box': Local Appropriation of Information and Communication Technologies (ICTs). Rome: Food and Agriculture Organization of United Nation's Communication for development Group, Extension, Education and Communication Services.

National Information and Communication Technology Policy (NICTP). (2003). *Ministry of Communication and Transport*. Retrieved from http://www.tanzania.go.tz/pdf/ictpolicy.pdf/

Ngwenyama, O., Andoh-Baidoo, K. F., Bollou, F., & Morawecynski, O. (2006). Is there a relationship between ICT, Heath, Education and Development? An empirical analysis of five West African Countries from 1997-2003. *The Electronic Journal on Information Systems in Developing Countries, 5*(23), 1–11.

Pigato, M. (2001). *Information and Communication Technology*. Poverty and Development in Sub-Saharan Africa and South Africa.

Rogers, E. M. (2003). Diffusion of innovations (5th ed.). New York: Free Press.

Sanga, C., Churi, A., & Tumbo, S. (2007). Status, Opportunities, Potential and Challenges of Technology-Mediated Open and Distance Education (Tech-MODE) for Agricultural Education and Improved Livelihoods: A Case Study of Tanzania. *Country Case Studies*, 113.

Sanga, C., Kalungwizi, V., & Msuya, C. (2013). Building agricultural extension services system supported by ICTs in Tanzania: Progress made, Challenges remain. *International Journal of Education and Development Using ICT, 9*(1), 80-99.

Sife, A. S., Kiondo, E., & Lyimo-Macha, J. G. (2010). Contribution of mobile phones to rural livelihoods and poverty reduction in Morogoro region, Tanzania. The Electronic Journal of Information Systems in Developing Countries, 42.

Tusubira, F. F., Kaggwa, I., & Ongora, J. (2005). Uganda. In *Towards an African e-Index; Household and individual ICT ACCESS AND USAGE across 10 African countries*. Research ICT Africa. The LINK Centre, Wits University School of Public and Development Management. Available: www.researchictafrica.net

United Nations Development Programme. (2011). *Making New Technologies Work for Human Development*. Retrieved from http://hdr.undp.org/reports/global/2011/en/

Vukanikinds-DTI. (2005). *An Analysis of the ICT Skills Audit for the dti and ISETT SETA*. Department of Trade and Industry (DTI)–South Africa. Available: http://www.isett.org.za/downloads/ICT_Skills_needs.pdf

Weema, M. T. (2005). A brief history of the development of an ICT Policy in Kenya. In E. F. Etta & L. Elder (Eds.), *At The Crossroads: ICT Policy Making in East Africa*. Nairobi: East African Educational Publishers.

World Bank Report. (2007). *Using Information and Communication Technologies (ICT) to support Rural Livelihood*. Evidence, Strategies and Tools, A Workshop for World Bank Staff. Retrieved from http://www.infodev.org/livelihoods on 11/23/12

ENDNOTES

1 www.tanzania.go.tz/pdt/ict policy

2 http://www.linkinglearners.net/downloads/RepICTstudy.pdf

Chapter 11
Doing Classic Grounded Theory Research in Information Systems
Trust in Emergence

Titus Tossy
Mzumbe University, Tanzania

Irwin Brown
University of Cape Town, South Africa

Andy Lowe
Grounded Theory Institute, USA

ABSTRACT

Grounded Theory (GT) is a latent pattern recognition research method that was developed by Barney Glaser and Anselm Strauss (1967). GT, which can use either quantitative or qualitative data, is a general research methodology and it produces empirically grounded but modifiable theory. Since the development of GT in 1967, some researchers have attempted to modify the original GT method, and remodelled GT approaches have been widely used in Information Systems (IS) research, often with unfortunate results due to disregard of the basic rules of the original GT. The continued use of remodelled GT methods has caused confusion and has prevented those in the IS research community from stimulating the development of new theory, which is the main benefit of using GT as it was originally outlined. Following publication of the original grounded theory by Glaser and Strauss (1967), Glaser (1978) went on to explain in detail how to operationalize the GT method. Despite this, some researchers mistakenly continued to classify GT as a subset of the qualitative data analysis research methodology. In doing so, they reveal that they have misunderstood both the purpose of GT and how to use it correctly. In this paper the author will concentrate on the authentic approach to GT, which is termed "Classic Ground Theory" (CGT), so as to differentiate it from any misconceived remodelled version of the theory. This paper explains how CGT can be carried out in an IS context by trusting in the emergence from the data of the required information, rather than forcing it.

DOI: 10.4018/978-1-5225-0539-6.ch011

INTRODUCTION

There are a number of theories from other disciplines–such as the social sciences– that are being adopted for use in IS (Gregor, 2006; Truex et al, 2006). While some are adopted as they are, others are modified and yet others are combined to fit within the IS discipline. Such combinations, when they include grounded theory approaches, erode the known procedures and methods of the Classic Grounded Theory (CGT) (Glaser, 1978, 1992). There is likely to be repetition of the same mistakes made in the original disciplines (Truex et al, 2006). The original, or Classic Grounded theory was developed from a combination of both quantitative and qualitative approaches to sociology (Glaser & Strauss, 1967). Many remodelled versions of CGT have been used in IS research projects, and the many approaches include evolved, mixed, and analytical versions (Matavire & Brown, 2011). This is an indication that there is confusion about how to use the original CGT in IS research projects.

CGT is a standalone general methodology and should not be used in a mixed method research design. While some IS researchers tend to mix CGT with other research methods to conduct their research, some simply use or borrow the vocabulary or analysis techniques of the CGT method to help them when analysing their data. Unfortunately, many of the claims for the use of CGT by IS researchers do not follow its original tenets (Tossy, 2012). Many researchers, for instance, mix CGT with other theories such as Quantitative Data Analysis (QDA). This is incorrect, since Glaser (2003:1-2) said that "Mixing QDA and GT downgrades or erodes the [classic] GT goal of conceptual theory through the process of default remodelling". CGT, in effect, becomes remodelled as another QDA method with all its descriptive baggage. This is because of confusion about the use of CGT and the evaluation of data using CGT. This paper fills a gap by outlining the necessary route to follow when using and evaluating CGT in IS.

The paper has four main sections: clarification as to what CGT is; an overview of the prior use of the GT theory method amongst the IS research community; how to conduct CGT research in an IS context; and the proposed principles for assessing the claim of a paper/thesis to have used CGT.

CLARIFICATION OF THE MEANING OF THE (CLASSIC) GROUNDED THEORY RESEARCH METHODOLOGY

Classic Grounded Theory methodology (CGTM) is faithful to the original formulation and follows all the original tenets of Grounded Theory as outlined by Glaser & Strauss (1967). The tenets have been further elaborated by Glaser (1978, 1992, 1998, 1999a, 1999b, 2001, 2003, 2005, 2006, 2008, 2009, 2011). CGTM is a general inductive research method designed to reveal deep seated latent patterns of human behaviour and how the main concerns of respondents studied are being continually resolved (Glaser, 1998). CGT provides IS researchers with a means to build theory relevant to the discipline, as explained in Matavire and Brown (2011). Lyytinen and King (2004) state that building theory for any field would help strengthen it. In the IS discipline, opportunities exist to build theory in both emerging fields and well-researched ones, as outlined by Matavire and Brown (2011). As Glaser (1992) states, in domains where a large body of knowledge already exists, new concepts and categories may not emerge, but CGT will still contribute to a better understanding of the basic social processes at play. Many authors, however, refer to having used CGT in their studies, when in fact what they mean is that they have been influenced by the idea of a grounded theory without following all the procedures (Bryant, 2002). This paper helps to avoid incorrect use of CGT, and enables the IS researcher to use CGT with skill and confidence.

The major source of confusion concerning the use of the GT method can be directly traced to a publication by Corbin and Strauss (1990), confusion that has probably been facilitated by the lack of clear understanding of all tenets of CGT. There have been various debates between the researchers following CGT and those following other GT approaches. The major source of debates is concerned with how the GT researcher should make the transition from low level substantive coding to higher level theoretical coding (Glaser, 1992). The authentic way to do this is by making use of a combination of constant comparison and sorting so that the way in which the lower level substantive codes relate to each other can emerge as a theoretical coding outcome. The other GT approaches, on the other hand, chose to introduce an intermediate stage between substantive and theoretical coding called axial coding. This incorrect and invented axial coding approach encourages the researcher to force the data into deductively derived patterns (selected by the researcher) rather than allowing the deep seated patterns to emerge. Using a shortcut when doing CGT research in IS, kills any prospect of revealing the core variable of the researched community. CGT is not a subset of any other GT approaches or methodology because it is designed to use any kind of data including qualitative and quantitative data. Any research methodology (including QDA as well as others) is a legitimate research method; but it should never be confused with CGT because CGT is a general research method.

An Overview of the Prior use of the Grounded Theory Research Methodology in Information Systems

It is not surprising that the IS research community have largely misunderstood the differences between CGT and the various remodelled versions of CGT. Firstly, such misunderstanding could be caused by the lack of literature in the IS community on conducting CGT research (Tossy, 2012). Secondly, the CGT method places several difficult demands on researchers that the remodelled CGT research methods do not. CGT researchers have to possess the twin characteristics of a high degree of tolerance for uncertainty and developed skills in conceptualisation. Other GT research approaches let the researcher off the hook by forcing the data instead of allowing it to emerge.

While CGT gives the appearance of being quite easy to conduct, in fact the reverse is true. This is because the CGT researcher always has to be mindful to avoid forcing the data and, instead, using the tenets of GT correctly, should allow the deep seated latent patterns of human behaviour to emerge. Researchers may claim to use CGT but they do not necessarily follow its tenets. A sample of the features of papers in IS that claim to use CGT but that actually have used their own remodelled versions is shown below in Table below.

Glaser (2007:41-42) explains the reasons for the persistence of a level of misunderstanding of the CGTM amongst many academics and researchers: they continue to fail to distinguish clearly between "conceptual generalisations" and "descriptive generalisations": he says,

The constant comparative method (at the heart of CGT) was originated in 1967 to generate concepts by constantly comparing indicators of a latent pattern. After several comparisons the result is conceptually naming the pattern and its properties. The category is abstract of time, place and people. The conceptual categories apply with emergent fit. The Formal CGT research generates categories that relate to each other as generalizations. In contrast descriptive comparisons are as old as research itself. They just compare differences and similarities and a sum of description of people's social actions in a unit of time. The concern is always accuracy, that is, the researcher has worrisome accuracy to constantly

Table 1. Analysis of IS Papers that Claimed to Use CGT but Used Remodelled Versions

Author	Why Method is Not Classic GT	Characteristic of Classic GT
Munro, M.C., Huff, S.L. & Moore, G. (1995)	• Mainly focused on doing hypothesis testing	• Hypothesis testing is not correct application of CGT
Gopal, A. & Prassad, P. (2000)	• Had preconceptions, using the symbolic interactions approach	• CGT is epistemologically free
Munro, M.C., Huff, S.L. & Moore, G. (1988)	• Hypothesis based focus: questionnaires were framed prior to meeting respondents, no emergent core variable	• Emergence of core variable
Zahedi, et al (2006)	• Pre-framed, an interpretive approach, meaning the research was QDA	• Not pre-framed
Ren, et al.(2008)	• Based on hypothesis testing, the utility of the concept "trajectory"	• Testing hypothesis is not correct way of doing CGT
Levina & Vaast (2006)	• Used ethnographic case study	• All is grounded in data
Watts & Henderson (2006)	• Testing hypothesis of innovative IT climates	• CGT never tests hypotheses, creates them
Lundell & Lings (2003)	• Used jargon of CGT but actually used a QDA approach	• In CGT you would never have 2-hour long single interviews with each respondent, and this will distract from creating meaningful theoretical memos
Kesseller (2008)	• Epistemologically constrained by making use of goal/question/ metric model	• CGT never tests hypotheses, creates them

contend with. Generalisations are difficult to make from one unit to another and even within the unit studied. These are the problems of accuracy of external and internal validity respectively of descriptive generalizations. Conceptual generalizations do not have this validity problem. They just apply within a context applied to with modification. For example, controlling clients by pseudo-friending simply varies for client types and client conditions.

Glaser (2009:15) also gives a robust response to the continuing misconceptions regarding CGT as articulated by Bryant and Charmaz (2007:120), in particular the "worrisome accuracy" associated with data analysis and the failure of researchers to appreciate the value of the conceptual level of CGT:

Data worries dominate the GT jargonizing of QDA issues in the Handbook. And why shouldn't it? Worrisome accuracy is the central issue of QDA. QDA research has to yield accurate description, which puts an emphasis on analysing the data used in any research every which way to see what, indeed, is being described and is verifiable. And furthermore how fast will the description get stale-dated and if it can be momentarily discursively generalized. This is a perennial, non-solvable problem. It totally neglects the conceptual level of CGT, which is free of and abstract of place, time and people and hence free of worrisome accuracy. Comparisons generate a CGT's constant concept modifiability.

Transparency of a fully grounded theory will be achieved only when the reader is able to clearly see how the researcher progressed from raw data to the final integrated core variable.

How to Do Classic Grounded Theory Research

Avoidance of Researcher Contamination

Classic Grounded Theory research focuses on the issues of concern to those being researched rather than issues of interest to those in the professional research community. CGT should not concern itself with issues of professional interest for any constituency. To do so would preconceive the outcome of the research.

The purpose of the CGT method is to discover robust but modifiable hypotheses which are grounded in data, not in the professional interest concerns. CGT's other objectives are twofold. Firstly, as mentioned, to discover the main concern(s) of those being researched. Secondly, to reveal how these main concerns are being continually resolved.

The researcher is advised not to research contexts with which they have great familiarity. This is because it is extremely difficult to do CGT research when the researcher's perspective is already contaminated by prior knowledge. Where it is not possible for the CGT researcher to not have any prior knowledge of the research context, a technique called "pre-understanding" needs to be employed. This means that the researcher, before doing any research, writes down all the issues that s/he believes are likely to be of concern to the respondent constituency involved in the CGT research project. The effect of this is to sensitise the CGT researcher both to his own "pre-understanding" and also to new issues which might arise.

Awareness of the Dangers of Using Remodelled GT Research Approaches and Methodologies

The GT research method has been compromised by many researchers due to their misunderstanding of what the method entails. Even in the *International Journal of Information Systems* there is evidence of confusion as to what the GT research method actually is. Walshaw (2006:324) states that, "Orlikowski (1993) describes the use of (C)GT including open and axial coding to generate a set of concepts from data". Yet Walshaw is quoting from a researcher, Orlikowski, who lacks the awareness that axial coding plays no part at all in the CGT method.

The axial coding concept was introduced by Corbin and Strauss (1990), and was not present in the original CGT in 1967. In fact, Glaser (1992) exposed this approach as a means of prematurely forcing the data into patterns dictated by the researcher rather than the data itself. At the heart of the CGT method is the necessity for the researcher to trust in emergence. This means that if the latent pattern is genuine it will be there whether the researcher reveals it or not. By following the tenets of the CGT research method the deep seated latent pattern will be revealed without any need for premature forcing.

Walshaw (2006:325) also incorrectly suggests that the concepts which arise in the GT method are chosen by the GT researcher himself:

In terms of learning from the data itself, grounded theory offers one way of doing this, although the coding is a subjective process, because the researcher chooses the concepts to focus on.

The CGT researcher does not choose the concepts to focus on. Correct use of the CGT method means that the researcher first develops substantive codes from the low level description of the respondents' narratives. Then the CGT method creates theoretical codes, which are a conceptualisation as to how the substantive codes relate to each other as a modelled, interrelated, multivariate hypothesis that accounts for the main concern of those being researched. Forcing of the data by the researcher is not correct application of grounded theory. Instead, the GT researcher has to make use of a combination of constant comparison, analysis and sorting, from which the core variable emerges. This is a mainly inductive rather than a deductive process. The problems defined above reveal just how important it is to disseminate the information in this paper so as to raise awareness about how to use CGT in IS.

Epistemological Stance of Classic Grounded Theory

CGT or the Classic Grounded Theory Methodology (CGTM) is known to be an ontologically- and epistemologically-free research methodology, for instance, by Christiansen (2005:81-84), who says that classic GT is an "ontology free and epistemology free methodology". Glaser (1978) reiterates that ontological assumptions have no place in the use of Classic GT since GT is grounded in data. Such arguments would consider epistemology (the theory of knowledge pertaining to knowing or 'recognizing'), as a activity and therefore as having little place in the application and use of CGT. To elaborate more on "CGT being free from ontological and epistemological assumptions", Glaser (2004:76) argues that "It is only a question of applying a rigorous and systematic method for discovering and explaining these patterns. Thus, just do it." Furthermore, Glaser (1978:45) explains that in the CGT methodology all ontological and epistemological suppositions are irrelevant and are not useful. As Glaser (1978:45) states "... logical conjectures or preconceptions, pre-frame the research, and pre-framing has to be avoided in any CGT work, in order to keep close to what the data has to say, and to discover and name emergent latent patterns in the data."

The CGT approach minimises the influence and potentially distorting effect of preconceptions, logical deductions and elaborations as well as ungrounded assumptions (Glaser, 1978). As a result, a researcher using CGT is free from both ontological & epistemological assumptions. However, if researchers choose to use other methodologies, ontological and epistemological diversity would be inevitable.

The misunderstanding identified by this paper in the use of CGT, is that those papers that were categorized as "GT" were then subdivided into either "interpretative" or "positivistic". Instead of categorizing, it is more important to say that several authors who had claimed to be using the CGT method in fact were either using either the QDA approach (interpretative) or hypothesis testing (positivistic).

Glaser (2004 :2) defines the GT research method thus: "GT is a latent pattern indicator based on index formation. The index formation model in turn is based on accumulation and summing of indicators [originally] from survey data to generate indexes or concepts is fundamental to CGT."

Whatever research methodology the researcher decides to employ it is always preferable for the researcher to stay within the tenets of that methodology instead of trying to introduce elements of other methods in order to validate their research (Glaser, 1978).

Basic Philosophy Underpinning Classic GT Methodology

As mentioned above, CGT was developed by Glaser and Strauss in 1967. At the time they considered this new research methodology to be a necessary challenge and alternative to the ungrounded and unquestioned "great men" theories being presented to students at academic institutions:

Many of our teachers converted departments ... into mere repositories of "great man" theories and taught these theories with a charismatic finality that students could seldom resist. Currently, students are trained to master "great-man" theories and to test them in small ways, but hardly to question the theory as a whole in terms of its position or manner of generation...some theories of our predecessors, because of their lack of grounding in data, do not fit, or do not work, or are not sufficiently understandable to be used and are therefore useless in research, theoretical advance and practical application" (Glaser & Strauss, 1967:10-11).

Glaser developed and expanded his theoretical position of the sixties in 1998, putting the case for a grounded theory which was more reliable than previous theories, was designed to minimise preconceptions and assumptions, and to address "the main concern",

A methodology was needed that could get through and beyond conjecture and preconception to exactly the underlying processes of what is going on so that professionals and laymen alike could intervene with confidence to help resolve the participants' main concern surrounding learning, pain and profit. In this sense, grounded theory has put vested social structures in some jeopardy, because the dependent variables are vital to work with; hence, it is hard to ignore grounded theory results. (Glaser, 1998:5)

Glaser (1998:115) argued that a researcher using (classic) Grounded Theory

... tries to understand the action in a substantive area from the point of view of the actors involved. This understanding revolves around the main concern of the participants whose behaviour continually resolves their main concern. Their continual resolving is the core variable.

The main objective of CGTM is, then, to discover a robust, empirically derived hypothesis or core variable through a specific process. The core variable articulates the main concern of the respondents. It is so called because the core variable is able to explain most of the variations. The particular feature of the CGTM is that it is transcendent of both time and place (Glaser, 1978, 1992, 1998, 2000, 2003, 2008, 2009). According to Glaser, this is made possible due to the fact that CGTM delivers a conceptual explanation rather than a low level descriptive context-based narrative. Thus the emerged GT pay its way by being able to demonstrate that it represents the main concern of those participants who are the subjects of the research.

Glaser (1992:15) outlines and explains the four central characteristics of CGT and emphasises the value to the research process of its "modifiability":

There are only four central criteria with which to evaluate CGT: fit, work, relevance and modifiability. If a grounded theory is carefully induced from the substantive area its categories and their properties will fit the realities under study in the eyes of the subjects, practitioners and researchers in the area. If a grounded theory works it will explain the major variations in behaviour in the area with respect to the processing of the main concerns of the subjects. If it fits and works the grounded theory has achieved relevance. The theory itself should not be written in stone or as a 'pet'; it should be readily modifiable when new data present variation in emergent properties and categories. The theory is neither verified nor thrown out; it is modified to accommodate by integration the new concepts. When these four criteria are met, then of course, the theory provides a conceptual approach to action and changes and accesses into the substantive area researched. (Glaser, 1992:15)

Stages and Procedures of the Grounded Theory Research Method

Although there are very specific processes through which the CGT researcher must go, they do not always occur in a linear format. This kind of flexibility in terms of sequence and process is necessary in order to reveal certain deep seated patterns of human behaviour. The structure of CGT will be outlined in the following section.

1. Generating Theoretical Sensitivity

It is essential that the researcher, prior to embarking on any CGT research, develop theoretical sensitivity. None of the advocates of remodelled GT make this a prerequisite. What this entails are explained by Glaser (1978): since there are already in existence literally hundreds of patterns of human behaviour, having a repertoire of these before embarking the research is absolutely essential. Some of the examples of theoretical codes are: Causal consequence models (cause, consequences, contexts, contingencies, co-variances and conditions); Process models (stages, phases, polarity, progressions, temporal pacing and cycling); Degree models (cutting points, probability, polarity, continuum and intensity); Asymptote models (mathematical models); Isomorphism (mathematical models showing how one established theory can trigger a new theory); Moment capture (business models which explain how many financial services operate)

If armed with a repertoire of theoretical codes, it is easier for the CGT researcher to start making sense of data.

2. Theoretical Sampling

This is a form a sampling in which it is both inappropriate and impossible, prior to doing the research, to state exactly what data will be needed (Glaser, 1978, 1998, 2006). This is because the sampling process can only cease once the core variable has emerged. The researcher samples data until s/he discovers the main concern of the respondents. The researcher then samples within a chosen group and demographic area for integrating into the research, the newly emerged concept. Once the researcher reaches the point where no further patterns emerge, the data is said to be saturated and the core variable to have emerged (Glaser, 1978, 1998).

According to the CGTM, theoretical sampling is a process of data collection for generating theory whereby the researcher simultaneously generates, codes, analyses and synthesizes data and uses this as a basis to decide what data to collect next and where to find them (Glaser, 1978). The researcher begins by selecting a context for the research. A series of short conversations with a small number of respondents is then set in motion. There are two reasons for this. Firstly, a high level of rapport must be established between the researcher and respondent(s) so that trust and respect can develop. This cannot happen easily with single encounters. Secondly, the researcher should not make use of any form of recording or note taking throughout the whole process of conversation/research with the respondent(s) (Glaser, 1978). This both deepens the researcher/respondent trust and increases the level of the researcher's concentration. Immediately following all encounters with respondents, the researcher should document the interaction using theoretical memos.

3. Theoretical Memos

Theoretical memos, as described by Glaser (1978:83), are "the theorising write up of ideas about codes and their relationships as they strike the analyst while coding". They are the means to abstraction and ideation, and can be used continually throughout the CGTM process. Initially a theoretical memo may consist of a short sentence but as the analysis proceeds, they are updated and developed and can be several pages long (Glaser, 1998). Theoretical memos are a crucial part of the GT method for six main reasons. Firstly, it is a device which allows the researcher to use both deductive and inductive logic in order to reveal the deep seated latent patterns of human behaviour which the GT method is designed to reveal. Secondly, theoretical memos encourage the researcher to have a dialogue with his data. This means that the GT researcher can interrogate the data to assess the veracity of emerging theoretical concepts. Thirdly, the theoretical memo enables the GT researcher to move from low level descriptive coding to higher level theoretical coding. The transition from description to conceptual coding cannot happen without theoretical memoing. Fourthly, theoretical memos are very important because they are the main way in which the GT researcher can understand how to carry out the most appropriate theoretical sampling strategy. Theoretical sampling is necessary in the GT research process because it ensures that the researchers look beyond the obvious descriptive narrative and understand what are the main concerns of those being researched. Fifthly, the GT method, for the researcher, is a delayed action phenomenon because it is impossible for any researcher to immediately understand all aspects of his data on a single examination. In other words, when the GT researcher is engaged in the simultaneous writing of several theoretical memos this allows the researcher's own thought processes to discover insights in partially completed theoretical memos that s/he is not currently involved in writing. This is because the mind is highly discursive and is always attempting to make sense of phenomena; but this takes time. Sixthly, when the researcher is engaged in the simultaneous writing of theoretical memos it stimulates the possibility of the cross fertilization of new ideas and concepts. This serendipity stimulates the discovery of new ideas and concepts.

In short, any research which claims to have used the GT research method and does not make full use of the theoretical memo will have failed to reveal the main concerns of those being researched.

4. Substantive Coding

These are the very first attempts at highlighting those data which the analyst believes may have importance for the research beyond the simple description of the context of the data (Glaser, 1978; 1998). These codes are labelled and often 'gerunded' (Glaser, 1978). In other words, the researcher converts the substantive code label into a verb ending in "ing" (e.g., "jargonizing"). The purpose of this is to sensitize the researcher to the processes and patterns which may be revealed at a later stage (Glaser, 1978; 1998:2006). Gerunding is effective because it introduces momentum into the data.

5. Theoretical Coding

This is a conceptual code. It arises from the synthesis of the substantive codes. To move from substantive codes to theoretical codes is a tricky and often elusive process (Glaser, 1978). This synthesis can be effected by examining the interplay between the theoretical memos. This process is known as the constant comparison method (Glaser, 1978). At this point the researcher has to engage both intellect and intuition to achieve a shift in perspective from low level context based description, to higher level conceptual abstraction.

6. The Constant Comparison Method

When using the constant comparison method, the researcher has to painstakingly examine all data in all theoretical memos to look for various types of indicators which might reveal previously hidden connections and patterns (Glaser, 1978; 1998). There is a hierarchy of processes involved in this research method. It starts with the recording of raw data in the theoretical memos and proceeds to substantive coding which is then fractured into categories and sub categories. These in turn have different properties (Glaser, 1998). The term "properties" is not strictly relevant in the context of explaining the constant comparison method. The expression "properties" is more relevant when writing theoretical memos. If the researcher believes that some elements of an emergent pattern might be on the brink of discovery, s/he could return to the data (or perhaps the respondent for another encounter) and try to establish what the properties are of a particular concept. So "properties" is another word for characteristics, and, as such, is not exclusively linked with the constant comparative method. Then, both the categories, as well as the properties, are constantly compared across all the theoretical memos. As this process begins to gather momentum, theoretical codes begin to emerge. What begins as a property of a category of a substantive code sometimes emerges as a theoretical code in its own right. Only through a process of constantly comparing the data for differences and similarities can theoretical codes emerge (Glaser, 1978). The exceptions to this are those rare occasions when the respondent is an articulate person able to conceptualize and is able to step back from personal context and give an authentic explanation of the main concern (Glaser, 1978; 1998).

7. Sorting

This is an iterative process which occurs at the later stages of the CGTM process. Sorting will ensure that the emerging theoretical codes have earned their place in the emergent process of theory creation (Glaser, 1978). If sorting does not take place several times it is unlikely that the research will move much beyond low level narrative description (Glaser, 1978; 1998). The principle aim of sorting is to ensure that the emerging core variable has been fully saturated and has earned its place in the research process (Glaser, 1978).

8. The Emergence of the Core Variable

The core variable is so called because it is able to account for most of the variation in the data (Glaser, 1978). It is the main concern of the respondents, translated and carried onto the conceptual level. There are frequently also sub-core variables which feed into the core variable (Glaser, 1978, 1998).

Respondent Encounters When Using Classic Grounded Theory

CGT allows the collecting of data through oral conversations, face to face, online and many other methods. The most important aspect of the interviews is that they should be of short duration, conducted without tape recordings, and followed immediately by detailed memo writing (Glaser & Strauss, 1967; Glaser, 1978). Several rounds of interviews are expected. For example, it is recommended that the initial round start from the respondent's perspective. Glaser (1978) recommends the following interviewing strategies: (1) no notes or tape recordings are recommended to be used in any of the respondent encounters; (2) no semi-structured interview questions. Instead, the researcher should concentrate in the initial conversations on establishing rapport and making the respondents feel comfortable and unthreatened. The rationale for such strategies is that the respondent encounters should be more like normal conversations than formal interviews; both parties should feel sufficiently relaxed to participate in a candid exchange of ideas. Immediately following each respondent encounter, the researcher should create very detailed notes, which later are converted into theoretical memos.

The CGT research method is known as a "delayed action phenomenon" (Glaser, 1978). This is because it takes some time for both the researcher and the respondent to consider what has been said. For this reason, it is essential not to attempt to collect all the data first and then proceed to attempt to analyse it. It is essential in CGTM that the researcher follows wherever the data leads. The GT researcher has to be able to move from the low level realm of contextual description to the higher level of theoretical explanation.

Prior to the next round of respondent encounters, more theoretical memos should be written. The focuses of the respondent encounters will shift and follow the issues emerging from the most recent iteration of theoretical memos. Although there is no standard template which all theoretical memos must adhere to, there are some basic issues which all theoretical memos must address. If these are not addressed, there will be no transparency in terms of how the researcher progresses from raw data to an emerged core variable. In order to meet transparency requirements, all memos should include: (1) Memo title; (2) Summary of any substantive codes, categories and properties; (3) Conceptual indicators (these are important links that form the bridge from substantive description to conceptual explanation); (4) Emergent theoretical codes, categories and properties; (5) A statement as to the type of data the researcher believes has been given to him or her by the respondent (Baseline, proper lined, vague, zero, or interpreted data). It does not matter when doing CGT what type of data the researcher has, because everything is considered to be data. If, for example, people are untruthful, in encounters, this would probably be more interesting and significant for the researcher than if they were being honest. What is of the utmost importance is that the researcher fully understands what type of data s/he has, rather than worry about accuracy. A research project should have: (1) A clear separation of empirical data from conjectural data. Both are important in CGT but they are very different types of data and need to treated and used accordingly; (2) Information on how the data has been fractured. Which conceptual theoretical perspectives has the researcher used to better understand the latent patterns embedded in the data? (3) Information on what links have been highlighted in other memos from the researcher's bank of memos. (4) Information on what unanswered questions arising from the data the researcher will use in his/her next meeting with the respondent, and (5) Constant comparison of the theoretical concepts in each memo, which should then be sorted. Information on what process has been used for sorting?

Once a critical mass of data had been generated the researcher should use a combination of constant comparison and analytic induction to tease out the main issues of concern as expressed by the respondents. Should two competing main issues or core variables emerge, the researcher should–following the specific guidance given by Glaser (1978:93-100)–re-enter the field and investigate which of the two variables proves to be more 'robust'. This exact situation occurred when Glaser and Strauss were writing *The Discovery of Grounded Theory* (1967). Two variables having emerged, Glaser and Strauss (1967) re-interviewed each respondent several times until latent patterns of behaviour did begin to emerge. Once these patterns emerged the second stage of interviewing could take place with more participants as well as other constituencies identified by the interviewees as being important people to speak with concerning the research agenda. As key concept indicators emerged, theoretical sampling becomes the feature of the second stage of interviewing. The interviewing and memo writing and sorting continues until no newer patterns emerge. When this happens the data is said to be 'saturated'. No more data is required because predictable patterns have begun to emerge.

Nature of the Respondent Encounters

The meetings with respondents should be scheduled before meeting them, to ensure a respondent will be free. Each meeting is introduced with an explanation. It should be made clear to the respondent that the research focus is to reveal, understand and explain what the main issues were which confronted a respondent in the execution of their role. The respondents should be encouraged to let the researcher know if they have felt uncomfortable about any line of enquiry. The absence of any tape recording or note taking during these encounters helps to make the respondents feel more relaxed, so increasing the possibility of transforming the nature of the encounters from an interview to something akin to a normal conversation during which both interviewer and respondent are able to pose questions to each other. All encounters should be kept brief so that the researcher is able to make detailed notes immediately after the conclusion of each meeting. The purpose of the series of meetings is to discover the respondents' main concerns regarding their professional workings, including the challenges, how they solve problems and what they consider to be achievements and successes in the course of their daily working lives. It is considered vital to schedule several meetings with each respondent as this will give each respondent the time and space to reflect before subsequent encounters, and it enables the researcher to encourage a respondent to think more deeply about what their real concerns are. Following this incremental process of having several interviews with each respondent ensures that it is the respondent's agenda that emerges rather than that of the researcher.

Choice of Interview's Location When Doing Classic Grounded Theory

The choice of locations is an important consideration. The idea is to try and ensure that the respondents feel relaxed. Thus a variety of locations other than their own offices is highly recommended when doing CGT. These include restaurants, university canteens, meeting rooms, public bars, walking in the street and driving in the car.

Types of Data When Doing Classic Grounded Theory

As part of the process of the grounded theory method, the researcher has immediately to be aware of the types of data which emerge from conversations with respondents. From a GT perspective there are five types of possible data to be derived in any encounter with respondents:

1. **Baseline Data:** The respondent is being as truthful as possible, as far as the interviewer can gauge.
2. **Interpreted Data:** The respondent is filtering his/her message, either to deliberately complicate or simplify her/his responses.
3. **Proper Lined Data:** The respondent deliberately edits the data so that it is not in conflict with the "official" line of argument from her/his employing authority.

4. **Vague Data:** The respondent deliberately gives an indistinct version of reality to try and confuse the researcher, obfuscate the issues, and throw the researcher 'off the scent'.
5. **Zero Data:** The respondent refuses to say anything at all in order to avoid saying anything which s/he might later regret.

From a GT perspective, it does not matter which type or types of data the respondent delivers because each type of data has the potential to contribute something of interest and value to the research.

Awareness of Ethical Issues

The GT research process will not be effective unless there is a very high level of rapprochement between the two parties. The researcher has to establish credibility by offering the respondent anonymity. This is made easier when using the GT research method because none of the respondent encounters are tape recorded. Note taking should also be avoided during the research process. This ensures that the encounters take the form of conversations rather than typical interviews that a researcher might when adopting other types of research methods. When the respondents are relaxed and comfortable they will be more likely to reveal issues which are of genuine concern.

Data Analysis

The GTRM is a general research method which treats everything as data, whether quantitative or qualitative Glaser (1992). GT primarily uses an inductive research approach in which the researcher has to be subordinate to the data. GT is not a sub-set of QDA–which values accuracy above all within a narrow research context; GT, on the contrary, is mainly concerned with the discovery of latent patterns of human behavior which are transcendent of the original context of the data. Glaser (1978, 1992) outlines the CGT Research process thus:

1. Theoretical sampling and substantive coding
2. Theoretical coding
3. Theoretical memo writing
4. Constant comparison between and within memos
5. Sorting codes
6. Data saturation
7. Emergence of the core variable.

Although the above are represented in a linear format, in practice the exact sequence varies according to what patterns seems to be emerging from conversations with respondents. The reason for this is that CGT is a process, not a unit based approach. The data is simultaneously, subsequently, and sequentially collected, analysed and synthesised. Once most of the variations in the data can be explained in terms of a key concept, the data is said to be saturated and there is therefore is no further need to collect more data since the core variable could be said to have emerged.

Coding Data for Emergent Fit

CGT is a latent pattern identifier. It is not evidence based but instead these patterns are revealed by the discovery of the main concerns of those being researched. These main concerns are continually being resolved by the respondents whether they have been revealed by research or not. The main concerns are not the respondents' narratives but rather their motivations behind their narratives. Because of this, the use of the CGT method means that emergent theory is transcendent of both time and place. The theoretical explanations of human behavior which CGT reveals will also be applicable across a wide range of different contexts. The product of any CGT research project is a robust hypothesis which can be modified as new data comes to light. As a result, CGTM has two basic models for studies:

(1) The first model is by pure discovery of all the concepts by going through the prescribed procedures; and (2) The second model is "starting from scratch." (Glaser, 1978)

The first model is the emergent fit model and applies the same procedures as the second model, but does not "start from scratch". It proceeds from a previously generated CGT and its substantive concepts. The second model is especially suited to an advanced user of a generated CGT who would want to scrutinise the extent to which a theory, which has been generated from a given set of data from a given substantive area, fits a researcher's data from the same or another substantive area.

Criteria for Evaluating Classic GT

The CGT is a general inductive methodology which can be applied to all types of data, both quantitative and qualitative, or the combination thereof as has been outlined by Glaser (1978, 1998). Glaser (1978, 1998, 2000, 2001) gives explicit examples of the four main criteria with which to evaluate grounded theory study: fit, workability, relevance and modifiability. These criteria are explored below:

- **Fit:** If a concept can be said to be a 'fit' it means that it adequately reflects the data that it purports to express. The categories of the emerged core variable must fit the data used to create the theory. Ungrounded assumptions and data which have been forced to fit into preconceived concepts derail fit in CGT, as well as invalidating the theory (Glaser, 1978). Although fit would appear to correspond to positivistic validity, it is fit in terms of CGT action and usage, not as a result of testing (Glaser, 1998). Unlike research based on preconceptions, such as research conducted in order to test a hypothesis, in CGT data which cannot be forced are discarded rather than used to correct the emerging categories. With CGT only data 'sorted' and used for developing theory can be said to constitute a fit (Glaser, 1978, 1998). The fit could be expressed as a 'refit' or an 'emergent fit'. During research categories emerge fast, so the need arises to refit them to the data as the research proceeds and to be sure that they fit all of the data they purport to indicate, thus increasing and ensuring transparency. Categories can emerge between data and pre-existent categories (Glaser, 1978).
- **Workability:** How the core variable accounts for the respondents' continual resolution of their main concern is a key issue. The emergent CGT must clearly explain what is happening, and the process of its happening, and by so doing should be able to predict future behavior (Glaser, 1978, 1998, 2001).

- **Relevance:** Also pertinent is how the emerged core variable has been received by the members of the constituency from which the data was drawn. Prior to the research it is very likely that respondents would be unable to articulate their main concern. Once they have read the research they should instantly be able to recognize the emerged core variable as being authentic (Glaser, 1998).
- **Modifiability:** The CGT is considered to be modifiable if it is easy for subsequent CGTM researchers to be aware of what research has been done so far in the area, and to proceed to modify or refine the theory as they collect and code new data (e.g., code new data for emergent fit), without invalidating the theory.

Since CGTM is an inductive research methodology it is appropriate to use a legitimate set of evaluation criteria as outlined by the CGT's co-originator Glaser (1978).

A Set of Principles for Assessing a Paper/Thesis Claim to use Classic Grounded Theory Research

Although Glaser (1978:1-2) specified the criteria for evaluating classic GT, there is persistent confusion among many researchers caused by the widespread use of remodelled versions of the CGT method. This raises the need for reiterating a set of principles for assessing a paper /thesis claim to use classic GT research. In this paper we propose a set of principles for the evaluation of an article/thesis claim to use CGT research in information systems and beyond. The source of these principles is Glaser (1978).

Like Myers and Klein (2011), citing Klein and Myers (1999), I use the word "principles" to emphasize that the general principles we propose are offered in the spirit of being fundamental ideas that may be helpful to CGT researchers in IS. The principles are culled from the philosophical writings of Glaser and Strauss (1967) and Glaser (1978, 1992, 2001, 2004, 2005, 2008, 2011). As noted by Myers and Klein (2011:24):

the principles should be helpful because they summarize important insights which as a whole are (as yet) not embedded in the current practices of [Classic Grounded theory], but could improve them if observed more systematically.

The Principle of the Emergence of Core Variable

This is the goal for doing classic grounded theory. As Glaser (1978:94) argues, the analyst should consciously look for a core variable when coding his data. In other words, while doing all you can do (comparing data, etc.), be alert to the one or two that are core. This means that a researcher looks for the main theme: for what, in their view, is the main concern or problem for the people in the setting; for what sums up in a pattern of behavior the substance of the what is going on in the data; for what is the essence of relevance reflected in the data; for gerunds which brings out process and change (Glaser, 1978:94). When the researcher starts coding, categories tend to emerge, giving way to finding core categories, and the researcher should be alert to these core variable.

Glaser (1978: 95) sums up the criteria by which a researcher can make judgements as to the core category:

Table 2. Summary of Principles for Assessing Paper/Thesis Claim to Use Classic GT Research

The fundamental principle of the emergence of core variable This principle suggests that the authentic GT outcome is a Core Variable which accounts for most of the variations in the data.
The principle of stages and processes in the discovery of core variable This principle insist on the necessity to follow all stages and processes in the discovery of the core variable, including developing theoretical sensitivity, identifying your substantive area, collecting data, open coding your data as you collect it, writing memos throughout the entire process, conducting selective coding and theoretical sampling, and sorting your memos and finding the theoretical code(s) which best organise your substantive codes-- the core variable will emerge.
The principle of core variable explains how the respondent's main concerns are continually resolved This principle means that categories must fit the data relevant to the respondent. Data should not be forced or selected to fit preconceived or pre-existent categories or discarded in favour of keeping an exact theory intact (Glaser, 1978:4)
The principle of research design involves the discovery of a hypothesis and does not test one GT exist for the discovery of hypotheses not to test them. Hypotheses should not be forced or selected, rather should be allowed to emerge from the data.
The principle of core variable is transcendent of the context of the data The theory should be able to explain what happened, predict what will happen, and interpret what is happening in an area of substantive or formal inquiry (Glaser, 1978:4)
The principle of emerged hypothesis allows modifiability as new data emerges in future research This principle allows for ready, quick modification to help explain surprising or new variations–such as developing new conditions–almost as new data appears. It is through this quick approach that the tractability of GT over social life is maintained, hence its relevance (Glaser, 1978:5)
The principle of researcher's theoretical sensitivity facilitates fully transparent transition from substantive to theoretical codes The principle allows the researcher to build necessary theoretical sensitivity by which they can render theoretically their discovered substantive, grounded categories.

- It is saturated: the core variable takes more time to saturate than other variables because it occurs frequently in the data
- It relates meaningfully and easily with other categories–such connections need not be forced
- A core category in a substantive study has clear and grabbing implication for formal theory
- It has carry-through
- It is a dimension of the problem
- It is completely variable
- It is grounded in the data

The principle of emergence of the core variable suggests that the authentic GT outcome is a core variable which accounts for most of the data. This forms the basis of the theory to come. This principle is fundamental to the other remaining principles.

The Principle of Core Variable Accountability for Most of the Variation in the Data

Glaser (1978) argues that CGT is grounded in data. Using the principle of emergence of the core variable, it then should occur frequently in the data. As it occurs frequently, it becomes accountable for most of the variation in the data. Its frequent occurrence will come to be seen as a stable pattern and will become increasingly to other variables. Glaser (1978:95) argues that if it does not recur frequently, this does not mean the category is uninteresting. It may be quite interesting in its own right, but it is not core.

The Principle of Stages and Processes in the Discovery of the Core Variable

The principles insist on following all stages and processes in the discovery of the core variable. These include the following: developing theoretical sensitivity, identifying the substantive area, and collecting data. Open coding should take please as data is collected, memos written throughout the entire process, conducting of selective coding and theoretical sampling, and sorting of memos and finding the theoretical code(s) which best organises your substantive codes-- the core variable will emerge.

The Principle of the Core Variable Explains How the Respondent's Main Concerns Are Continually Resolved

This principle means that categories must fit the data relevant to the respondent. Data should not be forced or selected to fit preconceived or pre-existent categories or discarded in favour of keeping an exact theory intact (Glaser, 1978:4). The core variable must be central, that is, related to as many of the other categories and their properties as possible and related more than are other candidates. This criteria of centrality is a necessary condition to make it core.

The Principle of Research Design Involves the Discovery of a Hypothesis and Does Not Test One

This principle explains how discovery of the core variable takes place. GT exists in order to discover hypotheses not to test them. Hypotheses should not be forced or selected: they emerged from data, data, and only data.

The Principle of the Core Variable Is Transcendent of the Context of the Data

The theory should be able to explain what happened, predict what will happen, and interpret what is happening in an area of substantive or formal inquiry (Glaser, 1978:4).

The Principle of an Emerged Hypothesis Ensures Modifiability as New Data Emerges in Future Research

This principle allows for ready, quick modification to help explain surprising or new variations, such as the development of new conditions, almost as new data appears. It is through this quick approach that the tractability of GT over social life is maintained, hence its relevance (Glaser, 1978:5).

The Principle of the Researcher's Theoretical Sensitivity Insures Full Transparent Transition from Substantive to Theoretical Codes

This principle allows the researcher to build necessary theoretical sensitivity by which they can render theoretically their discovered substantive, grounded categories.

The Principle of the Process of Both Sorting and Constant Comparison

The principle enables a researcher to discover a rich core variable. Firstly, analysis starts by comparing incident to incident. Secondly, research compares concept to more incidents. Lastly, while the first and second types of comparison continue throughout the study, the analyst also compares, thirdly, concept to concept (Glaser 1978:49-54).

Table 3. Operationalization of the outlined set of principles- overall assessment of the five articles claim to use Classic Grounded Theory

	Scott (2007)	**Orlikowski(1993)**	**Huff & Munro(1985)**	**Munro, Huff & Moore(1998)**	**Lundell &Lings(2003)**	**Adolph, Kruchten & Hall (2012)**
1. The principle of the emergence of core Variable (CV)	Temporal Integration	None Emerged	None Emerged	None Emerged	None Emerged	Reconciling perspectives
2. The principle of CV accountability for most of data	✓	x	x	x	x	✓
3. The principle of stages and processes in the discovery of CV	✓	No explanation to how they were revealed	No explanation to how they were revealed	x	x	✓
4. The principle of CV explains how the respondent's main concerns are continually resolved	✓	x	x	x	x	✓
5. The principle of research design involves the discovery of CV	✓	Hypothesis testing/QDA	Hypothesis testing/QDA	Hypothesis testing/QDA	Hypothesis testing/QDA	✓
6. The principle of core variable is transcendent of the context of data	✓	None, because no core variable	None, because no core variable	None, because no core variable	None, because no core variable	✓
7. The principle of emerged hypothesis is modifiability as new data emerges in future research	✓	No new concise thypothesis, therefore it is not modifiable	No new concise thypothesis, therefore it is not modifiable	No new concise thypothesis, therefore it is not modifiable	No new concise thypothesis, therefore it is not modifiable	✓
8. The principle of researcher's demonstrate full transparent transition from substantive to theoretical codes	✓	Opaque	✓	✓	✓	✓
9. The principle of process of both sorting and constant comparison	✓	No constant comparison stated	x	x	X	✓
OBSERVATIONS	**USED CGT**	**DID NOT USE CGT**	**DID NOT USE CGT**	**DID NOT USE CGT**	**DID NOT USE CGT**	**USED CGT**

CONCLUSION

This paper has set out to explain how a classic grounded theory research study can be carried out in an information systems context so as to facilitate good practice of the theory as well as to help prevent the misuse of classic grounded theory by the multitude of remodelled GT methods. The paper has responded to a call made by Matavire and Brown (2011) to build better understanding of the process and outcomes of following a CGT approach. We have detailed what an IS researcher should be aware of when using classic grounded theory: The researcher should be aware of the remodelled versions of the CGT method; the epistemological stance of CGT; the basic philosophy underpinning CGT; the stages and procedures of CGT; how to conduct interviews with respondents, including the nature of the meetings, choice of location, types of data collected, ethical issues, and special characteristics of CGT. Furthermore, an IS researcher should maintain a high level of transparency when using CGT. Finally, the researcher should be aware of the appropriate criteria for evaluating CGT as well as how to write papers that use the CGT method and that focus on previously published research in the IS field.

ACKNOWLEDGMENT

We thank Dr. Barney Glaser, the co-originator of classic grounded theory methodology, from whom we have learned much about conducting classic grounded theory research. We thank NRF for funding this research. Finally, we also thank various classic GT fellows in various countries around the globe for the promotion of the use of classic GT.

REFERENCES

Aldoph, S., Kruchten, P., & Hall, W. (2012). Reconciling perspectives: A grounded theory of how people manage the process of software development. *Journal of Systems and Software*, *85*(6), 1269–1286. doi:10.1016/j.jss.2012.01.059

Banville, C., & Landry, M. (1989). Can the field of MIS be disciplined? *Communications of the ACM*, *32*(1), 48–60. doi:10.1145/63238.63241

Baskerville, R., & Pries-Heje, J. (1999). Grounded action research: A method for understanding IT in practice, *Accounting. Management and Information Technology*, *9*(1), 1–23. doi:10.1016/S0959-8022(98)00017-4

Bryant, A. (2002). Re-grounding grounded theory. *Journal of Information Technology Theory and Application*, *4*(1), 25–42.

Chen, W., & Hirschheim, R. (2004). A paradigmatic and methodological examination of information systems research from 1991 to 2001. *Information Systems Journal*, *14*(3), 197–235. doi:10.1111/j.1365-2575.2004.00173.x

Glaser, B. (1964). *Organizational Scientists: Their Professional Careers*. Chicago: Aldine.

Glaser, B. (1965). The Constant Comparative Method of Qualitative Analysis. *Social Problems*, *12*(4), 436–445. doi:10.2307/798843

Glaser, B. (1978). *Theoretical Sensitivity: Advances in the Methodology of Grounded Theory*. Mill Valley, CA: Sociology Press.

Glaser, B. (1992). *Emergence vs Forcing: Basics of Grounded Theory Analysis*. Mill Valley, CA: Sociology Press.

Glaser, B. (1993). *Examples of Grounded Theory: A Reader*. Mill Valley, CA: Sociology Press.

Glaser, B. (Ed.). (1995a). *Grounded Theory 1984-1994, Volume One*. Mill Valley, CA: Sociology Press.

Glaser, B. (Ed.). (1995b). *Grounded Theory 1984-1994, Volume Two*. Mill Valley, CA: Sociology Press.

Glaser, B. (1996). *Gerund Grounded Theory: The Basic Social Process Dissertation*. Mill Valley, CA: Sociology Press.

Glaser, B. (1998). *Doing Grounded Theory: Issues and Discussions*. Mill Valley, CA: Sociology Press.

Glaser, B. (1999a). *The Discovery Of Grounded Theory: Strategies for Qualitative Research*. Mill Valley, CA: Sociology Press.

Glaser, B. (1999b). The Future of Grounded Theory: Qualitative Health Research. Oxford, UK: Sage Publications.

Glaser, B. (1999c). The Future of Grounded Theory, Keynote address at the 4th annual Qualitative Health Research Conference. *The Grounded Theory Review*, *1*(1), 1–9.

Glaser, B. (2001). *The Grounded Theory Perspective: Conceptualization Contrasted with Description*. Mill Valley, CA: Sociology Press.

Glaser, B. (2003). *The Grounded Theory Perspective II: Description's Remodeling of Grounded Theory*. Mill Valley, CA: Sociology Press.

Glaser, B. (2004). *Glaser's explanations on ontology and epistemology*. Academic Press.

Glaser, B. (2004). *Staying open*. Unpublished paper. Available at www.groundedtheoryinstitute.com/docs/staying_open.html

Glaser, B.. & Holton, J. (2005). Basic Social Processes. *The Grounded Theory Review*, *4*(3), 1–21.

Glaser, B. (2005a). *The Grounded Theory Perspective III: Theoretical Coding*. Mill Valley, CA: Sociology Press.

Glaser, B. (2005b). *The Roots of Grounded Theory*. Keynote presentation delivered to the 3rd International Qualitative Research Convention, Johor Bahru, Malaysia.

Glaser, B. (2006). *Doing Formal Grounded Theory: A Proposal*. Mill Valley, CA: Sociology Press.

Glaser, B. (2008). *Doing Quantitative Grounded Theory*. Mill Valley, CA: Sociology Press.

Glaser, B. (2011). *Getting Out of the Data: Grounded Theory Conceptualization*. Mill Valley, CA: Sociology Press.

Glaser, B., & Kaplan, D. W. (Eds.). (1996). *Gerund Grounded Theory: The Basic Social Process Dissertation*. Mill Valley, CA: Sociology Press.

Glaser, B., & Strauss, A. (1965). *Awareness of Dying*. Chicago: Aldine.

Glaser, B., & Strauss, A. (1967). *The Discovery of Grounded Theory: Strategies for Qualitative Research*. Hawthorne, NY: Aldine.

Glaser, B., & Holton, J. (2004). Remodeling Grounded Theory. *Forum: Qualitative Social Research, 5*(2), Article 4. Retrieved from http:// www.qualitative-research.net/fqstexte/2-04/2-04glaser-e.htm

Glaser, B. (2009). *Jargonizing: Using the Grounded Theory Vocabulary*. Mill Valley, CA: Sociology Press.

Gopal, A., & Prassad, P. (2000). Understanding GDSS in Symbolic Context: Shifting the Focus from Technology to Interaction. *Management Information Systems Quarterly, 24*(3), 509–546. doi:10.2307/3250972

Gregor, S. (2006). The Nature of Theory in Information Systems. *Management Information Systems Quarterly, 30*(3), 611–642.

Klein, H. K., & Myers, M. D. (2009). A Set Of Principles For Conducting And Evaluating Interpretive Field Studies In Information Systems. *Management Information Systems Quarterly, 23*(1), 67–94. doi:10.2307/249410

Levina, N., & Vaast, E. (2006). Turning a Community into a Market: A Practice Perspective on Information Technology Use in Boundary Spanning. *Journal of Management Information Systems, 22*(4), 13–37. doi:10.2753/MIS0742-1222220402

Lyytinen, K., & King, J. (2004). Nothing at the Center? Information Systems as a Reference Discipline? *Journal of the Association for Information Systems, 5*(6), 220–246.

Matavire, R., & Brown, I. (2008). Investigating the Use of 'Grounded Theory' in Information Systems Research. *Proceedings of the SAICSIT 2008 Conference*.

Matavire, R., & Brown, I. (2011, August). Profiling Grounded Theory Approaches in Information Systems Research. *European Journal of Information Systems, 30*. doi:10.1057/ejis.2011.35

Munro, M. C., Huff, S. L., & Moore, G. (1985). Information Technology Assessment and Adoption: A Field Study. *Management Information Systems Quarterly, 9*(4), 327–340. doi:10.2307/249233

Munro, M. C., Huff, S. L., & Moore, G. (1988). Expansion and Control of End-User Computing. *Journal of Management Information Systems, 4*(3), 5–27. doi:10.1080/07421222.1987.11517798

Myers, M. D., & Klein, H. K. (2011). A Set Of Principles For Conducting Critical Research In Information Systems. *Management Information Systems Quarterly, 35*(1), 17–36.

Ren, Y., Kiesler, S., & Fussell, S. R. (2008). Multiple Group Coordination in Complex and Dynamic Task Environments: Interruptions, Coping Mechanisms, and Technology Recommendations. *Journal of Management Information Systems, 25*(1), 105–130. doi:10.2753/MIS0742-1222250105

Truex, D., & Holmstrom, J., & Keil, M. (2006). Theorising Information Systems Research: A Reflective Analysis of the Adaptation of Theory in Information Systems Research. *Journal of the Association for Information Systems*, *7*(12), 797–821.

Watts, S., & Henderson, J. C. (2006). Innovative IT Climates: CIO Perspectives. *The Journal of Strategic Information Systems*, *15*(2), 125–151. doi:10.1016/j.jsis.2005.08.001

Zahedi, F. M., Van Pelt, W. V., & Srite, M. (2006). Web Documents' Cultural Masculinity and Femininity. *Journal of Management Information Systems*, *23*(1), 87–128. doi:10.2753/MIS0742-1222230103

Chapter 12
Information and Communication Technologies (ICTs) for Industrial Development:
Challenges and Opportunities

Felichesmi Selestine Lyakurwa
Mzumbe University, Tanzania

Joseph Sungau
Mzumbe University, Tanzania

ABSTRACT

Applicability of ICT has gained high research interests in both developed and developing countries due to the perceived social and economic benefits. With increased benefits of ICTs, many governments have invested heavily in the emerged technologies to gain competitive advantange over others in the business. Despite the high investment by many countries around the world, yet there are inadequate documentation about the cost - benefits of the massive ICT investment in developing countries especially Africa. Hence, developing an empirical study to reveal the perceived cost - benefits of ICT investment for the industrial development is critical. This chapter, presents a comprehensive review of various studies conducted in different countries of the world detailing the extent of ICTs, barriers hindering its use and the potential benefits. To date, there is existing empirical evidence to support the role of ICTs for industrial development in the developing countries, while the benefits of ICTs for industrial development in developing countries of Asia and Africa are not well documented.

DOI: 10.4018/978-1-5225-0539-6.ch012

INTRODUCTION

The adoption of the ICT facilities has attracted high research interests in many countries of the world due to its role in the support of the government development goals. The widespread use of ICTs by various production industries has created more chance for the constant changes in their operation and management models such that the industrial structures are converging in terms of the new business processes and organization structures. Besides, the ICTs ensures availability of valuable information to the people all over the world, including information about the improved farming techniques, markets for raw materials, prices of agricultural inputs, greener production techniques, markets for the industrial goods and products, access to information about health, education, government services to mention a few. However, access and use of ICT facilities might be hindered by inadequate telecommunications, electrical infrastructures and high cost of ICT appliances, e.g., computers, internet connectivity and little awareness on the benefits of ICT use for the development.

Tanzania as other developing countries strives to explore the benefits of ICT adoption in the development of the manufacturing and service industries. However, there are no empirical studies conducted to quantify cost - benefits of the ICT investment in the country. Nevertheless, considering the benefits of ICTs use, and mixed reactions on the benefits of ICTs for development, establishing empirical evidence is critical for future research. Hence, this chapter presents a comprehensive review of the literatures in relation to the adoption and use ICTs in both developed and developing countries and different questions to probe future research on ICT investment and their economic benefits. The challenges and opportunities of ICT use for industrial development, especially in developing countries are addressed, and the gaps of the ICT adoption between manufacturing and service industries are also covered that stimulates further research on ICTs and industrial development.

BACKGROUND

Information technology is critical towards the development of both manufacturing and service industries. In efforts to ensure availability of Information and Communication Technology (ICT), many developed and developing countries around the world have been invested and still investing heavily in ICT. The investments have been targeting at improving organizations efficiencies of their offerings. The countries have been investing in both human (staff) and physical resourses (machines and equipments). Not only that, but also, the countries have been investing in technological transfer issues to empower their countries to adopt available proven technology (Bankole et al., 2011).

Many scholars from different regions of the world, have defined technology in different ways and their definitions vary in terms of scope and perspectives but, yet they contain similar message. For instance, Scientists define technology as the branch of knowledge that deals with creation and use of technical means and their interrelation with life, society and the environment, drawing upon such subjects as industrial arts, engineering, applied science and pure science. Under the economists perspectives, technology is defined as the purposeful application of information in the design, production and utilization of goods and services and in the organization of human activities. And then, sociologists defined technology as the product or the outcome of human activity. However, in broader means, technology can be defined as the technical means through which people use to improve their surroundings and also the knowledge of using tools, machines and various scientific approaches to do tasks efficiently. Considering the definitions of technology given by different scholars, you can see the definitions are centered on applying different tools and techniques so as to simply work and achieve high productivity in the organization.

Technology has many importances to the society be it improved livelihood, safety at workplaces, timely access to information about health, education, markets and so on. Many countries have been investing in technologies in order to boost their economic growth. Technology has been diffusing from developed countries to less developed countries. Developing countries need to adopt the developed technologies. In this regard technology is very important to societies which need development. In oder to adopt the appropriate technology, the adopting organizations need to ask herself the following questions; how easy does the industrial technology diffused in both developed and developing countries? What obstacles hindered exploitation of ICT by industries and how these obstacles were resolved?

In the current world, many industries are investing in electronic (e) - business that provided easy access to markets, new product development and emerged technologies. E-business is the application of information and communication technology (ICT) in supporting of all organizational activities. While ICT is a term used to include all communication devices such as radio, television, cellular phones, computer and network hardware, satellite systems and others of a like. ICT is basically the use of technology to facilitate some processes or business processes. However, one may fail to draw a clear line to differentiate between technology and ICT since all the two words can be used interchangeably. Normally e-business is starting a change process in which industries should think about the operation models, organizational processes, and best means to meet and serve customers' needs and wants. In other words, E-business is a conduct of business processes on the internet. These electronic business processes include buying and selling products, supplies and services; servicing customers; processing payments; managing production control; collaborating with business partners; sharing information; running automated employee services; recruiting; and more.

E-business comprises a range of functions and services, ranging from the development of intranets and extranets to e-service, the provision of services and tasks over the internet by application service providers. Today, as major corporations continuously rethink their businesses in terms of the internet, specifically its availability, wide reach and ever-changing capabilities, they are conducting e-business to buy parts and supplies from other companies, collaborate on sales promotions, and conduct joint research. With the security built into today's browsers, and with digital certificates now available for individuals and companies from different countries, a certificate issuer must also give much of the early concerns about the security of business transaction on the websites and e-business by whatever means as accelerating. The e-business is not the same as e- commerce, although they might be used interchangeably. The "e", for both words is a short form for electronic. E-business and e-commerce is the application of electronic networks for organizational business activities. However, the e-commerce is a subset of e-business. In differentiating the two terms, e-commerce is any transaction that was made through an electronic system (or networked computers) such as transfer of ownership and right to use. In this regard, e-commerce includes all transactions made via internet, intranet, extranet, world wide web (WWW), by email and even by fax.

On the other hand, e-business includes transactions based on e-commerce and those which run in a traditional manner but caters for the online activities as well. An e-business runs internal processes online, including inventory management, risk management, finance, human resources. For a business to be e-commerce and e-business, it must both sell products online and handle other company activities or additional sales offline. The adoption and use of ICTs are substantial for the development of any business industry. The use of ICT improves organizational performance in both financial and non financial performance terms. ICT facilitates organizations to increase productivity and production efficiency, reduce maintenance time and cost, as well as optimization of materials used in the production systems (United Nations, 2011). With the emerged technologies, industries are in a better position for exploitation of the capacity of existing plants, diversification and quality improvement of products and substitution of locally available raw materials and machine parts with imported ones (Gholami et al., 2004). Despite the benefits of new ICT/technologies towards development of different manufacturing and service industries, the understanding of the extent to which industries adopt and are managed to exploit these available proven technologies is very necessary.

Over the past several decades, the extent to which industries utilize available technologies, including ICTs in both developed and developing countries of the world is relatively low (Selwood, 2004; Lyakurwa & Mangula, 2013). However, the reason for low extent might be due to managerial, technical, financial or resource problems, the dilemma that needs to be addressed through comprehensive literature reviews, case studies or experimental measurements. For the countries whose economies are low, especially developing countries of Africa, the benefits of ICT could further be exploited for industrial development through technology transfers (Fukasaku, 2005). The technology transfer meant here are imported experts, automated machines and equipments that need high costs for investments and payment of the experts. The capability of the industries in adopting technologies includes the ability to utilize the imported technologies effectively. However, in developing countries, this capability is questionable based on the level of developments in terms of infrastructures in ICTs, access to internet, expertise availability and financial position.

Based on the neoclassical formulation, technological change can be described as the shift in production function that affect the whole range of production techniques such that only major innovative points in the Schumpeterian sense are considered to have potential economic values (Gholami et al., 2004). Thinking about the kind of industries we have in developing countries of Africa including Tanzania with a view of technological change, all shall agree that, we are far behind in terms of utilizing ICTs for the development of our industries despite the great efforts made by our Government on ICT investments. One of the indicators is that most manufacturing industries are not operating, whereby the production of the country has been turned into a service industry, which cannot even be compared to service industries in the middle income countries of the world. So one of the hypotheses is that does ICT have a significant role in the development of industries in the least income countries including Tanzania? If yes, considering service and manufacturing industries, which best exploits the available ICTs. Is appropriateness of technology or mechanism of technology transfers; constitute one of the obstacles hindering the utilization of ICTs for industrial development? How does utilization of the ICT facilities for industrial development compare between developed and developing countries of the world?

The change in technology is usually generated within the productive systems. By using available technologies, productivity gains of the industry can increase without significant investments in terms of working capital, recruitment of new staffs or increase of working hours. This phenomenon was best explained by the Arrow's model that industrial productivity rises without significant economic investments because learning takes place as a byproduct of accumulating experiences in the production process (Arrows, 1969). Many studies of manufacturing and service industries revealed potential increase in productivity with fixed resources (manpower, money, machines and equipment) that indicate presence of accumulated experience, i.e., learning by doing approach (Alchian, 1963). The active and conscious forms of learning were considered to be responsible for the minor increase in the changes of technology in different industries on the technology transfer. For example, case studies of Argentine rayon and steel plant (United Nations, 1978) and a Brazilian steel mill (Dahlman & Fonseca, 1978) indicated that research and development (R&D) investments by these industries brought about significant technological changes which led to better exploitation of the capacity of manufacturing and service industries, diversification and quality improvement of manufactured products and serviced customers.

The ICT investment in the educational management is also necessary for the knowledge accumulation and skills to substitute the foreign technical assistance in the coming years (Selwood, 2004). Many countries of the world have invested heavily on information technology to increase efficiency of manufacturing processes, training of experts, and service provision to customers. To date, there is adequate evidence to support the role played by technologies towards improvements of teaching (teachers) - learning (students) environments. Most importantly include increased teaching standards and students' performance in primary and secondary schools, and higher learning institutions (e.g., Universities). Availability of ICTs to the training schools and institutions provide various benefits including effective leadership and management, provide teachers focus on their core tasks (e.g., teaching, research and consultancy), and provides effective collaboration between schools/Universities and their colleagues within and abroad, and greater learning chances for the students, teachers, their families and even the community (Selwood et al., 2001).

Usually all education sectors with clear ICT policies, covering issues related to the staff access to ICT resources, staff ICT training and encouragement of staffs ICT confidence and competence, has a better chance to develop in terms of students' performance, staff's competence and other productivity gains. Considering the governments need to quantify return on ICT investments in every education sector, such as schools, several questions should be addressed to collect qualitative evidence of ICT effectiveness in schools and higher learning institutions. Among the hypotheses include: (1) if there is significant access to ICT resources among school and University staffs; (2) there is significant ability level of ICT ability among staffs; (3) there is significant use of ICT resources among staffs; (4) there is a significant link between ICTs access, ability and use among the staffs in schools and Universities. The answers to these questions will provide useful inputs to the country policy making and planners that are necessary for the industrial development.

ADOPTION OF ICTS FOR THE INDUSTRIAL DEVELOPMENT

The adoption and use of ICT is the changing business process and the way in which the community live and works such that new innovations continue to emerge. The ICTs includes devices used for information processing, communications, display and the network services that transmit communication services and content through new and traditional media such as radios, television, fixed and mobile telephones, desktop computers and laptops. To date, there is adequate evidence that ICTs has contributed greatly towards economic growth of the developed countries, while the role of ICT industry towards development in developing countries is unclear. This might be caused by the fact that, developing countries are short of capital investment and knowledge know how such that developing countries lag far behind the industrialized nations in their ICT industry development and diffusion. Usually applicability of ICTs is constrained by the degree to which people have access to electricity be it mobile phones or battery powered devices that need to be charged. For example, about 80.5% of the global population have access to electricity, compared to less than 75% of the population in developing countries which have access to electricity. Besides, the distribution of electricity in rural versus urban areas of developing countries,

Figure 1. Mobile Cellular in 2013 (ITU, 2013)

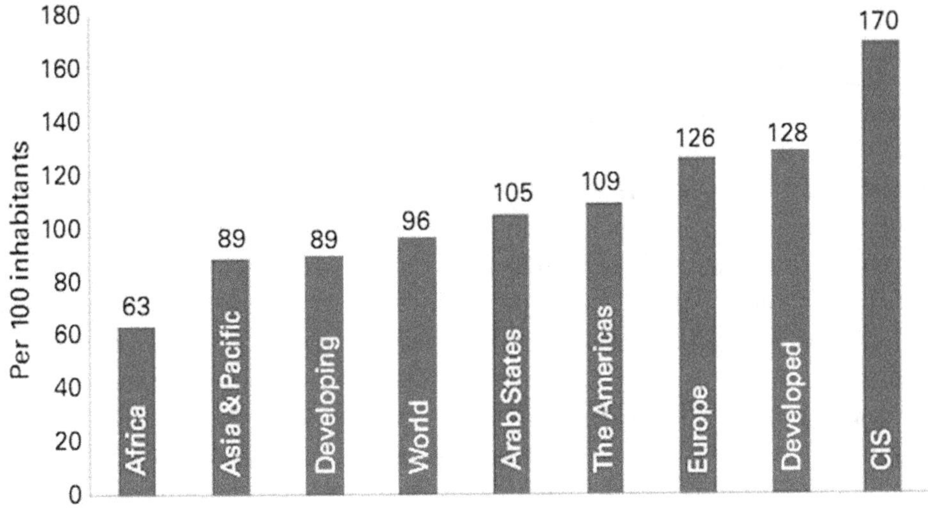

especially in Sub Saharan Africa is unevenly distributed such that studies indicated a total percent of the population with access to electricity in urban areas to be 59.9% and 14.2% in the rural areas (USAID, 2013). Despite the uneven distribution in terms of access to electricity between urban and rural areas, yet studies conducted to explore factors constraining distribution of electricity in rural areas of Sub Saharan Africa are inadequate.

To date, the global use of computers has been estimated to be 439 million, while 410 million computers were used in the USA, European Union and Asia, leaving 129 million computers used in the developing countries (Ssewanyana and Busler, 2007). Also, a study to assess a place where teachers' access computers in the UK revealed that about 84% of the teachers access computers at home, and 97% computer access at school (Selwood, 2004), while no study conducted to explore primary and secondary teachers access to computers at home and school in Sub Saharan Africa. Based on these projections for the adoption and use of ICTs, including computers, it is anticipated that the extent of ICT usage will increase significantly in the future. The internet access for example, and the rate of use was estimated to be 8 times in developed countries compared to the developing countries and then, by 2004 there were only 22 million internet users in Africa (ITU, 2004). Besides, the number of internet users in Sub Saharan Africa has increased significantly from 8 subscriptions per 100 inhabitants in 2001 to 38.8 per 100 inhabitants in 2013 (ITU, 2013) see figure 1.

In line with that, the number of mobile phone subscriptions has increased significantly such that the appropriate use of the same can improve the access to information about health, education, quality assurance of products, crop yields and others, which calls, a need for research on the role of mobile phones for the effective agriculture, industrial, health and education sector performances. Nevertheless, there is still a wide room for significant growth in the number of mobile phone subscribers in Africa, as Africa is being seen to have the worlds' largest working age of 2040, an indication of an economic potential with younger demographic (UNECA, 2012; AfDB, 2012; World Bank, 2012). The current infrastructure developments, through the undersea cables have sparked a stiff competition in Africa's broadband markets that led to lower telecommunication prices, forcing mobile phone subscribers announcing a significant reduction in the internet prices. Chavula (2013) documented that the median international bandwidth prices has fallen by more than 70% in the market, such that bandwidth supply in Sub Saharan Africa rose by 300% in 2010, leaving other countries raising their bandwidth intake in tenfold. In terms of telecommunications investment in Africa, for example, telecommunication investment rose at a compounding rate of 17% over the period of 1998 - 2007, i.e., from $3 billion in 1998 to 12 billion in 2007 the growth which outstrip the global infrastructure development.

In addition, the increase in the internet connection, web based applications, cellular networks, text message services and number of ICTs have improved the livelihood of the community, especially the rural poor such that with the ICTs, smallholder farmers who constitute more than 80% of the national population can access information about farming inputs, markets for their produce and weather changes. Recently, ICTs has introduced what is called "Networked Economy", whereby successful business organizations are linked with appropriate local and international suppliers, service providers, manufacturers and product shipping agents in real time. Moreover, many governments in developing countries, including Tanzania, are emphasizing the use of the internet for various online services as procurement of materials, marketing of products and services, access to greener means of industrial production (processes) as well as access to information necessary for the improved productivity of organizations that led to sustainable development of the country.

With ICTs, the coordination, communication, data and information processing costs of an organization are reduced significantly. The adoption and usage of ICTs in developing countries can be affected positively or negatively by the available ICT policy. For example, ICT policy introduced in East African countries including Tanzania and Uganda has paved a way towards the significant use of computers and internet connectivity in the education, health, agriculture and production sectors. The introduced policy by the governments was characterized by trade liberalization, privatization, civil service reform, financial sector reform and decentralization which have had an impressive economic performance (Ssewanyana and Busler, 2007; Omary et al., 2010). Absence of centralized policies in developing countries, i.e., policies that are specific to a certain sector, for example, mining, water and energy is one of the problems facing their implementation due to contradictions of one sector over the other, hence governments should introduce policies specific to ICT adoption and use so as to effect its usage and then, achieve the development goals.

OPPORTUNITIES AND CHALLENGES OF THE ICTS ADOPTION

Over the past two decades, the ICT sector has experienced a tremendous progress and improvements in comparison to other economic sectors in Africa caused by liberalization of the telecommunication market and developments of mobile technology in the region. With these developments, many industries, researchers and governments have developed a high research interest on the use of ICTs due to its impacts towards productivity improvements, effective collaborations within and abroad and timely access to relevant information (i.e., about new technologies, products and services, markets, materials). The effective use of ICTs by small and medium enterprises in Africa could lead into a significant increase in productivity in various ways, such as improved efficiency in resource allocation, reduced transaction costs and technological improvements. Besides, studies conducted to investigate the impact of ICTs in manufacturing industries in Europe and Asia (e.g., Iran), revealed a significant improvement in terms of productivity gains of manufacturing industries such that, the efficiency of production systems was found to be positive with significant productivity improvements (Greenan et al., 2001; Gholami et al.,

2004). According to Brynjolfsson and Hitts (2000) the ICT investments in US has brought significant improvements in labour productivity and more importantly, the multifactor productivity in at firm level and in the government.

Technological experts argued that, the access to inexpensive and rapid communication media lessens the disadvantages that arise in populations living in a remote and isolated region of the country. However, there exists a pessimistic view which assumes that the digital divide will increase and then, producers in developing countries and especially in rural areas will encounter even greater difficulties in comparison to their competitors in the developed countries (Wolf, 2001). Based on these arguments, try to think of how smallholder farmers, for example, use their mobile phones to access information about farming inputs or markets for their produce, the answer is a very few number of farmers use their phones for agriculture purposes. So access to ICT and the use of ICTs are two different things such that, developing an ICT awareness and personal change of attitudes towards the exploitation of ICT facilities is critical for sustainable development of any country especially developing countries of Africa.

Previous studies that utilize cross country data documented GDP per capita to be a key determinant factor for the diffusion of new technologies. The newly industrialized economies are in much more advanced level of exploitation of ICT facilities compared to the developing countries of South Asia, South East Asia and Pasific such that the diffusion of ICTs is characterized by a strong positive correlation with the per capita income of these countries (Gholami et al., 2004). Besides, studies documented that the number of internet hosts in Tanzania, Kenya and South Africa is higher than an average for the countries with similar GDP per capita levels, which reveal the fact that it is also true for countries with similar levels of per capita incomes to exhibit a wide variation in terms of ICT performances. According to the World Bank (1998) and Beardsley et al. (2010) the availability of ICT infrastructures has contributed greatly to the GDP growth in both developed and developing countries of the world. In addition, other inter-country studies revealed that, returns on ICT capital were positive and statistically significant for the industrialized countries only and not for the developing countries which qualifies previous statements on how income affects utilization of ICTs for industrial development.

However, there is a mix of information concerning the impact of ICTs towards growth of the manufacturing sectors in developing countries, that despite the intensive investment of ICT facilities, the contribution margin of industries in the national GDP has been reduced significantly compared to the production in 1970s'. One explanation of the above argument is that, reduced production or contribution margin of the manufacturing sectors of developing countries might be caused by the fact that most of the industries are not operating, whereas Sub Saharan Africa have been turned into market centers i.e., places where industrial produce from developed countries are sold. And, then developing a quantitative study that can explore the value for money investment on ICT facilities in developing countries will provide valuable information to the countries policy making and planning process. Based on the fact that, the ICT industry such as telecommunication, computer and software producers and electronic equipment manufacturers have played an important role in the growth of local and global economy, different governments have developed ICT policies that allow intensive investment of ICT facilities (Beardsley et al., 2010). Importantly, the predictions from previous studies revealed that the ICT industry can reduce carbon emissions, i.e., CO_2 (a current global concern) by 15% in 2020 (GeSI, 2008), and then, improve the livelihood of the community economically, socially and environmentally.

Hence, effective applications of the ICT facilities would create greener living environment through online shopping, teleworking, and remote communications made possible that is assisted by substantial change in human behavior. However, one could ask if at all we in the developing countries, we have had been able to see these opportunities and if not then, is it due to lack of expertise or reluctance of governments to use her people skilled enough in the area and educate the community through public presentations or using communication media such as radios, televisions, magazines and many others or else don't we see the benefits of ICT industry, of-course many governments have acknowledged the benefits earned, that's why there is massive investments in ICT and thus, did our governments document and implement some of the findings (i.e., value for money evaluation of ICTs investments in different sectors), which were obtained from the several studies based on information technology. Hence, effective utilization of ICTs in the country would bring potential benefits and massive expansion of different sectors in the country through the agricultural market information, micro-finance and m-banking services, health and nutrition services, low cost distance education and environmental facilities including CO_2 markets.

Despite the increased access to ICTs in developing countries, yet several barriers have been observed to hinder adoption and use of ICTs, especially by rural poor who constitute more than 80% of the population. The low rate of internet penetration and low bandwidth is still a challenge to most of the industries in developing countries such as Uganda and Tanzania, whereby only 62% of industries equipped with ICT facilities have the internet connection. According to the ITU (2007) report, by the year 2006, the internet penetration in Tanzania was 1% of the population, compared to 6%, 7.9% and 10.5% in Nigeria, Kenya and South Africa respectively. Similarly, poor ICT infrastructures that cannot deploy internet in many countries of the Sub Saharan Africa, has hampered the adoption of electronic services such as e-procurement and marketing, e-health care, e-banking and many others in the areas. Considering the majority of the population in developing countries are smallholder farmers who are poor and lives in the rural areas with no access to electricity, the adoption of ICTs have had been low despite the government efforts towards ICT adoption and use. High requirements in terms of the investment costs, and taxes on ICT facilities were also perceived to hinder the adoption and utilization of ICTs, especially by small industries in developing countries.

Besides, it has been established that the chance to adopt ICTs increases with the increase in the firm size such that business firms had a greater chance towards adoption of internet due to their capability in terms of skilled managers and workforce (OECD, 2000; Ssewanyana and Busler, 2007). The lack or the clarity of policies in place, to support ICT investment and adoption, which can be seen from presence of policies, regulations and laws that do not provide a framework necessary for the ICT adoption and use promotion activities. The reluctance of governments to take bold measures in fostering the necessary environment for the development of the ICT industry, be it due to finances, managerial or political concerns have contributed greatly to the low level of adoption and usage of ICT facilities in most of the developing countries. For example, pressures of unemployment that may arise from ICTs displacement effect since a large scale adoption of ICTs can reduce demand for low skilled labor has developed a negative attitude in some of the government leaders towards the strong developments of ICTs (Meng & Li, 2002). However, some studies documented that adoption and usage of ICTs have no significant displacement of employment, especially in the rural populations.

China, for example, before economic reforms in 1970's more than 80% of the country's population lived in the countryside, whereby the percentage was reduced to 69% in 1999, this entails that that dilemma of reduced employment due to ICT adoption and use does not hold water in all the conditions. In addition, most of the developing countries in the world are facing difficulties in the decision making process, especially when trying to establish the ICT promotion and overall national economic strategic goals. This comes into being due to lack of clear answers to the questions such as; can developing countries find competitive advantage upon building its ICT industry? Or shall ICT adoption and use increase the pressures of unemployment for the skilled or unskilled people? And can ICT development and use reduce disparity among nations? Thus, based on the literatures, yet the factors hindering the adoption and use of ICTs for industrial development cannot be explained precisely whether they are due to managerial, technical, financial or political, as quantitative studies are lacking.

The contribution margin of ICT goods and services towards country's economic development can be viewed from the production and utilization side, whereas those benefits earned from the usage side have been perceived to outweigh those of the production side (Kraemer & Dedrick, 2001). Considering the uneven distribution of the economic levels of different countries of the world, differences in per capita income of the population and population density of the countries, it becomes difficult to compare the economic benefits and diffusion of ICTs in the world, such that comparisons in terms of geographical areas are necessary.

GAPS FOR THE ADOPTION OF ICTS BY MANUFACTURING AND SERVICE INDUSTRIES

Over the past several decades, ICTs have been used in both manufacturing and service industries as means to increase the efficiency of production processes, improve quality of service and products manufactured, and reduce marketing, procurement and communication costs. Despite the benefits accrued from utilization of ICT facilities, yet most of the industries in in the Sub Saharan Africa, including Tanzania use old means for marketing her produce such as visiting the markets, radios and use of brokers in marketing products which lead to lower profits and lack of information about new technologies, markets and products. A study conducted in Uganda to explore adoption and use of ICTs in 40 industries, indicated that the adoption and use of ICT facilities are determined by the size of the company, such that most of the companies which uses computer and internet services in their daily operations were of larger sizes, i.e., both service and manufacturing industries. It was further documented that, the extent of adoption and use of ICTs in Uganda was extremely low compared to that of OECD countries whose usage was above 8% (Ssewanyana & Busler, 2007).

The human capital development, organization and proper management of the company operations were revealed to have significant impact towards the adoption and use of ICT facilities by manufacturing industries in Iran (Gholami et al., 2004). Based on these factors that contributed greatly to the low level of adoption and usage of ICT facilities in manufacturing industries, and the nature of the industries we have on developing nations impliedly the same factors might have the same impact to the service industries. Here you might ask why human capital development, management and organization were found to hinder the adoption of ICTs while our governments have invested heavily in educating her people in different professions. What might be the cause or do we really carry out needs assessment prior to any investment, be it training or procurements? When shall we wake up and discover where we went wrong, and is there any intention to rectify the cause so as to develop appropriate measures? The solution shall be made soon after our governments decide to use effectively, researchers from different Universities or utilize students' research reports available in University libraries, i.e., theses and dissertations in the planning and decision making process.

The adoption and use of ICTs in the different sectors in Tanzania have transformed the ways in which services are delivered. For instance, the adoption and use of ICTs in healthcare (e-healthcare) have been confirmed to have had significant impacts such as reduced human errors, reduced communication and service delivery costs, with improved efficiency and physician patient relationships (Omary et al., 2010). In the middle income countries such as China, the adoption and use of ICT facilities can be presented in the following order of importance, firstly, the farming, animal husbandry and fisheries. Secondly, the manufacturing industries and thirdly, encompasses other economic sectors of the country, whereby the rate of adoption and use for the first and second categories increased tremendously by two to seven fold respectively between 1992 and 1995, that was indicated by higher rates in these industries than the GDP growth (Meng & Li, 2002). Based on the rate of adoption and usage for the service and manufacturing sectors, you can see that the rate of adoption and use was favourable to the service industries compared to the manufacturing ones. The same scenario can be viewed in Tanzania, as most of the manufacturing industries are no longer functioning such that most of the business centers have been turned into warehouses and shopping centers to sell products manufactured from abroad such that, computers and internet connection is highly needed for the marketing and ordering of the products.

Considering the necessity of manufacturing industries to the economy of any nation, there is a high need for the quantitative research to establish the cause of the failures of many manufacturing industries in Tanzania and advise whether ICT usage by manufacturing industries can rescue failure of the few industries remained in operation. Based on the review, factors hindering the adoption and use of ICT facilities are determined by the per capita income of the country, readiness of the governments and availability of ICT infrastructures. Similarly, considering most of the manufacturing industries in developing countries of the world are not operating, there is quite few numbers of manufacturing industries that have exploited fully the benefits of ICTs, though quantitative research is needed to establish the extent of adoption and use of the ICT facilities.

SOLUTIONS AND RECOMMENDATIONS

Based on the challenges facing utilization of ICT facilities for industrial development, especially in developing countries, the following recommendations are made for improvements. Firstly, responsible organs in the specific countries has to use professionals and research results obtained by various academic institutions to ensure value for money investments on ICTs. Secondly, infrastructures to support the supply of alternative energy sources should be distributed, especially in rural areas so as to facilitate use of ICTs e.g., charging of mobile phone, and use of computers a mini-laptops in education, health and other production sectors. Thirdly, since many people in developing countries posses ICT facilities e.g., mobile phone but they just use them for social communications, educating the community on the need to use them for economic issues is necessary for the individual growth and nation at large.

Fourthly, considering most of the developing countries have been turned into service providers, i.e., places for collecting raw materials for the industries operating in developed countries, and places to sell products manufactured abroad, the use of ICT is important so as to get the information about the medium and long term impacts of their products. For example, risks accompanied with the massive importation of electrical and electronic products, e.g., mobile phone, computers, televisions and other electronic devices to human health and the environment. And lastly, with appropriate use of ICTs green growth of nations would be achieved through timely access of relevant information about green consumption of resources, greener design of products and services, greener procurement of raw materials and products and so on.

FUTURE RESEARCH DIRECTIONS

In view of the previous studies, the value for money ICT investments in the developing countries should consider that, the use of ICTs is significantly affected by the access and ability to use the ICT facilities and create awareness to the community especially the rural poor on the benefits of ICTs. This can be observed from the situation whereby most of the smallholder farmers in developing nations possess mobile phones, but they do not use them for obtaining information about agriculture inputs, markets or weather forecasts. In addition, the production engineers, for example, in the manufacturing industries do not have a culture of using the internet connection in their offices in search for new means of production or alternative materials that might produce quality products with less environmental pollution or proper recycling of materials so as to achieve good quality of performance for the manufactured products and many others.

The quantitative study to assess the benefits and challenges of ICT adoption and use by manufacturing and service industries in developing countries is inadequate, such that conducting studies to examine the impact of ICT adoption and use in terms of increase in productivity and efficiency over a certain period of time compared to productivity or efficiency prior to ICT adoption and use, is worth pursuing so as to advise on possible areas of investments. With tremendous increase in the number of people with mobile phone customization of the same with the appropriate information needed by people living in the rural and urban areas is critical to improve their living conditions. However, this would be effective by proper design of the awareness rising workshop or forums, to train all key stakeholders in a specific field, i.e., agriculture, health, education, business, mining and many others.

CONCLUSION

This chapter presents a comprehensive literature review on the ICT adoption and use in many countries of the world with a focus on the accrued benefits in developing countries. Different literatures concerning the adoption and use of the ICT facilities in various countries of the world were studied to reveal a global picture about ICT usage. Based on the existing body of knowledge, the ICT adoption and use is hindered by many factors such as the per capital income of the country, technology, readiness of the governments, education level of the people and finance, such that the rate of adoption and use between developing and developed countries are not comparable.

In the ancient times, technology transfers were found to be the best strategy to be adopted by industries that do not have specific technology to undertake some assignments that requires the advanced level of technology. However, the same strategy was practised in Tanzania in areas such as clean energy technologies, incubator programs and the like, whereby as in other developing countries, its adoption is still faced by inadequate infrastructures to support internet connectivity, access to electricity and financial capital.

REFERENCES

AfDB. (2012). *The transformational use of information and communication technologies in Africa.* World Bank, AfDB. Retrieved from http://siteresources.worldbank.org

Alchian, A. (1963). Reliability of progress curves in airframe production. *Econometrica, 31*(4), 679–693. doi:10.2307/1909166

Arrows, K. J. (1969). The organization of economic activity issues pertinent to the choice of market v/s non-markets allocation. Cambridge, MA: Belknap Press.

Bankole, F. O., Shiraz, F., & Brown, I. (2011). Investigating the impact of ICT investment on human development. *The Electronic Journal on Information Systems in Developing Countries, 48*(8), 1–19.

Beardsley, S. C., Enriquez, L., Bonini, S., Sandoval, S., & Brun, N. (2010). *Fostering the economic and social benefits of ICT.* The Global Information Technology Report, 2009 - 2010, World Economic Forum.

Brynjolfsson, E., & Hitts, L. (2000). Beyond computation: Information technology, organization transformation and business performance. *The Journal of Economic Perspectives, 14*(4), 23–48. doi:10.1257/jep.14.4.23

Chavula, H. K. (2013). The role of ICT in agriculture production in Africa. *Journal of Development and Agricultural Economics, 6*(7), 279–289.

Dahlman, C., & Fonseca, F. (1978). *From a technological dependence to technological development: The case of the USIMINAS steel plant in Brazil.* Working Paper 21, IBD/ECLA Research Program.

Fukasaku, Y. (2005). *Technology and industrial development in pre-war Japan.* Taylor and Francis e-Library.

GeSI. (2008). *SMART 2020: Enabling the low carbon economy in the information age.* GeSI, Global e-Sustainability Initiative, Creative Commons 2008 Attribution.

Gholami, R., Moshiri, S., & Lee, S. Y. T. (2004). ICT and productivity of manufacturing industries in Iran. *The Electronic Journal of Information Systems in Developing Countries, 19*(4), 1–19.

Greenan, N., Mairesse, J., & Bensaid, T. A. (2001). Information technology and research and development impacts on productivity and skills: Looking for correlations on Franch firm level data. In M. Pohjola (Ed.), *Information technology, productivity and economic growth: International evidence and implications for economic development* (pp. 119–148). Oxford, UK: Oxford University Press. doi:10.1093/acprof:oso/9780199243983.003.0006

ITU. (2004). *ITU world telecommunication indicators database.* International Telecommunication Union (ITU). Retrieved from http://www.itu.int/ITU-D/ict/statistics/index.html

ITU. (2007). *Telecommunications/ict markets and trends in Africa.* ITU.

ITU. (2013). ICT facts and figures 2013. ITU.

Kraemer, K. L., & Dedrick, J. (2001). Information technology and productivity: results and policy implications of cross-country studies. In M. Pohjola (Ed.), *Information technology, productivity and economic growth* (pp. 257–279). Oxford, UK: Oxford University Press. doi:10.1093/acprof:oso/9780199243983.003.0012

Lyakurwa, F. S., & Mangula, S. M. (2013). Delineation of the corporate use of environmental information systems: Seletcted cases of the corporate organizations in Tanzania. *Independent Journal of Management and Production*, *4*(2), 529–532. doi:10.14807/ijmp.v4i2.92

Meng, Q., & Li, M. (2002). New economy and ICT development in China. *Information Economics and Policy*, *14*(2), 275–295. doi:10.1016/S0167-6245(01)00070-1

OECD. (2003). *ICT and economic growth, evidence from OECD countries, industries and firms*. Paris: OECD.

Omary, Z., Lupiana, D., Mtenzi, F., & Wu, B. (2010). Analysis of the challenges affecting e-healthcare adoption in developing countries: A case of Tanzania. *International Journal of Information Studies*, *2*(1), 38–50.

Selwood, I. (2004). *Information technology and educational management in the knowledge society: Primary school teachers use of ICT for administration and management*. 6th International Working Conference on Information Technology in Educational Management (ITEM), Las Palmas de Gran Canaria, Spain.

Selwood, I., Smith, D., & Wishart, J. (2001). In A. C. W. Fung & M. A. Brown (Eds.), *Supporting UK teachers through the national grid for learning in Nolan. P* (pp. 159–171). Pathways to Institutional Improvement with Information Technology in Educational Management.

Ssewanyana, J. & Busler, M. (2007). Adoption and usage of ICT in developing countries: Case of Ugandan firms. *International Journal of Education and Development using Information and Communication Technology, 3*(3), 49-59.

UNECA. (2012). *Economic report on Africa 2012. Accelerating Africa's development through diversification*. Addis Ababa, Ethiopia: UNECA.

United Nations. (1978). *Technological change, economic development and intra and extra regional relations in Latin America. IDB/EC1A Science and Technolog*. Research Program in Latin America.

United Nations. (2011). Measuring the impacts of information and communication technology for development. UNICTAD current studies on science, technology and innovation.

USAID. (2013). *Assessing the role of information and communication technologies to enhance food systems in developing countries*. White Paper Series, USAID Higher Education Solutions Network.

Wolf, S. (2001). *Determinants and impact of ICT use for African SMEs: Implications for rural South Africa. Center for Development Research* (ZEF Bonn). 2001 Annual Forum at Misty Hills, Muldersdrift.

World Bank. (1998). *World development report 1998/99 - Knowledge for development*. Washington, DC: World Bank.

Chapter 13
Technology Access and Research Prolificacy:
An Econometric Analysis

Michael D'Rosario
Deakin University, Australia

ABSTRACT

Modern information and communication technologies enable research collaborations that were not conceivable a mere decade ago. Moreover, the diffusion of extant technologies (such as broadband, and cellular communication devices) more fully in both developing and developed nations has afforded more people access to key communications technologies, creating ever increasing networks and communities of researchers and collaborators. The current study seeks to more fully explicate the relationship between technology access, diffusion and research output. The study shall consider the impact of telecommunications tele-density, Broadband penetration, Computer access and ICT investment on research output and patent submissions. Both static and dynamic estimations are conducted, employing the Arellano & Bover systems estimator method. The findings suggest that telecommunications tele-density and ICT investment are key determinants of academic research output. They are however not significant determinants of patent submissions. These findings are robust for both static and dynamic estimations.

INTRODUCTION

Economic growth research represents amongst the most parsimonious inquiries for econometrics, given the interest in these and proximate matters. Theorists have considered the impacts of a number of key macroeconomic indicators on economic growth. Studies into these relations are as old as the discipline of economics itself. Increasingly research has considered the impact of ICT on economic activity. It is widely acknowledged that technology is a key driver of economic growth. The literature is replete with examples evidencing this set of relations. The relationship between telecommunications infrastructure and economic growth has been considered quite significantly in the extant literature, first by Jipp (1964), who posited the so called "Jipp curve" to define the underlying relationship. Therein, Jipp posited that

DOI: 10.4018/978-1-5225-0539-6.ch013

tele-density increases the greater the level of GDP. Much of the recent research into ICT and economic growth contends that information and communications technologies reduce the cost of transacting and as such improve the output of individual firms, therein improving aggregate production (Roller & Waverman, 2001, Madden & Savage 1998, 2000). What of the impact of such technologies on other sources of growth, there is a genuine dearth of research dealing with these matters? The impact of communications technologies on the diffusion of information and pertinently 'basic research' is of unquestionable importance and let little is known about this set of relations.

It may be self evident to many that innovations, or more simply put 'ideas' are pertinent to economic growth. Ideas are said to improve the technology of production (Jones, 2002). New ideas enable the factors of production to produce increased or better quality outputs. Utility is increased as more is produced or what is produced is better than before. These qualities are clearly growth enabling. Romer (1990) posits a worthwhile analogy. He asserts that ancient man used iron oxide based pigments to paint and decorate the walls of the caves they inhabited. Modern man then used these same pigments to 'paint' onto magnetised tape essential for the recording of video footage. The methods associated with the 'massification' of production developed by the Ford Motor Company revolutionised the manufacturing sector in the United States and abroad. The ideas developed by Ford were adopted by competitor firms improving the efficiency of automotive production across the sector.

The literature is replete with examples of innovations and technological advancements deriving from other disciplines and discourses. Nobel laureates Black and Scholes developed a formula for the valuation of options, their 'idea', a contemporary explication to value the uncertainty associated with options contracts was so useful that it is the approach adopted and taught within most business schools around the world and generally accepted by Wall St. The formula itself was based on another 'idea'; it was based on a heat diffusion formula developed in the engineering discourse years earlier.

The transitive quality of ideas, in particular basic research and its set or relations with economic growth is well espoused by Romer (1986). Romer (1986) formalised the conditions, by which ideas lead to growth. Romer asserts that the essence of ideas is that they are non-rivalrous, and by virtue of being non-rivalrous that the 'good' offers increasing returns to scale. Grossman and Helpman (1990) purport accurately that knowledge is a public good (or perhaps more accurately stated a near public good, to the extent that it is unrestricted) in that several parties may benefit from it simultaneously at no extra costs. The work of Romer (1986) is pertinent, Romer has asserted that so called 'knowledge spillovers' may cause investment in knowledge sectors to exhibit non decreasing returns to scale. This enables the innovation process to be sustainable in the long run (Grossman & Helpman, 1990).

Grossman and Helpman (1990) assert that when investment takes place in an economic environment with increasing returns to scale the marginal product of capital need not decline over time to the level of discount rate. Grossman and Helpman (1990) consider the role of Research and development asserting that it is a process essential to the knowledge generation process and consistent with Romer (1986).

As such ideas are profoundly different to other goods and services. Indeed the use of the manufacturing line by Ford did not preclude the use of a similar technique by competitor entities, nor did the application of six sigma methodology by GE preclude the use of similar managerial logic within other enterprises. While certain ideas can be made excludable, via patent and copyright legislation, this paper focuses on basic scientific research. This category of research is readily disseminated, with access relatively unrestricted. Basic research proffered in academic journals and periodicals provides other scholars and practitioners access to new ideas.

Table 1. Survey of the extant literature

Author(s)	Construct	Finding
Romer (1986)	Learning	Favourable
Lucas (1988)	Human Capital	Favourable
Romer (1990)	R&D	Favourable
Aghion & Howitt (1992)	R&D	Favourable
Barro (1990)	Public infrastructure	Favourable
Waverman (2001)	Public infrastructure	Favourable

Note: this table outlines the primary findings of the seminal studies within the extant literature relating to economic production, and innovation processes.

The extant literature evidences a recent trend whereby theoretical studies have sought to better enumerate the growth innovation relationship by attempting to fully endogenise innovation in the growth function. Broadly speaking the studies have considered four types of innovation process.

Empirical studies conducted in response to the theoretical work supports the posited relationship (see Coe & Helpman, 1993). There is however little national level research conducted into these matters. Geroski (1989) considering firm level data between 1976 and 1979 argued that innovations (employing the SPRU innovations database as a proxy) accounted for approximately 50% of growth for 30% of the entities considered. Bernstien & Mohnen (1994) investigate the impact of R&D in US and Japanese polities on TFP in the corresponding nation, the study finding that US R&D accounted for 60% of Japanese TFP while 20% of US TFP could be attributed to Japan.

Griliches & Lichtenberg (1984) shows evidence of spillovers between academic research and government research and development and the private sector. Pavitt et al (1987) assert that the relationship between firm size and innovation is non-linear, evidencing a strong associations between small and large firms and innovation and a weaker association between mid sized firms and innovation.

Academic science based research is shown to be a major contributor to productivity growth in Adams (1990) the author evidences a lag period of approximately 20 years. The work of Jaffe (1989) and Rosenberg (1994) is somewhat consistent with Adams (1990). Rosenberg offers a worthwhile example; consider the laser technologies so integral to medicine, manufacturing and technologies readily employed in contemporary society. The underlying scientific principles were espoused by Einstein in 1916; however industrial applications took some 44 years to be realized. Acs, Audretsch & Feldman (1992) observe that spillovers from university R&D can be quite significant with patent elastitities of approximately 10% for the corporate sector in respect of university based R&D.

This paper focuses on the impact of technology on the prolificacy of basic research, specifically basic science and technology based research published in academic periodicals. Modern day science has benefited greatly from the improved dispersion of research, enabling further research and innovation. The ability to access a periodical, to contact the contributing authors and understand the underpinning methodologies is made possible via contemporary communication technologies.

As asserted, it is a priori, readily evident that research enables capital to be employed more efficiently and effectively. Ceteris Paribas, research may enable greater a production quantum and superior outputs of production. There is significant extant research formalizing the relationship between economic growth and research and education. The seminal account of Mill (1900) tended to emphasise the role of scientific research and education in driving economic activity. Ricardo (1951) claimed that machinery was a key driver of per capita income growth, Romer (1986) emphasised the role of research in improving the factors or production, and consequently economic growth. Other significant contributions to the literature include those of Harrod (1939), Domar (1946), Solow (1956) and Swan (1956). It is acknowledged by much of the extant literature that different factors may be pertinent to different stages of economic growth. The provision and dispersion of technologies may be a key driver of growth in the future. This study is in the spirit of Mill (1900), Ricardo (1951) and Romer (1986) and acknowledges conceptually the role of knowledge and knowledge infrastructure in driving economic growth. Hitherto with this set of relations accepted, the focus consequently is a corollary consideration, the set of relations shared between research innovations (herein basic scientific research) and communications technologies (Fixed line/cellular architecture and broadband).

Grossman and Helpman (1990) claim that the lesser developed nations of the world may potentially benefit more from international relationships since in theory these nations can draw upon the significant knowledge capital accumulated in the industrial world. Herein the author offers a similar postulate, it may be asserted a priori that less developed nations may benefit disproportionately from technology adoption and diffusion (in terms of national research prolificacy) because they can benefit significantly from the stock of ideas, research, and technology already accrued in the developed world. The paper considers the impact of key ICT variables and ICT investment is considered in relation to research output and patent submissions, considering both developing and developed nations. Essentially the study is concerned with how significant communications technologies and infrastructures are to basic research prolificacy.

RELEVANT LITERATURE

Delong & Summers (1991) have evidenced that investment in capital equipment is associated with economic growth. Aschauer (1989) in his seminal paper showed that public capital and investment have positive and significant effects on productivity. This view is supported by Levine & Renalt (1992) who show that investment is a significant explanatory variable in relation to economic growth. Education is also shown to be positively related to economic growth (see Uzawa 1965, Lucas 1988, Barro & Sala-i-Martin 1995). Levine and Renelt show that secondary school enrolments are pertinent to economic growth. Barro and Sala-i-Martin show that human capital proxies and school access have positive effects on growth rates. Romer (1986) has focused on providing an endogenous explication for economic growth. Romer (1986) focuses on the role of positive externalities, specifically knowledge capital, as the key driver of economic growth.

The role of information and communications technologies in driving economic growth has been considered but what of other dynamic impacts, and second stage effects. Little has been said of the role of communications technologies in facilitating and nurturing innovation and research. This is clearly a worthwhile consideration, and is the impetus behind the current study. Given the seminal work of Mill (1900) and Ricardo (1951) and the contemporary accounts of Uzawa (1965), Lucas (1988), Delong and Summers (1991) and Romer (1986), we assume a positive association between economic growth and education, research and human capital. Herein, this study seeks to enumerate the relationship between Information and Communication Technologies and research. Specifically whether ICT drives innovation, and research output where research publications in science and technology proxy research, and patent submissions serve as a proxy for innovation.

Table 2. Variables and data sources

Variable	Description	Source
GDPCapita	Gross Domestic Product per capita	Worldbank
Tele100	Telecommunications lines per 100	ITU
SciTechCapita	Science and Tech Citations per capita	Worldbank
PatentCapita	Granted Patents per capita	ITU
Internet100	Internet access per 100	ITU
BroadSub100	Broadband Subscribers per 100	ITU

The study is a corollary of the accounts of Romer (1986) and Delong & Summers (1991), and seeks to determine the role that ICT plays in driving other posited drivers of economic growth, research and innovation. Much of the extant research considering proximate matters has incorporated static panel data approaches (see Madden & Savage 1998, Roller & Waverman 2001, Sridhar & Sridhar 2003). While this research has been apt, static explications have been limited to pooled OLS, Fixed effects (FE) and Random effects (RE) specifications. This study employs static and select dynamic estimation techniques to this intriguing new area of inquiry.

DATA AND BASIC CORRELATIONS

Before proceeding to the key estimations it is worthwhile to consider the basic correlations. The study utilizes consolidated country specific data across 166 developed and developing countries over 33 years between 1975 and 2008. The countries are listed in Table 1. This study incorporates data for both developed and developing countries. While the panel constructed was strongly balanced, far more data was available for developing countries than developed countries. The lack of Latin American and Caribbean data made regional panels for these groupings infeasible. The data gathered was procured from a number of different sources, the International Telecommunications Database 2008/9 (ITU), and the WorldBank WDI Database.

THE MODEL AND METHODOLOGY

The model contained herein is exogenous specification of research output (herewith RO). They are best described as falling into two general catagorisations. Firstly, the paper considers static exogenous models for research output. The key innovation of these model variants relates to the structure of the regressors and the dearth of research dealing with research output in economics and econometrics. The second model category considers research output and research innovation employing dynamic panel estimation techniques. The Arellano & Bond technique is employed. Given the potential complications associated with the static panel specifications using OLS, here both static and dynamic estimations are presented. When OLS is employed in a FE or RE specification, where a lagged dependent variable is included in the specification, the lagged variable may be correlated with the error term. The general static specifications are detailed below.

Table 3. Bivariate correlations (Pearson's)

	(1)	(2)	(3)	(4)	(5)	(6)	(7)	(8)
PatentApp (1)	1							
SciTech (2)	0.9566	1						
Tele100 (3)	0.154	0.174	1					
Broad100 (4)	0.1264	0.1219	0.622	1				
Internet100 (5)	0.1679	0.1794	0.8455	0.8016	1			
PCAcc100 (6)	0.2153	0.2393	0.8324	0.7307	0.9099	1		
Population (7)	0.5018	0.5142	-0.2333	-0.0903	-0.1828	-0.1546	1	
GDPCapita (8)	0.2429	0.2539	0.8121	0.6303	0.7998	0.8735	-0.125	1

$$\text{Roit} = \text{Bit} + \text{PCit} + \text{TELEFit} + \text{TELEMit} + \text{GDPCapitait} + \text{ICTCapitait} \quad (1)$$

Roit is the dependent research output where publications in science and technology serve as the principal proxy. Bit denotes broadband access. PCit denotes access to a computer per capita. TELEFit represents access to fixed-line telephones per 100. TELEMit represents access to mobile telephones per 100. GDPCapitait represents per capita GDP. Finally, ICTCapitait represents per capita ICT spending, where i indexes country classification and t indexes time.

It is commonly held that the dynamic specification is not limited by the problems of the static form, and as such enables a broader and more robust understanding of the intrinsic relationship. It is held by the author that this study is perhaps amongst the first studies to adopt the dynamic specification in this area of research. While the author holds the perspective that research growth and innovation are a consequence of a number of factors beyond those included in the primary iterations of the model specification, this study is proximately concerned with the deeper drivers of research output, but pertinently the ambit is limited to the set of relations shared between information and communication technologies and research output and research innovation. The dynamic specification adopted herein employs the GMM estimator developed by Arellano & Bond (1991). Consider the static specification denoted by model 1, by first differencing the specification detailed in model 1, we model growth by;

Table 4. Summary statistics

Variable	Obs	Mean	Std. Dev.	Min	Max
patentapp	2565	20579.39	93597.55	0	822037
Scitech	2883	12640.38	61174.76	0	582300
tele100	6216	25.15853	37.2403	0.011661	214.794
broad100	1026	4.411831	7.341149	0.000048	39.6972
internetac~0	2712	11.48164	18.30284	0.000018	91.6189
pcaccess100	2472	9.985926	15.43755	0.000032	94.3367
Population	6861	49900000	1.81E+08	12116	1100000000
GDPCapita	5744	6011.264	9899.843	62.8895	103042

$$D.Roit = aD.Riit + D.Bit + D.PCit + D.TELEFit + D.TELEMit + D.GDPCapitait + D.ICTCapitait \quad (2)$$

Where D is the first differenced operator and aD.Riit is the lagged first differenced dependent variable research innovation, and the remaining variables are the first differenced estimators as defined by model 1. In this specification levels of the dependent variable lagged one period and first differenced exogenous covariates are employed to estimate growth. The inclusion of this method provides a fuller explanation of growth, and a more robust series of findings. These model variants enable a number of the key challenges mounted against OLS static estimations to be accommodated. The specifications are strengthened as the paper proceeds from specification to specification, commencing with static OLS Pooled and FE estimations before considering the final specifications that employ the Arellano & Bond GMM estimator. Additionally the study incorporates a Worldbank dummy regressor, accounting for income level consistent with the Worldbank GNI classification method. Moreover, both pooled regression results and SUR estimations for each Worldbank classification are posited. This method enables the enumeration of income dependent effects on research output and research innovation.

One of the key challenges and limitations of the paper is the adequacy of data; the data limits the full replication of all static estimations employing the Arellano & Bond method (to afford a more robust response). Given the nature of the method requiring a lagged first differenced operator in the specification, the somewhat limited dataset cannot fully accommodate the inclusion of all desirable variables. The author acknowledges this limitation, though as stated this study is concerned with the relationship between ICT and both research innovation and research output, rather than affording a complete explication of research innovation and output. The dynamic estimations presented do in fact achieve this end. While an acknowledged limitation the dynamic estimations afforded in this piece provide a strong and compelling explication of the set of relations. The techniques employed are far more robust and rigorous that those employed in the extant research.

FINDINGS

Considering first the static estimations it would appear that PC Access, and broadband subscribership are positively associated with the science and technology research, significant at the 0.01 level and 0.05 levels respectively. GDP per capita is positively associated with research prolificacy across all but one of the specifications, though with differing levels of significance. The final iteration incorporates a broadband estimator (broadband tele-density per 100), broadband in positively and significantly associated with research prolificacy. The static estimations support the proposition that technology access is strongly associated with research prolificacy.

The dynamic estimations present far more robust findings. The incorporation of the lagged first differenced operator into the specification, addresses much of the frailty of the static form. However by virtue of the Arellano-Bond dynamic panel estimation technique, requiring the differencing of the country panel data, some of the data employed in the static estimation could not be incorporated into the dynamic form. This is not however a significant limitation.

The dynamic estimations present findings contrary to the findings of the static estimations. Teledensity is found to be positively associated with research prolificacy, significant at the 0.05 level across all iterations of the dynamic specification. GDP per capita is positively associated with research prolificacy

Table 5. Static models of ICT tele-density and research activity

SciTech Journals	Basic Telecom	Broadband	Internet Access	PC Access
PatentApp	0.6995498 (98.86)***	0.6659366 (58.05)***	0.6942416 (82.95)***	0.695378 (85.32)***
GDPCapita	0.0261948 (0.26)	0.3368124 (2.24)**	0.1674232 (1.75)*	-0.2119445 (-1.67)*
Tele100	41.39636 (1.53)			
Broad100		-776.1298 (-2.09)**		
Internet100			-59.99618 (-0.94)	
PCAccess100				342.5459 (3.73)***
Constant	1527.136 (1.56)	2113.24 (0.88)	2787.462 (2.05)**	2250.162 (1.78)*
Obs	1695	414	1056	1205
Adj R	86.49	89.87	87.7	86.93

The dependent variable is the number of Science and technology journal articles author at a national level. Robust standard errors. Model is a Simple Ordinary Least Squares model. The absolute t-statistics are in parentheses. * denotes significance at the 0.1 level, ** at the 0.05 level and *** at the 0.01 level.

Table 6. Dynamic panel models of Science and Technology research activity

SciTechCap	Broadband	Internet	PC Access	Tele-density
L1.	0.8232164 (17.1)***	0.8166327 (42.3)***	0.7513831 (45.74)***	0.6918871 (48.7)***
patentapp	0.0787386 (5.3)***	0.055022 (6.68)***	0.0732711 (8.52)***	
gdppercapita	0.0473923 (1.13)	0.0675614 (1.82)*	0.0493852 (1.18)	0.0968103 (12.27)***
broad100	26.10033 (0.62)			0.0911271 (2.35)**
Internet100		6.476783 (0.52)		
PCAccess100			31.61813 (1.61)	
tele100				23.38779 (3.9)***
constant	2316.117 (1.82)*	2696.544 (5.43)***	3730.513 (9.36)***	2633.851 (7.76)***
Obs.	324	934	1066	1371

The dependent variable is the number of Science and technology journal articles author at a national level. Robust standard errors. Model is a GMM System model based on Arellano & Blover/Blundell & Bond. The absolute t-statistics are in parentheses. * denotes significance at the 0.1 level, ** at the 0.05 level and *** at the 0.01 level.

across all specifications, though again with differing levels of significance. The final iteration incorporating the broadband estimator evidences that broadband is positively but not significantly associated with research prolificacy. The dynamic estimations support the proposition that telecommunications technology access is strongly associated with research prolificacy.

CONCLUSION

while there is strong evidence to support the proposition that communications technologies are drivers of economic growth, the literature replete with examples supporting this posited set of relations; it is conceivable that the impact of communications technologies on other growth drivers such as basic research may be equally significant. This paper evidences a strong and positive relationship between communications technologies and research prolificacy. While some may have made this determination a priori, this study is amongst the first to rigorously assess the set of relations employing dynamic estimation techniques. The study draws to light the need for further research into this pertinent relationship.

It is evident that research into the impact of ICT on research output in the developing world in particular would be apt. Given the role of research innovation in driving growth and development such analyses would be helpful in determining how limited ICT resources should be deployed. However limitations associated with the procurement of data render such analyses insatiable at present. Future research should also consider the differential impacts on country and regional groupings, through the development of balanced country and regional panels. Again such research questions cannot be addressed with data presently available.

The present study does again emphasise the importance of communications technologies beyond their relevance in conventional accepted applications. While communications technologies are generally considered growth drivers through direct applications of said technologies in promoting trade, reducing transaction costs and attenuating information asymmetries, their impacts are likely far more profound. Communications technologies enable the more fluid exchange of ideas, and research innovations, potentially resulting in improved productivity and quality.

Jequier (1984) claims that investment in the telecommunications sector has been seen by some in the development sector as not very fashionable, with development planners tending to take the position that such investment opportunities are generally less important than rural development, education, public health or urban rehabilitation. Moreover, many international development banks have tended to view the sector as an artifact or as a minor component of the development process remaining from the days where public infrastructures and utilities were built up (Jequier, 1984). To take such a view is both unhelpful and unproductive, as is evidenced within this and other cognate studies the sector has a key role to play in the development process, and is an important factor affecting the diffusion of ideas and research prolificacy.

REFERENCES

Acs, Z., Audretsch, D., & Feldman, M. (1994). R & D Spillovers and Recipient Firm Size. *The Review of Economics and Statistics*, *76*(2), 336–340. doi:10.2307/2109888

Adams, M. J. (1994). *Beginning to Read: A Summary*. Heinemann Educational Books - Primary Division.

Alleman, J., Madden, G., & Savage, S. (2000). Trade imbalance in international message telephone services. *Applied Economics*, *32*(10), 1311–1321. doi:10.1080/000368400404461

Arellano, M., & Bond, S. (1991). Some Tests of Specification for Panel Data: Monte Carlo Evidence and an Application to Employment Equations. *The Review of Economic Studies*, *58*(2), 277–297. doi:10.2307/2297968

Aschauer, D. A. (1990). *Public Investment and Private Sector Growth: The Economic Benefits of Reducing America's 'Third Deficit'*. Washington, DC: Economic Policy Institute, c.

B, J. A. (1986). Technological Opportunity and Spillovers of R&D: Evidence from Firms' Patents, Profits and Market Value. *American Economic Review*, *76*(5). doi:10.3386/w1815

Barro, R., & Sala-i-Martin, X. (1995). Technological Diffusion, Convergence, and Growth. *Journal of Economic Growth*, *2*. doi:10.3386/w5151

Bernstein, J., & Mohnen, P. (1994). International R&D Spillovers Between U.S. and Japanese R&D Intensive Sectors. *Journal of International Economics*, *44*. doi:10.3386/w4682

Coe, D., & Helpman, E. (1993). International R&D Spillovers. *Economic Review*, *39*. doi:10.3386/w4444

De Long, B., & Summers, L. (1990). Equipment Investment and Economic Growth. *The Quarterly Journal of Economics*, *106*. doi:10.3386/w3515

Domar, E. (1946). Capital Expansion, Rate of Growth, and Employment. *Econometrica*, *14*(2), 137–147. doi:10.2307/1905364

Geroski, P. (1989). Entry, Innovation and Productivity Growth. *The Review of Economics and Statistics*, *71*(4), 572–578. doi:10.2307/1928098

Griliches, Z., & Lichtenberg, F. (1984). Interindustry Technology Flows and Productivity Growth: A Reexamination. *The Review of Economics and Statistics*, *66*(2), 324–329. doi:10.2307/1925836

Grossman, G., & Helpman, E. (1990). Trade, Knowledge Spillovers, and Growth. *European Economic Review*, *35*. doi:10.3386/w3485

Harrod, R. (1939). An Essay in Dynamic Theory. *The Economic Journal*, *49*(193), 14–33. doi:10.2307/2225181

Jéquier, N. (1984). Telecommunications for development: Findings of the ITU-OECD project. *Telecommunications Policy*, *8*(2), 83–88. doi:10.1016/0308-5961(84)90028-4

Jipp, A. (1934). Telegraphenleitungen. *Moderne Telegraphie*, 113–190. doi:10.1007/978-3-662-32785-2_3

Levine, R. E., & Renelt, D. (1992). A Sensitivity Analysis of Cross-Country Growth Regressions. *The American Economic Review*, *84*(4).

Lucas, R. E. (1986). *On the mechanics of economic development*. Nanking, Taipei, Taiwan: Institute of Economics, Academia Sinica.

Madden, G., & Savage, S. (1998). Interconnection in international telecommunications. *Telecommunications Policy*, *22*(11), 953–961. doi:10.1016/S0308-5961(98)00062-7

Mill, J. S. (1900). *Principles of political economy, with some of their applications to social philosophy*. New York: Colonial Press.

Patel, P., & Pavitt, K. (1987). Is Western Europe losing the technological race? *Research Policy*, *16*(2-4), 59–85. doi:10.1016/0048-7333(87)90024-2

Ricardo, D., Sraffa, P., & Dobb, M. (1952). The Works and Correspondence of David Ricardo: Volume 8, Letters 1819-June 1821: v. 8: Letters 1819-June 1821. University Press for the Royal Economic Society.

Röller, L.-H., & Waverman, L. (2001). Telecommunications Infrastructure and Economic Development: A Simultaneous Approach. *The American Economic Review*, *91*(4), 909–923. doi:10.1257/aer.91.4.909

Romer, P. (1986). Increasing Returns and Long-Run Growth. *Journal of Political Economy*, *94*(5), 1002–1037. doi:10.1086/261420

Romer, P. (1990). Are Nonconvexities Important For Understanding Growth? *The American Economic Review*, *80*. doi:10.3386/w3271

Rosenberg, N. (1994). *Exploring the Black Box: Technology, Economics, and History* (1st ed.). Cambridge University Press. doi:10.1017/CBO9780511582554

Solow, R. (1956). A Contribution to the Theory of Economic Growth. *The Quarterly Journal of Economics*, *70*(1), 65–94. doi:10.2307/1884513

Sridhar, K. S. (2006). *Incentives for Regional Development: Competition Among Sub-National Governments*. Palgrave Macmillan.

Swan, A. (1956). Time-Study and Statistics. *OR*, *7*(4), 110–139. doi:10.2307/3007549

Uzawa, H. (1965). Optimum Technical Change in An Aggregative Model of Economic Growth. *International Economic Review*, *6*(1), 18–31. doi:10.2307/2525621

APPENDIX: COUNTRIES INCLUDED IN THE STUDY

A number of developed and developing nations were considered within this study. The majority of country level panels employed were strongly balanced.

Table 7.

Australia	El Salvador	Italy	Sweden
Austria	Finland	Japan	Syria
Canada	France	Kenya	Tanzania
Columbia	Greece	Malawi	Tunisia
Costa Rica	India	Malta	Turkey
Cyprus	Indonesia	Netherlands	United Kingdom
Denmark	Ireland	Norway	United States
Egypt	Israel	South Africa	Uruguay

Compilation of References

Aaltonen, K. (2013). The establishment of legitimacy: The case of international projects. *International Journal of Managing Projects in Business*, *6*(1), 13–35. doi:10.1108/17538371311291008

Aboh, C, L. (2008). Assessment of the frequency of ICT tools usage by Agricultural Extension agents in Imo State, Nigeria. *Journal of Agriculture and Social Research*, *8*(2).

Abouchedid, K., & Eid, G. (2004). E-learning Challenges in the Arab World: Revelation from a Case Study Profile. *Quality Assurance in Education*, *12*(1), 75–81. doi:10.1108/09684880410517405

Abou-Shouk, M., Lim, W. M., & Megicks, P. (2012). e-Commerce and small tourism businesses in developing countries: Drivers versus boundaries of adoption. *Tourism Planning & Development.*

Achterberg, J. S., Heng, M. S. H., & Gerrit, A. (1991). *Information systems research in the post-modern period.* Elsevier Science Publishers.

Acker, D., & Gasperini, L. (2008). Education for rural people: What have we learned. *Journal of International Agricultural and Extension Education*, *15*(1), 28–30. doi:10.5191/jiaee.2008.15103

Acs, Z., Audretsch, D., & Feldman, M. (1994). R & D Spillovers and Recipient Firm Size. *The Review of Economics and Statistics*, *76*(2), 336–340. doi:10.2307/2109888

Adams, J., & DeFleur, M. H. (2006). The acceptability of online degrees earned as a credential for obtaining employment. *Communication Education*, *55*(1), 32–45. doi:10.1080/03634520500343376

Adams, M. J. (1994). *Beginning to Read: A Summary*. Heinemann Educational Books - Primary Division.

Adebayo, E. L., & Adesope, O. M. (2007). Awareness, access and usage of information and communication technologies between female researchers and extensionists. *International Journal of Education and Development using Information and Communication Technology, 3*(1), 85-93.

Adesope, O. M., Asiabaka, C. C., & Agumagu, A. C. (2007). Effect of personal characteristics of extension managers and supervisors on information technology needs in the Niger Delta area of Nigeria. *International Journal of Education and Development Using ICT*, *3*(2).

Adolph, S., Hall, W., & Kruchten, P. (2008). A methodological leg to stand on: lessons learned using grounded theory to study software development. In *Proceedings of the 2008 Conference of the Center For Advanced Studies on Collaborative Research: Meeting of Minds*. ACM. http://doi.acm.org/10.1145/1463788.1463806

Adolph, S., Kruchten, P., & Hall, W. (2012). Reconciling perspectives: A grounded theory of how people manage the process of software development. *Journal of Systems and Software*, *85*(6), 1269–1286. doi:10.1016/j.jss.2012.01.059

Adrich, C. (2004). *Simulations and the Future of Learning.* London: Pfeiffer.

AfDB. (2012). *The transformational use of information and communication technologies in Africa.* World Bank, AfDB. Retrieved from http://siteresources.worldbank.org

Agwu, E. A., Uche-Mba, U. (2010). *Congruency, agreement and accuracy among researchers, extension workers and farmers on the role of ICTs in Nigeria's agricultural development.* Available at iaald2010.agropolis.fr/proceedings/national-experiences.html

Agwu, A. E., Uche-Mba, U. C., & Akinnagbe, O. M. (2008). Use of Information Communication Technologies (ICTs) among Researchers, Extension Workers and Farmers in Abia and Enugu States: Implications for a National Agricultural Extension Policy on ICTs. *Journal of Agricultural Extension, 12*(1), 37–49.

Ahmadpour, A., Mirdamadi, M., & Hossein, J. (2010). Factors influencing the design of electronic system in agricultural extension. *American Journal of Agricultural and Biological Sciences, 5*(2), 122–127. doi:10.3844/ajabssp.2010.122.127

Ahmed, A. (2004). Making Technology Work for the Poor: Strategies and policies for African sustainable development. *International Journal of Technology Policy and Management, 4*(1), 1–17.

Aizaki, H., Nakashima, Y., Ujie, K., Takeshita, H., & Tahara, K. (2010). Influence of Information about Agrochemicals provided via Internet –enabled mobile phones on Consumer Valuation of food. *Journal of Agricultural & Food Information, 11*(3), 209–211. doi:10.1080/10496505.2010.490509

Ajzen, I. (2008). *Theory of Planned Behaviour.* Retrieved on January 10, 2008, from http://hdppa.files.wordpress.com/2008/04/azjen.jpg

Ajzen, I. (1998). *Attitudes, Personality and Behaviour.* Chicago, IL: Dorsey Press.

Akabar, S. (2005). E-learning in developing countries: Challenges and opportunities Bangladesh perspectives. *Proceedings of the2nd International conference on e-learning for Knowledge-based society.*

Akar, E., Ozturk, E., Tuncer, B., & Wiethoff, M. (2004). Evaluation of a collaborative virtual learning environment. *Education + Training, 46*(6/7), 343-352.

Akhter, S., Jangjaimon, I., Chemin, Y., Uthayopas, P., & Honda, K. (2006). Development of a GRIDRPC tool for Satellite Images Parallel Data Assimilation in Agricultural Monitoring. *International Journal of Geoinformatics, 2*(3).

Akhter, S., Osawa, K., & Aida, K. (2007). Performance Evaluation of Distributed SWAP-GA Models with GridRPC. IPSJ SIG Technical Reports.

Akhter, S., Sakamoto, K., Chemin, Y., & Aida, K. (2010). Self-organizing GA for Crop Model Parameter Estimation using Multi-resolution Satellite Images. *International Journal of GeoInformatics, 6*(4).

Akhter, S., Aida, K., & Chemin, Y. (2010). GRASS GIS on High Performance Computing with MPI, OpenMP and Ninf-G Programming Framework. *Proceeding of ISPRS 2010.*

Akhter, S., Chemin, Y., & Aida, K. (2007). Porting a GRASS raster module to distributed computing. Examples for MPI and Ninf-G. *OSGeo Journal, 2*, 36–44.

Akhter, S., Honda, K., Chemin, Y., & Uthayopas, P. (2005). Distributed Pixel Method to speed-up the RS data assimilation of SWAP model. In *Proceedings of the MapAsia conference.*

Akhter, S., Honda, K., Chemin, Y., & Uthayopas, P. (2007). 2007a, Exploring Strategies for Parallel Computing of RS Data Assimilation with SWAP-GA. *Journal of Computer Science, 3*(1), 47–50. doi:10.3844/jcssp.2007.47.50

Akhter, S., Osawa, K., Nishimura, M., & Aida, K. (2008). Experimental Study of Distributed SWAP-GA Models on the Grid. *IPSJ Transactions on Advanced Computing Systems, 1*(2), 193–206.

Akkermans, N. (2010). *The Role of ICTs in Knowledge Sharing within Rural Communities in Ghana, Internship MA International Relations*. University of Groningen.

Alan, B. (2009). Handbook of Library Training Practice and Development. Abingdon, UK: Ashgate Publishing Group.

Alavi, M., & Leidner, K. (2001). Knowledge Management and Knowledge management systems: Conceptual Foundations and Research Issues. *Management Information Systems Quarterly*, *25*(1), 107–136. doi:10.2307/3250961

Albert, D., & Mori, T. (2001). Contributions of Cognitive psychology to the future of e-learning.*Bulleting of Graduate School of Education, Hiroshima University*, *50*(1), 25–34.

Alchian, A. (1963). Reliability of progress curves in airframe production. *Econometrica*, *31*(4), 679–693. doi:10.2307/1909166

Alexander, S. (2001). E-learning development and experience. *Education + Training*, *43*(4), 240-248.

Alleman, J., Madden, G., & Savage, S. (2000). Trade imbalance in international message telephone services. *Applied Economics*, *32*(10), 1311–1321. doi:10.1080/000368400404461

Allen, M. (2002). Michael Allen's Guide to E-Learning. Addison-Wesley.

Allens, M. (2007). *Designing successful e-learning*. John Wiley & Sons, Inc.

Al-Qirim, N. (2007a). The adoption of eCommerce communications and applications technologies in small businesses in New Zealand. *Electronic Commerce Research and Applications*, *6*, 462–473.

Al-Qirim, N. (2007b). A Research Trilogy into e-Commerce Adoption in Small Businesses in New Zealand. *Electronic Markets*, *4*, 263–285.

Andersson, A. S., & Grönlund, Å. (2009). A conceptual framework for e-learning in developing countries: A critical review of research challenges. *The Electronic Journal of Information Systems in Developing Countries*, *38*(8), 1–16.

Andrews, R., & Haythornthwaite, C. (2007). E-learning Journey. In R. Andrews & C. Haythornthwaite (Eds.), The Sage Handbook of E-learning research. Sage Publication.

Andrews, T. (2007, November). Reflections on 'The Discovery of Grounded Theory'. *The Grounded Theory Review*, 56.

Angehrn, A. (1997). Designing mature lnternet business strategies: The ICDT model. *European Management Journal*, *15*, 361–369.

Angelina, P. (2002a). Local initiatives in e-learning. *The Star Tech Plus*, *29*(Aug), 11.

Angelina, P. (2002b). E-learning: The alternative leaning process. *The Star Tech Plus*, *29*(Aug), 9–12.

Angello, C. (2010). The awareness and use of electronic information sources among livestock researchers in Tanzania. *Journal of Information Literacy*, *4*(2), 6–22. doi:10.11645/4.2.697

Anolina, A. (2005). Analysis of the terminologies used in the field of virtual learning. *Educational Technology and Society Journal*, *8*(3), 91-102.

Apache™ Hadoop. (2014) Retrieved from: http://hadoop.apache.org/

Appleton, J. V., & King, L. (2002). Journeying from the philosophical contemplation of constructivism to the methodological pragmatics of health services research. *Journal of Advanced Nursing*, *40*(6), 641–648. doi:10.1046/j.1365-2648.2002.02424.x PMID:12473042

Aranda, N. (2007). *A Brief History of E-learning and Distance Education*. Retrieved March 23, 2009, from http://ezinearticles.com/?A-Brief-History-of-E-learning-and-Distance-Education&id=496460

ArcGIS. (2012). *ArcGIS as a System for Emergency/Disaster Management, ArcGIS as a System for Emergency/Disaster Management*. Author.

Arellano, M., & Bond, S. (1991). Some Tests of Specification for Panel Data: Monte Carlo Evidence and an Application to Employment Equations. *The Review of Economic Studies*, *58*(2), 277–297. doi:10.2307/2297968

Arokoyo, T. (2010). *ICTs in the transformation of Agricultural Extension: The case of Nigeria*. Available at www.fao.org

Arokoyo, T., Chikwendu, D., & Ogunbameru, K. (2002). *A study of the access of Rural Women to Public and Private Extension Service in Nigeria*. Report of a study commissioned and funded by CTA.

Aroyo, L., & Dicheva, D. (2004). The New Challenges for E-learning: The Educational Semantic Web. *Journal of Educational Technology & Society*, *7*(4), 59–69.

Aroyo, L., & Kommers, P. (1999). Intelligent Agents for Educational Computer-Aided Systems. *Journal of Interactive Learning Research*, *10*(3–4).

Arrows, K. J. (1969). The organization of economic activity issues pertinent to the choice of market v/s non-markets allocation. Cambridge, MA: Belknap Press.

Aschauer, D. A. (1990). *Public Investment and Private Sector Growth: The Economic Benefits of Reducing America's 'Third Deficit'*. Washington, DC: Economic Policy Institute, c.

Ashley, C., & Carney, D. (1999). *Sustainable Livelihoods: Lessons from early experience*. London: Department for International Development.

Ashrafi, R., & Murtaza, M. (2008). Use and Impact of ICT on SMEs in Oman. *Electronic Journal Information Systems Evaluation*, *11*(3), 125–138.

Asian Development Bank. (2004). Building e-community centers for rural development. Report of the Regional Workshop, Bali, Indonesia.

Asian Productivity Organization (APO). (2002). *Report of a study meeting on Integration of Agricultural Research and Extension*. APO.

ASTER Image Webpage. (2009). Retrieved from http://asterweb.jpl.nasa.gov/

B, J. A. (1986). Technological Opportunity and Spillovers of R&D: Evidence from Firms' Patents, Profits and Market Value. *American Economic Review*, *76*(5). doi:10.3386/w1815

Baker, R. K. (2003). A Framework for Design and Evaluation of Internet-Based Distance Learning Courses Phase One – Framework Justification, Design and Evaluation. *Online Journal of Distance Learning Administration*, *6*(2), 43–51.

Baldwin, D. A. (1978). Power and social exchange. *The American Political Science Review*, *72*(4), 1229–1242. doi:10.2307/1954536

Ball State University. (1997). *A Brief History of Distance Education*. Available at http://www.seniornet.org/edu/art/history.html

Ballantyne, P. (2009). Accessing, Sharing and Communicating Agricultural Information for Development: emerging trends and issues. *Sage Journal*, *25*(4), 260-271. Available at http://www.sagepublications.com

Banard, D., & Vonk, T. (2003). *Information and Communication Technologies in African Local Governments*. Freiburg, Germany: EDITOSIA – Electronic Distance Training on Sustainability in African Local Governments. ICLEI European Secretariat GmbH.

Banathy, B. H. (1991). *Systems designs of education: A journey to create the future*. Englewood Cliffs, NJ: Educational Technology Publications.

Bankole, F. O., Shiraz, F., & Brown, I. (2011). Investigating the impact of ICT investment on human development. *The Electronic Journal on Information Systems in Developing Countries*, *48*(8), 1–19.

Banville, C., & Landry, M. (1989). Can the field of MIS be disciplined? *Communications of the ACM*, *32*(1), 48–60. doi:10.1145/63238.63241

Barbara, A. (2002). *E-learning and teaching in Library and information services*. Neal-schuman publishers.

Barbara, A. (2004). *Project Management: Tools and Technique for Today's LIS professionals*. Neal-schuman publishers.

Barron, J. (2006). Top Ten Secretes of Effective E-learning. *Industrial and Commercial Training*, *38*(7), 360–364. doi:10.1108/00197850610704552

Barro, R., & Sala-i-Martin, X. (1995). Technological Diffusion, Convergence, and Growth. *Journal of Economic Growth*, *2*. doi:10.3386/w5151

Barry, B. (2002). *ISD and the e-learning framework*. Available online at http://www.wit.ie/library/webct/isd.html

Baskerville, R., & Pries-Heje, J. (1996). Information Technology Diffusion: Building Positive Barriers. *Proceedings of the 4thEuropean Conference on Information Systems*. AIS.

Baskerville, R., & Pries-Heje, J. (2001). Racing the E-Bomb: How the Internet Is Redefining Information Systems Development Methodology. In B. FitzGerald, N. Russo, & J. DeGross (Eds.), Realigning Research and Practice in Is Development: The Social and Organisational Perspective (pp. 49-68). New York: Kluwer.

Baskerville, R., & Myers, M. (2002). Information Systems as a Reference Discipline. *Management Information Systems Quarterly*, *26*(1), 1–14. doi:10.2307/4132338

Baskerville, R., & Pries-Heje, J. (1999). Grounded action research: A method for understanding IT in practice, *Accounting. Management and Information Technology*, *9*(1), 1–23. doi:10.1016/S0959-8022(98)00017-4

Batchelor, S., Evangelista, S., Hearn, S., Sugden, S., & Webb, M. (2003). *ICT for development-Contributing to the Millennium Development Goals: Lessons learned from Seventeen infoDev Projects*. Washington, DC: World Bank. Available on line: www.infodev.org/symp2003/publications/CaseStudies.pdf

Bates, A. (2009). *Is e-learning failing in higher education?* Available at www.tonybates.ca/2009/06/is-e-learning-failing-in-higher-education/

Bates, A. (2005). *Technology, e-learning and Distance Education*. London: Routledge. doi:10.4324/9780203463772

Bates, T. (2000). *Managing Technological Change: Strategies for College and University Leaders*. Jossey Bass.

Bates, T. (2001). *National Strategies for e-Learning in Post-Secondary Education and Training*. UNESCO.

Bates, T. (2007). *Strategic Planning for E-learning in Polytechnic*. Tony Bates Associates Ltd. doi:10.4018/978-1-59140-950-2.ch004

Bayes, A., Von Braun, J., & Akhter, R. (1999). *Village Pay Phones and Poverty Reduction: Insights from a Grameen Bank, Initiative in Bangladesh.* ZEF Discussion Papers on Development Policy No. 8. Centre for Development Research. Available on line: http://www.apdip.net/publications/iespprimers/eprimer-pov.pdf

Beardsley, S. C., Enriquez, L., Bonini, S., Sandoval, S., & Brun, N. (2010). *Fostering the economic and social benefits of ICT.* The Global Information Technology Report, 2009 - 2010, World Economic Forum.

Becker, H. J. (2000). Findings for the teaching, learning and computing survey: Is Larry Cuban right? School Technology Leadership conference of the Council of Chief State Schools officers, Washington, DC.

Bednarek-Michalska, B., & Wolodko, A. (2007). E-learning model for Polish libraries: BIBWEB. *The Electronic Library, 25*(1), 80–89. doi:10.1108/02640470710729137

Beebe, M. (2003). Partnerships, Alliances and Networks for E-learning. In AfricaDotEdu. IT Opportunities and Higher Education in Africa. New Delhi: Tata McGraw Hill.

Beebe, M. (2004). *Impact of ICT Revolution on the African Academic Landscape.* CODESRIA Conference on Electronic Publishing and Dissemination, Dakar, Senegal. Retrieved from: http://www.codesria.org/Links/conferences/el-publ/beebe.pdf

Beebe, M. et al. (2003). *AfricaDotGov: IT Opportunities and Higher Education in Africa.* New Delhi: Tata McGraw-Hill Publishing Company, Ltd.

Beebe, M. (2003). *Partnerships, Alliances and Networks for eLearning.* New Delhi: Tata McGraw-Hill Publishing Company, Ltd.

Bennane, A. (2013). Adaptive Educational Software by Applying Reinforcement Learning. *Informatics in Education, 12*(1), 13–28.

Bennett, C. (2004). E-learning in Scotland: The North Lanarkshire Experience. *New Library World, 105*(11/12), 410–416. doi:10.1108/03074800410568743

Berners-Lee, T., Hendler, J., & Lassila, O. (2001). The Semantic Web. *Scientific American, 284*(5), 34–43. doi:10.1038/scientificamerican0501-34 PMID:11396337

Bernstein, J., & Mohnen, P. (1994). International R&D Spillovers Between U.S. and Japanese R&D Intensive Sectors. *Journal of International Economics, 44.* doi:10.3386/w4682

Bertolini, R. (2004). *Strategic thinking: making information and communication technologies work for food security in Africa.* Available online: http://www.ifpri.org/pubs/ib/ib27.pdf

Beverland, M. B., Ewing, M. T., & Matanda, M. J. (2006). Driving-market or market-driven? A case study analysis of the new product development practices of Chinese business-to-business firms. *Industrial Marketing Management, 35,* 383–393.

Bhatnagar, S., & Schware, R. (2000). *Information and communication technology in development: Cases from India.* Sage Publications India Pty, Ltd.

Bheenick, K., & Brizmohunr, G. (2003). *The scope of information and communication technology application in agriculture extension in Mauritius.* Available on line: www.Uom.ac.mu/faculty/foa/ais/amas/htm

Bigus, O. E. (1972). The Milkman and his Customer: A Cultivated Relationship. *Journal of Contemporary Ethnography, 1*(2), 131–165. doi:10.1177/089124167200100201

Bird, J., & Morgan, C. (2003). Adults Contemplating University Study at a Distance: Issues, themes and concerns. *International Review of Research in Open and Distance Learning*, *4*(1), 24–36.

Bitektine, A. (2008). Exploring slow institutional variables in healthcare context: A study of electronic health record in Canada. *ASAC*, *29*(28), 15–39.

Bleimann, U. (2004). Atlantis University: A new pedagogical approach beyond e-learning. *Campus-Wide Information Systems*, *21*(5), 191–195. doi:10.1108/10650740410567536

Block, J. (1971). *Mastery Learning: Theory and Practice*. New York: Holt, Rinehart and Winston.

Blomqvist, K. (2002). *Partnering in the dynamic environment: The role of trust in asymmetric technology partnership formation*. (Doctoral Thesis). Lappeenranta University of Technology.

Blomqvist, K., Hurmelinna, P., & Seppänen, R. (2005). Playing the collaboration game right—balancing trust and contracting. *Technovation*, *25*(5), 497–504. doi:10.1016/j.technovation.2004.09.001

Bloom, B. S., Engelhart, M. D., Furst, E. J., Hill, W. H., & Krathwohl, D. R. (1956). *Taxonomy of educational objectives: Handbook I: Cognitive domain*. New York: David McKay.

Bloom, B. (1968). Learning for Mastery. *Evaluation Comment*, *1*(2), 1–5.

BLR-Employee Training Centre. (2012). *The Most Effective Training Techniques*. Available at http://trainingtoday.blr.com/employee-training-resources/How-to-Choose-the-Most-Effective-Training-Techniques

Blyth, A. (2011). *Brilliant online marketing. How to use the Internet to market your business*. Pearson Education Limited.

Boaden, R., & Lockett, G. (1991). Information technology, information systems and information management: Definitions and developments. *European Journal of Information Systems*, *1*(1), 23–32. doi:10.1057/ejis.1991.4

Bonk, C. J., & Reynolds, T. H. (1997), Learner-centered Web instruction for higher-order thinking, teamwork and apprenticeship. In B. H. Khan (Ed.), Web-based instruction (pp. 167-178). Englewood Cliffs, NJ: Educational Technology Publications.

Boone, K.M., & Paulson, C.E., & Barrick, R.K. (1993). Graduate Education in Agricultural Communication: The need and Role. *Journal of Applied Communications*, *77*(1), 16–26.

Boone, K., Meisenbach, T., & Tucker, M. (2000). *Agricultural Communications: Changes and Challenges*. Iowa State University Press.

Bose, K. (2003). An E-learning Experience: A written Analysis Based on my Experience with Primary School Teachers in an E-learning Pilot Project. *International Review of Research in Open and Distance Learning*, *4*(2).

Bourne, J. R., Harris, D. A., & Mayadas, A. F. (2005). Online Engineering Education: Learning Anywhere, *Anytime. The Journal of Engineering Education*, *94*(1), 131–146. doi:10.1002/j.2168-9830.2005.tb00834.x

Bower, M., Hedberg, J. G., & Kuswara, A. (2010). A framework for Web 2.0 learning design. *Educational Media International*, *47*(3), 177–198. doi:10.1080/09523987.2010.518811

Brause, R. S. (2000). *Writing Your Doctoral Dissertation: Invisible Rules for Success*. London: Farmers Press.

Breckenridge, J., & Jones, D. (2009). Demystifying theoretical sampling in grounded theory research. *The Grounded Theory Review*, *8*, 113–126.

Broadbent, B. (2003). *Championing e-learning*. Retrieved from http://www.elearninghub. com/articles/chamioning.htm#Pro%20and%20cons%20of%20%20e-learning

Brooks, C., & McCalla, G. (2006). Towards flexible learning object metadata. *International Journal of Continuing Engineering Education and Lifelong Learning*, *16*(1/2), 50–63. doi:10.1504/IJCEELL.2006.008917

Brown, A. D. (1995). Managing understandings: Politics, symbolism, niche marketing and the quest for legitimacy in IT implementation. *Organization Studies*, *16*(6), 951–969. doi:10.1177/017084069501600602

Brusilovsky, P., Eklund J., & Schwarz, E. (1998). Web-based education for all: a tool for development adaptive courseware. *Computer Networks and ISDN Systems, 30*(1-7), 291–300.

Brusilovsky, P. (1996). Methods and techniques of adaptive hypermedia. *User Modeling and User-Adapted Interaction*, *6*(2–3), 87–129. doi:10.1007/BF00143964

Brusilovsky, P. (2003). Developing adaptive educational hypermedia systems: From design models to authoring tools. In *Authoring Tools for Advanced Technology Learning Environment* (pp. 377–409). Dordrecht: Kluwer Academic Publishers. doi:10.1007/978-94-017-0819-7_13

Brusilovsky, P., & Millan, E. (2007). User Models for Adaptive Hypermedia and Adaptive Educational Systems. In P. Brusilovsky, A. Kobsa, & W. Neidl (Eds.), Lecture Notes in Computer Science: Vol. 4321. *The Adaptive Web: Methods and Strategies of Web Personalization* (pp. 3–53). Springer. doi:10.1007/978-3-540-72079-9_1

Bryant, A., & Charmaz, K. (2007). Grounded theory research: Methods and practices. In A.Bryant & K. Charmaz (Eds.), The Sage handbook of grounded theory (pp. 1-29). London: Sage Publications Limited.

Bryant, A. (2002). Re-grounding grounded theory. *Journal of Information Technology Theory and Application*, *4*(1), 25–42.

Bryant, A. (2002). Re-grounding Grounded Theory. *Journal of Information Technology Theory and Application*, *4*(1), 25–42.

Bryant, A. (2009). Grounded theory and pragmatism: The curious case of Anselm Strauss. *Forum Qualitative Sozial Forschung*, 10.

Bryant, A., & Charmaz, K. (Eds.). (2007). *The Sage Handbook of Grounded Theory*. London: Sage Publications.

Brynjolfsson, E., & Hitts, L. (2000). Beyond computation: Information technology, organization transformation and business performance. *The Journal of Economic Perspectives*, *14*(4), 23–48. doi:10.1257/jep.14.4.23

BUA News. (2010). *South Africa: Limpopo farmers get ICT boost*. Author.

Burawoy, M. (1991). Reconstructing social theories. In M. Burawoy (Ed.), Ethnography unbound. Power and resistance in the modern metropolis (pp. 8-27). Berkeley, CA: University of California Press.

Buriano, L., Marchetti, M., Carmagnola, F., Cena, F., Gena, C., & Torre, I. (2006). The role of ontologies in context-aware recommender systems. In *Proceedings of 7th International Conference on Mobile Data Management (MDM) 2006*. IEEE. Available at: http://creative.eun.org/

Burke, M. E. (2007). Making Choices: Research Paradigms and Information Management (IM): Practical Application of Philosophy in IM Research. *Library Review*, *58*(6).

Burney, S. M. A., & Mahmood, N. (2006). A Brief History of Mathematical Logic and Applications of Logic in CS/IT. *Karachi University Journal of Science*, *34*(1), 61–75.

Burrell, G., & Morgan, G. (1979). *Sociological Paradigms and Organisational Analysis*. London: Heinemann.

Burrell, G., & Morgan, G. (1979). *Sociological Paradigms and Organizational Analysis: Elements of the Sociology of Corporate Life*. London: Heinemann.

Burton-Jones, A. (2009). Minimizing Method Bias through Programmatic Research. *Management Information Systems Quarterly*, *33*(3), 445–471.

Calder, J., & McCollum, A. (1998). *Open and flexible learning in vocational education and training*. London: Kogan Page.

Calder, J., & McCollum, A. (1998). *Open and Flexible Learning in Vocational Education and Training*. London: Kogan Page.

Calvert, J. (2001). Deakin University: Going online at a dual mode university. *International Review of Research in Open and Distance Learning*, *1*(2), 12.

Calvert, J. (2005). Distance Education at the Crossroads. *Distance Education*, *26*(2), 227–238. doi:10.1080/01587910500168876

Campbell, D. T., & Russo, M. J. (2001). *Social Measurement*. Thousand Oaks, CA: Sage Publications. doi:10.4135/9781412985857

Carliner, S. (2002). *Designing e-Learning*. Alexandria, VA: American Society for Training and Development.

Carliner, S., & Shank, P. (2007). *The e-Learning Handbook: Past Promises, Present Challenges*. San Francisco: Wiley & Sons.

Carr, N. G. (2003, May). IT Doesn't Matter. *Harvard Business Review,* 41-49.

Carrier, C. A., Davidson, G. V., Williams, M. D., & Kalweit, C. M. (1986). Instructional options and encouragement effects in a microcomputer-delivered concept lesson. *The Journal of Educational Research*, *79*(4), 222–229. doi:10.1080/00220671.1986.10885681

Carr, J. (1999). The Role of Higher Education in the Effective Delivery of Multimedia Management Training to Small and Medium Enterprises. *Journal of Educational Technology & Society*, *2*(2), 31–38.

Casey, T., & Wilson-Evered, E. (2012). Predicting uptake of technology innovations in online family dispute resolution services: An application and extension of the UTAUT. *Computers in Human Behavior*, *28*, 2034–2045.

Castells, M. (1996). *The Rise of the Network Society* (Vol. 1). Malden, MA: Blackwell.

Chacha, E. (2009). *Understanding the Tools of the Social Learning Landscape*. Available online at http://www.elearninglearning.com/social-media/delicious/2009/?query=chacha

Chacha, F. (2009). Annual E-learning Report on East Africa.*Annual East African University Council Meeting*, Arusha, East African University Council.

Chadha, G., & Kumail, S. M. N. (2002). *E-Learning: An Expression of the Knowledge Economy*. New Delhi: Tata McGraw-Hill Publishing Company Limited.

Chambers, R., & Conway, G. R. (1992). *Sustainable Rural Livelihoods: Practical Concepts for the 21st century, Institute of Development Studies*. Sussex, UK: IDS.

Charmaz, K. (2003). Grounded theory - Objectivist and constructivist methods. In N. K. Denzin & Y. S. Lincoln (Eds.), Strategies of Qualitative Inquiry (pp. 249-291). London: Sage.

Charmaz, K. (2008a). Grounded theory. In J. A. Smith (Ed.), Qualitative psychology: A practical guide to research methods (pp. 81-110). Los Angeles, CA: SAGE.

Charmaz, K. (2008b) Constructionism and the grounded theory method. In Handbook of Constructionist Research. Guilford Press.

Charmaz, K. (1983). The grounded theory method: An explication and interpretation. In R. Emerson (Ed.), *Contemporary field research: A collection of readings* (pp. 109–126). Boston, MA: Little Brown Company.

Charmaz, K. (1990). 'Discovering' chronic illness: Using grounded theory. *Social Science & Medicine*, *30*(11), 1161–1172. doi:10.1016/0277-9536(90)90256-R PMID:2360052

Charmaz, K. (2006). *Constructing grounded theory: A practical guide through qualitative analysis*. London: Sage Publications Limited.

Chavula, H. K. (2013). The role of ICT in agriculture production in Africa. *Journal of Development and Agricultural Economics*, *6*(7), 279–289.

CHEMS (Ed.). (2000). *Cases, Issues and Themes in Higher Education Distance e-Learning: a Study on Market Issues for the Proposed e-University*. Higher Education Academy.

Chen, H., & Chen, T. J. (2002). Asymmetric strategic alliances: A network view. *Journal of Business Research*, *55*(12), 1007–1013. doi:10.1016/S0148-2963(02)00284-9

Chen, W., & Hirschheim, R. (2004). A paradigmatic and methodological examination of information systems research from 1991 to 2001. *Information Systems Journal*, *14*(3), 197–235. doi:10.1111/j.1365-2575.2004.00173.x

Chen, Y., Wu, C., Guo, X., & Wu, J. (2012). Semantic Learning Service Personalized. *International Journal of Computational Intelligence Systems*, *5*(1), 163–172. doi:10.1080/18756891.2012.670528

Chiemelie, I. B. (2012). *Disadvantaged of traditional methods of training*. Available at http://bnetwork.us/showthread.php?337-Disadvantaged-of-traditional-methods-of-training

Chilimo, W. (2008). Information and communication technologies and sustainable livelihoods: a case of selected rural areas of Tanzania (Doctoral dissertation, University of KwaZulu-Natal, Pietermaritzburg, South Africa).

Chin, K. L., & Kon, P. N. (2003), Key Factors for a Fully Online E-learning Mode: a delphi study. In Crisp G., D. Thiele, I. Scholten, S. Barker, & J. Baron (Eds.), *Interact, Integrate, Impact:Proceedings of the 20th Annual Conference of the Australasian Society for Computers in Learning in Tertiary Education*.

Chizari, M., Lashgarara, F., Mirdamadi, S. M., & Hosseini, S. J. F. (2009). Role of information and communication technologies in improving food availability of Iranian rural households. *International Journal of Education and Development using Information and Communication Technology*, *5*(3), 193-205.

Choi, T. Y., Barratt, M., & Li, M. (2011). Qualitative case studies in operations management: Trends, research outcomes, and future research implications. *Journal of Operations Management*, *29*, 329–342.

Chong, S. (2006). An empirical study of factors that influence the extent of deployment of electronic commerce for small-and medium sized enterprises in Australia. *Journal of Theoretical and Applied Electronic Commerce Research*, *1*(2), 45–57.

Christensen, Ó. (2007). A simpler Understanding of Classic GT: How it is a fundamentally different methodology. *The Grounded Theory Review*, *6*(3), 39–61.

Christiansen, Ó. (2005). The theory of "opportunizing" and the sub-process of "conditional befriending". *Journal of Business & Economic Research Journal*, *3*(4), 73–88.

Christiansen, Ó. (2006). Opportunizing: A classic grounded theory study on business and management. *The Grounded Theory Review*, *6*(1), 109–133.

CIRTL Network. (2012). *Lecturing: Advantages and Disadvantages of the Traditional Lecture Method.* Available at http://www.cirtl.net/node/2570

Clarke, C. (2003), *Towards a Unified E-learning Strategy*. A consultant document submitted to the Department for Education and Skills, United Kingdom. Available online at http://www.education.gov.uk/consultations/downloadableDocs/towards%20a%20unified%20e-learning%20strategy.pdf

Clarke, C. (2003). Towards a Unified E-learning Strategy, A consultant Document submitted to the. Department for Education and Skills. Available online at http://www.education.gov.uk/consultations/downloadableDocs/towards%20a%20unified%20e-learning%20strategy.pdf

Clarke, T., & Hermens, A. (2001). Corporate developments and strategic alliances in e-learning. *Journal of Education and Training, 43*(4/5), 256–267. doi:10.1108/00400910110399328

Clark, R. C., & Mayer, R. E. (2011). *E-Learning and the Science of Instruction: Proven Guidelines for Consumers and Designers of Multimedia Learning* (3rd ed.). Hoboken, NJ: Pfeiffer. doi:10.1002/9781118255971

Coe, D., & Helpman, E. (1993). International R&D Spillovers. *Economic Review, 39*. doi:10.3386/w4444

Coffield, F., Moseley, D., Hall, E., & Ecclestone, K. (2004). *Learning styles and pedagogy in post-16 learning. A systematic and critical review*. Learning and Skills Research Centre.

COL. (2002). *The Commonwealth of Learning, "An Introduction to Open and Distance Learning*". Available online at http://www.col.org/ODLIntro/introODL.htm

Coldeway, D. O. (1982). Recent Research in Distance Learning. In J. S. Daniel, M. A. Stroud, & J. R. Thompson (Eds.), *Learning at a Distance: A World Perspective* (pp. 29–37). Edmonton, Alberta, Canada: Alberta University.

Collins, C., Buhalis, D. & Peters, M., A. (2003). Enhancing SMTEs' Business Performance through the Internet and Elearning platforms. *Education + Training, 45*(8/9), 483-494.

Collins, G. R. (2002). Case Study: A Satellite-based Internet Learning System for the Hospitality Industry. *Online Journal of Distance Learning Administration, 5*(4), 23–34.

Collins, J. C., & Jerry, I. P. (2000). *Built to Last – Successful Habits of Visionary Companies*. London: Random House.

Collis, B. (1997). *Implementing ICT in the Faculty: Letting a 1000 Flowers bloom or managing change?* Keynote presentation for the national conference, Studying in Digital Learning Environments, University of Utrecht. Retrieved from http://education2.edte.utwente.nl/teletophomepage.nsf/PapersNLViewForm?readform

Collis, B. (1996). *Tele-learning in a digital world-the future of Distance Learning*. London: International Thomson Computer Press.

Collis, B., & Moonen, J. (2001). *Flexible Learning in a Digital World: Experiences and Expectations*. London: Kogan Page.

Collis, J., & Hussey, R. (2009). *Business Research: A practical guide for undergraduate & postgraduate students* (3rd ed.). Palgrave Macmillan.

Commons, C. (2012). *e-Learning Fundamentals*. Available at http://www.leerbeleving.nl/wbts/1/history_of_elearning.html

Conole, G., & Fill, K. (2005). A learning design toolkit to create pedagogically effective learning activities. *Journal of Interactive Media in Education, 2005*(1), 9. doi:10.5334/2005-8

Cosier, R. A., & Ruble, T. L. (1981). Research on conflict-handling behavior: An experimental approach. *Academy of Management Journal, 24*(4), 816–831. doi:10.2307/256179

Cowie, N., & Sakui, K. (2013). It's never too late: An overview of e-learning. *ELT Journal*, *67*(4), 459–467. doi:10.1093/elt/cct037

Cramer, S., Krasinki, S., Crutchfield, M. D., Sackmary, B., & Scalia, L. (2000). Using Collaboration and the Web to Implement the new CEC Standards. *Teaching Exceptional Children*, *32*(5), 12–19. doi:10.1177/004005990003200502

Creswell, J. (2003). *Research Design: Qualitative, Quantitative and Mixed Method Approaches* (2nd ed.). Thousand oaks, CA: SAGE publishers.

Cross, J. (2004). An Informal History of eLearning. *On the Horizon*, *12*(3), 103–110. doi:10.1108/10748120410555340

Crotty, M. (1998). *The foundations of social research: Meaning and perspective in the research process*. London: Sage Publications Limited.

Cruthers, M. (2008). *Education technology gives teachers a wider reach*. Available from http://www.etni.org.il/etnirag/issue5/mark_cruthers.htm

CTL. (2001). *A vision of e-learning for America's workforce*. Available online at http://www.nga.org/cda/files/ELEARNINGREPORT.pdf

Cuban L. (1993). Computers Meet Classroom: Classroom wins Teachers College Record. *Volunteer, 95*(2), TCR ID 82.

Cupchik, G. (2001). Constructivist Realism: An Ontology That Encompasses Positivist and Constructivist Approaches to the Social Sciences. *Forum: Qualitative Social Science*. Available online at http://www.qualitative-research.net/index.php/fqs/article/view/968/2112

Customer-Essay. (2012). *Dissertation Structures: Some Important Things to Consider*. Available at http://custom-essay.org/article/dissertation-structures

Cutcliffe, J. R. (2004). Adapt or adopt: Developing and transgressing the methodological boundaries of grounded theory. *Journal of Advanced Nursing*, *51*(4), 421–428. doi:10.1111/j.1365-2648.2005.03514.x PMID:16086811

Dabbagh, N. H., Bannan-Ritland, B., & Silc, K. (2000). Pedagogy and Web-based Course Authoring Tools: Issues and implications. In B. H. Khan (Ed.), Web-based training (pp. 343-354). Englewood Cliffs, NJ: Educational Technology Publications.

Dagger, D., Wade, V., & Conlan, O. (2002). Towards a standards-based approach to e-Learning personalization using reusable learning objects. In *World Conference on E-Learning in Corporate, Government, Healthcare, and Higher Education.*

Dahlman, C., & Fonseca, F. (1978). *From a technological dependence to technological development: The case of the USIMINAS steel plant in Brazil*. Working Paper 21, IBD/ECLA Research Program.

Dam, N. (2004). *The E-learning Fieldbook. Implementation Lessons and Case Studies from Companies That Are Making E-learning Work*. New York, NY: McGraw-Hill.

Dana, L. P. (2001). Networks, Internationalization & Policy. *Small Business Economics*, *16*(2), 57–62. doi:10.1023/A:1011199116576

Davis, F. D. (1989). Perceived Usefulness, Perceived Ease of Use, and Users` Acceptance of Information Technology. *Management Information Systems Quarterly*, *13*(3), 318–340. doi:10.2307/249008

Davis, F. D. (1993). Users` acceptance of information technology: System characteristic, user perceptions and behavioral impacts. *International Journal of Man-Machine Studies*, *38*(3), 475–487. doi:10.1006/imms.1993.1022

De Long, B., & Summers, L. (1990). Equipment Investment and Economic Growth. *The Quarterly Journal of Economics*, *106*. doi:10.3386/w3515

Dearnley, C. (2003). Student Support in Open Learning: Sustaining the Process. *International Review of Research in Open and Distance Learning*, *4*(1), 67–74.

Dhanarajan, G. (2001). *Open and Flexible Learning: Commonwealth Experience and Indian's Challenge.* Inaugural Remarks made at the 76th Annual Meeting of the Association of Indian Universities, Central University of Chandigarh, Chandigarh, India. Available at http://www.col.org/resources/speeches/2001presentations/pages/2001-12-08.asp

Dobson, P. J. (2002). Critical Realism and Information Systems Research: Why Bother with Philosophy?. *Information Research*, *7*(2). Available at http://InformationR.net/ir/7-2/paper124.html

Dodds, T. (1999). *Non-Formal and Adult Basic Education Through Open and Distance Learning in Africa*. Academic Press.

Domar, E. (1946). Capital Expansion, Rate of Growth, and Employment. *Econometrica*, *14*(2), 137–147. doi:10.2307/1905364

Dorji, M. (2003). *Integration of SWAP Model and SEBAL for Evaluation of on Farm, Irrigation Scheduling with Minimum Field Data*. Enschede: ITC.

Doughty, P. L., Spector, M., & Yonai, B. A. (2003). Time, Efficacy and Cost Considerations of e-Collaboration in Online University Courses. *Brazilian Review of Open and Distance Learning*. Available at http://www.abed.org.br/publique/cgi/cgilua.exe/sys/start.htm?UserActiveTemplate=1por&infoid=746&sid=70

Dreyfus, H. L. (1999). Anonymity versus Commitment: The Dangers of Education on the Internet. *Ethics and Information Technology*, *1*(1), 15–21. doi:10.1023/A:1010010325208

Dringus, L. P. (2003). From both sides now: On being an Online Learner and Online Instructor. *E-Learn Magazine*. Available at URL:http://www.elearnmag.org/subpage/sub_page.cfm?section=3&list_item=1&page=1

Driscoll, M. (2002). *Web-based Training: Designing E-learning Experiences*. Jossey-Bass.

Drucker, P. (2000). *Need to Know: Integrating e- Learning with High Velocity Value Chains*. A Delphi Group White Paper.

Drucker, P. (1994, November). The Age of Social Transformation. *Atlantic Monthly*, 68.

Dube, L., & Pare, G. (2003). Rigor in Information Systems Positivist Case Research: Current Practices, Trends and Recommendations. *Management Information Systems Quarterly*, *27*(4), 597–636.

Duchaste, A. (1996). Design for web-based learning.*Proceedings of the WebNet-96 World Conference of the Web Society*.

Duncan, E. A. S., & Nicol, M. M. (2004). Subtle realism and occupational therapy: An alternative approach to knowledge generation and evaluation. *British Journal of Occupational Therapy*, *67*(10), 453–456. doi:10.1177/030802260406701006

EAC. (2001). *Retreat to establish East African Community*. East African Community. Available at http://www.eac.int/treaty/

EAC. (2001). *Treaty Establishing the East African Community.* Available at http://www.eac.int/treaty/

e-Commerce Awards. (2013). *e-Commerce Awards Source*. Retrieved May 2, 2013, from http://www.ecommerceawards.co.za/2012/nominations/travelground

Economic Commission for Africa (ECA). (2005). *Emerging issues in science and technology for Africa's development: science, technology and innovation for meeting key MDGs*. Addis Ababa: ECA Sustainable Development Division.

Edland, M. (2011). *ELearning and the Benefits That It Offers*. Ezine@Articles. Available at http://ezinearticles.com/?ELearning-and-the-Benefits-That-It-Offers&id=5715586

Edmundson, A. (2007). *Globalized e-learning cultural challenges*. London: Idea Group. doi:10.4018/978-1-59904-301-2

Eggleston, K., Jensen, R., & Zeckhauser, R. (2002). Information and Communication Technologies, Markets and Economic Development. In *The Global Information Technology Report: Readiness for the Networked World* (pp. 62–63). New York: University Press.

Eisenhardt, K. M. (1989). Building theories from case study research. *Academy of Management Review*, *14*(4), 532–550.

E-learning Africa. (2012). *Perspectives on ICT and Education in Africa*. Available from http://www.elearning-africa.com/eLA_Newsportal/E298tanzania-beyond-tomorrow%E2%80%99-soon-ready/

E-learning in India. (2001). *E-learning in India*. Presented in the E-Learning in India Conference. Available at http://www.slideshare.net/missan/e-l-india-finl-presentation

El-Gohary, H. (2012). Factors affecting E-Marketing adoption and implementation in tourism firms: An empirical investigation of Egyptian small tourism organisations. *Tourism Management*, *33*, 1256–1269.

Ellington, H. (1995). Flexible learning, your flexible friend. In C. Bell, M. Bowden & A. Trott (Eds.), Implementing flexible learning (pp. 3-13). London: Kogan Page.

Ellington, H. (1995). Flexible learning, your flexible friend. In C. Bell, M. Bowden, & A. Trotts (Eds.), Implementing flexible learning (pp. 3-13). London: Kogan Page.

El-Tigi, M. A., & Khan, B. H. (2001). Web-based Learning Resources. In B. H. Khan (Ed.), Web-Based Training (pp. 59-72). Englewood Cliffs, NJ: Educational Technology Publications.

Engelhard, R. (2001). *Wireless: a help line for agricultural development*? Report of the 4th Consultative Expert meeting of CTA"s Observatory on ICTs. CTA Working Document-8024. Available: http://www.agricta.org/observatory

Ennew, C., & Fernandez-Young, A. (2006). Weapons of Mass Instruction? The rhetoric and reality of online learning. *Marketing Intelligence & Planning*, *24*(2), 148–157. doi:10.1108/02634500610654008

ESRF. (2007). *Tanzania Country Report: Enhancing Livelihoods of the Rural Poor: The Role of Information and Communication Technologies (ICT) – The Final Report*. ESRF, Dar-es-Salaam.

Ettinger, A., & Holton, V. (2005). E-learning lessons from the sharp end: The BBC. *Training & Management Development Methods*, *19*(2), 7–13.

Ettinger, A., Holton, V., & Blass, E. (2006a). E-learner experiences: Key questions to ask when considering implementing e-learning. *Industrial and Commercial Training*, *38*(3), 143–147. doi:10.1108/00197850610659409

Ettinger, A., Holton, V., & Blass, E. (2006b). E-learner experiences: What is the future for e-learning? *Industrial and Commercial Training*, *38*(4), 208–212. doi:10.1108/00197850610671991

EUN. (n.d.). Available at http://www.eun.org/

Evans, J.R. & Haase, I.M. (2001). Online Business Education in the Twenty-First Century: An Analysis of Potential Target Markets. *Internet Research: Networking*.

Evans, C., & Fan, J. P. (2002). Lifelong Learning through the Virtual University. *Campus-Wide Information Systems*, *19*(4), 127–134. doi:10.1108/10650740210438810

FACET. (2010). *ICT to enhance farm extension services in Africa*. Available on line: www.facet.org

FAO. (2000). *The role of information and communication technologies in rural development and food security*. Rome: FAO. Available on line: http://www.fao.org/sd/CDdirect/CDre0055.htm

FAO. (2003). *Virtual extension and research communication network.* Available on line: http://www.fao.org/sd/2003/KN10053_en.html

FAO. (2004). *Institution Building to strengthen Agriculture Extension.* 27th FAO Regional Conference for Asia and the Pacific, Beijing, China. Available on line: www.fao.org

FAO. (2006). *The Internet and Rural and Agricultural Development – An Integrated Approach.* Food and Agriculture Organisation. Available on line: http://www.fao.org/docrep/w6840e/w6840e05.htm

Farrington, J., Chapman, R., & Slaymaker, T. (2001). *Sustainable Livelihoods approaches in practice: Potentials and constraints.* Paper prepared for the SIDA Poverty Workshop, London, UK.

Felder, R.M. (1996). Matters of Style. *ASEE Prism, 6*(4).

Felder, R. M., & Silverman, L. K. (1988). Learning and teaching styles in engineering education. *English Education, 78*(7), 674–681.

Fendt, J., & Sachs, W. (2008). Grounded Theory Method in Management Research. *Organizational Research Methods, 11*(3), 430–455. doi:10.1177/1094428106297812

Fernández, M., Gómez-Pérez, A., Pazos, A., & Pazos, J. (1999). Building a Chemical Ontology Using METHONTOLOGY and the Ontology Design Environment. *IEEE Intelligent Systems, 14*(1), 37–46. doi:10.1109/5254.747904

Fiorini, P. M., & Lipsky, L. R. (2012). Search marketing traffic and performance models. *Computer Standards & Interfaces, 34*, 517–526.

Fitzgerald, B. & Howcroft, D. (1998b). Competing Dichotomies in IS Research and Possible Strategies for Resolution. *ICIS 1998 Proceedings.*

Fitzgerald, B., & Howcroft, D. (1998). Towards the dissolution of the IS debate. From polarisation to polarity. *Journal of Information Technology, 13*(4), 313–326. doi:10.1057/jit.1998.9

Fleming, N. (2006). Teaching and learning styles: VARK (2nd ed.). Academic Press.

Fleming, N. D. (1995). I am different; not dumb. Modes of presentation (V.A.R.K.) in the tertiary classroom. In A. Zelmer (Ed.), *Research and development in higher education.Proceedings of the1995 annual conference of the higher education and research development society of Australia* (HERDSA).

Fleming, S. (2012). Information and Communications Technologies (ICTs) and Democracy Development in the South: Potential and Current Reality. *EJISCD, 3*(10), 1–10.

Flinn K. & Lawrence, M. (2003). *E-learning Development in Sub-Saharan Africa: Issues and Challenges.* Paper Presented to National ICT policy round table.

Flor, A. G., & Hazelman, M. (2004). *Regional Prospects and Initiatives for Bridging the Rural Digital Divide.* AFITA/WCCA Joint Congress on IT in Agriculture. Available online www.upou.edu.ph/downloads/2004AFITA.pdf

Flynn, D., & Hussain, Z. (2001). *Using Structuration Theory To Explain Information Systems Development And Use In A Public Health Organization.* Paper presented to the 9th European Conference on Information Systems, Bled, Slovenia. Available at http://sdaw.info/asp/aspecis/20010034.pdf

Flynn, D., & Hussain, Z. (2004). Seeking Legitimation for an Information System: A Preliminary Process Model. *ECIS 2004 Proceedings.* Available online at http://aisel.aisnet.org/ecis2004/27

Frempong, G., & Stork, C. (2005). Towards an African e-Index: Household and individual ICT access and usage across 10 African countries. In *Research ICT Africa*. The LINK Centre, Wits University School of Public and Development Management. Available: www.researchictafrica.net

Fresen, J. W., & Boyd, L. G. (2005). Caught in the Web of Quality. *International Journal of Educational Development*, *25*(3), 317–331. doi:10.1016/j.ijedudev.2004.12.002

Fry, K. (2000). *Forum Focus and Overview. The Business of E-learning: Bringing your Organization in the Knowledge Economy, Telcam Group*. Sydney: University of Technology.

Fukasaku, Y. (2005). *Technology and industrial development in pre-war Japan*. Taylor and Francis e-Library.

Fulton, K. (1996). *Technology Infusion and School Change: Perspective and Practices*. Available at http://www.ncrel.org/sdrs/areas/issues/methods/technlgy/te600.pdf

Galagan, P. (2000). Getting Started With E-learning. *Journal of Training and Development*, *54*(4), 62–64.

Gardner, H. (1999). *Intelligence Reframed: Multiple Intelligences for the 21st Century*. Basic Books.

Garrison, D. R., & Shale, D. (1987). Mapping the Boundaries of Distance Education: Problems in Defining the Field. *American Journal of Distance Education*, *1*(1), 7–13. doi:10.1080/08923648709526567

Gasson, S. (1998). Framing Design: A Social Process View of Information System Development. In *Proceedings of the Nineteenth International Conference on Information Systems* (pp. 224 – 236). Atlanta, GA: Association for Information Systems (AIS).

Gauch, S., Speretta, M., Chandramouli, A., & Micarelli, A. (2007). User profiles for personalized information access. In The adaptive Web (pp. 54–89). Springer Berlin Heidelberg. doi:doi:10.1007/978-3-540-72079-9_2 doi:10.1007/978-3-540-72079-9_2

GBR Institute for Interactive Technologies. (2006). *E-Learning Concepts and Techniques*. Available from http://iit.bloomu.edu/Spring2006_eBook_files/chapter1.htm

Gelb, E., Gal, B., & Wolfson, D. (2008). *Information and Communication Technologies(ICT) for Agricultural Extension –An Overtime Israeli Perspective*. Israel: Ministry of Agriculture and Rural Development.

George, T., & Mcgee, M. K. (2003). Educational Advantage. *Information Week*, 57–58.

Geroski, P. (1989). Entry, Innovation and Productivity Growth. *The Review of Economics and Statistics*, *71*(4), 572–578. doi:10.2307/1928098

GeSI. (2008). *SMART 2020: Enabling the low carbon economy in the information age*. GeSI, Global e-Sustainability Initiative, Creative Commons 2008 Attribution.

Ghobakhloo, M., Arias-Aranda, D., & Benitez-Amado, J. (2011). Adoption of e-Commerce applications in SMEs. *Industrial Management & Data Systems*, *8*, 1238–1269.

Gholami, R., Moshiri, S., & Lee, S. Y. T. (2004). ICT and productivity of manufacturing industries in Iran. *The Electronic Journal of Information Systems in Developing Countries*, *19*(4), 1–19.

Giddens, A. (1984). *The Constitution of Society: Outline of the Theory of Structuration*. Cambridge: Polity Press.

Gilbert, P. K. (2002). *The Virtual University an Analysis of Three Advanced Distributed Leaning Systems*. Available at http://gseacademic.harvard.edu/~gilberpa/homepage/portfolio/research/pdf/edit611.pdf

Gilmore, A. (2003). *Service Marketing and Management*. London: SAGE Publications Ltd.

GIS Development. (2006). *GIS application*. Available online: http://www.gisdevelopment.net/application/agriculture/overview/agrio0011dhtm

Glaser, B. (1996). *Gerund Grounded Theory: The Basic Social Process Dissertation*. Mill Valley, CA: Sociology Press.

Glaser, B. (1996). *Gerund Grounded Theory: The Basic Social Process Dissertation*. Sociology Press.

Glaser, B. (1999b). The Future of Grounded Theory: Qualitative Health Research. Oxford, UK: Sage Publications.

Glaser, B. (2004). *Glaser's explanations on ontology and epistemology*. Academic Press.

Glaser, B. (2004). *Staying open*. Unpublished paper.

Glaser, B. (2004). *Staying open*. Unpublished paper. Available at www.groundedtheoryinstitute.com/docs/staying_open.html

Glaser, B. (2004a). *Glaser's explanations on ontology and epistemology*. Academic Press.

Glaser, B. (2004b). *Staying open*. Unpublished paper.

Glaser, B. (2005a). The Impact of Symbolic interactionism on GT. *GT Review*, *4*(2), 1-23.

Glaser, B. (2005b). *The Roots of Grounded Theory*. Keynote presentation delivered to the 3rd International Qualitative Research Convention, Johor Bahru, Malaysia.

Glaser, B. (2005c). *The Roots of Grounded Theory*. Keynote presentation delivered to the 3rd International Qualitative Research Convention, Johor Bahru, Malaysia.

Glaser, B. (2009). *Jargonizing: Using the Grounded Theory Vocabulary*. Mill Valley, CA: Sociology Press.

Glaser, B., & Holton, J. (2004). Remodeling grounded theory. *Forum: Qualitative Social Research*, *5*(2), Article 4. Retrieved from http:// www.qualitative-research.net/fqstexte/2-04/2-04glaser-e.htm

Glaser, B., & Holton, J. (2004). Remodeling Grounded Theory. *Forum: Qualitative Social Research*, *5*(2), Article 4. Retrieved from http:// www.qualitative-research.net/fqstexte/2-04/2-04glaser-e.htm

Glaser, B., & Kaplan, D. W. (Eds.). (1996). *Gerund Grounded Theory: The Basic Social Process Dissertation*. Mill Valley, CA: Sociology Press.

Glaser, B. (1964). *Organizational Scientists: Their Professional Careers*. Chicago: Aldine.

Glaser, B. (1965). The Constant Comparative Method of Qualitative Analysis. *Social Problems*, *12*(4), 436–445. doi:10.2307/798843

Glaser, B. (1978). *Theoretical Sensitivity: Advances in the Methodology of Grounded Theory*. Mill Valley, CA: Sociology Press.

Glaser, B. (1992). *Basics of grounded theory analysis*. Mill Valley, CA: Sociology Press.

Glaser, B. (1992). *Emergence Vs Forcing: Basic of Grounded Theory Analysis*. Mill Valley, CA: Sociology Press.

Glaser, B. (1992). *Emergence vs Forcing: Basics of Grounded Theory Analysis*. Mill Valley, CA: Sociology Press.

Glaser, B. (1993). *Examples of Grounded Theory: A Reader*. Sociology Press.

Glaser, B. (1998). *Doing Grounded Theory: Issues and Discussions*. Mill Valley, CA: Sociology Press.

Glaser, B. (1999). *The Discovery Of Grounded Theory: Strategies for Qualitative Research*. Mill Valley, CA: Sociology Press.

Glaser, B. (1999b). *The Future of Grounded Theory, Qualitative Health Research. Sage Publications*, *9*(6), 836–845.

Glaser, B. (1999c). The Future of Grounded Theory, Keynote address at the 4th annual Qualitative Health Research Conference. *The Grounded Theory Review*, *1*(1), 1–9.

Glaser, B. (2001). *The Grounded Theory Perspective: Conceptualization Contrasted with Description*. Mill Valley, CA: Sociology Press.

Glaser, B. (2002). Constructivist grounded theory? *Forum Qualitative Sozial Forschung*, 3.

Glaser, B. (2003). *The Grounded Theory Perspective II: Description's Remodeling of Grounded Theory*. Mill Valley, CA: Sociology Press.

Glaser, B. (2004c). Naturalist inquiry and grounded theory. *Forum Qualitative Sozial Forschung*, 5.

Glaser, B. (2005). *The Grounded Theory Perspective III: Theoretical Coding*. Mill Valley, CA: Sociology Press.

Glaser, B. (2005b). *The grounded theory perspective III: Theoretical coding*. Mill Valley, CA: Sociology Press.

Glaser, B. (2006). *Doing Formal Grounded Theory: A proposal*. Mill Valley, CA: Sociology Press.

Glaser, B. (2006). *Doing Formal Grounded Theory: A Proposal*. Mill Valley, CA: Sociology Press.

Glaser, B. (2008). *Doing Quantitative Grounded Theory*. Mill Valley, CA: Sociology Press.

Glaser, B. (2009). *Jargonizing: Using the grounded theory vocabulary*. Sociology Press.

Glaser, B. (2011). *Getting out of the data: Grounded theory conceptualisation*. Mill Valley, CA: Sociology Press.

Glaser, B. (2011). *Getting Out of the Data: Grounded Theory Conceptualization*. Mill Valley, CA: Sociology Press.

Glaser, B. (Ed.). (1993). *Examples of Grounded Theory: A Reader*. Mill Valley, CA: Sociology Press.

Glaser, B. (Ed.). (1995a). *Grounded Theory 1984-1994, Volume One*. Mill Valley, CA: Sociology Press.

Glaser, B. (Ed.). (1995b). *Grounded Theory 1984-1994, Volume Two*. Mill Valley, CA: Sociology Press.

Glaser, B., & Holton, J. (2004). Remodelling Grounded Theory. *Forum Qualitative Sozial Forschung*, *5*(2), 1–17.

Glaser, B., & Holton, J. (2005). Basic Social Processes. *The Grounded Theory Review*, *4*(3), 1–21.

Glaser, B., & Strauss, A. (1965). *Awareness of Dying*. Chicago: Aldine Publishing Company.

Glaser, B., & Strauss, A. (1967). *The Discovery of Grounded Theory: Strategies for Qualitative Research*. Hawthorne, NY: Aldine De Gruyter.

Glendenning, C. J., Babu, S., & Asenso-Okyere, K. (2010). *Review of Agricultural Extension in India, Are farmers' information needs being met?* IFPRI Discussion Paper 01048. Eastern and Southern Africa Regional Office.

Goodear, L. (2001). *Cultural Diversity and Flexible Learning*. Presentation of Findings 2001 Flexible Learning Leaders Professional Development Activity. South West Institute of TAFE. Australia. Available at http://www.flexiblelearning.net.au/leaders/events/pastevents/2001/statepres/papers/lyn-handout.pdf

Google BigQuery. (2014). Retrieved from: https://developers.google.com/bigquery/

Gopal, A., & Prassad, P. (2000). Understanding GDSS in Symbolic Context: Shifting the Focus from Technology to Interaction. *Management Information Systems Quarterly*, *24*(3), 509–546. doi:10.2307/3250972

Gotschall, M. (2000). E-learning Strategies for Executive Education and Corporate Training. *Fortune*, *141*(10), S5–S59.

Goulding, C. (1998). Grounded theory: The missing methodology on the interpretivist agenda. Qualitative Market Research. *International Journal (Toronto, Ont.)*, *1*(1), 50–57.

Goulding, C. (2001). Grounded theory: A magical formula or a potential nightmare. *The Marketing Review*, *2*(1), 21–33. doi:10.1362/1469347012569409

Grange, L. (2004). E-learning: Some Critical Thought, South Africa. *The Journal of Higher Education*, *18*(1), 87–97.

Grant, D., & Cliff, O. (1996). *Metaphors and Organization*. London: Sage Publications.

Grant, L. K., & Spencer, R. E. (2003). The Personalized System of Instruction: Review and Applications to Distance Education. *International Review of Research in Open and Distance Learning*, *4*(2), 28–37.

Grasha, A. (1996). *Teaching with Style*. Alliance Publishers.

Grass GIS. (2009). *Geographic Resources Analysis Support System*. Retrieved from http://grass.itc.it/

Gray, B. (1985). Conditions facilitating interorganizational collaboration. *Human Relations*, *38*(10), 911–936. doi:10.1177/001872678503801001

Gray, B. (1989). *Collaborating: Finding Common Ground for Multiparty Problems*. San Francisco: Jossey-Bass.

Greenan, N., Mairesse, J., & Bensaid, T. A. (2001). Information technology and research and development impacts on productivity and skills: Looking for correlations on Franch firm level data. In M. Pohjola (Ed.), *Information technology, productivity and economic growth: International evidence and implications for economic development* (pp. 119–148). Oxford, UK: Oxford University Press. doi:10.1093/acprof:oso/9780199243983.003.0006

Gregor, S. (2006). The nature of theory in information systems. *Management Information Systems Quarterly*, *30*(3), 611–642.

Gregor, S. (2006). The Nature of Theory in Information Systems. *Management Information Systems Quarterly*, *30*(3), 611–642.

Griliches, Z., & Lichtenberg, F. (1984). Interindustry Technology Flows and Productivity Growth: A Reexamination. *The Review of Economics and Statistics*, *66*(2), 324–329. doi:10.2307/1925836

Grix, J. (2002). Introducing students to the generic terminology of social research. *Politics*, *22*, 175–186.

Grooms, L. D. (2003). Computer-Mediated Communication: A Vehicle for Learning. *International Review of Research in Open and Distance Learning*, *4*(2), 70–76.

Grossman, G., & Helpman, E. (1990). Trade, Knowledge Spillovers, and Growth. *European Economic Review*, *35*. doi:10.3386/w3485

Gruber, T. R. (1993). A translation approach to portable ontologies. *Knowledge Acquisition*, *5*(2), 199–220. doi:10.1006/knac.1993.1008

Gruninger, M., & Fox, M. (1995). Methodology for the Design and Evaluation of Ontologies. In *Proceedings of the IJCAI'95.Workshop on Basic Ontological Issues in Knowledge Sharing*.

Gudza, L. D. (2010). *Podcasts can inform poor farmers.* Available on line http//www.scidev.net/en/agriculture-and-environment/farming-practices/opinions/podcasts-can-inform poor farmers.html

Gunga, S., & Ricketts, I. (2007). Facing the Challenges of E-learning Initiatives in African Universities. *British Journal of Educational Technology*, *38*(5), 896–906. doi:10.1111/j.1467-8535.2006.00677.x

Gurstein, M. (2000). Community Informatics: Enabling Communities with information and communication technologies. Idea Group Publishing. doi:doi:10.4018/978-1-878289-69-8 doi:10.4018/978-1-878289-69-8

Habiba, M., & Akhter, S. (2012). MAS Workflow Model and Scheduling Algorithm for Disaster Management System. *International Conference On Cloud Computing Technologies, Applications And Management* (ICCCTAM-12).

Habiba, M., & Akhter, S. (2013). A Cloud Based Natural Disaster Management System. *The 8th International Conference on Grid and Pervasive Computing* (GPC 2013). doi:10.1007/978-3-642-38027-3_16

Haigh, J. (2004). Information Technology in Health Professional Education: Why IT Matters. *Nurse Education Today*, *24*(7), 547–552. doi:10.1016/j.nedt.2004.07.008 PMID:15465170

Halawani, F., Abdullah, M. M. B., Rahman, M. S., & Halawani, Y. (2013). A Proposed Framework for E-Commerce Usage and Competitive Advantage on Small and Medium Tourism Enterprises (SMTES) in Lebanon. *Journal of Social and Development Sciences*, *6*, 258–267.

Hall, B. (2001). E-Learning: Building Competitive Advantage Through People and Technology, A Special Section on Elearning. *Forbes Magazine*. Available at http://www.forbes.com/specialsections/elearning/

Hall, B. (1997). *Web-based training cookbook*. New York: Wiley.

Halse, M.L., & Mallinson, B.J. (2009). Investigating popular Internet applications as supporting e-learning technologies for teaching and learning with Generation Y. *International Journal of Education and Development using Information and Communication Technology*, *5*(5), 58-71.

Hameed, A. M., Counsell, S., & Swift, S. (2012). A Conceptual Model for the Process of IT Innovation Adoption in Organizations. *Journal of Engineering and Technology Management*, *29*, 358–390.

Hamid, A. A. (2002). E-Learning - Is it the “e” or the Learning that Matters. *The Internet and Higher Education*, *4*(1), 311–316.

Harper, G. W., & Salina, D. D. (2000). Building collaborative partnerships to improve community-based HIV prevention research: The university-CBO collaborative partnership (UCCP) model. *Journal of Prevention & Intervention in the Community*, *19*(1), 1–20. doi:10.1300/J005v19n01_01

Harris, R. (2004). *Information and communication technologies for poverty alleviation*. Asia-Pacific Development Information Programme.

Harrod, R. (1939). An Essay in Dynamic Theory. *The Economic Journal*, *49*(193), 14–33. doi:10.2307/2225181

Hartman, H. J. (2001). *Metacognition in learning and instruction*. Academic Press.

Harvey, B., & Beards, D. (2004). E-learning in Scottish Further and Higher Education. *Education + Training, 46*(6/7), 353-360.

Hawaii Public Schools. (2012). *How Can Technology Benefit Our Students?*. Available from http://www.k12.hi.us/~teono/teri/benefits_of_technology.htm

Heath, H., & Cowley, S. (2004). Developing a Grounded Theory Approach: A Comparison of Glaser and Strauss. *International Journal of Nursing Studies*, *41*(2), 141–150. doi:10.1016/S0020-7489(03)00113-5 PMID:14725778

Heeks, R. (1999). *The Tyranny of Participation in Information Systems: Learning from Development Projects.* Institute for Development Policy and Management, University of Manchester. Available on line: http://knownetgrin.honeybee.org/honeybee.html

Henze, N., & Nejdl, W. (2004). A logical characterizing of adaptive education hypermedia. *Journal of New Review in Hypermedia and Multimedia, 10*(1).

Hernandez, C. A. (2006). *Are there two methods of grounded theory?: Demystifying the debate.* Paper presented at the Grounded Theory Seminar, New York, NY.

Hernandez, T. (2006). Next Generation E-Learning: A Buyer's Market. *Certification Magazine*. Available at http://www.certmag.com/issues/jul00/feature_hernandez.cfm

Herrmann, N. (1996). *The Whole Brain Business Book*. McGraw-Hill.

Herselman, M. (2005). An Investigation into E-learning Practices Implication for Providers of Education and Training. *South African Journal of Higher Education, 19*(2), 393–410.

Hiltz, S. R. (1997). Impacts of College Level Courses via Asynchronous Learning Networks: Some Preliminary Results. *Journal of Asynchronous Learning Networks, 1*(2), 1–18.

Hinson, R. (2006). The Internet for Academics: Towards a Holistic Adoption Model. *Online Information Review, 30*(5), 542–554. doi:10.1108/14684520610706415

Hirschheim, R. (1985). Information systems epistemology: An historical perspective. In E. Mumford, R. Hirschheim, & R. Fitzberald et al. (Eds.), *Research Methods in Information Sysfems* (pp. 13–38). Amsterdam: North-Holland.

Hirschheim, R., & Klein, H. (1989). Four paradigms of information systems development. *Communications of the ACM, 32*(10), 1199–1216. doi:10.1145/67933.67937

Hitt, M. A., Hoskisson, R. E., & Kim, H. (1997). International Diversification: Effects on Innovation and Firm Performance in Product-Diversified Firms. *Academy of Management Journal, 40*(4), 767–798. doi:10.2307/256948

Hive. (2014). Retrieved from: http://hive.apache.org/

Ho, C., Lin, M., & Chen, H. (2012). Web users' behavioural patterns of tourism information search: From online to offline. *Tourism Management, 33*, 1468–1482.

Hofmann, J. (2004). *The Synchronous Trainer's Survival Guide*. Pfeiffer.

Holton, J. A. (2007). The Coding Process and Its Challenges. In The Sage Handbook of Grounded Theory. Sage Publications.

Holton, J. A. (2009). Qualitative tussles in undertaking a grounded theory study. *The Grounded Theory Review, 8*, 37–50.

Holton, W. K. (2000). *Designing Web-based Training: How to Teach Anyone Anything Anywhere Anytime.* John Wiley & Sons.

Homan, G., & Macpherson, A. (2005). E-learning in Corporate Universities. *Journal of European Industrial Training, 29*(1), 75–90. doi:10.1108/03090590510576226

Honey, P. (2001). E-learning: A performance appraisal and some suggestions for improvement. *The Learning Organization, 8*(5), 200–202. doi:10.1108/EUM0000000005913

Honey, P., & Mumford, A. (1992). *The Manual of Learning Styles* (3rd ed.). Maidenhead, UK: Peter Honey.

Honey, P., & Mumford, A. (2000). *The learning styles helper's guide.* Maidenhead, UK: Peter Honey Publications Ltd.

Hong, W., & Zhu, K. (2006). Migrating to internet-based e-commerce: Factors affecting e-commerce adoption and migration at the firm level. *Information & Management*, *43*, 204–221.

Horrocks, I., Patel-Schneider, P. F., & Van Harmelen, F. (2003). From SHIQ and RDF to OWL: The making of a web ontology language. *Web Semantics: Science, Services, and Agents on the World Wide Web*, *1*(1), 7–26. doi:10.1016/j.websem.2003.07.001

Horton, W. (2001). *Leading e-Learning: American Society for Training and Development.* Available at http://www.elearninggurus.com/articles.html

Hosseini, S. J. F., Niknami, M., & Chizari, M. (2009). To determine the challenges in the application of ICT's by the agricultural extension service in Iran. *Journal of Agricultural Extension and Rural Development, 1*(1), 27-30.

Hotspot Service Area. (2004). *Telekom Malaysia.* Available at http://isp.tm.net.my/hotspot/service-klangv.shtml

Howard, H. (2003). *Individual Preferences in E-Learning.* Gower Publishing Limited.

Howcroft, D., & Trauth, E. M. (2005). Choosing Critical IS Research. In Handbook of Information Systems Research: Critical Perspectives on Information Systems Design, Development and Implementation. Edward Elgar Publishing.

Howell, S. L., Williams, P. B., & Lindsay, N. K. (2003). Thirty-two Trends Affecting Distance Education: An Informed Foundation for Strategic Planning. *Online Journal of Distance Learning Administration*, *6*(3).

Hsbollah, H., & Idrisk, K. (2009). E-learning Adoption: The Role of Relative Advantages, Trialability and Academic Specialization. *Campus-Wide Information Systems*, *26*(1), 23–37.

Huang, Z., & Benyoucef, M. (2013). From e-commerce to social commerce: A close look at design features. *Electronic Commerce Research and Applications*, *12*, 246–259.

Husain, W., & Dih, L. (2012). A Framework of a Personalized Location-based Traveler Recommendation System in Mobile Application. *IJMUE*, *7*(3), 11–18.

Hussain, Z., Taylor, A., & Flynn, D. (2004). A case study of the process of achieving legitimation in information systems development. *Journal of Information Science*, *30*(5), 408–417. doi:10.1177/0165551504046725

Iacono, J. C., Brown, A., & Holtham, C. (2011). The use of the Case Study Method in Theory Testing: The Example of Steel eMarketplaces. *The Electronic Journal of Business Research Methods*, *9*(1), 57–66.

iCOPER D3.1. (2009). *ICOPER Report on the Standardized Description of Instructional Models. Deliverable 3.1.* Available at: http://nm.wu.ac.at/nm/icoper

Iivari, J., Hirschheim, R., & Klein, H. K. (1998). A Paradigmatic Analysis Contrasting Information Systems Development Approaches and Methodologies. *Information Systems Research*, *9*(June), 164–193. doi:10.1287/isre.9.2.164

Im, I., Hong, S., & Kang, M. S. (2011). An international comparison of technology adoption Testing the UTAUT model. *Information & Management*, *48*, 1–8.

INACOL. (2011). *Use of ICT in Learning and Teaching*. The International Association for K-12 Online Learning. Available at http://www.inacol.org

Ines, A. V. M. (2004). *Improved Crop Production Integrating GIS and Genetic Algorithms.* (PhD Thesis). Asian Institute of Technology (AIT), KhlongLuang, Bangkok, Thailand.

InfoDev. (2005). Knowledge Map: ICT in Education. Author.

InfoSphere Streams. (2014) Retrieved from: http://www.ibm.com/developerworks/bigdata/streams/

International Fund for Agricultural Development (IFAD). (2002). *Challenges in rural development and poverty alleviation in Eastern and Southern Africa: the role of civil society and development institutions*. Available on line: http://www.ifad.org/ngo/events/pf/reports/session3.pdf

International Institute for Communication and Development (IICD). (2006). *ICTs for agricultural livelihoods: impact and lessons learned from IICD supported activities*. The Hague: IICD.

Irani, Z., Alshawi, S., & Missi, F. (2011). Organisational, technical and Data Quality Factors in CRM Adoption-SMEs Perspective. *Industrial Marketing Management*, *40*, 376–383.

iTEC. (n.d.). Available at: http://itec.eun.org/web/guest

ITU. (2004). *ITU world telecommunication indicators database*. International Telecommunication Union (ITU). Retrieved from http://www.itu.int/ITU-D/ict/statistics/index.html

ITU. (2007). *Telecommunications/ict markets and trends in Africa*. ITU.

ITU. (2013). ICT facts and figures 2013. ITU.

IUCEA. (2001). ICT for Learning and Training. *Proceeding during roundtable meetings of The Inter-University Council of East Africa (IUCEA)*. Available at http://www.iucea.org/resources/ict_for_learning_and_training/

Ivergard, T., & Hunt, B. (2005). Towards a Learning Networked Organization: Human Capital, Compatibility and Usability in E-Learning Systems. *Applied Ergonomics*, *36*(2), 157–167. doi:10.1016/j.apergo.2004.09.006 PMID:15694069

Jadhav, A., & Sonar, R. (2009). An integrated rule-based and case-based reasoning approach for selection of the software packages. In Information Systems, Technology and Management (pp. 280–291). Springer Berlin Heidelberg. doi:doi:10.1007/978-3-642-00405-6_30 doi:10.1007/978-3-642-00405-6_30

Jamal, T. B., & Getz, D. (1995). Collaboration theory and community tourism planning. *Annals of Tourism Research*, *22*(1), 186–204. doi:10.1016/0160-7383(94)00067-3

James-Gordon, Y., Young, A., & Bal, J. (2003). External Environment Forces Affecting E-learning Provider'. *Marketing Intelligence & Planning*, *21*(3), 168–172. doi:10.1108/02634500310474984

Jennings, M. M. (1994, November 28). Business Schools' Formula for Irrelevance. *Wall Street Journal, Manager's Journal*.

Jéquier, N. (1984). Telecommunications for development: Findings of the ITU-OECD project. *Telecommunications Policy*, *8*(2), 83–88. doi:10.1016/0308-5961(84)90028-4

Jere, N. R., Thinyane, M., & Terzoli, A. (2010). Enhancing e-commerce services for a rural community: online customer negotiation application. ZA_WWW 2010, Graduate School of Business (UKZN).

Jeyaraj, A., & Sabherwal, R. (2008). Adoption of information systems innovations by individuals: A study of processes involving contextual, adopter, and influencer actions. *Information and Organization*, *18*, 205–234.

Jipp, A. (1934). Telegraphenleitungen. *Moderne Telegraphie*, 113–190. doi:10.1007/978-3-662-32785-2_3

Johnston, J. (1987). *Electronic Learning: From Audio Tape to Videodisc*. London: Lawrence Erlbaum Associates.

Jones, G. E. (1986). Investing in Rural Extension: Strategies and Goals. Elsevier Applied Science Publishers.

Jones, M. G., & Farquhar, J. D. (1997). User Interface Design for Web-Based Instruction. In B. H. Khan (Ed.), Web-based instruction (pp. 239-244). Englewood Cliffs, NJ: Educational Technology Publications.

Juhary, J.B. (2005). A step towards e-learning: some pedagogical issues. *International Journal of Pedagogies and Learning*, *1*(1), 48-58.

Juhary, J. (2005). A Step Towards E-learning: Some Pedagogical Issues. *International Journal of Pedagogies and Learning*, *1*(1), 48–58. doi:10.5172/ijpl.1.1.48

Juma, M. (2001). *The Establishment of a Higher Education Open and Distance Learning Knowledge Base for Decision Makers in Kenya*. Available online at http://www.UNESCO.org/education/studyingabroad/highligt/od/kb/kenya.doc

Juma, M. (2002). *The Impact of Implication of Open Courseware for Higher Education in Developing Countries*. Available online at http://www.wcet.info/unesco/juma_kenya.pdf

Juma, M. (2003). An evolution of an existing institution. In *The university – Current challenges and opportunities*. Nairobi: African Virtual University. Retrieved February 29, 2009, from http://www.unesco.org/iiep/virtualuniversity/home.php

Juskeviciene, A., & Kurilovas, E. (2014). On Recommending Web 2.0 Tools to Personalise Learning. *Informatics in Education*, *13*(1), 17–30.

Kabanda, S. (2011). "E-Commerce Institutionalization is not for us": SMEs perception of E-Commerce in Tanzania. *The African Journal of Information Systems*, *1*, 1–16.

Kaltan, G. (1983). *Introduction to Survey Sampling*. Beverly Hills, CA: Sage Publications.

Kamble, B., & Chemin, Y. H. (2006). *GIPE in GRASS Raster Add-ons*. Retrieved from http://grass.gdf-hannover.de/wiki/

Kao, D., Tousignant, W., & Wiebe, N. (2000). A Paradigm for Selecting an Institutional Software. In *Proceedings IS-ECON 2000*. AITP Foundation for Information Technology Education.

Kapoor, B., & Sharma, S. (2010). A Comparative Study Ontology Building Tools for Semantic Web Applications. *International journal of Web & Semantic Technology*, *1*(3), 1–13.

Karanasios, S. S. (2008). *An e-Commerce framework for small tourism enterprises in developing countries*. (Unpublished Doctoral of Philosophy Thesis). Victoria University.

Karon, R. L. (2000). Bankers go Online: Illinois Banking Company Learns Benefits of E-training. *E-learning*, *1*(1), 38–40.

Kasonde, T. (2009). *Zambia: Promoting Agriculture through the use of ICT's*. Available on line: www.lusaka times.com

Kathawala, Y., Abdou, K., & Elmulti, D. S. (2003). The Global MBA: A Comparative Assessment for its Future. *Journal of European Industrial Training*, *26*(1), 14–23. doi:10.1108/03090590210415867

Kauffman, R. J., & Kumar, A. (2005). *A Critical Assessment of the Capabilities of Five Measure for ICT Development*. University of Minnesota.

Kavulya, J. (2007). Digital Libraries and Development in Sub-Saharan Africa: A Review of Challenges and Strategies. *The Electronic Library*, *25*(3), 299–315. doi:10.1108/02640470710754814

Kay, J. (2001). Learner Control. *User Modeling and User-Adapted Interaction*, *11*(1/2), 111–127. doi:10.1023/A:1011194803800

Keable, M., Landry, M., & Banville, C. (1998). The legitimacy gap between IS professionals and users. *Communications of the ACM*, *41*(5), 146–156. doi:10.1145/276404.276406

Kearsley, G. (1996). The World Wide Web: Global Access to Education. *Educational Technology Review*, (5), 26-30.

Kearsley, G. (2000). *Teaching & Learning in Cyberspace*. Available at http://home.sprynet.com/~gkearsley/chapts.htm

Kearsley, G., & Shneiderman, B. (1999). *Engagement Theory: A Framework for Technology-Based Teaching and Learning*. Available online at http://home.sprynet.com/~gkearsley/engage.htm

Ke, B., & Zhang, X. (1999). *The development of ICT and its impacts on rural economy in China*. Beijing: RCRE.

Keefe, J. (1979). Learning style: an overview. In *NASSP's Student Learning Styles: Diagnosing and Prescribing Programs*.

Keegan, D., et al. (2007), E-learning Initiatives that did not Reach Targeted Goals. Megatrends Project 2007.

Keegan, D., Lõssenko, J., Mázár, I., Michels, P. F., Paulsen, M. F., Rekkedal, T., ... Zarka, D. (2007). *E-learning initiatives that did not reach targeted goals*. Bekkestua: NKI Forlaget.

Keegan, D. (1986). *The Foundation of Distance Education*. Routledge, Kegan & Paul.

Keirsey, D. (1998). *Please Understand me II: Temperament Character Intelligence*. Prometheus Nemesis Books.

Kelle, U. (2005). Emergence vs forcing of empirical data? A crucial theory of grounded theory reconsidered. *Forum: Qualitative Social Research, 6*(2), Article 27. Retrieved from http://www.qualitativeresearch.net/fqs-texte/2-05/05-2-27-e.htm

Kelly, K. B., & Bonner, K. (2005). Digital Text, Distance Education and Academic Dishonesty: Faculty and Administrator Perceptions and Responses. *Journal of Asynchronous Learning Networks*, *9*(1), 43–52.

Kember, D. (1995). *Open Learning Courses for Adults. A model of Student Progress*. Englewood Cliffs, NJ: Educational Technology Publications.

Kember, D. (1999). Integrating Part-time Study with Family, Work and Social Obligations. *Studies in Higher Education*, *24*(1), 109–124. doi:10.1080/03075079912331380178

Kember, D., Armour, R., Jenkins, W., Lee, K., Leung, D. Y. P., & Li, N. et al. (2001). Orientation to Enrollment of Part-time Students: A Classification System Based upon their Perceived Lifelong Learning Needs. *Higher Education Research & Development*, *20*(3), 265–280. doi:10.1080/07294360120108359

Kember, D., Ying, C. K., Wan, C. S., & Yung, C. S. (2005). How students cope with part time study: An analysis of coping mechanisms through an on-line forum. *Active Learning in Higher Education*, *6*(3), 230–242. doi:10.1177/1469787405057662

Kerrigan, K., Lindsey, G., & Novak, K. (1994). Computer networking in International Agricultural Research Experience of the CGNET. New Information technologies in Agriculture. *Quarterly Bulletin of the International Association of Agricultural Information Specialists, 39*(1-2), 182-193.

Khalid, M. S., Rongbutsri, N., & Buus, L. (2012). *Facilitating adoption of Web tools for problem and project based learning activities*. Academic Press.

Khalil, M., Dongier, P., & Qiang, C. Z. W. (2009). Overview in Information and communications for development (IC4D) Information and communications for development (IC4D). Washington, DC: The World Bank.

Khan, B. H. (1997a). Web-based Instruction: What is it and why is it? In B. H. Khan (Ed.), Web-based instruction (pp. 5-18). Englewood Cliffs, NJ: Educational Technology Publications.

Khan, B. H. (2001a). Virtual U: A Hub for Excellence in Education, Training and Learning Resources. In B. H. Khan (Ed.), Web-Based Training (pp. 491-506). Englewood Cliffs, NJ: Educational Technology Publications.

Khan, B. H., & Ealy, D. (2001). A Framework for Web-Based Authoring Systems. In B. H. Khan (Ed.), Web-based Training (pp. 355-364). Englewood Cliffs, NJ: Educational Technology Publications.

Khan, B. H., Waddill, D., & McDonald, J. (2001a). Review of Web-Based Training Sites. In B. H. Khan (Ed.), Web-Based Training (pp. 367-374). Englewood Cliffs, NJ: Educational Technology Publications.

Khan, B. (2005). *Managing E-learning Strategies: Design, delivery, implementation and Evaluation.* London: Idea group. doi:10.4018/978-1-59140-634-1

Khan, B. H. (2001). *A framework for e-learning*. LTI Magazine.

Khan, B. H. (2001b). Web-Based Training: An Introduction. In B. H. Khan (Ed.), *Web based Training*. Englewood Cliffs, NJ: Educational Technology Publications.

Khan, B. H. (2005a). *E-Learning QUICK Checklist.* Hershey, PA: Information Science Publishing. doi:10.4018/978-1-59140-812-3

Khan, B. H. (2007a). Introduction to open, flexible and distributed learning. In B. H. Khan (Ed.), *Flexible Learning*. Englewood Cliffs, NJ: Educational Technology Publications.

Khan, B. H. (2007b). Obstacles Encountered During Stages Elearning Process. In B. H. Khan (Ed.), *Flexible Learning*. Englewood Cliffs, NJ: Educational Technology Publications.

Khan, B. H., & Smith, H. L. (2007c). Student Evaluation Instrument for Online Programs. In B. H. Khan (Ed.), *Flexible Learning*. Englewood Cliffs, NJ: Educational Technology Publications.

Kilby, T. (2001). The direction of web-based training: A practitioner's view. *The Learning Organization*, *8*(5), 194–199. doi:10.1108/EUM0000000005912

Kim, D., & Lee, J. (2013). Development of Intelligent Instruction System for Mathematical Computation. *Informatics in Education*, *12*(1), 93–106.

Kizito, J. B., & Bijan, K. (2006). *An Empirical study on Education Strategies to E-leaning in Developing Countries.* Emerald Group Publishing Limited.

Klein, D. E., Mallory, C. A., & Safstrom, D. W. (1997). Analysis, Design, And Implementation of a web-Based Training System for Multi-Criteria decision Support, Integrating Hypertext. In Multimedia-Based Case Studies and Training Software. Monterey, CA: Naval Postgraduate School.

Klein, H. K., & Hirschheim, R. A. (1989). Legitimation in information systems development: A social change perspective. *Information Technology & People*, *5*(1), 29–46.

Klein, H. K., & Myers, M. D. (1999). A Set of Principles for Conducting and Evaluating Interpretive Field Studies in Information Systems. *MIS Quarterly*, *23*(1), 21–29.

Klein, H. K., & Myers, M. D. (2009). A Set Of Principles For Conducting And Evaluating Interpretive Field Studies In Information Systems. *Management Information Systems Quarterly*, *23*(1), 67–94. doi:10.2307/249410

Klein, H. K., Nissen, H. E., & Hirschheim, R. (1991). *A Pluralist Perspective of the Information Systems Research Arena. In Information Systems Research: Contemporary Approaches & Emergent Traditions*. Amsterdam: North-Holland.

Kohli, R., & Kettinger, W. J. (2004). Informating the clan: Controlling physicians' costs and outcomes. *Management Information Systems Quarterly*, *28*(3), 363–394.

Kohli, R., & Kettinger, W. J. (2004). Informating the Clan: Controlling Physicians' Costs and Outcomes. *Management Information Systems Quarterly*, *28*(3), 363–394.

Kolb, D. A. (1985). *LSI Learning- Style Inventory.* McBer & Company.

Kolesinski, M. T., Nelson-Weaver, E., & Diamond, D. (2013). *Digital Solidarity in Education: Promoting Equity, Diversity, and Academic Excellence through Innovative Instructional Programs*. Routledge.

Kothari, C. R. (2002). *Research Methodology: Methods & Techniques* (2nd ed.). New Delhi, India: K.K Gupta.

Koza, M., & Lewin, A. (2000). Managing partnerships and strategic alliances: Raising the odds of success. *European Management Journal, 18*(2), 146–151. doi:10.1016/S0263-2373(99)00086-9

Kraemer, K. L., & Dedrick, J. (2001). Information technology and productivity: results and policy implications of cross-country studies. In M. Pohjola (Ed.), *Information technology, productivity and economic growth* (pp. 257–279). Oxford, UK: Oxford University Press. doi:10.1093/acprof:oso/9780199243983.003.0012

Krantz, L. (2001). *The Sustainable Livelihood Approach to Poverty Reduction: An introduction*. Stockholm: Division for Policy and Socio-Economic Analysis.

Krauth, B. (1998). Distance Learning: The Instructional Strategy of the Decade. In G. P. Connick (Ed.), The Distance Learner's Guide. Upper Saddle River, NJ: Prentice Hall.

Krejcie, R. V., & Morgan, D. W. (1970). Determining sample size for research activities. *Educational and Psychological Measurement, 30*, 607–610.

Kruse, K. (2006). *The Benefits and Drawbacks of e-Learning*. Available at http://74.125.155.132/scholar?q=cache:DJhiHmX0drMJ:scholar.google.com/&hl=tr&assdt=2000

Kshetri, N. (2007). Barriers to e-Commerce and Competitive Business Models in Developing Countries: A Case Study. *e-Commerce Research and Applications, 6*, 443–452.

Kuchi, R., Gardner, R., & Tipton, R. (2003). *A Learning Framework for Information Literacy and Library Instruction Programs at Rutgers University Libraries, Recommendations of the Learning Framework Study Group*. Rutgers University Libraries.

Kumar, R., Rangan, U. S., & Rufín, C. (2005). Negotiating complexity and legitimacy in independent power project development. *Journal of World Business, 40*(3), 302–320. doi:10.1016/j.jwb.2005.05.006

Kurilovas, E., & Dagiene, V. (2010). *Evaluation of Quality of the Learning Software. Basics, Concepts, Methods, Monograph*. Saarbrücken, Germany: LAP LAMBERT Academic Publishing.

Kurilovas, E., Kubilinskiene, S., & Dagiene, V. (2014). Web 3.0 – Based Personalisation of Learning Objects in Virtual Learning Environments. *Computers in Human Behavior, 30*, 654–662. doi:10.1016/j.chb.2013.07.039

Kushner, J., & Chong, P. (2004). *Conducive environments for promoting community e-centers.Asian Development Bank Regional Workshop*, Bali, Indonesia.

Kwan, F. R., & Joseph. (2005). Web-Based Learning. *Proceedings of the 4th International Conference, 'Technology and Pedagogy'*. River Edge, NJ: World Scientific Publishing Co.

Kypreos, T. (2003). Building a Business Case for Elearning. *E-Learn Magazine*. URL:http://www.elearnmag.org/subpage/sub_page.cfm?section=3&list_item=13&page=1

Lahn, L.L. (2004). Dilemmas in the Development of Elearning at Work. *Journal of Workplace Learning, 16*(8), 466-478.

Laird, D. (1985). *Approaches To Training and Development* (2nd ed.). Reading, MA: Addison-Wesley.

Land, S. M., & Hannafin, M. J. (1997). Patterns of Understanding with Open-Ended Learning Environments: A Qualitative Study. *Educational Technology Research and Development, 45*(2), 47–73. doi:10.1007/BF02299524

Larson, A., Teisberg, E. O., & Johnson, R. R. (2000). Sustainable Business: Opportunity and Value Creation. *Interfaces*, *30*(3), 1–12. doi:10.1287/inte.30.3.1.11658

Lashgarara, F., Mirdamadi, S. M., & Hosseini, S. J. F. (2010). Determining the causal model role of ICT in improving food availability of Iranian rural households. *International Journal of Advanced Science and Technology*, *24*, 27–40.

Lau, M. P. (2002). *Online Teaching & Learning*. Nanyang Siang Pau's New Century.

Laurillard, D. (1995). Multimedia and the Changing Experience of the Learner. *British Journal of Educational Technology*, *26*(3), 179–189. doi:10.1111/j.1467-8535.1995.tb00340.x

Lave, J., & Wenger, E. (1991). *Situated Learning: Legitimate Peripheral Participation (Learning in Doing: Social, Cognitive and Computational Perspectives)*. Cambridge, UK: Cambridge University Press. doi:10.1017/CBO9780511815355

Laws, R. D., Howell, S. L., & Lindsay, N. K. (2003). Scalability in Distance Education: "Can we Have Our Cake and Eat it Too?". *Online Journal of Distance Learning Administration*, *6*(4), 80–89.

Lee, B. (1997). *The Power Principle – Influence with Honor*. New York: Simon & Schuster.

Lee, C. S., & Al-Hawamdeh, S. (2001). New Mode of Course Delivery for Virtual Classroom. *Aslib Proceedings*, *53*(6), 238–242. doi:10.1108/EUM0000000007057

Lee, F. (2002a). *Post Keynesian Price Theory*. Cambridge, UK: Cambridge University Press.

Lee, F. (2002b). Theory Creation and the Methodological Foundation of Post Keynesian Economics. *Cambridge Journal of Economics*, 26789–26804.

Lee, F. (2005). Grounded Theory and Heterodox Economics. *The Grounded Theory Review*, *3*(1), 70–86.

Leeuwis, C. (2004). *Communication for Innovation: Rethinking Agricultural Extension* (3rd ed.). Blackwell Science Ltd.

Lehaney, B. (2004). *Beyond Knowledge Management*. Idea Group publishing. doi:10.4018/978-1-59140-180-3

Leitch, S. (2006). *Prosperity for all in the global economy - world class skills*. London: HM Treasury.

Levina, N., & Vaast, E. (2006). Turning a Community into a Market: A Practice Perspective on Information Technology Use in Boundary Spanning. *Journal of Management Information Systems*, *22*(4), 13–37. doi:10.2753/MIS0742-1222220402

Levine, R. E., & Renelt, D. (1992). A Sensitivity Analysis of Cross-Country Growth Regressions. *The American Economic Review*, *84*(4).

Levy, S. (2003). Six Factors to Consider when Planning Online Distance Learning Programs in Higher Education. *Online Journal of Distance Learning Administration*, *6*(1), 35–43.

Li, L., Tang, S., Fang, L., Xiao, R., Deng, X., Xu, Y., & Xu, Y. (2007). Voeditor: a visual environment for ontology construction and collaborative querying of semantic Web resources. In *31st Annual International Computer Software and Applications Conference* (COMPSAC). IEEE. doi:doi:10.1109/COMPSAC.2007.237 doi:10.1109/COMPSAC.2007.237

Li, B. (2009). The Use of E-learning in Pre-Service Teacher Education. *Campus-Wide Information Systems*, *26*(2), 132–136. doi:10.1108/10650740910946855

Lie, R. (2006). *Background paper: ICTs for Agricultural Development. An exercise in Interdisciplinarity, A final report*. Hampton Press.

Lisewski, B., & Joyce, P. (2003). Examining the five-stage e-moderating model: Designed and emergent practice in the learning technology profession. *Alt-J*, *11*(1), 55–66. doi:10.1080/0968776030110106

Litto, F. M. (2002). The Hybridization of Distance Learning in Brazil – An Approach Imposed by Culture. *International Review of Research in Open and Distance Learning*, *2*(2), 54–64.

Litvin, S. W., Pan, B., & Goldsmith, R. E. (2008). Electronic word-of-mouth in hospitality and tourism management. *Tourism Management*, *29*, 458–468.

Locke, J. (2001). *An Essay Concerning Human Understanding* (A. C. Fraser, Ed., Vol. 1). Oxford, UK: Clarendon.

Locke, K. (2001). *Grounded Theory in Management Research*. London: Sage.

Lowe, A. (1996). *An Explanation of Grounded Theory* (Working paper). Glasgow, UK: University of Stratchlyde, Department of Marketing.

Lowe, A. (1996). *An Explanation of Grounded Theory*. Glasgow, UK: University of Stratchlyde. Unpublished manuscript.

Lowe, A. (2005). *Trust in Emergence*. Keynote presentation delivered to the 3rd International Qualitative Research Convention, Johor Bahru, Malaysia.

Lowe, A. (1995). The Basic Social Processes of Entrepreneurial Innovation. *International Journal of Entrepreneurial Behaviour and Research*, *1*(2), 54–76. doi:10.1108/13552559510090622

Lowe, A. (1998). Managing the Post-Merge Aftermath by Default Remodeling. *Management Decision*, *36*(2), 102–110. doi:10.1108/00251749810204188

LRE APv4.7. (2011). *Learning Resource Exchange Metadata Application Profile v.4.7*. Available at: http://lreforschools.eun.org/c/document_library/get_file?p_l_id=10970&folderId=12073&name=DLFE-1.pdf

LSL. (n.d.). Available at: http://lsl.eun.org/

Luambano, I., & Nawe, J. (2004). Internet Use by Students of the University of Dar es salaam. *Library Hi Tech News*, *2*(10), 13–17. doi:10.1108/07419050410577550

Lucas, R. E. (1986). *On the mechanics of economic development*. Nanking, Taipei, Taiwan: Institute of Economics, Academia Sinica.

Lujara, S. K. (2008). *Development of e-Learning Content and Delivery for Self Learning Environment: Case of Selected Rural Secondary Schools in Tanzania*. Blekinge Institute of Technology. Available online at http://www.bth.se/tks/teknovet.nsf/bilagor/Suzan%20Lujara%20Lic%20manuscript_pdf/$file/Suzan%20Lujara%20Lic%20manuscript.pdf

Lyakurwa, F. S., & Mangula, S. M. (2013). Delineation of the corporate use of environmental information systems: Seletcted cases of the corporate organizations in Tanzania. *Independent Journal of Management and Production*, *4*(2), 529–532. doi:10.14807/ijmp.v4i2.92

Lynch, M. M. (2001). *The online educator: a guide to creating the virtual classroom*. London: Routledge.

Lytras, M. D., Pouloudi, A., & Poulymenakou, A. (2002). Knowledge Management Convergence – Expanding Learning Frontiers. *Journal of Knowledge Management*, *6*(1), 40–51. doi:10.1108/13673270210417682

Lyytinen, K., & King, J. (2004). Nothing at the center? Information systems as a reference discipline? *Journal of the Association for Information Systems*, *5*(6), 220–246.

Lyytinen, K., & King, J. (2004). Nothing at the Center? Information Systems as a Reference Discipline? *Journal of the Association for Information Systems*, *5*(6), 220–246.

Mabe, L. K. (2005). *The Role of Women in Agriculture and other non-farm activities in three regions of the North-West Province*. (MSc. Thesis). North-West University, Mafikeng Campus.

MacFadzean, S., & Newton, B. (2005). Models of ELearning: Technology Promise vs Learner Needs Literature Review. *The International Journal of Management Education*, *3*(1), 43–55.

MacKeogh, K., & Fox, S. (2009). Strategies for Embedding e-Learning in Traditional Universities: Drivers and Barriers. Electronic Journal of e-Learning, 7(2), *147 - 154.*

Macpherson, A., Homan, G., & Wilkinson, K. (2005). The Implementation and Use of E-learning in the Corporate University. *Journal of Workplace Learning*, *17*(1/2), 33–48. doi:10.1108/13665620510574441

Madden, G., & Savage, S. (1998). Interconnection in international telecommunications. *Telecommunications Policy*, *22*(11), 953–961. doi:10.1016/S0308-5961(98)00062-7

Mafu, S. (2004). From the Oral Tradition to the Information Era: The Case of Tanzania. *International Journal on Multicultural Societies*, *6*(1), 53–78.

Maliyamkono, T. L & Mason H. (2006). *The Promise.* TEMA Publishers Company Ltd and Siyaya Publishing (pty).

Mallaliev, K. I. (2010). *Selecting Sustainable ICT Solutions for Pro-Poor intervention, Department of Electrical and Computer Engineering, The University of west indies. St.* Augustine, Trindad and Tobago.

Malopinsky, L., Kirkley, J., Stein, R., & Duffy, T. (2000). *An Instructional Design Model for Online Problem Based Learning (PBL) Environments: The Learning to Teach with Technology Studio.* Paper Presented at Association for Educational Communications and Technology Conference (AECT), Denver, CO.

Manning, S. (2008). Embedding projects in multiple contexts–a structuration perspective. *International Journal of Project Management*, *26*(1), 30–37. doi:10.1016/j.ijproman.2007.08.012

Manville, B. (2004). *Organizing Enterprise-Wide eLearning and Human Capital Management.* Chief Learning Officer.

Mapeshoane, T. J., & Pather, S. (2012). Adoption of e-commerce in typical developing country context: Lesotho tourism industry.*Proceedings of the 14th Annual Conference on World Wide Web Applications.*

Marker, P., McNamara, K., & Wallace, L. (2002). *The Significance of Information and Communication Technologies for Reducing Poverty, The Think! Programme.* Development Policy Department, Department of International Development.

Martins, C., Oliveira, T., & Popovi, A. (2013). Understanding the Internet banking adoption: A unified theory of acceptance and use of technology and perceived risk application. *International Journal of Information Management.*

Martin, V. B. (2006). The postmodern turn: Shall classic grounded theory take that detour? *The Grounded Theory Review*, *5*, 119–128.

Maswera, T., Dawson, R., & Edwards, J. (2008). e-Commerce Adoption of Travel and Tourism Organisations in South Africa, Kenya, Zimbabwe and Uganda. *Telematics and Informatics*, *25*, 187–200.

Matavire, R., & Brown, I. (2008). Investigating the Use of 'Grounded Theory' in Information Systems Research. *Proceedings of the SAICSIT 2008 Conference.*

Matavire, R., & Brown, I. (2008). Investigating the use of 'Grounded Theory' in Information Systems research. *Proceedings of the SAICSIT 2008 Conference, Wilderness.*

Matavire, R., & Brown, I. (2008). Investigating the use of 'Grounded Theory' in Information Systems research. *Proceedings of the SAICSIT 2008 Conference.*

Matavire, R., & Brown, I. (2013). Profiling grounded theory approaches in information systems research. *European Journal of Information Systems*, *22*(1), 119–129. doi:10.1057/ejis.2011.35

Materi, R. R., & Fahy, P. J. (2004). Interim Report: A Case Study of Internet-Based Distance Education Program Development in Vietnam. *International Review of Research in Open and Distance Learning*, *5*(1), 7–16.

Matthews, D. (1999). The Origins of Distance Education.T.H.E. Journal, 27(2), 56-66.

Mayer, R. (2003). Elements of a science of learning. *Journal of Educational Computing Research*, *29*(3), 297–313. doi:10.2190/YJLG-09F9-XKAX-753D

McAdam, D. (2003). *E-Learning for Management and Marketing in Libraries*. IFLA Publications Series No. 115.

McCalla, G. (2004). The ecological approach to the design of e-learning environments: purpose-based capture and use of information about learners. Journal of Interactive Media in Education.

McCraken, H. (2004). Extending Virtual Access: Promoting Engagement and Retention through Integrated Support Systems. *Online Journal of Distance Learning Administration*, *6*(1), 44–52.

McKnight, M. (2000). Distance Education: Expressing emotions in video-based Classes. In *Conference on College Composition and Communication*, Minneapolis, MN.

McPherson, M., & Nunes, J. (2005). Critical Issues for E-learning; What May Seem Obvious is not Always Put Into Practices. *Journal of Computer Assisted Learning*, *5*(2), 34–42.

Meera, S. N., Jhamtani, A., & Rao, D. U. M. (2004). Information and communication technology in agricultural development: A comparative analysis of three projects from India. *Agricultural Research and Extension Network, 135.*

Mello, R. (2002). 100 Pounds of Potatoes in a 25-Pound Sack: Stress, Frustration, and Learning in the Virtual Classroom. Teaching *With Technology Today*, *8*(9). Available at http://www.elearningmag.com/elearning/article/articleDetail.jsp?id=2031

Meng, Q., & Li, M. (2002). New economy and ICT development in China. *Information Economics and Policy*, *14*(2), 275–295. doi:10.1016/S0167-6245(01)00070-1

Meredith, S., & Newton, B. (2003). Models of e-Learning: Technology promise vs. learner needs–literature review. *The International Journal of Management Education*, *4*(3), 43–56. doi:10.3794/ijme.33.73

Mersham, G., & Viviers, D. (2007). Looking at E-learning Through the Lens of a Communication Model. *SOLSTICE 2007 Conference*. Edge Hill University.

Meyen, E. (2002). E-learning: A Programmatic Research Construct for the Future. *Journal of Special Education Technology*, *17*(3), 37–46.

Meyen, E. L. (2000). *Using Technology to Move Research to Practice: The Online Academy.* New York: National Center for Learning Disabilities.

Meyer, K. A. (2003). Face-To-Face Versus Threaded Discussions: The Role of Time and Higher-Order Thinking. *Journal of Asynchronous Learning Networks*, *7*(3), 55–65.

Mgendi, M. F. (2010). *Introducing Web-Based E-learning Platforms at An African University.* Available http://www.isprs.org/proceedings/XXXVIII/part6/ papers/Mgendi/Mgendi_ISPRS_VI_symposium B1 D.pdf

Michiels, S. I., & Van Crowder, L. (2007). Discovering the 'Magic Box': Local Appropriation of Information and Communication Technologies (ICTs). Rome: Food and Agriculture Organization of United Nation's Communication for development Group, Extension, Education and Communication Services.

Middleton, S. E., Roure, D. D., & Shadbolt, N. R. (2009). Ontology-based recommender systems. In S. Staab & R. Studer (Eds.), Handbook on Ontologies, International Handbooks on Information Systems (pp. 779–796). Springer. do i:doi:10.1007/978-3-540-92673-3_35 doi:10.1007/978-3-540-92673-3_35

Mihhailova, G. (2006). E-learning as Internationalization Strategy in Higher Education: Lecturer's and student's perspectives. *Baltic Journal of Management*, *1*(3), 270–284. doi:10.1108/17465260610690926

Miles, M., & Huberman, M. (1994). *Qualitative Data Analysis*. Thousand Oaks, CA: Sage.

Miller, M. D., & Corley, K. (2001). The Effect of E-Mail Messages on Student Participation in the Asynchronous On-Line Course: A Research Note. *Online Journal of Distance Learning Administration*, *4*(3), 17–24.

Mill, J. S. (1900). *Principles of political economy, with some of their applications to social philosophy*. New York: Colonial Press.

Mingers, J. (2001). Combining IS research methods: Towards a pluralist research methodology. *Information Systems Research*, *12*(3), 240–259. doi:10.1287/isre.12.3.240.9709

Ministry of Education and Culture. (2001). *Basic Education Master Plan (BEMP): Medium Term Strategic and Programme Framework 2000-2005*. Available from www.mwanza.go.tz/kurasa/nyaraka_mbalimbali/BEMP.pdf

Ministry of Education and Vocational Training. (2007). *Information & Communication Technology (ICT) Policy For Basic Education*. Available from http://planipolis.iiep.unesco.org/upload/Tanzania/UR/Tanzania_ICT_Policy_for_BasicEducation_2007.pdf

Ministry of Education and Vocational Training. (2011). *Higher Education Baadhi ya Taarifa*. Available from http://www.moe.go.tz/pdf /HE_Organogram.pdf

Ministry of Education and Vocational Training. (2011). *Takwimu za Kielimu kwa Baadhi ya Miaka ya Uhuru wa Tanzanganyika*. Available from http://www.moe.go.tz/pdf/TAKWIMUTANGUUHURUHADIT2011.pdf

Ministry of Works. (2003). *Housing and communications*. Uganda: National ICT Policy.

Minshall, T., Mortara, L., Valli, R., & Probert, D. (2010). Making "asymmetric" partnerships work. *Research-Technology Management*, *53*(3), 53–63.

Mitchell, A., & Honore, S. (2007). Criteria for Successful Blended Learning. *Industrial and Commercial Training*, *39*(3), 143–149. doi:10.1108/00197850710742243

Mmbengeni, M. E., & Mokoka, M. S. (2002). *Provincial Report on Education and Training for Agriculture and Rural Development in Limpopo Province*. Limpopo Department of Agriculture.

MOCT. (2003). *National Information and Communications Technologies Policy*. Available from http://www.tanzania.go.tz/pdf/ictpolicy.pdf

MODIS Image Webpage. (2009). Retrieved from http://modis.gsfc.nasa.gov/

Moeng, B. (2004). *IBM Tackles Learning in the Workplace, IBM*. Management Development Solutions.

Mohan, P., & Brooks, C. (2003). Learning Objects on the Semantic Web.*Proc. Int'l Conf. Advanced Learning Technologies*. doi:10.1109/ICALT.2003.1215055

Molla, A., & Licker, P. S. (2005). eCommerce adoption in developing countries: A model and instrument. *Information & Management*, *42*, 877–899.

Moore, M. G. (1998). Introduction. In C. C. Gibson (Ed.), *Distance learners in higher education*. Madison, WI: Atwood Publishing.

Morgan, G. (2001). Thirteen "Must Ask" Questions About Elearning Products and Services. *The Learning Organization*, *8*(5), 203–210. doi:10.1108/EUM0000000005914

Morrison, J. L., & Khan, B. H. (2003). *The Global E-learning Framework: An Interview with Badrul Khan. The Technology Source.* A Publication of the Michigan Virtual University. Available at http://ts.mivu.org/default.asp?show=article&id=1019#options

Morse, J. M., Stern, P. N., Corbin, J., Bowers, B., Charmaz, K., & Clarke, A. E. (2009). *Developing Grounded Theory: The Second Generation*. Walnut Creek, CA: Left Coast Press.

MoST. (2003). *National Information and Communication Technologies Policy*. Tanzania: Ministry of Science and Technology of Tanzania.

MoST. (2006). *The Use of ICT in National Development, an Opening Remark given during the Inaugural of "Tume ya Sayansi na Tekinoloji", (2-4 May) by Minister of Ministry of Science and Technology*. Available at www.most.gov.tz/resources/speeches/2006/tume_ya_sayansi_na_tekinolojia/

Motik, B., Sattler, U., & Studer, R. (2005). Query answering for OWL-DL with rules. *Web Semantics: Science, Services, and Agents on the World Wide Web*, *3*(1), 41–60. doi:10.1016/j.websem.2005.05.001

Mpazanje, F., Sewchurran, K., & Brown, I. (2013). Rethinking information systems projects using Actor-Network Theory–Perspectives from a developing country. *The Electronic Journal of Information Systems in Developing Countries*, *58*(7), 1–37.

MUC. (2012a). *Academic Programmes*. Available at http://www.makumira.ac.tz/index.php/academic-programmes

MUC. (2012b). *Photo gallery*. Available at http://www.makumira.ac.tz/zindex.php/photo-gallery

Mudannayake, I. (2006). *Forging relationships between Libraries and Extension Services for Improved Access to Agriculture Information: A case study in Sri Lanka*. World Library and Information Congress: 72nd IFLA General Conference and Council, Seoul, Korea.

Mulkeen, A. (2003). What Can We do to Encourage ICT Integration? Evidence from the Irish School System. Education Department, NNI Maynooth, Ireland.

Mulvenna, M. D., Anand, S. S., & Buchner, A. G. (2000). Personalization on the Net using Web mining: Introduction. *Communications of the ACM*, *43*(8), 122–125. doi:10.1145/345124.345165

Mungania, P. (2003). *The Seven E-learning Barriers Facing Employees*. Available at http://www.aerckenya.org/docs/ElearningReport.pdf

Munro, M. C., Huff, S. L., & Moore, G. (1985). Information Technology Assessment and Adoption: A field study. *Management Information Systems Quarterly*, *9*(4), 327–340. doi:10.2307/249233

Munro, M. C., Huff, S. L., & Moore, G. (1988). Expansion and Control of End-User Computing. *Journal of Management Information Systems*, *4*(3), 5–27. doi:10.1080/07421222.1987.11517798

Munyua, H. (2000). Application of ICTs in Africa's Agrcultural Sector: A gender perspective. International Development Research Center (IDRC).

Munyua, H. (2008). *Report on ICTs and small - scale Agriculture in Africa: A scoping study*. IDRC.

Muyinda, P. (2007). M-learning: Pedagogical, Technical and Organisational Hypes and Realities. *Campus-Wide Information Systems*, *24*(2), 97–104. doi:10.1108/10650740710742709

Myers, M. (1997b). Interpretive Research in Information Systems. In J. Mingers & F. Stowell (Eds.), *Information Systems: An Emerging Discipline?* (pp. 239–266). London: McGraw-Hill.

Myers, M. D. (1997). Qualitative research in information systems. *Management Information Systems Quarterly*, *21*(2), 241–242. doi:10.2307/249422

Myers, M. D. (2009). *Qualitative Research in Business & Management (3rd ed.).* London: Sage Publications.

Myers, M. D., & Klein, H. K. (2011). A Set Of Principles For Conducting Critical Research In Information Systems. *Management Information Systems Quarterly*, *35*(1), 17–36.

Myers, M. D., & Newman, M. (2007). The Qualitative Interview in IS Research: Examining the Craft. *Information and Organization*, *17*(1), 2–26. doi:10.1016/j.infoandorg.2006.11.001

Myers, M. D., & Walsham, G. (1998). Exemplifying Interpretive Research in Information Systems: An Overview. *Journal of Information Technology*, *13*(4), 233–234. doi:10.1057/jit.1998.2

Mysite. (2012). *History of Distance Education*. Available at http://mysite.du.edu/~kkeairns/de/Text/Lessons/Lesson1.pdf

NACTE. (2012). *List of Accredited Institutions as of June 2011*. Available at http://www.nacte.go.tz/registered_institution.php?lang=kis

Nagarsi, K. (2000). *Assess of creative potential and practice of extension personnel*. New Delhi: IARI Pusa.

Naidu, S. (2006). *E-Learning: A Guidebook of Principles, Procedures and Practices*. Available from http://cemca.org/e-learning_guidebook.pdf

Naidu, S. (2006). *E-learning: A Guidebook of Principles, Procedures and Practices*. New Delhi: CEMCA.

Nanas, N. (2001). *Literature Review: Information Filtering for Knowledge Management*. The Open University.

Nasirin, S., Birks, D. F., & Jones, B. (2003). Re-examining fundamental GIS implementation constructs through the grounded theory approach. *Telematics and Informatics*, *20*(4), 331–347. doi:10.1016/S0736-5853(03)00012-1

National Information and Communication Technology Policy (NICTP). (2003). *Ministry of Communication and Transport*. Retrieved from http://www.tanzania.go.tz/pdf/ictpolicy.pdf/

Nazrov, E. (2011). *Emergency Response management in Japan*. Final Research report. ASIAN Disaster Reduction Center: FY2011A Program.

Ndag, I., Sanusi, R. A., & Aigbekaen, E. O. (2008). Comparative analysis of Information and Communication Technology (ICT) use by agricultural extension workers in South-west and North-central Nigeria. *Proceedings of the 19th Annual International Information Management Association.*

Ndume, V., Tilya, F.N., & Twaakyondo, H. (2006). Challenges of Adaptive E-learning at Higher Learning Institutions: A Case Study in Tanzania. *International Journal of Computing and ICT Research, 2*(1).

Ndume, V., Tilya, F. N., & Twaakyondo, H. (2008). Challenges of Adaptive eLearning at Higher Learning Institutions: A Case Study in Tanzania, International. *Journal of Computing and ICT Research*, *2*(1), 47–59. Available online at http:www.ijcir.org/volume1-number2/article6.pdf

Ndume, V., Tilya, F., & Twaakyondo, H. (2008). Challenges of Adaptive elearning at higher learning institutions: A case study in Tanzania. *International Journal of Computing and ICT Research*, *2*(1), 47–59.

Ndume, V., Tilya, F., & Twaakyondo, H. (2008). Challenges of adaptive e-learning at higher learning institutions: A case study in Tanzania. *International Journal of Computing and ICT Research*, *2*(1), 47–59.

Nettland, G. (2007). E-learning in a Large Organisation: A Study of the Critical Role of Information Sharing. *Journal of Workplace Learning*, *19*(6), 392–411. doi:10.1108/13665620710777129

Neuman, W. L. (1997). *Social Research Methods* (3rd ed.). Allyn and Bacon.

Newman. (2008). *Advantages and Disadvantages of E-learning*. Available at http://www.newman.ac.uk/Students-Websites/~m.m.friel/

Newton, R. (2003). Staff Attitudes to the Development and Delivery of E-learning. *New Library World*, *104*(1193), 412–425. doi:10.1108/03074800310504357

Ngugi, C. (2006). Report on the AVU (African Virtual University). *ODeL Seminar: Supporting Open, Distance and e-Learning in African Universities.*

Ngwenyama, O., Andoh-Baidoo, K. F., Bollou, F., & Morawecynski, O. (2006). Is there a relationship between ICT, Heath, Education and Development? An empirical analysis of five West African Countries from 1997-2003. *The Electronic Journal on Information Systems in Developing Countries*, *5*(23), 1–11.

Nichols, M. (2012). *International Forum of Educational Technology & Society: Formal Discussion Initiation-A theory for eLearning*. Available at http://ifets.ieee.org/discussions/discuss_march2003.html

Nicholson-Crotty, S. (2005). Bureaucratic Competition in the Policy Process. *Policy Studies Journal: the Journal of the Policy Studies Organization*, *33*(3), 341–361. doi:10.1111/j.1541-0072.2005.00119.x

Njenga, J. K., & Fourie, L. C. H. (2010). The myths about e-learning in higher education. *British Journal of Educational Technology*, *41*(2), 199–212. doi:10.1111/j.1467-8535.2008.00910.x

Njenga, J., & Fourie, L. (2008). The Myths About E-learning in Higher Education. *British Journal of Educational Technology*, *3*(1), 1–14.

Noe, R. A. (2005). *Traditional Training Methods*. Available from http://www.stamfordonline.com.my/courses/dhrm/dhr304/DHR%20304%20Lecture%207.pdf

Nunan, T. (1996). Flexible Delivery - What is it and Why is it a Part of Current Educational Debate? Different Approaches: Theory and Practice in Higher Education. *Proceedings HERDSA Conference 1996*. Available at http://www.herdsa.org.au/confs/1996/nunan.html

O`Brien, J. A. (2004). *Management Information Systems: Managing Information Technology in the Business Enterprise* (6th ed.). New York: McGraw-Hill/Irwin.

O'Malley, J., & McGraw, H. (1999). Students Perceptions of Distance Learning, Online Learning and the Traditional Classroom. *Online Journal of Distance Learning Administration*, *2*(4), 23–36.

O'Regan, K. (2003). Emotion and E-learning. *Journal of Asynchronous Learning Networks*, *7*(3), 78–92.

Obayelu, A., & Ogunlade, I. (2006). Analysis of the uses of Information & Communications Technology for Gender Empowerment & Sustainable Poverty Alleviation in Nigeria. *International Journal of Educational Development.*

OECD. (2003). *ICT and economic growth, evidence from OECD countries, industries and firms*. Paris: OECD.

Oladele, O. I. (2011). Effect of Information Communication Technology on Agricultural Information Access Among Researchers, Extension Agents, and Farmers in South Western Nigeria. *Journal of Agricultural and Food Information, 12*(2), 167- 176.

Olaniran, B. (2007). Challenges to Implementing E-learning in Lesser-Developed Countries. In A. Edmundson (Eds.), Globalized E-learning Cultural Challenges. Idea group.

Omary, Z., Lupiana, D., Mtenzi, F., & Wu, B. (2010). Analysis of the challenges affecting e-healthcare adoption in developing countries: A case of Tanzania. *International Journal of Information Studies, 2*(1), 38–50.

Omwenga, E. I., Waema, T. M., & Wagacha, P. W. (2004). A Model for Introducing and Implementing E-learning for Delivery of Educational Content Within the African Context, *African Journal of Science and Technology. Science and Engineering Series, 5*(1), 34–46.

Oomen-Early, J., & Murphy, L. (2008). Self-actualization and e-learning: A qualitative investigation of university faculty's perceived needs for effective online instruction. *International Journal on E-Learning, 8*(2).

Open University of Tanzania. (2011). *PROSPETUS 2011/2012*. Retrieved from http://www.out.ac.tz/announcements/Prospectus 2011-2012 for printing3.pdf

OpenLayer. (2009). *OpenLayers: Free Maps for the Web*. Retrieved from http://openlayers.org/

OPPAPERS. (2012). *Traditional Training Methods*. Available from http://www.oppapers.com/essays/Traditional-Training-Methods/685211

Orlikowski, W. J. (1993). CASE Tools as Organizational Change: Investigating Incremental and Radical Changes in Systems Development. *Management Information Systems Quarterly, 17*(3), 37–45. doi:10.2307/249774

Orlikowski, W. J., & Baroudi, J. J. (1991). Studying Information Technology in Organizations: Research Approaches and Assumptions. *Information Systems Research, 2*(1), 1–28. doi:10.1287/isre.2.1.1

Pachnowski, L. M., & Jurczyk, J. P. (2003). Perceptions of Faculty on the Effect of Distance Learning Technology on Faculty Preparation Time. *Online Journal of Distance Learning Administration, 6*(3), 89–102.

Pade, C., Plamer, R., Kavhai, M., & Gumbo, S. (2009). *Siyakula Living Lab: Baseline study report*. (Unpublished report). Grahamstown: Rhodes University.

Palloff, R., & Pratt, K. (1999). *Making the Transition: Helping Teachers to Teach Online*. In the 15th Annual Conference on Distance Teaching and Learning, Madison, WI.

Palmer, A. (2005). *Principle of Services Marketing* (4th ed.). McGraw-Hill Education.

Pandey, S., Karunamoorthy, D., & Buyya, R. (2011). *Workflow Engine for Clouds,CloudComputing: Principles and Paradigms*. New York: Wiley Press.

Papalewis, R. (2003). *What are the Advantages /Disadvantages of Technology in the classroom?* Available at http://answers.yahoo.com/question/index?qid=20080712164252AA1iB7Y

Pashler, H., McDaniel, M., Rohrer, D., & Bjork, R. (2008). Learning styles concepts and evidence. *Psychological Science in the Public Interest, 9*(3), 105–119. PMID:26162104

Patalong, S. (2003). Using the Virtual Learning Environment WebCT to Enhance Information Skills Teaching at Coventry University. *Library Review, 52*(3), 103–110. doi:10.1108/00242530310465906

Patel, P., & Pavitt, K. (1987). Is Western Europe losing the technological race? *Research Policy*, *16*(2-4), 59–85. doi:10.1016/0048-7333(87)90024-2

Paulsen, A. (2003). *Online Education Obituaries*. Available online at www.studymentor.com/studymentor/Obituaries.pdf

Pergler, C., & Littlejohn, A. (2007). *Preparing for Blended E-learning*. New York: Routledge.

Petter, S., DeLone, W., & McLean, E. (2008). Measuring information systems success: Models, dimensions, measures, and interrelationships. *European Journal of Information Systems*, *17*, 236–263.

Phang, J. (2004). Impact of ICT Innovations: The Islamic Context. *The Star in Tech*, *8*(April), 18.

Piccoli, G., Ahmad, R., & Ives, B. (2001). Web-based virtual learning environments: A research framework and a preliminary assessment of effectiveness in basic IT skills training. *Management Information Systems Quarterly*, *25*(4), 401–426. doi:10.2307/3250989

Pig. (2014). Retrieved from: https://pig.apache.org/

Pigato, M. (2001). *Information and Communication Technology*. Poverty and Development in Sub-Saharan Africa and South Africa.

Pig, L. C. (2003). Exploring Critical Aspects of Information Technologies Integration in Singapore Schools. *Australasian Journal of Educational Technology*, *19*(1), 1–24.

Piskurich, G. M. (2002). *AMA Handbook of E-Learning: Effective Design, Implementation, and Technology Solutions*. AMACOM Books.

Popescu, E. (2009). Diagnosing students' learning style in an educational hypermedia system. In Cognitive and Emotional Processes in Web-based Education: Integrating Human Factors and Personalization. IGI Global.

Porter, L. R. (1997). *Creating Virtual Classroom: Distance Learning with the Internet*. New York: J. Wiley & Sons.

Powell, W. W. (1990). Neither market nor hierarchy: Network forms of organization. In B. M. Staw & L. L. Cummings (Eds.), *Research in Organizational Behavior* (pp. 295–336). JAI.

Purnomo, S. H., & Lee, Y. H. (2010). An Assessment of Readiness and Barriers towards ICT Programme Implementation: Perceptions of Agricultural Extension Officers in Indonesia. *International Journal of Education and Development using Information and Communication Technology*, *6*(3), 19-36.

Puteh, M. (2008). E-learning implementation in Malaysian Universities: The Universiti Teknologi Malaysia Experience. *Proceedings of the 3rd international conferences on e-learning*. University of Cape Town.

Puteh, M. (2008). E-learning Implementation in Malaysian Universities: The Universiti Teknologi Malaysia Experience. *Proceedings of the 3rd International Conference on E-learning*. University of Cape Town.

PyWPS. (2009). *Python Web Processing Service*. Retrieved from http://pywps.wald.intevation.org/

Qiang, C. W., Clarke, G. R., & Halewood, N. (2006). The role of ICT in doing business. 2006 information and communications for development: global trends and policies. Washington, DC: World Bank.

Raabe, K. (2008). *Reforming the agricultural extension system in India: What do we know about what works where and why?* IFPRI Discussion Paper 00775. Washington, DC: International Food Policy Research Institute.

Rafea, A. (1999). *Managing agricultural knowledge: role of information and communication technology*. American University in Cairo.

Ray, K., & Day, J. (1998). Students` attitudes towards electronic information resources. *Information Research, 4*(2). Available at http://informationr.net/ir/4-2/paper54.html

Reigeluth, C. M., & Khan, B. H. (1994). *Do Instructional Systems Design (ISD) and Educational Systems Design (ESD) Really Need each other?* Paper presented at the Annual Meeting of the Association for Educational Communications and Technology (AECT), Nashville, TN.

Relan, A., & Gillani, B. B. (1997). Web-Based Instruction and Traditional Classroom: Similarities and Differences. In B. H. Khan (Ed.), Web-based instruction (pp. 41-46). Englewood Cliffs, NJ: Educational Technology Publications.

Ren, Y., Kiesler, S., & Fussell, S. R. (2008). Multiple Group Coordination in Complex and Dynamic Task Environments: Interruptions, Coping Mechanisms, and Technology Recommendations. *Journal of Management Information Systems, 25*(1), 105–130. doi:10.2753/MIS0742-1222250105

Resources.com. (2012). *Research Methodology - Key Concepts of the Scientific Method.* Available from www.experiment-resources.com/research-method

Ricardo, D., Sraffa, P., & Dobb, M. (1952). The Works and Correspondence of David Ricardo: Volume 8, Letters 1819-June 1821: v. 8: Letters 1819-June 1821. University Press for the Royal Economic Society.

Ricci, F., Rokach, L., & Shapira, B. (2011). Introduction to recommender systems handbook. Springer US. doi:doi:10.1007/978-0-387-85820-3 doi:10.1007/978-0-387-85820-3

Richards, T. (1998). The Emergence of Open Standards for Learning Technology. Macromedia Interactive Learning Division.

Richardson, D. (1997). *The internet and rural and agricultural development – an integrated approach.* Rome: FAO. Available online: http://www.fao.org/docrep/w6840e/w6840e00.htm

Richardson, D. (2003). *Agricultural extension transforming ICTs? Championing universal access* (Background Paper for CTA's ICT Observatory). Available on line: http://www.cta.int/fr/content/download/2792/20425/version/1/file/backgroundpaper2003.pdf

Richardson, D. (2006). *ICTs – Transforming Agricultural Extension.* Report of the 6th Consultative Experts Meeting of CTA's Observatory on ICT's, The ACP-EU Centre for Agrcultural and Rural Cooperation (CTA).

Richardson, D. (2005). How can agricultural extension best harness ICTs to improve rural livelihoods in developing countries. In E. Gelb & A. Offer (Eds.), *ICT in Agriculture: Perspectives of Technological Innovation.* Jerusalem: Hebrew University of Jerusalem, Centre for Agricultural Economics Research.

Ritchie, D. C., & Hoffman, B. (1997). Incorporating Instructional Design Principles with the World Wide Web-Stride Handbook. In B. H. Khan (Ed.), Web-based instruction (pp. 135-138). Englewood Cliffs, NJ: Educational Technology Publications.

Ritchie, J., & Lewis, J. (Eds.). (2004). *Qualitative Research Practice.* London: Sage.

Rivera, J. C., & Rice, M. L. (2002). A Comparison of Student Outcomes & Satisfaction between Traditional & Web Based Course Offerings. *Online Journal of Distance Learning Administration, 5*(3), 75–83.

Rivera, W. M., & Sulaiman, R. V. (2009). Extension: Object of reform, engine for innovation. *Outlook on Agriculture, 38*(3), 267–273. doi:10.5367/000000009789396810

Rodriguez, J. (2008), Development of an ICT Training Model for In-service Teachers in Nicaragucen Schools. *23rd International Technical Conference on Circuits/system, Computers and Communication* (ITC-CSCC 2008).

Roffe, I. (2002). E-learning: Engagement, Enhancement and Execution. *Quality Assurance in Education, 10*(1), 40–50. doi:10.1108/09684880210416102

Rogers, E. M. (2003). Diffusion of innovations (5th ed.). New York: Free Press.

Rogers, D. (2000). A Paradigm Shift: Technology Integration for Higher Education in the New Millennium. *Educational Technology Review, 6*(3), 19–27.

Rogers, E. M. (2003). *Diffusion of innovations* (5th ed.). New York: Free Press.

Rogerson, C. M. (2005). Unpacking Tourism SMMEs in South Africa: Structure, Support Needs and Policy Response. *Development Southern Africa, 22*, 623–642.

Röller, L.-H., & Waverman, L. (2001). Telecommunications Infrastructure and Economic Development: A Simultaneous Approach. *The American Economic Review, 91*(4), 909–923. doi:10.1257/aer.91.4.909

Romer, P. (1986). Increasing Returns and Long-Run Growth. *Journal of Political Economy, 94*(5), 1002–1037. doi:10.1086/261420

Romer, P. (1990). Are Nonconvexities Important For Understanding Growth? *The American Economic Review, 80*. doi:10.3386/w3271

Romiszowski Alexander, J. (2004). How's the E-learning Baby? *Factors Leading to Success or Failure of an Education Technology Innovation.* Available at http://www.itslifejimbutnotasweknowit.org.uk/files/elearning_failure_study-romiszowsky.pdf

Romiszowski, A. J. (2004). How's the E-learning Baby? Factors Leading to Success or Failure of an Educational Technology Innovation. *Educational Technology, 44*(1), 5–27.

Rosenberg, M. J. (2001). *E-Learning: Strategies for Delivering Knowledge in the Digital Age.* New York: McGraw-Hill.

Rosenberg, N. (1994). *Exploring the Black Box: Technology, Economics, and History* (1st ed.). Cambridge University Press. doi:10.1017/CBO9780511582554

Rossett, A., & Sheldon, K. (2001). *Beyond The Podium: Delivering Training and Performance to a Digital World.* San Francisco: Jossey-Bass/Pfeiffer.

Rourke, L., & Anderson, T. (2002). Using Web-based, Group Communication Systems to Support Case Study Learning at a Distance. *International Review of Research in Open and Distance Learning, 3*(2), 101–112.

Roux, L. l. (2005). *Qualitative research: method in the madness?* A Working paper, University of Stellenbosch Business school.

Rovai, A. P., & Jordan, H. M. (2004). Blended Learning And Sense Of Community: A Comparative Analysis With Traditional And Fully Online Graduate Courses. *International Review of Research in Open and Distance Learning, 5*(2), 65–71.

Roy, M.H. (1996). Collaborating Across Campuses: The Virtual University Experience. *Campus Wide Information Systems, 13*(4), 6-10.

Roy, S. (2005). *Globalisation, ICT and Developing Nations: Challenges in the information age*. New Delhi, India: Sage Publications Ltd.

Rumble, G. (2000). The Globalisation of Open and Flexible Learning: Considerations for Planners and Managers. *Online Journal of Distance Learning Administration, 3*(3), 1–15.

Rutter, D. J. (2005). *From Diversity to Convergence: British Computer Networks and the Internet, 1970-1995*. (Unpublished Doctoral thesis). University of Warwick, Warwick.

Said, A., Rafea, A., El-Beltagy, S., & Hassan, H. (2009). Automatic Generation of Explanation for Expert Systems implemented with Different Knowledge Representations. *WSEAS Transactions on Systems, 8.*

Salau, E. S., & Saingbe, N.D. (2008). Access and Utilization of Information and Communication Technologies (ICTs) Among Agricultural Researchers and Extension Workers in Selected Institutions in Nasarawa State of Nigeria. *PAT, 4*(2), 1-11. Available online: www.patnsukjournal.com/currentissue

Salmon, G. (2002). Pedagogic Requirements of VLEs: Pets and Planets. *UCISA TLIG-SDG User Support Conference*. Available at http://sstweb.open.ac.uk:8282/oubs/gilly

Salmon, G. (2000). *E-Moderating The Key to Teaching and Learning Online*. Sterling, VA: Kogan Page.

Salomon, G. (1993). No Distribution without Individuals' Cognition. In G. Salomon (Ed.), *Distributed Cognitions: Psychological and education consideration* (pp. 111–138). New York: Cambridge University Press.

Saltiel, I. M. (1998). Defining collaborative partnerships. *New Directions for Adult and Continuing Education, 1998*(79), 5–11. doi:10.1002/ace.7901

Saltzbert, S., & Polyson, S. (1995). Distributed learning on the world wide web. *Syllabus, 9*(1), 10–12.

Saltzbert, S., & Polyson, S. (1995). Distributed Learning on the World Wide Web. *Syllabus, 9*(1), 10–12.

Sambrook, S. (2003). E-learning in Small Organizations. *Education + Training, 45*(8/9), 506-516.

Sampson, D. (2002). Current and Future Research and Technology Developments in e-Learning. *Industry and Higher Education Journal, 16*(4).

Sampson, D., & Karagiannidis, C., & Kinshuk. (2002). Personalised Learning: Educational, Technological and Standardisation Perspective. *Interactive Educational Multimedia, 4*, 24–39.

Sanga, C., Churi, A., & Tumbo, S. (2007). Status, Opportunities, Potential and Challenges of Technology-Mediated Open and Distance Education (Tech-MODE) for Agricultural Education and Improved Livelihoods: A Case Study of Tanzania. *Country Case Studies*, 113.

Sanga, C., Kalungwizi, V., & Msuya, C. (2013). Building agricultural extension services system supported by ICTs in Tanzania: Progress made, Challenges remain. *International Journal of Education and Development Using ICT, 9*(1), 80-99.

Sangineto, E., Capuano, N., Gaeta, M., & Micarelli, A. (2008). Adaptive course generation through learning styles representation. *Journal of Universal Access in the Information Society, 7*(1), 1–23. doi:10.1007/s10209-007-0101-0

Sarasin, L. (1998). *Learning Style Perspectives: Impact in the Classroom*. Atwood Publishing.

Satyanarayana, K. V., Reddy, M. N., Balasubramani, N., Sharma, V. P., & Bhaskar, G. (2008). *National Institute of Agricultural Extension Management*. Available at www.manage.com

Saunders, M., Lewis, P., & Thornhill, A. (2009). *Research Methods for Business Students* (5th ed.). New York: FT/ Prentice Hall.

Schank, R. C. (2001). *Designing World-class E-learning*. McGraw-Hill.

Schifter, C. C. (2004). Compensation Models in Distance Education: National Survey Questionnaire Revisite'. *Online Journal of Distance Learning Administration, 7*(1), 54–63.

Schmitz, J. G. (2003). Agricultural Extension on the Web. University of Illinois.

Schott, M., Chernish, W., Dooley, K. E., & Linder, J. R. (2003). Innovations in Distance Learning Program Development and Delivery. Online Journal of Distance Learning Administration, 6(2), 81-92.

Schreiber, D. A., & Berge, Z. L. (1998). *Distance Training: How Innovative Organizations are Using Technology to Maximize Learning and Meet Business Objectives*. San Francisco: Jossey-Bass.

Schunk, D. (1996). *Learning Theories: An Educational Perspective*. Prentice Hall.

Scott, H. (2004). *Design for Collaborative Learning Online: The Integrated Pedagogical and Community Model.* Paper presented at the Seamless Learning: Integrating Teaching, Research & Technology, Stony Brook State University of New York.

Scott, H. (2007). *The Temporal Integration of Connected Study into a Structured Life: a Grounded Theory*. (PhD Thesis). Portsmouth University.

Scribd. (2005). *Traditional Training Methods*. Available at http://www.scribd.com/doc/35921013/Traditional-Training-Method

Seabra, C., Abrantes, J. L., & Lages, F. L. (2007). The Impact of Using Non-media Information Sources on the Future Use of Mass Media Information Sources: The Mediating Role of Expectations Fulfilment. *Tourism Management, 28*, 1541–1554.

Secretariat of the Pacific Community (SPC). (2011). *Pacific Community*. Retrieved from: http//www.spc.int

Selin, S., & Chaves, D. (1995). Developing a Collaborative Model for Environmental Planning and Management. *Environmental Management, 19*(2), 189–195. doi:10.1007/BF02471990

Selwood, I. (2004). *Information technology and educational management in the knowledge society: Primary school teachers use of ICT for administration and management.* 6th International Working Conference on Information Technology in Educational Management (ITEM), Las Palmas de Gran Canaria, Spain.

Selwood, I., Smith, D., & Wishart, J. (2001). In A. C. W. Fung & M. A. Brown (Eds.), *Supporting UK teachers through the national grid for learning in Nolan. P* (pp. 159–171). Pathways to Institutional Improvement with Information Technology in Educational Management.

Sife, A. S., Kiondo, E., & Lyimo-Macha, J. G. (2010). Contribution of mobile phones to rural livelihoods and poverty reduction in Morogoro region, Tanzania. The Electronic Journal of Information Systems in Developing Countries, 42.

Sife, A., Lwoga, E. & Sanga, C. (2007). New technologies for teaching and learning: Challenges for higher learning institutions in developing countries. *International Journal of Education and Development Using ICT, 53*(2), 57-67.

Sife, A., Lwoga, E., & Sanga, C. (2007). New Technologies for Teaching and Learning: Challenges for Higher Learning Institutions in Developing Countries. *International Journal of Education and Development using ICT, 3*(2). Available: http://ijedict.dec.uwi.edu/viewarticle.php?id=246

Sifuna, N. D. (1998). *Crisis in the Public Universities in Kenya*. Available online at http://ahero.uwc.ac.za/index.php?module=cshe&action=downloadfile&fileid=81806115511732720564415

Sifuna, D. (2007). The challenge of increasing access and improving quality: An analysis of universal primary education interventions in Kenya and Tanzania since the 1970s. *International Review of Education, 53*(5-6), 687–699. doi:10.1007/s11159-007-9062-z

Simonson, M., Smaldino, S., Albright, M., & Zvacek, S. (2000). *Teaching and Learning at a Distance: Foundations of distance education*. Upper Saddle River, NJ: Prentice Hall.

Simpson, O. (2003). *Student Retention in Online, Open and Distance Learning*. Kogan Page.

Simpson, O. (2004). The impact on retention of interventions to support distance learning students. *Open Learning*, *19*(1), 79–96. doi:10.1080/0268051042000177863

Singh, H. (2003). Building Effective Blended Learning Programs. *Educational Technology*, *44*(1), 5–27.

Sirin, E., Parsia, B., Grau, B. C., Kalyanpur, A., & Katz, Y. (2007). Pellet: A practical owl-dl reasoner. *Web Semantics: Science, Services, and Agents on the World Wide Web*, *5*(2), 51–53. doi:10.1016/j.websem.2007.03.004

SJCET. (2012). *School of Information Technology*. Available at http://www.sjcet.in/sit.html

Skiera, B., Eckert, J., & Hinz, O. (2010). An analysis of the importance of the long tail in search engine marketing. *Electronic Commerce Research and Applications*, *9*, 488–494.

Skiera, B., & Nabout, N. A. (2012). Return on Quality Improvements in Search Engine Marketing. *Journal of Interactive Marketing*, *26*, 141–154.

Skiera, B., Nabout, N. A., Stepanchuk, T., & Gerstmeier, E. (2012). An analysis of the profitability of fee-based compensation plans for search engine marketing. *International Journal of Research in Marketing*, *29*, 68–80.

Smith, A.D., & Rupp, W.T. (2004). Managerial Implications of Computer-Based Online/Face-to-Face Business Educational Case study. *Online Information Review*, *28*(2), 100-109.

Smulders, D. (2003). Designing for Learners, Designing for Users. *E-Learn Magazine*. Retrieved from http://www.elearnmag.org/subpage/sub_page.cfm?section=3&list_item=11&page=1

Soloman, B., & Felder, R. M. (1998). *Index of learning styles questionnaire*. Available at: http://www.engr.ncsu.edu/learningstyles/ilsweb.html

Solow, R. (1956). A Contribution to the Theory of Economic Growth. *The Quarterly Journal of Economics*, *70*(1), 65–94. doi:10.2307/1884513

Sorokin, P. A. (1937). *Social and Cultural Dynamics*. New York: American Book Company.

Sorokin, P. A. (2006). *Social and cultural dynamics: a study of change in major systems of art, truth, ethics, law, and social relationships*. London: Transaction Publishers.

Sparks, B. A., & Browning, V. (2011). The impact of online reviews on hotel booking intentions and perception of trust. *Tourism Management*, *32*, 1310–1323.

Sridhar, K. S. (2006). *Incentives for Regional Development: Competition Among Sub-National Governments*. Palgrave Macmillan.

Ssewanyana, J. & Busler, M. (2007). Adoption and usage of ICT in developing countries: Case of Ugandan firms. *International Journal of Education and Development using Information and Communication Technology, 3*(3), 49-59.

Starkey, L. (2011). Evaluating learning in the 21st century: A digital age learning matrix. *Technology, Pedagogy and Education*, *20*(1), 19–39. doi:10.1080/1475939X.2011.554021

Steiner. (2005). *African tertiary institution connectivity survey*. Nairobi: African Virtual University. Available online at http://www.avu.org/documents/partnership%20connectivity%20report-%20revised%2017-10-04.pdf

Stepanyan, K. (2008). *Challenges and opportunities for practicing e-learning globally*. Buckinghamshire Cilterns University College.

Stockley, D. (2003). *E-learning definition and Explanation*. ECampus.com.au.

Stockley, D. (2006). *E-learning Definition and Explanation*. Retrieved from http://derekstockley.com.au/e-learning-definition.html

Stockley, D. (2006, February). Implementing e-learning successfully: Building a successful e-learning strategy. *The EI Magazine*, *2*(7), 4–5.

Strauss, A., & Corbin, J. (1989). Grounded Theory's Applicability to Nursing Diagnostic Research. In Monograph of the Invitational Conference of Research Methods for Validating Nursing Diagnoses, Palm Springs, CA.

Strauss, H. (2003). Getting all of our Courses Online: A Euphoric State Case Study. *E-Learn Magazine*. Retrieved from http://www.elearnmag.org/subpage/sub_page.cfm?section=3&list_item=4&page=3

Strauss, A. (1987). *Qualitative Analysis for Social Scientists*. Cambridge, UK: Cambridge University Press. doi:10.1017/CBO9780511557842

Strauss, A. L., & Corbin, J. (1990). *Basics of qualitative research: Grounded theory procedures and techniques*. Thousand Oaks, CA: Sage.

Strauss, A., & Corbin, J. (1990). *Basics of Qualitative Research*. Sage.

Strauss, A., & Corbin, J. (1994). Grounded Theory Methodology – An Overview. In N. K. Denzin & Y. S. Lincoln (Eds.), *Handbook of Qualitative Research*. Sage.

Strauss, A., & Glaser, B. (1970). *Anguish*. Mill Valley, CA: Sociology Press.

Suchman, M. (1995). Managing legitimacy: Strategic and institutional approaches. *Academy of Management Review*, *20*(3), 571–610.

Suddaby, G., & Milne, J. (2008). Coordinated Collaborative and Coherent: Developing and Implementing E-learning Guidelines Within a National Tertiary Education Systems. *Campus-Wide Information Systems*, *25*(2), 114–122. doi:10.1108/10650740810866602

Sulaiman, R. (2003). *Agricultural Extension: Involvement of Private Sector*. Mumbai, India: National Bankfor Agricultural and Rural Development.

Sure, Y., Staab, S., & Studer, R. (2003). On-to-knowledge methodology (OTKM). In Handbook on Ontologies. Springer.

Swan, A. (1956). Time-Study and Statistics. *OR*, *7*(4), 110–139. doi:10.2307/3007549

Swanson, B. E. (2008). Global review of good agricultural extension and advisory service practices. Rome: Food and Agriculture Organization of the United Nations.

Swanson, B. E., & Rajalahti, R. (2010). Strengthening Agricultural Extension and Advisory Systems: Procedures for Assessing, Transforming and Evaluating Extension Systems. The International Bank for Reconstruction and Development/The World Bank.

TCU. (2010). *Statistics for Universities and University Colleges Enrollment of Students*. Available at http://www.tcu.go.tz/uploads/file/Statistics%20for%202009-2010.pdf

Telematica Instituut E-learninggroep. (2001). E-learning, voordelen en kansen op Organisatie-niveau. *M&I, 2001*(3), 5.

The eLearning Guild. (2002). *The E-Learning Development Time Ratio Survey*. Available online at http://www.elearningguild.com/pdf/1/time%20to%20develop%20Survey.pdf

Thomas, D. R. (2003). A General Inductive Approach for Qualitative Data Analysis. University of Auckland. Available online at www.health.auckland.ac.nz/hrmas/resources/qualdatanalysis.html

Thomson, W. (2001). *A Guide for the Young Economist: Writing and Speaking Effectively*. Available at www.leaonline.com/doi/abs/10.1207/S15327736ME1501_7

Thomson, A. M., & Perry, J. L. (2006). Collaboration processes: Inside the black box. *Public Administration Review*, *66*(1), 20–32. doi:10.1111/j.1540-6210.2006.00663.x

Tinto, V. (1993). *Leaving College: Rethinking the Causes and Cures of Student*. Academic Press.

Tinto, V. (1997). *Classrooms as Communities: Exploring the Educational Character*. Academic Press.

Tinto, V. (1975). Dropout from Higher Education: A Theoretical Synthesis of Recent Research. *Review of Educational Research*, *45*(1), 89–125. doi:10.3102/00346543045001089

TRA. (2012). *Tax Exemption for deemed capital goods*. Available at www.tra.go.tz/index.php?option=com_content&task=view&id=238&Itemid=76

TravelGround. (2013). *The easiest way to find and book accommodation across South Africa*. Retrieved May 2, 2013, from: http://www.travelground.com

Tregurtha, N., & Vink, N. (2002). *E-commerce and the South African horticultural export industry: current status and future directions*. Available on line: http://www.gapresearch.org/production/EcommercefinalreportAug2002.pdf

Trochim, W. M. K. (2006). *Philosophy of Research: Induction and Deduction, Research Methods Knowledge Base*. Available at http://www.socialresearchmethods.net/kb/dedind.php

Troussas, C., Virvou, M., & Alepis, E. (2014). Collaborative Learning: Group Interaction in an Intelligent Mobile-Assessed Multiple Language Learning System. *Informatics in Education*, *13*(2), 279–292. doi:10.15388/infedu.2014.08

Truex, D., & Holmstrom, J., & Keil, M. (2006). Theorising Information Systems Research: A Reflective Analysis of the Adaptation of Theory in Information Systems Research. *Journal of the Association for Information Systems*, *7*(12), 797–821.

Truex, D., Holmstrom, J., & Keil, M. (2006). Theorising information systems research: A reflective analysis of the adaptation of theory in information systems research. *Journal of the Association for Information Systems*, *7*(12), 797–821.

Tsuji, G. Y., Uehara, G., & Salas, S. (1994). *DSSAT v3.0*. Honolulu, HI: University of Hawaii.

Tusubira, F. F., Kaggwa, I., & Ongora, J. (2005). Uganda. In *Towards an African e-Index; Household and individual ICT ACCESS AND USAGE across 10 African countries*. Research ICT Africa. The LINK Centre, Wits University School of Public and Development Management. Available: www.researchictafrica.net

Tyler, K., Tan, J., & Manica, A. (2007). Business-to-business adoption of e-Commerce in China. *Information & Management*, *44*, 332–351.

Tynjala, P., & Hakkinen, P. (2005). E-learning at Work: Theoretical Underpinnings and Pedagogical Challenges. *Journal of Workplace Learning*, *17*(5/6), 2005. doi:10.1108/13665620510606742

UDOM. (2012). *Call For Admission into Undergraduate Degree Programmes for Equivalent Applicants*. Available at http://udom.ac.tz/index.php/ admission/undergraduate-application

UDOM. (2012). *Organisation Structure*. Available at http://udom.ac.tz/index.php/admission/undergraduate-application

UDOM. (2012). *University History*. Available at http://udom.ac.tz/index.php/about-udom/history

UNDP. (2001). *Information Communications Technology for Development. Essentials ICTD 31. Synthesis of Lessons Learned*. Evaluation Office N0. 5. Retrieved from http://www.undp.org/eo

UNECA. (2012). *Economic report on Africa 2012. Accelerating Africa's development through diversification*. Addis Ababa, Ethiopia: UNECA.

United Nations Development Programme. (2004). *Asia-Pacific Development Information Programme (UNDP-APDIP)*. Available on line: www.apdip.net

United Nations Development Programme. (2011). *Making New Technologies Work for Human Development*. Retrieved from http://hdr.undp.org/reports/global/2011/en/

United Nations. (1978). *Technological change, economic development and intra and extra regional relations in Latin America. IDB/EC1A Science and Technolog*. Research Program in Latin America.

United Nations. (2011). Measuring the impacts of information and communication technology for development. UNICTAD current studies on science, technology and innovation.

University of Florida. (2012). *The History of Distance Education*. Available at http://iml.jou.ufl.edu/projects/spring01/declair/history.html

UN-OHRLLS. (2009). *Least Developed Countries: about LCDS*. The United Nations Office of the High Representative for Least Developed Countries. Available online at http://www.unohrlls.org/en/ldc/25/

Unwin, T. (2008). Survey of e-learning in Africa- based on a questionnaire survery of people on the e-learning Africa. *Database*, 2007.

Urdan, T. A., & Weggen, C. C. (2000). *Corporate E-learning: Exploring a new Frontier*. WR Hambrecht + Co. Available at http://www.wrhambrecht.com/research/coverage/elearning/ir/ir_explore.html

Urquhart, C. (1997). Exploring Analyst-Client Communication: Using Grounded Theory Techniques to Investigate Interaction in Informal Requirements Gathering. In Information Systems and Qualitative Research. London: Chapman & Hall.

Urquhart, C., & Fernandez, W. (2006). Grounded theory method: the researcher as blank slate and other myths. *ICIS 2006 Proceedings*. Retrieved from http://aisel.aisnet.org/icis2006/31

Urquhart, C. (1998a). Analysts And Clients In Conversation: Cases In Early Requirements Gathering. *Proceedings of theNineteenth International Conference on Information Systems*.

Urquhart, C. (1998b). Strategies for Conversation and Systems Analysis in Requirements Gathering: A Qualitative View of Analyst-Client Communication. *Qualitative Report*, *4*(1/2). Available at http://www.nova.edu./ssss/QR/QR4-1/urquhart.html

Urquhart, C. (1999a). Reflections On Early Requirements Gathering – Themes From Analyst- Client Conversation. In T. J. Larsen, L. Levine, & J. I. DeGross (Eds.), *Information Systems: Current Issues and Future Changes*. Laxenburg, Austria: IFIP.

Urquhart, C. (1999b). Themes In Early Requirements Gathering – The Case Of The Analyst. *The Client And The Student Assistance Scheme. Information Technology & People*, *3*(1), 25–34.

USAID. (2013). *Assessing the role of information and communication technologies to enhance food systems in developing countries*. White Paper Series, USAID Higher Education Solutions Network.

Uschold, M., & King, M. (1995). Towards a Methodology for Building Ontologies. *Proceedings of the IJCAI'95.Workshop on Basic Ontological Issues in Knowledge Sharing*.

Utsumi, Y. (2005). *Connect the World official launch*. Available online at http://www.itu.int/partners/pdf/SG_preso.pdf

Uys, P. M. (2003). Critical Success Factors in the Infusion of Instructional Technologies for Open Learning in Development Settings: The Case of the University of Botswana. *International Review of Research in Open and Distance Learning*, *4*(2), 33–39.

Uys, P. M. (2004). A Syntagm of Networked Educational Management: Case Study – University of Botswana. *Campus-Wide Information Systems*, *21*(1), 22–28. doi:10.1108/10650740410512293

Uzawa, H. (1965). Optimum Technical Change in An Aggregative Model of Economic Growth. *International Economic Review*, *6*(1), 18–31. doi:10.2307/2525621

Van Audenhove, L. (2003). *Towards an integrated information society policy in South Africa: an overview of political rhetoric and policy initiatives 1994 – 2000*. Cape Town: HSRC Publishers. Available on line: www.hsrcpress.ac.za

Van Dam, J. C., Kroes, J. G., Groenendijk, P., Hendriks, R. F. A., & Jacobs, C. M. J. (1997). *SWAP Model*. Retrieved from: http://www.swap.alterra.nl/

Van Dam, N. (2004). *The E-Learning Fieldbook*. New York: McGraw-Hill.

Van Niekerk, J. C., & Roode, J. D. (2009). Glaserian and Straussian grounded theory: similar or completely different? *Proceeding SAICSIT '09 Proceedings of the 2009 Annual Research Conference of the South African Institute of Computer Scientists and Information Technologists*. Available at http://dl.acm.org/ft_gateway.cfm?id=1632163&ftid=694294&dwn=1&CFID=81873259&CFTOKEN=70405415

Van Veldhuizen, L., Waters-Bayer, A., & De Zeeuw, H. (1997). *Developing Technology with Farmers: A Trainers Guide for Participatory Learning*. Zed Books Ltd.

VanBrakel, P., & Chisenga, J. (2003). Impact of ICT-based distance learning: The African Story. *The Electronic Library*, *21*(5), 476–486. doi:10.1108/02640470310499867

Vega-Gorgojo, G., Bote-Lorenzo, M. L., Gomez-Sanchez, E., Asensio-Perez, J. I., Dimitriadis, Y. A., & Jorrn-Abellan, I. M. (2008). Ontoolcole: Supporting educators in the semantic search of CSCL tools. *Journal of Universal Computer Science*, *14*(1), 27–58.

Venkatesh, V., Morris, M. G., Davis, G. B., & Davis, D. F. (2003). User acceptance of information technology: Toward a unified view. *Management Information Systems Quarterly*, *2*, 425–478.

Vernon, M. (2002). Understanding E-learning. *Financial Times*. Available at http://specials.ft.com/elearning/FT35PSCM2ZC.html

Vesin, B., Ivanović, M., Klašnja-Milićević, A., & Budimac, Z. (2012). Protus 2.0: Ontology-based semantic recommendation in programming tutoring system. *Expert Systems with Applications*, *39*(15), 12229–12246. doi:10.1016/j.eswa.2012.04.052

Vesin, B., Ivanović, M., Klašnja-Milićević, A., & Budimac, Z. (2013). Ontology-based architecture with recommendation strategy in java tutoring system. *Computer Science and Information Systems*, *10*(1), 237–261. doi:10.2298/CSIS111231001V

Viviers, D., & Mersham, G. (2007). Looking at E-learning through the lens of a Communication Model. *SOLSTICE 2007 Conference*. Edge Hill University.

Volkoff, O., Strong, D. M., & Elmes, M. B. (2005). Understanding enterprise systems-enabled integration. *European Journal of Information Systems*, *14*(2), 110–120. doi:10.1057/palgrave.ejis.3000528

Vukanikinds-DTI. (2005). *An Analysis of the ICT Skills Audit for the dti and ISETT SETA*. Department of Trade and Industry (DTI)–South Africa. Available: http://www.isett.org.za/downloads/ICT_Skills_needs.pdf

Wagner, E. (2000). *Strategies for leveraging learning objects: PowerPoint presentation delivered at the TechLearn2000 conference*. Available at http://www.techlearn.net/layout/default.cfm?page=sessiondetail&session=1236

Wallden, S., & Makinen, E. (2014). Educational Data Mining and Problem-Based learning. *Informatics in Education*, *13*(1), 141–156.

Wang, H. C., & Huang, T. H. (2013). Personalized e-learning environment for bioinformatics. *Interactive Learning Environments*, *21*(1), 18–38. doi:10.1080/10494820.2010.542759

Wang, M. Y. (2003). The Strategic Role of Digital Libraries: Issues in E-Learning Environments. *Library Review*, *52*(3), 111–116. doi:10.1108/00242530310465915

Wang, Y., & Qualls, W. (2007). Towards a Theoretical Model of Technology Adoption in Hospitality Organizations. *Hospital Management*, *26*, 560–573.

Wanhill, S. (2002). *Sustaining Tourism SMEs.VII Congreso Internacional del CLAD sobre la Reforma del Estado y de la Administración Pública*, Lisboa, Portugal.

Wassenaar, D. R. (2006). Ethical issues in social science research. In *Research in practice: Applied methods for the social sciences*. Cape Town: UCT Press.

Watson, J. F., et al. (2004). *Education evolution: The need to keep pace with development of K-12 online learning*. Available from www.learningpt.org/expertise/ schoolimprovement/pdf

Watts, S., & Henderson, J. C. (2006). Innovative IT climates: CIO perspectives. *The Journal of Strategic Information Systems*, *15*(2), 125–151. doi:10.1016/j.jsis.2005.08.001

WB. (2002). *World Bank Population growth Report*. New York: WB.

Weema, M. T. (2005). A brief history of the development of an ICT Policy in Kenya. In E. F. Etta & L. Elder (Eds.), *At The Crossroads: ICT Policy Making in East Africa*. Nairobi: East African Educational Publishers.

Wentling, T. L., Waight, C., Gallaher, J., La Fleur, J., Wang, C., & Kanfer, A. (2000). *Elearning -A Review of Literature*. Available at http://learning.ncsa.uiuc.edu/papers/elearnlit.pdf

Werthner, H., & Ricci, F. (2004). e-Commerce and Tourism. *Communications of the ACM*, *47*, 101–105.

Wesley, D. (2002). A Critical Analysis on the Evaluation of E-learning. *International Journal on E-Learning*, *2*(1), 45–56.

White, C. (2005). Contribution of Distance Education to the Development of Individual Learners. *Distance Education*, *26*(2), 165–181. doi:10.1080/01587910500168835

White, K. W., & Weight, B. H. (2000). *The Online Teaching Guide*. Canada: Tony Bates Associates Ltd.

Wiersma, W. (1995). *Research methods in education* (6th ed.). Boston: Allyn and Bacon.

Wiersma, W., & Jurs, S. G. (2009). *Research Methods in Education* (9th ed.). Pearson Education Inc.

Wikipedia. (2009). *Least developed countries*. Wikimedia Foundation, Inc. Available online at http://en.wikipedia.org/wiki/Least_Developed_Countries

Wild, R.H., Griggs, K.A., & Downing, T. (2002). A Framework for E-learning as a Tool for Knowledge Management. *Industrial Management & Data Systems*, *102*(7), 371-380.

Wiley, D. A. (2000). *Connecting Learning Objects to Instructional design Theory: a definition, a Metaphor, and a Taxonomy*. Utah State University. Available at: http://www.reusability.org/read/

Willging, P. A., & Johnson, S. D. (2004). Factors that influence students' decisions to drop out of online courses. *Journal of Asynchronous Learning Networks*, *8*(4), 105–118.

Williams, P., & Nicholas, D. (2004). *What Literature Tell us About Distance Education*. Available from http://site.ebrary.com/lib/coventry/docDetail.actiondocID=10085710&p00=informatione-learning

Willis, B. (Ed.). (1994). *Distance Education: Strategies and Tools*. Englewood Cliffs, NJ: Educational Technology Publications.

Wolf, S. (2001). *Determinants and impact of ICT use for African SMEs: Implications for rural South Africa. Center for Development Research* (ZEF Bonn). 2001 Annual Forum at Misty Hills, Muldersdrift.

Wolodko, A., & Bednarek-Michalska, J. (2007). E-learning Models for Polish Libraries. *BIBWEB, the ElectronicLlibrary*, *25*(1), 80-89.

Woodley, A. (2004). Conceptualizing student dropout in part-time distance education: Pathologizing the normal? *Open Learning*, *19*(1), 47–64. doi:10.1080/0268051042000177845

Woodley, A., de-Lange, P., & Tanewski, G. (2001). Student Progress in Distance Education: Kember's model re-visited. *Open Learning*, *16*(2), 63–71. doi:10.1080/02680510123105

Woodward, H., & Pilling, S. (1993). *The international serials industry: The International Serials Industry*. Aldershot, UK: Gower Publishing.

Woody, C. (1927). Research Methodology. *Journal of Social Sciences*.

World Bank Report. (2007). *Using Information and Communication Technologies (ICT) to support Rural Livelihood*. Evidence, Strategies and Tools, A Workshop for World Bank Staff. Retrieved from http://www.infodev.org/livelihoods on 11/23/12

World Bank. (1998). *World development report 1998/99 - Knowledge for development*. Washington, DC: World Bank.

World Business Education. (2012). *E-learning: Adoption Rates and Barriers*. Available at http://www.internettime.com/itimegroup/FrmRep1.pdf

Worldwidelearn. (2012). *E-learning Essentials: Benefits of E-learning*. Available from http://www.worldwidelearn.com/elearning-essentials/elearning-benefits.htm

Wright, J. (2008). *Primary ICT and E-Learning Co-Coordinator's Manual Kit*. SAGE Publications Inc.

Wymer, S. A., & Regan, E. A. (2005). Factors Influencing e-commerce Adoption and Use by Small and Medium Businesses. *Electronic Markets*, *15*(4), 438–453.

Xiang, Z., & Gretzel, U. (2010). Role of social media in online travel information search. *Tourism Management*, *31*, 179–188.

Xiang, Z., & Pan, B. (2011). Travel queries on cities in the United States: Implications for search engine marketing for tourist destinations. *Tourism Management*, *32*, 88–97.

Yang, P., Wu, Y., & Tao, Y. (2008). The use of unified theory of acceptance and use of technology to confer the behavioral model of 3G mobile telecommunication users. *Journal of Statistics & Management Systems*, *11*, 919–949.

Yick, A., Patrick, P., & Costin, A. (2005). Navigating distance and traditional higher education: Online faculty experiences. *International Review of Research in Open and Distance Learning*, *6*(2).

Yieke, F.A. (2005). Towards Alternatives IN Higher education: the benefits and challenges of E-learning in Africa. *CODESRIA Bulletin*, (3-4), 73.

Yong, A. (2003). Success Factors in E-learning Implementation. *The Star*, *19*, 93.

Yoo, S., Kim, V., & Park, S. (2013). An Educational tool for Browsing the Semantic Web. *Informatics in Education*, *12*(1), 143–151.

Yorke, M. (2004). Retention, persistence and success in on-campus higher education, and their enhancement in open and distance learning. *Open Learning*, *19*(1), 19–32. doi:10.1080/0268051042000177827

Young-Ju, J., Bong, M., & Choi, H. J. (2000). Self-Efficacy for Self-Regulated Learning, Academic Self-Efficacy, and Internet Self-Efficacy in Web-Based Instruction. *Educational Technology Research and Development*, *48*(2), 5–13. doi:10.1007/BF02313398

Youn, S., & McLeod, D. (2006). *Ontology Development Tools for Ontology-Based Knowledge Management. Encyclopedia of E-Commerce, EGovernment and Mobile Commerce*. Idea Group Inc.

Yum, J. C. K., Kember, D., & Siaw, I. (2001). Study Examines Coping Methods in Part-time Students. *The Newsletter of the National Resource Center for the First Year Experience and Students in Transition*, *14*(1), 4–5.

Zahedi, F. M., Van Pelt, W. V., & Srite, M. (2006). Web Documents' Cultural Masculinity and Femininity. *Journal of Management Information Systems*, *23*(1), 87–128. doi:10.2753/MIS0742-1222230103

Zahm, S. (2000). No Question About it – E-learning is here to Stay: A Quick History of the E-learning Evolution. *E-learning*, *1*(1), 44–47.

Zhang, J., Khan, B. H., Gibbons, A. S., & Ni, Y. (2001). Review of Web-Based Assessment Tools. In B. H. Khan (Ed.), Web-Based Training (pp. 137-146). Englewood Cliffs, NJ: Educational Technology Publications.

Zijp, W. (1994). Improving the transfer and use of agricultural information: A guide to information technology. The International Bank for Reconstruction and Development.

About the Contributors

Titus Tossy is a Lecturer of ICT at Mzumbe University Faculty of Science and Technology. Dr. Tossy has 14 years of teaching experiences in higher learning institutions in Africa and UK. Dr. Tossy's main research interest is in understanding the uses of ICT for national Development as well as how people, span organizational, professional, and cultural and other boundaries in the process of building and using new technology. He uses qualitative and quantitative methods and a range of social and organizational theories in his work including classic grounded methodology. In 2007, he received a SPIDDER grant to establish national e-learning at the open university of Tanzania. In 2012, he was a founder of Tanzania Business School. Dr. Tossy received his Bsc (Hons) in Computer Science and Statistics from university of Dar es salaam and MBA in Information Technology Management from Coventry University. He received his Doctor of Philosophy in Information Systems from University of Cape Town. Due to his future career ambition, Dr. Tossy now is doing Bachelor of Education (Policy and Management) at the Open University of Tanzania (OUT).

* * *

Md. Shamim Akhter received his Ph.D. in Information Processing from Tokyo Institute of Technology (TokyoTech), M.Sc. in Computer Science and Information Management from Asian Institute of Technology (AIT) and B.Sc. in Computer Science from American International University Bangladesh (AIUB) in 2009, 2005 and 2002 respectively. He joined AIUB as a Lecturer in 2002 and since 2005 he has been an Assistant Professor. He was also a JSPS Post Doctoral Research Fellow in National Institute of Informatics (NII) from FY 2009-2011, Visiting Researcher in Tokyo Institute of Technology, Japan from FY 2009-2011, Research Associate at the RS and GIS FoS, Asian Institute of Technology, Thailand in 2005 and Global COE Research Assistant from Sep 2007~ Aug 2009 in Tokyo Institute of Technology, Japan. He was awarded "The Excellent Student of The Year, FY2008", Global COE Program, Photonics Integration-Core Electronics (PICE), Tokyo Institute of Technology, Japan and Magna-Cum Laude for academic excellence from American International University Bangladesh in 2002. His research interests are Artificial Intelligent, Evolutionary Algorithms and Models for their Parallelization, Remote Sensing (RS) and GIS applications, High Performance Computing (HPC), Algorithm and Complexity Analysis, Wireless Communications and Mobile Computing. He is a senior member of IEEE and member of JARC-Net.

Pearson Broome is a lecturer in Political Science and the Programme Coordinator for the MSc. e-Governance for Developing Countries in the Department of Government, Sociology and Social Work at the University of the West Indies, Cave Hill Campus, Bridgetown, Barbados. He has diverse working experience in the Caribbean and served as a Consultant for UNDP, the Commonwealth Secretariat and CSME Unit, CARICOM Secretariat. His research interests include: EGovernance/eGovernment/eCommerce/eDemocracy, with particular emphasis on the theoretical approaches and the institutionalization of ICTs; ICTs and the globalisation process focusing on outsourcing, trade, investment and technology transfer flows; the evaluation of national ICT Policies and National Systems of Innovation, the politics of science and technology policy, deregulation and Telecommunications policy, new public management and the restructuring of state bureaucracies, and public policy and its application to public sector management in developing countries.

Eugenijus Kurilovas is Senior Research Scientist in Vilnius University Institute of Mathematics and Informatics and Associate Professor in Vilnius Gediminas Technical University. He has published about 100 scientific papers, is a member of 35 scientific journals' and conferences Editorial boards and Programme committees. Dr. E. Kurilovas is honoured TOP 100 Scientist 2014 by International Biographical Centre (IBC), Cambridge, and his biographical records are included in: Who is Who in the World – 2014 (31st Edition) and 2015 (32nd Edition); 2000 Outstanding Intellectuals of the XXIst Century – 2014 (8th Edition); Who is Who in Science and Engineering – 2016-2017 (12th Edition); and Dictionary of International Biography – 2014 (37th Edition) and 2015 (38th Edition) by International Biographical Centre (IBC), Cambridge.

Felichesmi S. Lyakurwa is a Lecturer in the Department of Engineering Management Studies, Mzumbe University.

Camilius Sanga is an Associate Professor of Informatics in the Department of Informatics at Sokoine University of Agriculture (SUA), Tanzania. He is a head of Department of Informatics in the Faculty of Science, SUA. He has PhD in Computer Science from University of the Western Cape, South Africa. Also, he holds MSc. Computer Science and BSc. in Computer Science from Osmania University and University of Dar es Salaam respectively. His research interest is in the area of Information and Communication Technology for Development (ICT4D). He has published papers in a number of International journals. He has also published articles in local and International conferences in the field of ICT4D. Furthermore, he has co-authored two books as well as a number of book chapters. Some of the research projects which Sanga has been involved are: (i) Development of Monitoring and Evaluation system for Projects under Enhancing Pro-poor Innovations in Natural Resources and Agricultural Value-chains (EPINAV) at SUA, (ii) Promoting participation of female students in Science, Technology, Engineering and Mathematics using e-learning and (iii) Farmer Voice Radio (FVR) Project - Building a radio - based, impact driven small farmer extension service system. Currently, he is an assistant project leader for the research titled "The role of mobile phones towards improving coverage of agricultural extension services: a case study of maize value chain". Lastly, he is involved in the project titled "Assessing the impacts of climate variability and change on agricultural systems in Eastern Africa while enhancing the region's capacity to undertake integrated assessment of vulnerabilities to future changes in climate" (2012-2014).

Joseph Sungau is a Lecturer in the Department of Engineering Management Studies, Mzumbe University.

Index

A

B

C

D

E

F

G

H

I

K

L

M

P

R

S

T

W

www.ingramcontent.com/pod-product-compliance
Lightning Source LLC
LaVergne TN
LVHW060406300926
849162LV00006B/1361